D1612358

LIVERPOOL JMU LIBRARY
3 1111 01470 1021

Port Designer's Handbook

Third edition

If you copy or steal from one,

it is called plagiarism.

If you copy or steal from two,

it is called evaluation.

If you copy or steal from many,

it is called research.

Wilson Mizner

This book is dedicated to my family, in acknowledgement of their understanding and encouragement that helped me so much in my research. Without their support I would neither have been able to start nor finish this book.

Carl A. Thoresen

Port Designer's Handbook

Third edition

Carl A. Thoresen

Published by ICE Publishing, One Great George Street, Westminster, London SW1P 3AA.

Full details of ICE Publishing sales representatives and distributors can be found at: www.icevirtuallibrary.com/info/printbooksales

Other titles by ICE Publishing:

Coasts, Marine Structures and Breakwaters: From Sea to Shore – Meeting the Challenges of the Sea (2 volumes). ISBN 978-0-7277-5975-7
Coasts, Marine Structures and Breakwaters: Adapting to Change (2 volumes). ISBN 978-0-7277-4129-5
Construction Health and Safety in Coastal and Maritime Engineering. ISBN 978-0-7277-3345-0
Piers, Jetties and Related Structures Exposed to Waves. ISBN 978-0-7277- 3265-1

www.icevirtuallibrary.com

A catalogue record for this book is available from the British Library

ISBN 978-0-7277-6004-3

ICE Publishing is a division of Thomas Telford Ltd, a wholly-owned subsidiary of the Institution of Civil Engineers (ICE).

Commissioning Editor: Jo Squires
Production Editor: Vikarn Chowdhary
Market Development Executive: Catherine de Gatacre

Typeset by Academic + Technical, Bristol
Index created by Indexing Specialists (UK) Ltd
Printed and bound in Great Britain by TJ International, Padstow

Contents

Foreword

This third edition of the handbook on modern design and construction of port and harbour engineering will, in my opinion, give harbour colleagues around the world an opportunity not only to study the Norwegian practices and solutions in the design and construction from traditional berth structures to complicated oil and gas berths, but also gain knowledge and a general understanding of the design of port and harbour constructions.

The invitation to write the foreword to the third edition of the *Port Designer's Handbook*, which has been updated to cover new port developments and standards, provides the ideal opportunity to express my belief that harbour experts around the world should write and share their knowledge and information more frequently with their colleagues.

This new edition will contribute and make available new information based on the author's long experience and knowledge to others within the harbour and port sector, so that the field can continue to develop.

I feel, therefore, that this handbook will be invaluable to practising harbour and port engineers and to postgraduate and senior university students of the state of the art.

Kirsti L. Slotsvik
Director General of the Norwegian Coastal Administration

Preface and acknowledgements

The purpose of the third edition of the *Port Designer's Handbook* is to update its coverage, particularly with regards to the berthing of large vessels and the design and construction of berth structures due to new developments in the construction and use of new materials. Over the last 5–10 years there has been considerable improvement and much new thinking in the design and construction of port and harbour structures. The main aim of the third edition is not to explain the full scope of port design and construction but to give guidelines and recommendations to try to deal with some of the main new items and assumptions in the layout, design and construction of modern port structures, and the forces and loadings acting on them.

The use of new concrete technology for berth structures in the marine environment is dealt with in detail, as well as the types of deterioration and methods of repair of these structures. Safety considerations and maintenance problems have also been covered in detail.

The book is mainly based on the author's more than 45 years of experience gained from practical engineering and research from more than 800 different small and large port and harbour projects both in Norway and abroad. He worked as the chief engineer and was technically responsible for the port and harbour division in the largest Norwegian consulting company Norconsult AS. After retiring, he started his own office, PORT-CAT.

The book includes material from many lectures held by the author over the years on berth and port structures at the Norwegian University of Science and Technology, and at the Chalmers University of Technology, Sweden, as well as on postgraduate courses arranged by the Norwegian Society of Chartered Engineers.

Over the last 10 years or so, the growing interest in the design of port and harbour structures has produced a huge number of research reports and technical papers and amount of information, especially from the International Navigation Association (PIANC). Much of this latest information has been evaluated and summarised in this handbook.

The author hopes that the contents of the book will make it a readable and useful handbook that provides practical guidance for port engineers who are responsible for the design of port and harbour structures, and that it will contribute to further development of the subject. Many of the subjects mentioned in the text are worthy of further study, and there are further reading lists at the end of each chapter to direct the reader to other detailed sources of information.

Finally, the author would like to express his deepest thanks to the many friends and colleagues who have contributed with helpful encouragement, information, comments and suggestions to this third edition. Particular and sincere thanks go to the following persons, whom the author has drawn upon for their experience and knowledge and for their input to the following chapters, and therefore deserve to be mentioned for their contribution to the book:

Senior Geotechnical Engineer Bjørn Finborud, Norconsult AS, Norway: Chapter 1.
Senior Geotechnical Engineer Odd Gregersen, Norwegian Geotechnical Institute, Norway: Chapters 1 and 11.
Senior Port Consultant George R. Steel, Scott Wilson Group, UK: Chapters 1 and 13 and co-author to Chapter 20.
Special Adviser Arne Lothe, Norconsult AS, Norway: Chapters 1 and 2.
Professor Shigeru Ueda, Tottori University, Japan: Chapters 4, 14 and 21.
Project Manager Sergi Ametller, Panama Canal Enlargement, Grupo Unidos por el Canal, Panama: Chapter 2.
Professor Knut V. Høyland, The Norwegian University of Science and Technology, Norway: Chapter 2
Professor Han Ligteringen, Delft University of Technology, Netherlands: Chapter 2.
Project Manager Frederic J. L. Hannon, TOTAL Tour Atlantique, France: Chapter 3.
Manager Andy Dogherty, OCIMF (Oil Companies International Marine Forum), UK: Chapters 1 and 3.
Senior Marine Advisor Graeme Robertson, Shell Shipping Technology, UK: Chapters 3 and 4
Dr Mark McBride, HR Wallingford, UK: Chapters 3, 4 and 21.
Consulting Engineer Svein Ove Nyvoll, Nyvoll Consult AS, Norway: Chapters 6, 18 and 19.
Sector Head Structures Steinar Leivestad, Norwegian Council for Building Standardization: Chapter 6.
Senior Structural Engineer Bjørn Vik, Norconsult AS, Norway: Chapter 6.
Consulting Engineer Hans Petter Jensen, Sivilingeniør Hans Petter Jensen AS, Norway: Chapters 9, 10 and 11.
Senior Geotechnical Engineer Asmund Eggestad, Norconsult AS, Norway: Chapter 10.
Technical Director Dr Svein Fjeld, Dr.techn. Olav Olsen AS, Norway: Chapters 3, 10 and 15.
Senior Adviser and Strategy and Research Manager Per G. Rekdal, Oslo Port Authority, Norway: Chapter 12
Manager Marc Sas, IMDO International Marine and Dredging Consultants, Netherlands: Chapter 15.
Senior Hydraulic Engineer Henk J. Verheij, Delft Hydraulics, Netherlands: Chapter 15.
Port Director Tore Lundestad, Borg Port Authority, Norway: co-author to Chapter 13, and for his general input to the book.
Technical Director Stephen Cork, HR Wallingford, UK: Chapter 13.
Director Gisli Viggosson, Icelandic Maritime Administration, Iceland: Chapter 14.
Senior Adviser Reidar Kompen, Norwegian Public Road Administration, Norway: Chapters 17, 18 and 19.
Concrete Technologist Bernt Kristiansen, AF Gruppen ASA, Norway: Chapters 17, 18 and 19.
Chief Engineer Trygve Isaksen, Norconsult AS, Norway: Chapter 19.

In addition, thanks go to the following friends and colleagues whom I have known for many years and have had the privilege of exchanging experiences with in the field of port and harbour engineering for this book:

Senior Port Engineer Hans Magnus Rossebø, Norconsult AS, Norway, who has been a close colleague for many years and for his valuable comments to the book.
Professor Alf Tørum, Sintef Coastal and Ocean Engineering, Norway.
President Harbour and Marine Construction Lars Gunnar Andersen, Altiba Construction Company AS, Norway.
Professor Øivind A. Arntsen, The Norwegian University of Science and Technology, Norway.
Project Manager Harbour Facilities Rikard Karlstrøm, Multiconsult AS, Norway.
Senior Advisor J. U. Brolsma, Rijkswaterstaat, the Netherlands.
Director Jose Llorca Ortega, Department of Technology and Standards, Puertos del Estado, Spain.
Technical Director Per Øivind Halvorsen, Oslo Port Authority, Norway.
Technical Director Svein-Inge Larsen, Kristiansand Port Authority, Norway.
Chief Engineer Roar Johansen, Borg Port Authority, Norway.
Senior Port Engineer Riyad Zen Al-Den, Oslo Port Authority, Norway.
Managing Director S. Ghosh, Consulting Engineering Service, India.

PIANC is thanked for letting me use information and materials from their publications.

Thanks are also due to ICE Publishing for all their help and assistance during the preparation of the typescript for the third edition of the *Port Designer's Handbook*.

About the author

Carl A. Thoresen graduated from the University of Strathclyde, Glasgow, in 1963. During more than 45 years as a professional consulting engineer he has been involved in practical engineering and research for over 800 different small and large port and harbour projects both in Norway and overseas as the Chief Engineer and responsible consultant for the largest Norwegian consulting company, Norconsult AS. For the last decade he has run his own company, PORT-CAT. In addition, he has worked for the Norwegian Agency for International Development, the World Development Bank, the Asian Development Bank and the African Development Bank.

The projects he has overseen include waterways, complicated oil and gas harbours, commercial multi-purpose and container terminals, breasting and mooring dolphins, lighthouses, breakwaters, ferry berths, fishing harbours, marinas and small craft harbours. His involvement on these projects has included planning, technical and economic evaluations, design, preparations of tender documents, tender evaluations and negotiation, construction supervision, maintenance and rehabilitation work.

He is the author of the books *Port Design: Guidelines and Recommendations* (published 1988) and *Port Designer's Handbook* (first edition in 2003, reprint in 2006 and a second edition in 2010), and the co-author of *Norwegian Recommendations for Waterfront Structures* and the *Norwegian Recommendations for Concrete Construction in Water.*

He has been a member of many official technical committees for harbour design, constructions and waterfront structures, and actively involved in the PIANC's various working groups, including:

Working Group 145, 'Berthing Velocity and Fender Design'
Working Group 135, 'Design Principles for Container Terminals in Small and Medium Ports'
Working Group 116 (chairman), 'Safety Aspects Affecting the Berthing Operations of Large Tankers to Oil and Gas Terminals'
Working Group 48, 'Guidelines for Port Construction Related to Thrusters'
Working Group 33, 'Guidelines for the Design of Fender System 2002'.

Over the years he has presented many lectures on berth and port structures at the Norwegian University of Science and Technology, the Chalmers University of Technology, Sweden, and on postgraduate courses arranged by the Norwegian Society of Chartered Engineers.

He has been an active participant in around 30 international conferences. He has written some 50 papers and articles on port and harbour planning, berth and fender design and construction, and marine repair and rehabilitation works.

Port Designer's Handbook
ISBN 978-0-7277-6004-3

http://dx.doi.org/10.1680/pdh.60043.001

Chapter 1
Port planning

1.1. Introduction

The advantages and disadvantages of various berth alternatives for accommodating all types of ship in a port cannot be assessed in detail without well-developed and well-defined port plans. All port plans represent a set of compromises between several goals. This chapter evaluates the activities necessary for the preparation of a detailed port plan, and discusses the criteria that form the basis of the planning, from the open sea, through the approach channel, the harbour basin, the berth and the terminal, as indicated in Figure 1.1.

The port authority and its consulting engineer should identify the activities necessary to establish the terms of reference for the engineering planning and to specify the work to be executed by, for example, the consulting engineer, the contractors and the port operator, within the fixed margin of expenditure.

1.2. Planning procedures

There are many activities, all of which have to be recorded, clarified and assessed. Essential basic information includes, among other things, data on the physical and technical conditions in the development area and information obtained from experienced port users. A checklist for the planning of port developments should cover at least the following main items:

(a) resolution by the port authority to start planning
(b) selection of the consulting engineer
(c) scope of work:
  (i) introduction
  (ii) background
  (iii) scope of project
  (iv) basic data
(d) registration of users:
  (i) public
  (ii) private
(e) recording of users' needs:
  (i) types of port and berth structures
  (ii) traffic statistics
  (iii) types and specifications of ships
  (iv) coastal areas and maritime conditions
  (v) berth and land area requirements
  (vi) growth factors
(f) impact study
(g) site evaluation:

Figure 1.1 The activities that need to be investigated

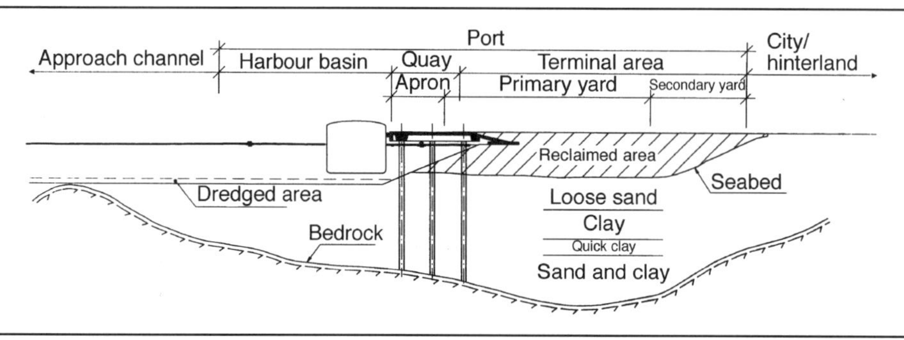

    (i) existing areas
    (ii) potential areas
    (iii) natural conditions
    (iv) relationship with neighbours
(h) layout plan
(i) economic analysis
(j) work schedule.

The items listed above are outlined in the following sections, in order to describe the various activities that require closer study and assessment in connection with proper port planning. But one should always remember that ports often define their own needs. Some ports are predominantly bulk ports, others are high-value cargo ports and others are multi-purpose ports, etc. Depending on the character of existing traffic and expectations about future potential, the port's needs and future capacities will vary. A port usually exists in a dynamic business and social environment, and therefore the needs of the port can change rapidly over short periods of time.

1.2.1 Resolution to start planning

After engaging a consultant, but before the planning starts, it is essential that the client or the authority concerned has prepared a project plan stating clearly the conditions and target of the planning or the work to be done.

The planning and implementation of a project for a new port or for a major port extension can be subdivided into the following main phases:

(a) project identification study
(b) preliminary planning study:
    (i) reconnaissance mission
    (ii) fact-finding mission
    (iii) feasibility study
    (iv) appraisal mission and study
(c) detailed planning work:
    (i) inception planning
    (ii) interim planning
    (iii) final planning and report

(d) pre-engineering work:
    (i) design criteria and structural specifications
    (ii) preliminary cost estimate
    (iii) final pre-engineering report
(e) detailed engineering work:
    (i) design calculations
    (ii) tender drawings (formwork drawings)
    (iii) technical specification for construction
    (iv) bill of quantity
    (v) tender evaluation
(f) construction work:
    (i) construction drawings
    (ii) construction supervision
(g) project completion report.

One of the most important tasks is probably the preliminary planning study, which serves the purpose of verifying whether a suggested project is really justified from an economic point of view and whether it can be implemented at a reasonable cost under safe technical conditions. The most convenient site for the suggested works should be tentatively selected, or alternative locations suggested. A preliminary plan of the port, an approximate cost estimate and an economic and financial evaluation should form the final part of a preliminary study report.

The results of the planning study for the port development should, therefore, always be summarised in an action-oriented programme containing an evaluation of the following:

(a) operational analysis
(b) technical analysis
(c) economic analysis
(d) financial analysis.

The fate of the project will depend on the conclusions of the preliminary report. The general character of the port, the layout of the port facilities, and their capacity and extent are determined in the preliminary plan, notwithstanding such modifications or corrections that may be made afterwards. The preparation of a preliminary study should, therefore, be entrusted to port planners with the widest possible range of experience, both in technical planning and in port operation under various conditions, and a thorough understanding of economic and transportation problems. When the general conclusions of the preliminary study have been approved and its recommendations accepted, the next predominantly technical phase of the planning will include all necessary field investigations and the detailed design.

1.2.2 Selection of the consulting engineer (planners)

It is a fact of life that the competition for consultancy business in the harbour sector is now tougher than it has ever been. It is therefore necessary for consultancy companies to be highly specialised in the use of the latest technical skills and development tools. The company's past experience and performance as a whole, the experience of the leading personnel who will be involved in the project, and the company's proposed methodology are important factors for the client to consider when selecting a consulting engineer.

As a basis for selection, the consulting engineers must enumerate and describe the projects they have undertaken, naming previous employers for reference. They should also indicate the general manpower available (e.g. graduate engineers), whether they can step-up planning and design if so desired by the client, and whether they can mobilise divers (frogmen), an underwater camera, a diver's telephone outfit, etc., to carry out underwater investigations and supervision.

A client should always make sure that the personnel named in a proposal will also form the project team working on the project. If a team member is replaced, the client should always demand that the new team member have at least the same qualifications as the original team member.

The Federation International des Ingenieurs-Conseils, **FIDIC**, has the following policy statement:

> A Consulting Engineer provides a professional service. A Client, in selecting a Consulting Engineer, is selecting a professional adviser. The Consulting Engineer's role is to put expert knowledge at the disposal of his Client. On engineering matters, he serves his Client's interests as if they were his own. It is essential that he should have the necessary ability. It is equally important that the Client and Consulting Engineer should proceed on the basis of mutual trust and co-operation. In the professional relationship, the Consulting Engineer identifies with his Client's aims.

It is in the client's interest to select the most qualified and experienced company and to negotiate a fair price for the consultancy services. Saving 1 or 2% of the project cost on engineering is penny-wise and pound-foolish.

Payment for consultancy services can be defined in the following ways:

(a) payment on a time basis
(b) payment of a lump sum, based on either:
    (i) the consulting engineer's estimate of the work involved, or
    (ii) a generally accepted fee scale
(c) payment as a percentage of the cost of the works.

Direct expenses, such as travel costs, hotels, etc., are normally reimbursed separately. The fee for the consultancy service itself is normally invoiced at agreed intervals.

Some international development banks select the consulting engineer after what was previously called **the two envelope system**. The consulting engineer is requested to submit its proposal for consulting services in one technical envelope and one financial envelope. The technical proposal should be placed in a sealed envelope clearly marked 'technical proposal' and the financial proposal should be placed in a sealed envelope clearly marked 'financial proposal'.

The **technical proposal** should contain the following:

(a) The consulting company's general expertise for doing the work. If the company does not have the full expertise, the company can be associated with another company. For work in a developing country it is considered desirable to associate with a local company.
(b) Any comments or suggestions on the terms of reference (TORs) and a description of the methodology (work plan) the consulting engineer proposes for executing the services, illustrated with bar charts of activities and the graphics of the type used in the critical path method (CPM) or programme evaluation review technique (PERT).
(c) The estimated number of key professional staff required to execute the work according to the TORs. The composition of the proposed engineering team, and the task which would be

assigned to each member. Estimates of the total time effort supported by bar charts showing the time proposed for each team member.
(d) Curricula vitae (CVs) for the proposed key team members. The majority of the team members should be permanent employees of the company.
(e) If the TORs specify training as a major component of the assignment, the proposal should include a detailed description of the proposed methodology, staffing, budget and monitoring.

A maximum of 100 points is given to the technical proposal and the points are usually divided as follows:

(a) specified experience of the company related to the assignment: 10 points
(b) adequacy of the proposed work plan and methodology in responding to the TORs: 30 points
(c) qualifications and competence of the consulting team: 60 points.

The **financial proposal** should contain and list the costs associated with the assignment, and include the following:

(a) the remuneration for the staff assigned to the team, either foreign or local, both in the field and at headquarters
(b) subsistence per diem, housing, etc.
(c) transportation, both international and local, for the team
(d) mobilisation and demobilisation
(e) services and equipment, such as vehicles, office equipment, furniture, printing, etc.
(f) the tax liability and the cost of insurances.

The **evaluation of the proposals** will usually be carried out according to the alternatives described below:

(a) When the proposals have been invited on a fixed budget basis, the client and/or the evaluation committee will select the firm that submitted the highest ranked technical proposal within the indicated budget (this is normally supplied with the invitation to tender). Proposals that exceed the indicated budget or which do not get the minimum score, usually 75 out of 100 points, will be rejected. The company having the best technical proposal will then be invited to contract negotiations.
(b) When the selection is to be made on a quality basis, the highest ranked firm on the basis of both the technical and the financial proposal is invited to negotiate a contract. Under this procedure, the lowest priced financial proposal is given a financial score of 100 points. To obtain the overall score the financial proposal is typically counted as 20% and the technical proposal 80% of the overall score.

1.2.3 Scope of work

The consulting engineer should assist the client in defining the assignment. The following items should, therefore, be clearly specified.

Introduction

(a) The client and client's contact persons
(b) the type of project (preliminary or final design)

(c) the geographical position and boundaries of the project
(d) who and what will be affected by the project (people, firms, etc.).

Background

(a) Project background (existing infrastructure, traffic increase, development restrictions, old installations, excessive maintenance costs at existing facilities, etc.).

Scope of project

(a) The project area and boundaries
(b) project involvement (activities, nature of work, scope of project, etc.)
(c) the schedule for execution of the project.

Basic data

(a) Which reports and data can be used as a basis? When were the reports prepared and by whom?

1.2.4 Registration of users

Experience has shown that it can be difficult to register all port users who may influence the preparation of a port plan, i.e. the present and potential users of the harbour facilities. It is advisable to register users in the following two groups.

Public users

(a) Port authorities, municipality, district, state, etc.

Private users

(a) Shipping companies
(b) private industry, service industry (charterer, stevedores, etc.)
(c) clubs (marinas, etc.).

1.2.5 Recording users' needs

The rapid growth in regional and international trade has placed increasing demands on the shipping and port sectors. Therefore, in order to achieve a realistic port plan it is essential to record the needs of users which may have a bearing on the plan and to obtain relevant data on this subject. The recording of users' requirements must be carried out in close cooperation with the client or port authority. This also includes the organisation of the port itself. The following are different approaches to organising the port:

(a) **Resource (tool) port**: the port owns the land, infrastructure and fixed equipment, provides common-user berths and rent-out equipment and space on a short-term basis to cargo-handling companies and commercial operators.
(b) **Operating (service) port**: the port provides berths, infrastructure and equipment, together with services to ships and their cargo.
(c) **Landlord port**: for larger ports this is the most common system, wherein the port owns the land and basic infrastructure and allows the private sector to lease out berths and terminal areas.

In order to obtain a general view of users' needs, data should be recorded as outlined below.

Type of port facilities

The type of port facilities should be evaluated and registered as follows:

(a) Evaluate and register the various types of port facilities that exist and identify those that would meet the future requirements (commercial port, bulk-cargo port, industrial port, fishing port, supply port, ferry berths, marinas).
(b) Register which of the facilities would be public or privately owned.

Recording of traffic data cargo volumes

The planner should have access to statistics compiled either by private companies or by port authorities on traffic, traffic density and the volume/tonnage of goods handled in the harbour area. Unfortunately, the statistics seldom specify the type, size, weight and other details of individual consignments. It may therefore be necessary to carry out additional recording and research for a limited period of time in order to obtain annual, weekly and daily averages of the port traffic and to identify peaks. This recording, or more detailed research, must be oriented towards the port plan objectives. The general pattern of the cargo flow which can be expected through the terminal area of a port is illustrated in Figure 1.2.

Figure 1.2 Cargo flow through a port

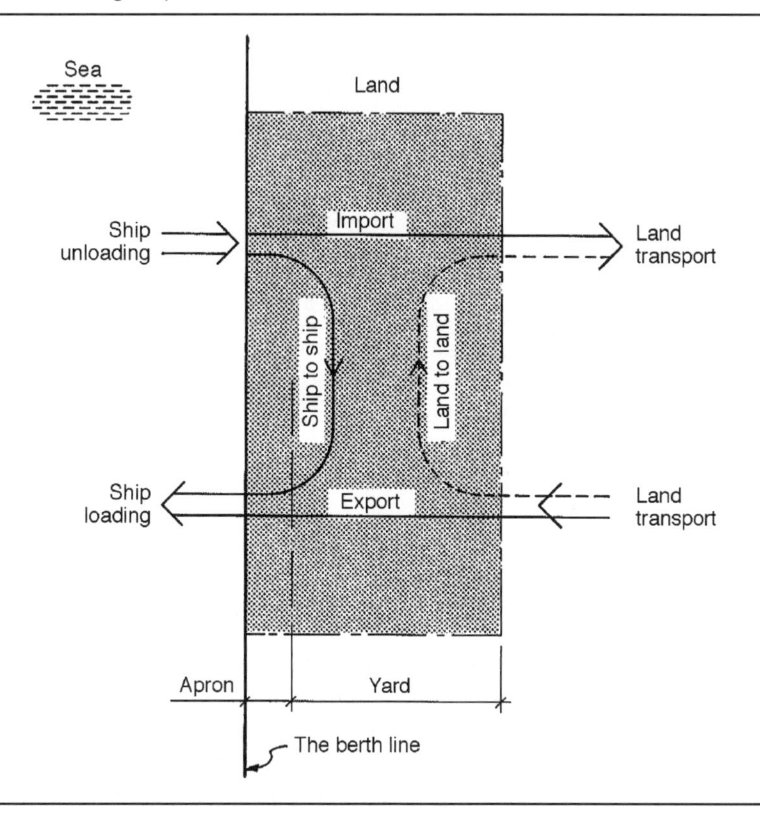

The recording of **traffic densities** and cargo volumes should give a detailed account of cargo and passenger handling by day of week, hour of day, and mode of transportation to and from the port, for the following:

(a) ocean-going tramp ships
(b) foreign liner ships
(c) domestic liner ships
(d) ferries
(e) trucks
(f) buses
(g) railways
(h) possibly aircraft.

The **annual turnover** in the port should, if possible, be subdivided into the following categories:

(a) bulk/general cargo
(b) transhipment ship/ship
(c) transhipment ship/rail
(d) goods carried by coastal ships/tramps
(e) goods/general cargo handled at terminal
(f) storage time
(g) type of storage
(h) customs clearance.

Commodities should be described in detail:

(a) type of cargo
(b) present and potential cargo tonnage and volume
(c) frequency of cargo arrival
(d) origin and destination of cargo
(e) times of loading and discharging
(f) space requirements for cargo
(g) cargo-handling rate/time of storage
(h) commodity classification
(i) cargo-handling operations analysis
(j) storage requirements (cold or warm).

It is essential to specify if the goods require **special handling equipment**, such as:

(a) loading and unloading equipment
(b) capacities of cranes (mobile or stationary)
(c) fork-lift trucks.

Based on the above-mentioned data it is possible to evaluate the optimum storage time. Generally, owners prefer goods to remain in the harbour until required. This means that, for the owner, unnecessary loading/unloading should be avoided. On the other hand, port authorities want goods to leave the harbour as quickly as possible so as to clear the harbour for new arrivals. Therefore, the design criteria to be adopted for the harbour area should be based on actual shipping statistics and cargo volumes.

For the majority of ports the storage time in the port area is one of the most critical factors in evaluating the port capacity. Nowadays, with faster loading and unloading of ships, it is very seldom the berthing capacity (the number of berths) that reduces port efficiency. As a rule, it is the limitation of the storage area behind the berth that is the determining factor. Roughly, one can say that if the storage time can be halved, the harbour capacity can be doubled. This would ensure the best use of invested capital and would, at the same time, result in lower cost per ton handled.

Types and specifications of ships

During the past 20 years in particular, trends in shipping have had a great impact on port and harbour development. The larger tankers, container ships and cargo ships require deeper water and highly mechanised cargo-handling equipment and systems. The rapid growth in, for example, containerisation has had a great effect on the handling equipment, the layout of the yard and the size of the berth. For this reason the following must be studied closely:

(a) ship types (fishing, cargo, roll-on/roll-off, load-on/load-off, tankers, warships, etc.)
(b) ship sizes
(c) frequency of arrivals and times of day
(d) ships' origins and destinations
(e) analysis of future conditions.

Based on ship parameters, one can analyse the demand for berth facilities and determine the required water depths at the various berth structures.

Coastal areas and maritime conditions

Port basin requirements, as well as local maritime conditions, have sometimes been neglected in port planning. A port plan should include information about the following:

(a) the general conditions of navigation between open sea and the berth facilities, tugs required, anchorage grounds, waiting area
(b) the length, width and depth of the access channel and basin area, and depths together with the location of submarine cables, etc.
(c) restrictions in manoeuvring conditions due to wind and current, and possibly in terms of waiting time for better weather conditions
(d) the need for shelter
(e) requirements regarding pilot services, beacons, safety zones, tugboat assistance, etc.
(f) collision possibilities and other dangers such as height obstructions (bridges, high-voltage lines)
(g) possible restrictions with regard to berthing and departure times.

Berth and land area requirements

Port owners should be aware of the danger of not forward planning and should always be conscious of possible future requirements. Therefore, one of the most important requirements of a long-term port plan is that it should have built-in flexibility. A good port plan is one in which the basic strategy remains intact even when some of the details need to be adjusted.

In addition, probably the most important function of the port plan is to reserve at least 25% more land for the port to develop and expand to meet future growth in traffic or changes in technology. Therefore the following information should be obtained:

(a) Location of the port area in relation to:
  (i) regional conditions
  (ii) local conditions
  (iii) local traffic (railways, trucks, bicycles, passenger traffic, pedestrians, ferry traffic with parking space for waiting vehicles, etc.)
  (iv) car parks (public and private).
(b) Location of berths:
  (i) general requirements
  (ii) relationship to port area
  (iii) natural conditions
  (iv) neighbourhood relationship.
(c) Size of berth:
  (i) types and numbers of berths (general cargo, containers, roll-on/roll-off, load-on/load-off, bulk, etc.)
  (ii) length and surface of each berth, depth alongside
  (iii) dolphins
  (iv) special mooring facilities
  (v) loads
  (vi) utilisation of the port facilities
  (vii) utilisation of berth capacities for separate berths.
(d) Land area:
  (i) present and future needs for land areas
  (ii) indoor and outdoor storage capacity
  (iii) access roads
  (iv) development of land area (gravel, asphalt, type of traffic, etc.)
  (v) area restrictions (building lines, cables, power lines, etc.)
  (vi) facilities for dockers, stevedores, service functions
  (vii) local authorities' demands and benefits.
(e) Demand for auxiliary services and installations:
  (i) electricity, water and telephone connections
  (ii) lighting
  (iii) mooring facilities, fenders
  (iv) lifesaving equipment and ladders
  (v) refuse collection and disposal, cleaning of tanks, waste-water tanks, oil-protection and fire-fighting equipment
  (vi) water and fuel bunkering
  (vii) maintenance facilities
  (viii) repair workshop, slip.

When evaluating potential sites, it is advisable to divide the port into activity zones.

1.2.6 Growth factors

During the recording of available data to be used in the port planning process, the planner must bear in mind the consequences that the port development will have on society:

(a) population increase locally and regionally
(b) economic growth
(c) traffic growth and modified transport modes

(d) industrial developments
(e) environmental problems.

1.2.7 Impact study

In recording users' requirements, the planner must assess whether the existing activities in the harbour can, or have to, be relocated and what the impact will be for a specific user. Would such relocation represent great expense to the user and thus make him less competitive? Or would the relocation to more developed areas make the user more competitive?

At an early stage one must assess the negative impacts that may arise. For instance, increased ship traffic would also increase the possibility of ships colliding. This could be disastrous, especially if ships carrying dangerous cargo, such as gas, ammunition, etc., were to collide. In other words, an impact study must consider all risks involved.

1.2.8 Site evaluation

Potential sites for harbour development must be examined to ensure that they meet the specific functions of the port and the needs of various port users. A detailed investigation is often not necessary, but the examination should provide sufficient information for the evaluation. The following items should be considered.

Natural conditions

It is often difficult to visualise the effect that the natural conditions may have, but problems can be solved by means of model studies. The following must be evaluated:

(a) Topographical and maritime conditions:
   (i) description of land area/topographical conditions
   (ii) hydrographical conditions
   (iii) manoeuvring and navigation conditions.
(b) Geotechnical conditions:
   (i) stability and load-bearing capacity
   (ii) choice of type of berth structure
   (iii) location of berth based on geotechnical evaluation
   (iv) seabed conditions
   (v) dredging and blasting conditions
   (vi) dumping authorisation
   (vii) sounding, acoustic profiles.
(c) Geological conditions:
   (i) structure and composition of strata.
(d) Water-level recordings:
   (i) tidal variations
   (ii) depth references.
(e) Water quality:
   (i) quality of water (pH value, salt content, etc.)
   (ii) degree of pollution
   (iii) visibility
   (iv) corrosion characteristics (corrosion, deterioration of concrete, attacks by marine borers).
(f) Wind:
   (i) wind force, directions and durations (compass card)

   (ii) critical wind forces and directions.
(g) Waves:
   (i) wave heights caused by wind, significant wave length, maximum wave height,
(h) Wave direction
   (i) swell
   (ii) waves from passing vessels.
(i) Climatic conditions:
   (i) air temperatures (maximum, minimum)
   (ii) air humidity
   (iii) water temperature.
(j) Current:
   (i) strength, direction and duration
   (ii) erosion and siltation, sea bottom conditions.
(k) Ice:
   (i) thickness, duration, extent
   (ii) possibility of ice-breaking assistance.
(l) Visibility conditions:
   (i) fog and number of foggy days
   (ii) topographical conditions
   (iii) need for navigational aids, lighthouse, radar, radio.
(m) Evaluation of natural resources:
   (i) impact of development on environment.
(n) Model testing:
   (i) stability, protection measures and facilities
   (ii) erosion and sedimentation frequencies.
(o) Materials:
   (i) investigation of available local construction materials.
(p) Contractor's equipment:
   (i) assess type, availability and capacity of equipment for marine structures.

The choice of type of berth structure and layout should be based on a thorough knowledge of the natural conditions and the market circumstances. Superficial investigations of these factors could result in severe negative economic consequences.

Relationship with neighbours

The following should be assessed to determine the impact of a port plan on the neighbourhood:

(a) Existing properties:
   (i) recording of private ownership of land, of port installations and plant which would be affected by the port development
   (ii) power lines and cables in the area, water supply and sewerage systems, etc.
(b) Local traffic:
   (i) by land
   (ii) by sea (size of ships and frequency of calls)
   (iii) hindrance and delay caused by traffic from neighbours
   (iv) anchorage grounds required by neighbours
   (v) manoeuvring needs of neighbours
   (vi) tugboat use by neighbours.

(c) Cargoes:
   (i) existing and future volumes and tonnages of cargo to be transported to adjacent premises by land or sea.
(d) Offshore traffic:
   (i) ship traffic outside the development area, size of ships and frequency, range and sailing channel.
(e) Future traffic:
   (i) forecast of seaborne and overland transport in the area
   (ii) frequency of calls at neighbour berth and movement of vessels offshore.
(f) Damages and drawbacks:
   (i) damages, drawbacks and consequences to neighbours due to their own development and traffic
   (ii) damages and drawbacks caused by a third party's development (waves and nuisance due to increased traffic)
   (iii) traffic conditions (queues, etc.).
(g) Possibilities of expansion:
   (i) acquisition of neighbouring sea and land areas
   (ii) acquisition costs.

1.2.9 Layout plan

Based on records of users, needs of users, assessment of consequences, location of site, etc., a port layout plan can be prepared to cater for the various activities in the area. Such a layout plan can, in addition to being used to select the most convenient technical solutions, include various political solutions, which will not be dealt with here.

1.2.10 Economic analysis

All port projects can be broadly divided into the following three main groups:

(a) rehabilitation of existing port facilities
(b) expansion of port facilities
(c) development of a new port.

Therefore, in order to obtain a picture of what a project would cost, it is advisable to show the costs of the planning phase and of the development phase separately:

(a) The planning phase expenditures will depend on how much the client (port authorities, municipalities, government, etc.) is prepared to invest in planning.
   In order to assess the expenditures involved in the planning, one must know how much data collection, site investigations, etc. will or can be carried out by the port authorities, municipal engineers, port users, etc. To obtain a realistic design, geotechnical evaluations of the areas concerned should be carried out. This would facilitate the allocation of areas to the various activities.
(b) The construction-phase expenditures will mainly comprise the construction expenses plus the consulting engineer's study expenses. The latter usually amount to 3–5% of the construction expenses.
   In connection with planning, there will almost always be alternative development options, which can influence the choice of berth structure. This requires an adequate knowledge of

engineering techniques and an understanding of the requirements to ensure a satisfactory cost/benefit ratio for the project.

(c) There should always, if possible, be competitive tendering between the construction contractors, because without competition there can be no true cost comparison. A lack of competition inevitably leads to a low level of efficiency, and political decision-makers may be unable to see possibilities to save on investment or operating costs, even when these possibilities are within reach.

In principle, the port development should earn a satisfactory financial rate of return, but this could sometimes be difficult to achieve during the first years after the completion of the development, due to the fact that it might take some years for port traffic to build up. In the development of a new port, some construction items, such as a new breakwater, large dredging works, etc., will have long service lives and might be adequate for further port expansion in 20 or 30 years. Therefore, during the first years of operation it may not be possible for the traffic of a new port to be strictly cost-based.

1.2.11 Work schedule

A complete port plan includes a programme for a staged development of the port. One must first of all try to record when the various parts of the port should be made ready for the respective users. This work schedule will also indicate when construction work should start on the various new areas.

1.3. Subsurface investigations

1.3.1 General

One of the most important tasks in the planning and design process is to get detailed knowledge of the geotechnical and the subsurface conditions of the port area. This is an important basis for selection of the most suitable port structures, efficient design of the facilities and a smooth construction process. Construction difficulties and cost overruns of port projects are commonly caused by unexpected soil or rock conditions. Sufficient site investigations can, therefore, be considered as a relatively cheap 'insurance premium' to reduce the risks of encountering technical and economic problems during implementation.

The investigations should provide data for various engineering and construction issues, such as:

(a) foundation of onshore and marine structures
(b) settlement of reclaimed areas
(c) stability of filled and dredged slopes
(d) nature of soils to be dredged and suitable dredging methods
(e) contaminated sediments
(f) use of dredged materials
(g) sources and properties of natural construction materials such as embankment fills.

The required extent of the investigations and the depth of boreholes will depend on the type and size of project, the complexity of the site conditions and the information that is already available. Investigation depth should be deeper than any potential stability failure surface, and deeper than the possible penetration of foundation piles or sheet piles or to bedrock. British Standard BS 6349-1: 2000 (BSI, 2001, Article 49.6) recommends a minimum boring depth below the base of the structure of 1.5 times the structure width, and in dredging areas to hard strata a minimum of 5 m below dredging level.

The first step in the search for data on the soil and rock conditions is the collection of existing information, the main sources of which are:

(a) geological information from maps, reports and study of aerial photos
(b) reports from earlier physical investigations in the area, collected from official or private files
(c) previous history and use of the site, including experience from construction works in the vicinity and any defects and failures attributable to ground conditions.

The next step is to analyse the project and the need for additional information, in order to establish a programme for necessary site investigations. For large projects, such investigations will normally be done in steps corresponding to the requirements of the different planning stages. The most detailed and most costly investigations will be made in connection with the design and construction phases. Sufficient data should be available before a construction contract is signed.

The port authority or client and his consultant normally determine the necessary extent and types of investigations but there may also be minimum requirements stated in local codes, such as building codes in the USA.

1.3.2 Organisation of the site investigations

Methods and procedures for undertaking site investigations vary from one country to another, depending on local practice and available resources. It is generally advisable to follow local practice if it has proved successful in the past.

In some countries or areas, authorities have their own equipment for basic investigations. The normal procedure is that the employer or consulting engineer engages a firm to carry out the site investigations. Such firms could be consultants or contractors having a range of investigation equipment, or specialist firms with a narrow product line, for instance geophysical investigations.

Offshore investigation can be performed from a barge properly anchored and positioned, or a jack-up platform as shown in Figure 1.3. Also available are vessels and jack-up platforms specially equipped for subsurface investigations, which are used on large projects, at great water depths and/or to avoid disturbance from waves. Survey for positioning is usually done using a global positioning system (GPS).

A normal procurement procedure is that qualified firms are invited to submit competitive tenders. A contract is established between the client or engineer and the successful firm(s) based on recognised local or international contract conditions, for instance the FIDIC Model Sub-consultancy Agreement or, for large assignments, the FIDIC Red Book (FIDIC, 1999).

The extent of supervision that the client or engineer will need to do will depend on the reliability of the selected firm, but some monitoring and supervision of the investigations is highly recommended. Field investigations and laboratory testing should be done according to recognised procedures, which for most methods are stated in international standards or codes of practice, such as the American Society for Testing and Materials (ASTM) and American Association of State Highway and Transportation Officials (AASHTO) standards, British standards or codes (BS) or European standards (Eurocodes).

The contractor is normally required to submit progress records and boring logs during the course of the work, and a final report together with documentation of all investigation results.

The investigation methods may be divided into the following types or groups:

(a) geophysical methods

Figure 1.3 A jack-up platform. (Courtesy of Altiba AS, Norway)

(b) soundings or simple borings
(c) in situ tests
(d) soil and rock sampling
(e) field trials
(f) laboratory tests.

The methods most commonly used are described in the following sections. Different methods may provide similar information. The selection of a method will, to a large extent, be determined by availability and costs. Each method has certain advantages and limitations, which should be taken into consideration.

1.3.3 Geophysical methods

Geophysical methods are suitable for initial surveys offshore because they can provide a significant amount of information at a modest cost. Such methods do not provide accurate data on soil and rock conditions, and need to be calibrated using information obtained from boring, sampling and/or in situ tests.

The commonly used methods are seismic reflection and seismic refraction.

Seismic reflection (acoustic profiling)

Acoustic profiling is performed using equipment towed by a vessel at low speed (2–5 knots), working according to the same principle as an echo sounder used for depth recording, but operating at a different frequency that allows penetration into the seabed. A high-energy acoustic pulse passes

Figure 1.4 Acoustic survey

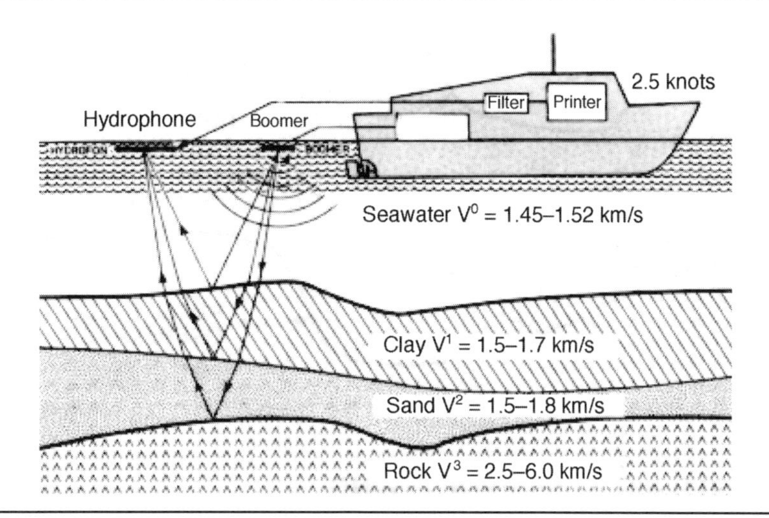

through the water and into the soil (Figure 1.4). The pulse is reflected back by the seabed and by subsoil interfaces corresponding to changes in density and sonic velocity. A profile of the return times from the various reflecting horizons is continuously recorded and transformed into electric signals that are visualised by printing a 2D diagram (echogram), which needs to be interpreted by a specialist.

Different types of equipment are available, with different characteristics and capabilities. The main difference is the frequency, which determines the maximum penetration depth and the resolution accuracy. Low-frequency pulses achieve deep penetration but have a low resolution. The main types are:

(a) Penetrating echo sounder (pinger): frequency 3–10 kHz, penetration depth up to 25 m and resolution around 0.5 m.
(b) Boomer, consisting of a metal plate placed on a small catamaran: frequency 0.5–2.5 kHz, penetration depth 100 m and resolution 1 m.
(c) Sparker, consisting of a metal frame with electrodes, which is towed slightly below the water level: frequency 0.05–1.0 kHz, penetration depth 500 m and resolution 5 m.

The indicated penetration depths are valid for sediments of clay, silt and sand. The results obtained from such a survey are the thickness of soil above bedrock and borders between different soil types. Determination of soil layering is uncertain unless there are large and distinct changes in the seismic velocities.

The main advantage of the method is that a rough picture of the subsoil conditions over large areas can be obtained quickly and at a modest cost. With profiles measured at intervals of 10 m, an area of around 20 km^2 can be covered in one day. The disadvantage is that the results are not accurate and need to be checked and calibrated by comparison with data obtained from boring and or sampling.

Seismic refraction

A seismic refraction survey can be used on land and in the sea. A system of seismometers (geophones) is placed along a profile and connected by cables to a recorder. A shock wave is initiated by explosives or

Figure 1.5 The difference between the reflection method and the refraction method

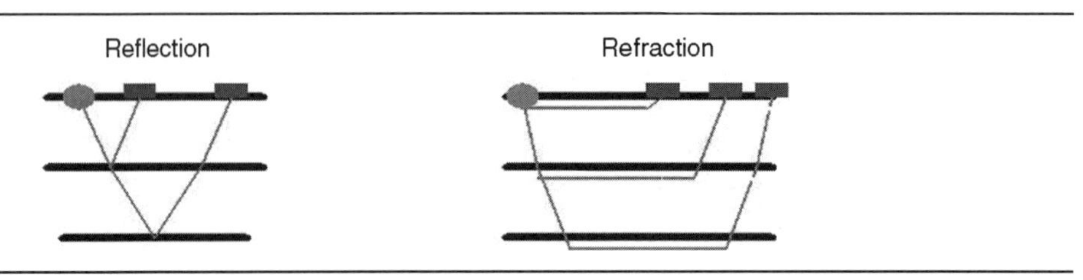

by a hammer blow. The wave travels faster in more consolidated soil than in soft soil, and fastest in compact bedrock. By recording the travel time of the direct and reflected waves to the geophones, a profile of soil boundaries and depth to bedrock can be obtained.

Seismic refraction is more accurate than a reflection survey (Figure 1.5) but the capacity is less and it is therefore more costly. Seismic refraction is used for preliminary investigations over relatively limited areas and to supplement acoustic surveys. The refraction results also need to be checked and calibrated against data obtained from borings and/or sampling.

1.3.4 Soundings or simple borings (probings)

Soundings are performed by inserting a rod with a special tip into the ground by rotation, hammering or by other mechanical means. The methods are named according to the method of advancing the rod, such as rotary sounding, ram sounding and motorised sounding. A range of equipment from hand-held engines to rock-drilling plant is available.

The recordings consist of the soil resistance plotted against penetration:

(a) Rotary sounding: number of rotations per unit of penetration (for instance every 20 cm).
(b) Ram sounding: the number of blows per unit of penetration.
(c) Rotary/pressure sounding: continuous (automatic) registration of resistance at a specified penetration speed.
(d) Motorised sounding or drilling: the time spent per unit of penetration and/or recorded resistance.

Different types of equipment are used around the world. The penetration capacity depends on the strength of the rod, the type and size of the tip, and the energy of the driving medium. Heavy rock-drilling equipment (Figure 1.6) has very good penetration ability but needs a special flashing technique for deep penetration in soils. It is a suitable method for determining the boundary between soil and hard rock.

Sounding is a simple technique, which provides information on the relative density, layering and thickness of soils. Sounding is suitable for preliminary investigations but needs to be supplemented by data from other investigations for calibration and to provide engineering data. Sounding is also a cheap method of obtaining additional data to interpolate measures between sampled boreholes.

1.3.5 Borings with in situ tests

In situ tests are used to measure certain characteristics or engineering properties of the ground. These methods avoid the disturbance or scale effect associated with testing of small samples in the laboratory.

Figure 1.6 Rock control drilling with water flashing

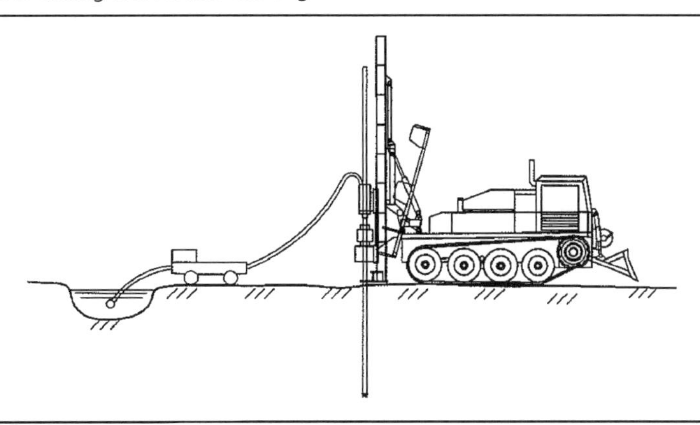

Standard penetration test (SPT)

The standard penetration test (SPT) is historically the most common investigation method used worldwide, particularly in countries influenced by the USA. This test is undertaken in boreholes. A small-diameter tube fitted with a cutting edge (split-spoon sampler) is driven into the soil below the base of a casing by means of a 65 kg hammer with a falling head of 700 mm. The sampling tube is driven 450 mm into the soil and the penetration resistance, or the *N* value, is the number of blows required to drive the tube the last 300 mm. During driving, a disturbed sample is obtained which can be used for soil classification and testing of index properties.

Although there is no direct relation between the *N* value and engineering properties, a range of empirical relationships have been established, which are valid for different soil types and site conditions.

The SPT method has a good penetration capacity in hard layers. Other advantages are the wide availability of the equipment, and the fact that the test is well known and understood by construction contractors.

Cone penetration test (CPT or CPTU)

The cone penetration test (CPT) provides a direct measurement of the resistance to the penetration of a thin rod with a conical tip. The resistance is measured in terms of the end bearing and side-friction components of the tip. The CPT penetrometer is advanced into the ground by a thrust machine placed at the surface or on a floating rig. There is also available CPT equipment designed to operate from the seabed, where the reaction force is provided by a heavy template; this is normally used offshore at large water depths.

The resistance to penetration can be measured mechanically by pressure gauges at the surface at intervals of, say, 200 mm, or continuously by electric load cells mounted in the tip. Electric penetrometers can also be equipped to measure the pore water pressures induced in the soil as the tip advances and the reduction in excess pressure with time during a break in the penetration (CPTU).

Theoretical and empirical relationships between the penetration results and soil types and engineering properties have been developed. Computer programs are also available for the interpretation. Interpretation of CPT/CPTC data requires experience.

The main advantage of this method is that it provides data that are otherwise available only through undisturbed sampling and laboratory testing, and normally obtains these data at a lower cost. Disadvantages are a limited penetration capability in hard layers, and a limited availability of and experience in the method in many regions.

Shear vane test

The shear vane test is used to determine the undrained shear strength of cohesive soils. The vane consists of two blades fixed at right angles and attached to a rod, which is pushed into the soil from the surface, or from the bottom of a borehole. A calibration torque head is used to apply an increasing turning force to the rod until failure occurs.

The measured strength may need to be corrected to allow for the effect of soil friction on the rod, the anisotropy of the soil, certain scale effects and the soil plasticity. In particular, results of trial embankments show that in high-plasticity clays the actual shear strength is less than the vane strength measured in situ, and empirical correction factors have been established.

1.3.6 Soil and rock sampling

Different boring methods are available to provide samples of the ground. The choice of method depends on the type of materials, the penetration depth required, and the requested size and quality of the samples. The methods can be divided into disturbed and undisturbed sampling.

Disturbed sampling

There is a great variety of methods and equipment types available worldwide, and only the main types are mentioned below. The structure of the natural soil may be disturbed to a considerable degree by the action of the boring tool or excavation equipment, but the sample should be representative, i.e. keep a similarity to the material in situ. In addition to the SPT, mentioned under in situ tests (see Section 1.3.5), the following methods provide disturbed samples.

Test pits Excavation of pits by hand or machine is the simplest form of sampling, which is usually limited in depth to 3–6 m. Advantages are that the layering and behaviour of the soil can be observed in the field, and large samples can be taken for laboratory testing.

Auger borings Hand-operated or mechanical augers are cheap means of sampling in favourable soils that do not contain stones.

Percussion methods Soil sampling can be performed using various methods of advancing a borehole into the ground, such as:

(a) Percussion, cable tool, or shell and auger borings, where the soil is loosened by various types of tool and the soil is removed by shell or augers.
(b) Wash boring, where a stream of water or drilling mud removes the soil. To obtain samples, the soil, which is in suspension, needs to be settled in a pond or tank. This sampling procedure is unreliable because certain soil fractions may be lost and the soil structure is lost.

Borehole stability is normally secured by a steel casing, which is pressed or driven down in steps as the hole is advanced. Uncased holes may be used if the walls are stable, such as in stiff clays, or if the walls are stabilised by a drilling mud. A tripod is usually used to operate the casing and drilling tools.

Shallow offshore sampling may be done by gravity or vibro coring, which are rapid and cost-effective methods of investigating soft soils below the sea bottom:

(a) Gravity corers, penetrating by their own weight, and operated from a vessel.
(b) Vibro corers consisting of a frame carrying a sample tube, normally 75 or 100 mm diameter and up to 6 m long, operated by a crane from a support vessel. The sample tube is vibrated into the sea bottom and withdrawn, while the sample is retained by a spring. Larger diameter and more sophisticated vibro corers have also been developed.

Rotary drilling Various types of rotary drills are used to advance boreholes and to provide samples in hard soils and rock. Sampling is done by means of a tube (core barrel), fitted with diamond or tungsten cutting bits, which is rotated by a drilling rig and advances into the soil or rock by cutting an annulus into the material. A core enters the barrel and is retained by a spring. The bit is cooled and lubricated by pumping water or a drilling fluid down the hollow drilling rods and into the bit. Core diameters are typically 50–80 mm, while the length of the core barrels can vary from 1 to 3 m. Cores can be taken at intervals or continuously. They are normally placed in core boxes for inspection, description/classification and further testing.

Undisturbed sampling

Undisturbed samples represent, as closely as is practicable, the true in situ structure and water content of the soil. The usual sampling method is to push or drive a thin-walled tube for its full length into the soil, and then to withdraw the tube with its contents. The typical sample diameter is 50–90 mm. The retracted tube is sealed at both ends and transported to the laboratory for opening, soil classification and testing. Special equipment has been developed to reduce the disturbance when pushing out the sample from the tube.

There are basically two types of such sampler:

(a) Open tubes (Shelby tubes), which are operated from the bottom of uncased or cased boreholes. The upper part of the samples may be disturbed to some degree due to the effects of advancing the borehole.
(b) Various types of piston samplers, where a closed sampler is carried into the soil, and the tube is pushed into the soil at the actual depth. This type is preferable for use in soft clay.

The length of the sample tubes is usually 50–80 cm. Undisturbed samples are typically taken at intervals of 1–2 m.

1.3.7 Field trials

Field trials can provide field data that are not obtained through boring methods, and are used to test foundation methods or field procedures prior to or during the full-scale construction work.

Test piling and test loading Test piles can be driven in order to check the pile-driving resistance the penetration depth and/or the bearing capacity. Bearing capacity can be recorded by analysis of driving data through a device called the pile driving analyser (PDA). This procedure was developed in the early 1970s by a research team at the Case Western Reserve University, Ohio, USA ('Case method'). The data measured are the hammer impact force and the acceleration during driving by use of strain transducers and accelerometers. The data are analysed using a computer program, originally known as the CAse Pile Wave Analysis Program (CAPWAP). The measured data include efficiency of the pile hammer, pile-bearing capacity along the pile and pile integrity. Any structural damage to the pile is indicated, as is the depth of such a deficiency.

PDA equipment is generally available throughout the world, and is often used routinely during production piling by piling contractors. PDA measurements have, to a large extent, replaced full-scale static test loadings, because the PDA, besides being reliable, can be used on a great number of piles and is less costly.

Static pile test loading may be required where there are complex soil conditions, and may be mandatory under certain codes such as those that pertain in the USA.

Trial dredging Data provided by ground investigation methods cannot be transformed directly into a measure of the performance of dredging plant. Therefore, for large dredging undertakings, where geological conditions are complex or at the borderline of the capabilities of certain plant, trial dredging may be advisable. Conditions triggering trial dredging include those where it has to be decided whether hard soil or rock can be dredged directly with available equipment or has to be pretreated, for instance by blasting. Trial dredging is expensive but, if properly carried out and the data properly recorded, provides the best type of information.

Trial embankments Trial embankments are sometimes constructed in areas to be reclaimed in order to provide field data on the rate and magnitude of settlements and to test the effect of ground improvement methods. Trial embankments have historically been used in many cases to test vertical drains, because the effect of such drains with respect to the acceleration of settlements in complex soils cannot be calculated with certainty.

1.3.8 Laboratory tests

Laboratory tests on soil and rock samples can be divided into five groups:

(a) soil description and index (or routine) tests
(b) strength tests
(c) consolidation tests
(d) other physical tests
(e) mineralogical and chemical tests.

Index tests Index tests are generally simple and inexpensive, and provide basic information about the physical characteristics of the material, such as density, porosity, grain size, permeability, organic content, water content, plasticity, etc. The index properties are used for soil classification, and to obtain information on engineering properties through empirical relationships.

Description of the soil is often based on a standard classification method such as, for example, the AASHTO method or the Unified Soil Classification System. Such standard classification should be supplemented by a visual description of the soil. For dredging, PIANC has presented a Classification of Soils to be dredged, which is based on the British standard. Classification of rock cores is normally done by visual description by an engineering geologist.

Strength tests Strength tests are used to determine the strength properties of soil and rock in compression, tension and shear. The most common test is the unconfined compression test (UCT), which can be used for clay and rock samples. Undrained strength in clay can also be tested by means of a falling cone or a laboratory vane. More sophisticated testing to provide effective soil strength parameters (cohesion and friction angle) is done using triaxial apparatus, either as drained or undrained triaxial tests, where the sample is subjected to a confining pressure. Various types of shear box are also used.

Consolidation tests Consolidation tests in oedometers are used to determine parameters for calculating the magnitude and rate of settlements caused by embankment fills and structural loads.

Other physical tests Compaction tests are used to determine the compaction properties and requirements of materials used in fills and pavements. Various material tests are performed to establish the strength and durability of pavement materials, such as the Los Angeles Abrasion and Sodium Sulphate Soundness tests.

Mineralogical and chemical tests Rock samples are subjected to mineralogical and petrographical tests. Chemical tests are used to track the content of salts and other deleterious substances in concrete aggregates, for instance to assess the risk of alkali–silica reactions.

1.4. Hydraulic laboratory studies

The use of hydraulic models should, in terms of port planning and design, generally be a standard part of all important port and harbour projects, where complex interactions between the berth structures within the mooring evaluation and the waves and the seabed and coastline are involved. Although great strides have been, and will continue to be, made in the mathematical and numerical modelling of such processes, laboratory models or field studies are still required to calibrate such methods, and the use of physical models in combination with mathematical models is often preferred as a more expedient way of obtaining reliable information.

Physical laboratory modelling and testing are commercial services that are provided by many commercial hydraulic laboratories, research institutions and universities around the world.

It is convenient to distinguish between the 2D and the 3D models.

The **3D models** are models that cover a large area of the sea bottom and shoreline, and where the width and breadth of the basin are of approximately the same magnitude, typically 10–40 m. The point of interest for a port, port entrance, breakwater or other type of structure is built at one end of the basin and the wave generators are placed at the opposite end where near-deep-water conditions prevail. Where an ocean or river current is required, this is usually introduced by letting water in at one end and draining it at the other.

Figure 1.7 shows a 3D model at a scale of 1 : 100 of a fishing port at Sirevaag in south Norway. The water depth in the model is 0.55 m. The laboratory model was used to determine wave-height distributions along the proposed new breakwater (background), and to determine wave heights and ship motions at berth inside the new harbour basin. The wave generators at the far end of the model basin produce wave spectra that were calculated using a numerical refraction model.

Such tests may be used for a number of tasks, such as:

(a) determining the wave-height distribution along a breakwater
(b) assessing the effects on wave agitation of different types of structure or civil works in a port, such as breakwaters, dredging, piers or artificial beaches

Figure 1.7 A 3D model of a fishing port. (Photo courtesy of SINTEF, Norway)

(c) assessing the quality of planned berths by measuring wave heights and (preferably) ship motions at berth
(d) locating wave-breaking zones.

The **2D models** are models where one horizontal dimension is minimised so that the basin has the shape of a canal or a flume. When applying waves in a 2D model, it is assumed that the direction has been established by other means, so that wave height, length and breaking are the only parameters. General wave flumes may be up to several hundred metres in length, but more typical facilities for port engineering purposes are 30–100 m long and 1–10 m wide, with depths of 1–4 m.

Figure 1.8 shows an almost 2D model of wave attack on a breakwater and industrial area at Gismeroy, Norway. The breakwater was modelled at a scale of 1 : 50 in a wave canal which was 4.5 m × 60 m × 1.5 m.

Examples of tasks that are resolved using wave flumes within a 2D model are:

(a) stability testing of rubble-mound breakwaters and concrete-block breakwaters
(b) wave run-up and overtopping of breakwaters
(c) wave forces on structures such as breakwaters, pilings, pipelines, etc.

Scale and model effects are the results of factors that are either not reproduced correctly in the model, or are due to factors that have been introduced into the model that are not present in nature. A typical model effect is the introduction of basin walls. These walls will create reflected waves that are not present in nature, and action must be taken to either minimise these waves through passive or active absorbers, or account for their effect when analysing the final results. Scale effects include, for example, viscosity and surface tension. While these effects can be assumed

Figure 1.8 A 2D model of a breakwater. (Photo courtesy of SINTEF, Norway)

to be negligible when considering wave attack on a several ton armour unit in a breakwater, these effects become more important as the model scale decreases. As there is no practical alternative to using plain water in the model tests, there is no way of eliminating these effects. The preferred method of reducing and controlling the scale effects is to keep the model scale as large as the equipment in the laboratory permits.

Wave generators are used to create waves. A number of types and methods exist, but the simplest and most commonly used principle is a flat paddle, which is moved horizontally to create the waves. Monochromatic wave generators (i.e. capable of only one wave period at a time) are obsolete and not widely used.

Modern facilities have random wave generators that are hydraulically, pneumatically or electrically powered. Depending on the type, the wave generator may also be fitted with software and sensors that automatically compensate for reflected waves, so that the operator may specify a given wave height, which will then be produced from the generator. If no such compensation exists, a calibration procedure must be followed.

Three-dimensional wave generators create waves that closely replicate real sea states by combining waves of different heights, periods and directions in one sea state. Such waves are, however, of limited use in port design studies, because the field in which the whole array of waves exists is very limited. These wave generators are commonly used to study wave impact in very small areas, such as wave forces on a slender structure (Figure 1.9).

An example of a 3D model used to investigate wave loading on a breakwater, wave agitation in the approach channel and wave agitation inside a port is as shown in Figure 1.9. The ship model was used for illustration purposes only, and was not part of the testing programme. The scale was 1 : 100, and the water depth at a distance of 2 km from the port was in the range 8–10 m (8–10 cm

Figure 1.9 A 3D model for investigating wave loading. (Photo courtesy of SINTEF, Norway)

in the model). Thus, in order to generate proper waves, the entire model was built on a platform that lets the waves form and be generated in a 50 cm model water depth. The stability of the breakwater was tested in a separate 1 : 50 model.

The guidelines on commonly used values for the most requested port-related studies are shown in Table 1.1.

Table 1.1 Guidelines on commonly used values

Type of study	Type of model	Scale range	Primary parameters	Other
Wave loading on breakwaters, distribution	3D	1 : 80–1 : 120	Wave height, period and direction, water depths, water level	Waves must be allowed to generate and form in near-deep-water conditions
Wave agitation inside ports	3D	1 : 80–1 : 100	Wave height, period and direction, water depths, water level	
Breakwater stability, behaviour of individual stones in a breakwater	2D, 3D if space permits	1 : 20–1 : 60	Wave height and period, breaker type, water level	
Wave forces on structures, seawalls, piles, etc.	2D	1 : 10–1 : 40	Wave height and period, breaker type, water level	

1.5. Life-cycle management

The overall maintenance policy of a port should be to maintain all the facility assets to the extent that the level of expenditure is justified, in order to maintain the design life of the assets and for reasons of safety and security. It therefore follows that to avoid unexpected large-scale rehabilitation measures and costly downtime as a result of neglected maintenance, systematic planning and budgeting of maintenance activities will be necessary.

To this end it is recommended from the outset that the principles of life-cycle management (LCM) be embraced in port design. Two International Navigation Association (PIANC) reports, *Inspection, Maintenance and Repair of Maritime Structures Exposed to Damage and Material Degradation Caused by the Salt Water Environment*, published in 1991 and revised in 2004, and *Life Cycle Management of Port Structures – General Principles*, published in 2007, provide a useful background to the application of LCM.

LCM and its precursor whole-life costing (WLC) will make a realistic contribution to the maintenance policy of a port, including decision-making, planning, budgeting, and funding of inspection and repair activities during the lifetime of the facilities.

1.6. Safety management and risk assessment

Marine operations conducted within a port should be controlled by policies and procedures within a safety management system (SMS), which is based on a structured and formal assessment of risk within the port. Such a risk assessment system should be an integral part of an SMS, and used by the harbour authority to manage all port operations and activities.

The aim of an SMS is, therefore, to manage and minimise risk. Risk assessment methods should be used to identify priorities and to set objectives for eliminating hazards and reducing risks. If risks cannot be eliminated, they are minimised by physical controls or, as a last resort, through systems of work and personal protection equipment.

Objectives

The establishment of an SMS requires the development of the following elements:

(a) definition of the safety policy
(b) definition of the organisation and personnel roles within it
(c) setting standards
(d) performance monitoring
(e) audit and review system.

Tools

The above objectives should be achieved by developing port-specific functional procedures, which may typically include:

(a) ship movement (navigation management) procedures
(b) vessel traffic system (VTS)
(c) port control procedures
(d) pilotage procedures
(e) towage guidelines
(f) conservancy procedures

(g) emergency procedures
(h) administrative procedures, including:
  (i) training and certification
  (ii) management of change
  (iii) risk assessment
  (iv) incident and near miss reporting
  (v) audit and review.

In a large port it is likely that separate procedures will be developed for each of the above; in a small port, some procedures may be combined. In addition, there are significant areas of overlap between the above functions, and items which appear in one of these documents in one port may be found in another in a different port.

Safety management system performance

The SMS should include proper recording procedures and self-monitoring to ensure that the system is functioning. This should cover premises, plant, substances, people, procedures and systems, including individual behaviour and performance.

The system should include provision for systematic review of performance based on the information obtained from monitoring and from independent audits of the whole system. The port authority should work towards continuous improvement through the constant development of policies, systems and techniques for risk control.

If controls fail, or if near misses or incidents occur, these should be investigated to determine the root causes, and corrective action taken where appropriate; this may include amendment of the SMS or procedures.

Risk assessment system

The risk assessment comprises the first step in the above SMS cycle, and is the first step in application of the SMS. The aim of the risk assessment is to establish the risks that need to be managed in the port, and identify means to control them at acceptable levels.

In order to do so, the process must identify the hazards, together with the events or circumstances that may give rise to their realisation; determine the risk posed by them, and identify the barriers that can be put in place to control the risk, by preventing the realisation of the hazard and/or mitigating its effect if it does occur.

The risk assessment process consists of five parts:

(a) data gathering
(b) hazard identification
(c) risk analysis
(d) assessment of existing measures
(e) identification of risk control measures/options.

(a) Data gathering

The data-gathering process will aim to establish an initial list of hazards. In essence, data gathering involves familiarisation with all aspects of the existing port operations.

(b) Hazard identification
The process of hazard identification attempts to list all the hazards that currently exist within the port as a result of operations conducted therein. This includes and builds upon the hazards identified in the data-gathering process.

(c) Risk analysis
Risk can be defined as the product of the probability of an event occurring and the consequences flowing from it. Thus an event that occurs infrequently and has a low level of consequence constitutes a lower risk than one which occurs more frequently and has a higher consequence. The analysis for each hazard requires the establishment of probability of occurrence, and the consequences reasonably expected to be associated with that level of probability.

(d) Assessment of existing measures
Existing control measures and defences identified in the data-gathering and hazard identification stages should be reviewed. Additional control measures may be identified to address gaps, or where enhancement of measures is indicated as required by the analysis. There may be areas where risk control measures are disproportionately high, considering the risk involved, and may be reduced with subsequent benefit to resource allocation.

(e) Risk control
This stage identifies the specific control measures to be put in place to achieve the risk profile required by the port's safety policy and/or other relevant legislation/standards. This will include consideration of all identified risk control options, together with the resource requirements, benefits and other consequences of their implementation.

Once the risk assessment has been completed, with risk control measures selected, the SMS can be established (or modified). The operation of these measures then becomes part of the SMS. As the SMS includes performance measurement, and an audit and review process, the control measures adopted will be checked, audited and reviewed on a regular basis. The frequency of these reviews may be fixed, or vary depending on the degree of risk identified.

Figure 1.10 shows the relationship between the SMS and risk assessment. The SMS manages the risk, and the risk assessment defines the risk.

1.7. The International Ship and Port Facility Security (ISPS) Code and the Container Security Initiative (CSI)

1.7.1 Introduction

The International Maritime Organisation (IMO) has established an international framework of measures to enhance maritime security, through which ships and port facilities can cooperate to detect and deter acts that threaten security in the maritime transport sector. Once within the port terminal, it should be possible to identify, check and accept temporary storage in a restricted area all cargo while waiting for shipment.

The comprehensive security regimen came into force in July 2004, with the intention of strengthening maritime security to prevent and suppress acts of terrorism against shipping.

The International Ship and Port Facility Security (ISPS) Code and the Container Security Initiative (CSI) represent the culmination of work by the International Maritime Organisation's Maritime

Figure 1.10 Relationship between the SMS and risk assessment

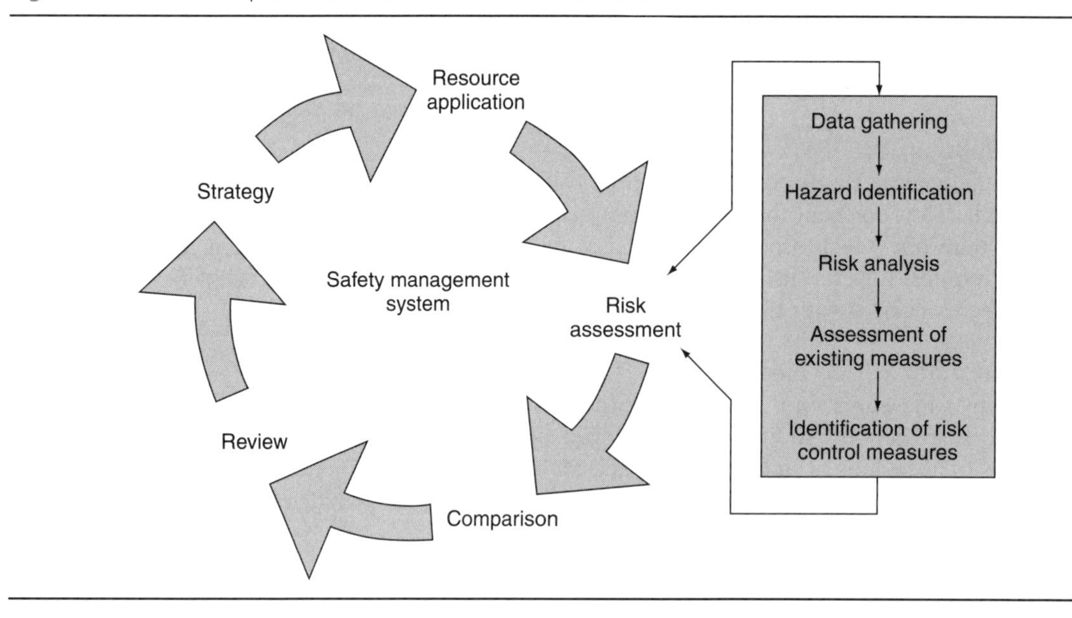

Safety Committee and the United States Custom and Border Protection Service in the aftermath of the terrorist atrocities in the USA in September 2001.

The ISPS Code takes the approach that the security of ships and port facilities is basically a risk management activity, and that in order to determine what security measures are appropriate an assessment of the risks must be made for each particular case. The CSI, on the other hand, seeks to use non-intrusive inspection and radiation detection technology to screen high-risk containers before they are shipped to the USA.

1.7.2 Application of the ISPS Code

The purpose of the code is to provide a standardised, consistent framework for evaluating risk, enabling ports and governments to offset changes in threat against changes in vulnerability to ships and port facilities.

The code applies to:

(a) The following types of ships engaged on international voyages:
    (i) passenger ships, including cruise ships and high-speed craft
    (ii) cargo ships, including high-speed craft of 500 gross tonnes and above
    (iii) mobile offshore drilling rigs.
(b) Port facilities serving such ships on international voyages.

To begin the process, each contracting government must conduct a port facility security assessment, which will have three essential components. First it must identify and evaluate the asset and infrastructure that are critical to the facility, as well as those structures that, if damaged, could cause significant loss of life or damage to the facility. Second, the assessment must identify the actual threats to those critical assets and infrastructure in order to prioritise security measures. Finally, the assessment must address the vulnerability of the facility by identifying its weaknesses in terms of physical security, structural integrity, protection systems, procedural policies, communications systems, transportation infrastructure, utilities and other areas that may be a likely target. Once this assessment has been completed, the contracting government can accurately evaluate the risks.

The risk management concept is embodied in the code through a number of minimum functional security requirements for ships and port facilities.

For port facilities the requirements will include:

(a) a facility security plan
(b) security personnel
(c) security equipment.

In addition, the requirements will include:

(a) monitoring and controlling access
(b) monitoring the activities of people and cargo
(c) ensuring security communications are readily available.

As each port facility presents different risks, the method by which a facility will meet the specific requirements of the code will be determined and approved by the contracting government on a case-by-case basis.

1.7.3 Application of the Container Security Initiative (CSI)

The three core elements of the CSI are:

(a) identification of high-risk containers using automated targeting tools based on advance information and strategic intelligence
(b) pre-screening and evaluation of containers before they are shipped, generally at the port of departure
(c) use of large X-ray and/or gamma-ray technology to ensure that screening can be done rapidly without slowing down the movement of trade.

To be eligible for participation in the CSI, each candidate nation must commit to the following minimum standards:

(a) The customs administration must be able to inspect cargo originating, transiting, exiting or being transhipped through a country. Non-intrusive inspection equipment with gamma- or X-ray imaging capabilities and radiation-detection equipment must be utilised for conducting such inspections without disrupting the flow of legitimate trade.
(b) The port must have regular, direct and substantial container traffic to ports in the USA.
(c) The port must establish a risk management system to identify potentially high-risk containers, including a mechanism for validating threat assessments, targeting decisions and identifying best practices.
(d) Commit to sharing critical data, intelligence and risk management information with the United States Customs and Border Protection and develop an automatic mechanism for these exchanges.

(e) Conduct a thorough port assessment to ascertain vulnerable areas.
(f) Commit to maintaining programmes to prevent lapses in employee integrity.

1.7.4 ISPS maritime security levels

In order to communicate the threat at a port facility or for a ship, the contracting government must set the appropriate level of security. Security levels 1, 2 and 3 correspond to normal, medium and high threat situations, respectively. The security level sets the link between the ship and the port facility, as it triggers the implementation of appropriate security measures for both the ship and shore facility.

The maritime security levels are defined as follows:

- Level 1: normal: the level at which ships and port facilities normally operate.
- Level 2: heightened: the level applying for as long as there is a heightened risk of a security incident.
- Level 3: exceptional: the level applying for the period of time when there is probable or imminent risk of a security incident.

1.7.5 Port facility security assessment

The port facility security assessment is fundamentally a risk analysis of all aspects of a port facility's operation in order to determine which parts of it are more susceptible and/or more likely to be the subject of attack. Security risk is seen as a function of the threat of an attack coupled with the vulnerability of the target and the consequence of an attack.

On completion of the analysis it will be possible to produce an overall assessment of the level of risk. The assessment will help determine which port facilities are required to appoint security personnel and also assist in the preparation of a port facility security plan. This plan should indicate the operational and physical security measures that the facility should take to ensure that it always operates at security level 1. The plan should also indicate the additional, or intensified, security measures the facility needs to adopt to move to and operate at security level 2 when instructed to do so. It should also indicate the actions the facility would need to take to allow prompt response to the instructions that may be issued for security level 3.

1.7.6 Security measures

Once the security measures have been established they will be implemented according to the following security levels that apply to the facility.

Security level 1: normal

This is the level at which the ships and port facilities normally should operate. The security should, in principle, include fencing and guarding of the terminal, routine checking of cargo, cargo transport and storage of all cargo entering the terminal. The checking of the cargo may be accomplished by visual and physical examination, use of scanning detection equipment, mechanical devices, dogs, etc.

This level is, therefore, the minimum security level. This threat level indicates the possibility of a (general) threat against facilities and shipping, and the security provisions for this level must always be maintained during normal working conditions.

(a) Maintain 24-hour guard procedures.
(b) Positively identify anyone accessing the facility and their vehicles for any reason.

(c) Undertake searches of persons, personal effects and vehicles.
(d) Monitor restricted areas.
(e) Supervise cargo loading, including stored/staged cargo operations.
(f) Require advance notice for arrival of ships, vehicles carrying stores, supplies, or repair items/replacement parts, and non-routine vendors/visitors.
(g) Composition of stores, driver and vehicle registration.
(h) Search of stores/supplies delivery vehicles.
(i) Ensure that communication is fully established when vessel(s) are berthed alongside.
(j) Maintain a high awareness and report suspicious activity or abnormal behaviour.
(k) Maintain awareness and report any suspicious activity or abnormal behaviour to the senior security officer and/or necessary authorities.
(l) Turn on appropriate lighting during hours of darkness.
(m) Ensure all unused accesses are locked.

Protective measure	Security level		
	1	2	3
All facility personnel must review and exercise their security duties and responsibilities through drills, training and exercises	×	×	×
Provide security information to all facility and security personnel that addresses the security level and any specific threat information	Optional	×	×
Security officer(s) must communicate with vessels alongside to coordinate protective measures	×	×	×
Security manual is in place and implemented	×		
Security officer(s) and security personnel are assigned	×		
Security plan is written, revised and working	×		
Restricted areas are marked and monitored	×		
Access to restricted areas is denied contingent on normal safety	×		
Vulnerable points are locked	×		
Accesses are secured	×		
Personnel and visitor IDs are worn	×		
Security equipment, lighting and surveillance are operational	×		
Normal liaison with the port security committee is continued and the security level status is acknowledged	×		
High situational awareness for suspicious activity is maintained	×		
All suspicious activity is reported to the security officer(s), then the port security committee and the contracting government	×		
Cargo operations are supervised	×		
Communications at level 1 are established	×		

Security level 2: heightened

This is the level applying for the period of time during which there is a heightened risk of a security incident. At this level the security is intensified to detailed checking of all cargo and operations inside the terminal.

Security level 2 must, therefore, be established when intelligence from a reliable source has been received indicating that a threat to ports/terminals has been identified within an operating region, area, or port or type of vessel, although no 'specific' target has been identified.

(a) Assign additional personnel to guard access points and patrol perimeter barriers.
(b) Pre-approve/screen in advance any persons requesting access to the facility.
(c) Increase safeguards for ship stores to include coordination with the ship security officer and escorts of vehicles.
(d) Limit physical access to the facility and its sensitive areas (e.g. restricted areas).
(e) Increase package/supply/stores screening.
(f) Increase awareness of suspicious activity.
(g) Verify the inventory of cargo in storage at the facility.

Protective measure	Security level		
	1	2	3
All measures listed for security level 1		×	
Security briefings are provided to all security personnel on any specific threats and on the need for vigilance and reporting of suspicious persons, objects or activities		×	
Visitors are allowed only by appointment. Visitors will be escorted and allowed into restricted areas by authorisation only			
Random personnel searches are conducted by security personnel at gate access points using supplied security equipment		×	
Increase search of facility and ship stores		×	
Restrict movement of ship crews through the facility		×	
Manpower is augmented for guard and patrol proposes, if the threat persists		×	
Liaison with ship security officers and the port security committee to coordinate response		×	
Communications at level 2 are established		×	

Security level 3: exceptional

This level can apply for the period of time when there is the probable or imminent risk of a security incident, and may include restriction or suspension of cargo movements or operation within all or part of the terminal.

Security level 3 therefore represents a high-level security threat against a specific target and represents the highest threat level based on reliable intelligence services. This level indicates that a specific vessel or port facility has been identified as a 'target' and that the threat is highly probable or imminent.

(a) Maximise use of lighting and surveillance equipment.
(b) Prohibit non-essential access to the facility.
(c) Secure all access points to the facility.
(d) Consider ceasing all cargo-handling/transfer operations.
(e) Follow government authority and company instructions regarding facility operations.
(f) Revise ship schedules until the threat is eliminated.

(g) Ensure that vessels berthed alongside and/or entering or leaving the facility are informed.
(h) Secure all barge hatches to prevent unauthorised cargo discharge.
(i) Consider waterside surveillance.
(j) Prepare to evacuate or partially evacuate the facility.
(k) Implement specific/additional protective measures/actions ordered by the appropriate government authorities.

Protective measure	Security level		
	1	2	3
All measures listed for security levels 1 and 2			×
Additional lighting is turned on			×
Essential visitors only are allowed entry. All visitors are escorted			×
100% personnel searches are conducted			×
Cargo operations are ceased			×
Fire hoses are laid out and connected			×
Ship crew and stores are prohibited from transiting the facility			×
Liaison with the port security committee and law enforcement authorities			×
Provide waterside surveillance			×
Communications at level 3 are established			×

All ports should have a port facility security plan (PFSP), which should indicate the operational and physical security measures the port should take to ensure that it always operates at security level 1. The security plan should indicate the additional security measures the port must take without delay to move to and operate at security level 2, and the plan should also indicate the possible preparatory actions the port should take to allow prompt response to security level 3.

REFERENCES AND FURTHER READING

Agerschou H (2004) *Planning and Design of Ports and Marine Terminals*. Thomas Telford, London.

Agerschou H, Lundgren H, Sørensen T, Ernst T, Korsgaard J, Schmidt LR and Chi WK (1983) *Planning and Design of Ports and Marine Terminals*. Wiley, Chichester.

Alderton PM (2005) *Port Management and Operations*. Lloyd's Practical Shipping Guides. LLP, London.

Asian Development Bank (2000) *Developing Best Practices for Promoting Private Sector Investment in Infrastructure*. Asian Development Bank, Manila.

BSI (British Standards Institution) (2000) BS 6349-1: 2000. Maritime structures. Code of practice for general criteria. BSI, London.

Bruun P (1990) *Port Engineering*, vols 1 and 2. Gulf Publishing, Houston, TX.

Carmichael J (2000) *The Global Change? International Ports Congress 1999*. Thomas Telford, London.

FIDIC (Federation Internationale des Ingenieurs-Conseils) (1999) *Conditions of Contract for Construction, General Conditions*. FIDIC, Lausanne.

Frankel EG (1987) *Port Planning and Development*. Wiley, Chichester.

Institution of Civil Engineers (1985) *Port Engineering and Operation*. Conference. Thomas Telford, London.

PIANC (International Navigation Association) (1998) *Life Cycle Management of Port Structures. General Principles. Report of Working Group 31*. PIANC, Brussels.
PIANC (2000) *Site Investigation Requirements for Dredging Works. Report of Working Group 23*. PIANC, Brussels.
PIANC (2000) *Dangerous Cargoes in Ports. Report of Working Group 35*. PIANC, Brussels.
PIANC (2004) *Inspection, Maintenance and Repair of Maritime Structures Exposed to Damage and Material Degradation Caused by the Salt Water Environment*. PIANC, Brussels.
PIANC (2007) *Life Cycle Management of Port Structures: Recommended Practice for Implementation. Report of Working Group 42*. PIANC, Brussels.
PIANC (2010) *The Safety Aspect Affecting the Berthing Operations of Tankers to Oil and Gas Terminals. Report of Working Group 55*. PIANC, Brussels.
Quinn A De F (1972) *Design and Construction of Ports and Marine Structures*. McGraw-Hill, New York.
Takel RE (1974) *Industrial Port Development*. Scientechnia, Bristol.
UNCTAD (United Nations Conference on Trade and Development) (2003) *Modern Port Management, Port Training Programme*. UNCTAD, Geneva.
United Nations (1978) *Port Development, A Handbook for Planners in Developing Countries*. United Nations, New York.

Port Designer's Handbook
ISBN 978-0-7277-6004-3

http://dx.doi.org/10.1680/pdh.60043.037

Chapter 2
Environmental forces

2.1. General

Consideration should be given to the environmental forces and conditions at all stages of a berth structure, both during the construction phase and during the service life. One should bear in mind the variable and often unpredictable character of all environmental forces that can act on a berth structure; it would be unrealistic to expect substantial cost savings by attempting to design the structure for a shorter design life.

Generally, it is not easy to forecast what the maximum environmental forces on a berth, a berth with a moored ship or a ship will be. It is dependent on the direction of the wind and waves, the current, the size and type of ship, tugboat assistance, and whether the ship is loading or unloading, etc.

It is always recommended that, when dealing with environmental forces on important berth structures or forces on larger ships, a hydraulic institute with experience in coastal engineering, for example, be consulted.

When evaluating the environmental forces, it is necessary to obtain estimates of the expected extreme conditions that may act on the port site area. These estimates can be obtained by observations and calculations of high wind speeds, and are then applied to a forecasting technique for later use in a mathematical or physical model.

2.2. Wind

Figure 2.1 is an example of a windrose diagram showing the yearly distribution of wind directions in decadegrees and the forces as a percentage of time. The dominating or prevailing wind directions are, in this case, southeasterly and northwesterly. In Figure 2.2 the same wind forces are shown as the frequency of the yearly wind forces for each Beaufort interval.

The wind forces are classified in accordance with the Beaufort wind scale. The **Beaufort** range of intensity from 0 to 12 is shown in Table 2.1.

The mean wind velocity and direction should, in accordance with the Beaufort scale, be recorded 10 m above the mean sea level and should be based upon the 10 min averages of the wind velocity and direction.

The Norwegian Petroleum Directorate has recommended that the maximum wind speed, averaged over short time periods, may be obtained by multiplying the actual 10 min mean wind speed by the gust factors, as shown in Table 2.2.

Figure 2.1 An example of a windrose for the yearly distribution of wind and forces as a percentage of time

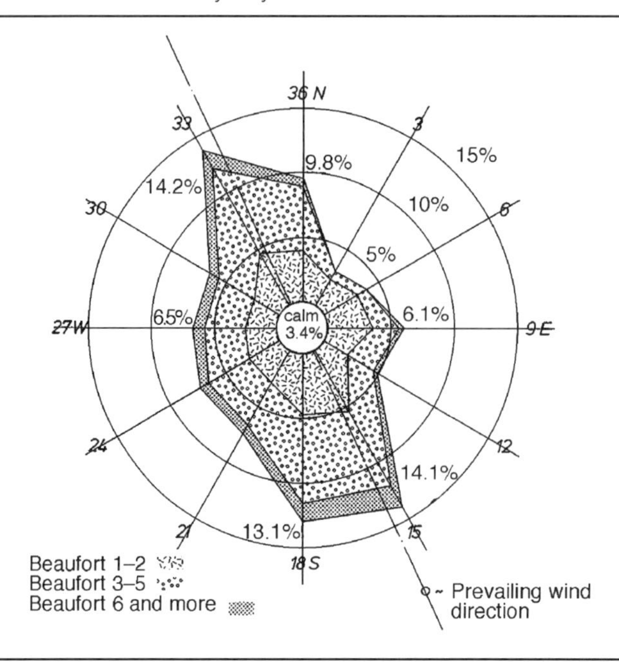

Provided that no specified information on the gust ratio is available (i.e. the ratio between the short-period wind speeds to the mean wind speed for the wind conditions at the site), the International Navigation Association (PIANC) recommends that Table 2.2 could tentatively be applied for higher wind velocities. Table 2.3 shows the relationship between the 1 h mean wind speed and the associated maximum speeds for a range of shorter mean durations.

Figure 2.2 The frequency of the yearly wind forces

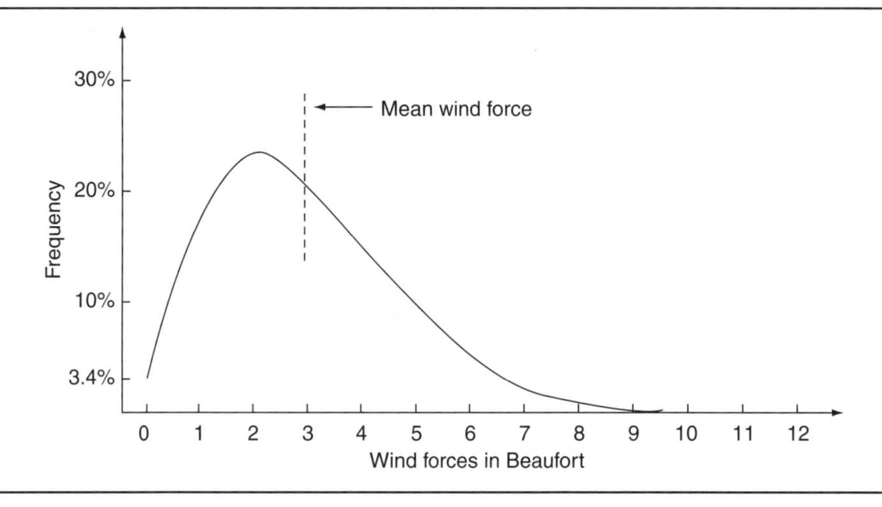

Table 2.1 The Beaufort wind scale

Beaufort	Description	Velocity	
		m/s	knots
0	Calm	0.0–0.2	0–0.5
1	Light air	0.3–1.5	1–3
2	Light breeze	1.6–3.3	4–6
3	Gentle breeze	3.4–5.4	7–10
4	Moderate breeze	5.5–7.9	11–16
5	Fresh breeze	8.0–10.7	17–21
6	Strong breeze	10.8–13.8	22–27
7	Near gale	13.9–17.1	28–33
8	Gale	17.2–20.7	34–40
9	Strong gale	20.8–24.4	41–47
10	Storm	24.5–28.4	48–55
11	Violent storm	28.5–32.6	56–63
12	Hurricane	32.7–	64–

Table 2.2 Relationship between 10 min mean wind speed and the gust factor

Wind duration	Gust factor
3 s mean	1.35
10 s mean	1.30
15 s mean	1.27
30 s mean	1.21
1 min mean	1.15
10 min mean	1.0

Table 2.3 Relationship between 1 h mean wind speed and the gust factor

Wind duration	Gust factor
3 s mean	1.56
10 s mean	1.48
1 min mean	1.28
10 min mean	1.12
30 min mean	1.05
1 h mean	1.00

The gust factor will depend upon the topographical conditions around the harbour basin and the port location. If there is a lack of proper wind information in the area, use of the gust factors in Table 2.3 is recommended.

When considering the wind forces acting on a ship, the focus should be on the wind gust that is capable of overcoming the ship inertia. For a given ship (characterised by its length, shape, etc.), such wind can be defined by two dependent variables: wind duration and velocity.

For port and ship operation it is claimed that gust durations shorter than about 1 min will be of secondary importance, but at some oil terminals in the North Sea it has been observed that wind durations down to 20–30 s may affect tankers in a ballasted condition.

In the northern and southern hemispheres there can appear what are known as **polar lows** or arctic hurricanes, which are defined as small but intense cyclones. The wind speed varies from Beaufort 7 (or around 15 m/s) up to Beaufort 10, and thunder storms and even waterspouts are possible. The temperatures drop sharply with the passage of the storm centre, causing an extremely high risk of icing over sea.

2.2.1 Wind forces

The wind forces acting on a ship may vary considerably, as do the current forces, with both the type and size of the ship, and should therefore best be established by testing in a hydraulic institute. This is especially so for the wind forces acting upon a ship with a large windage sided area; for example, fully loaded container ships or large passenger ships are influenced greatly, while very large oil tankers have large variations in the longitudinal forces depending upon the shape or design of the bow. Generally, the wind effects on port and harbour operations are more important than those of the wave and current.

In this chapter, different methods and national standards and regulations for calculation of the wind forces are compared against each other. It should be noted that in important evaluations these standards and regulations should only be used as a guide to the magnitude of the forces on the ship.

The magnitude of the wind velocity V_w to be applied in design varies from place to place, and has to be assessed in each case. The design wind velocity should correspond to the maximum velocity of the gusts that will affect the ship, and not only to the average velocity over a period of time. A 30 s average wind velocity is recommended for use in the wind force equations for mooring analyses.

These gust velocities can be about 20% higher than the average velocity. In the case of moored ships, the gust duration must be sufficient for the full mooring line or fender strains to develop, taking into account the inertia of the ship. This can lead to a reduced design wind speed. It should also be taken into account that the wind area is not symmetrical about the midship line, which implies the development of a moment of rotation.

Figure 2.3 shows the relationship between wind pressure and wind velocity over a 10 min period. When reading the wind pressure, the curve with the gust factor $V + 20\%$ should be selected to account for the wind gust factor.

A wind velocity lower than 30 m/s – after the Beaufort scale – with a gust factor of 1.2 should not be assumed for the design of berth structures (i.e. the minimum wind pressure should be 0.81 kN/m^2). If

Figure 2.3 Wind pressure and wind velocity

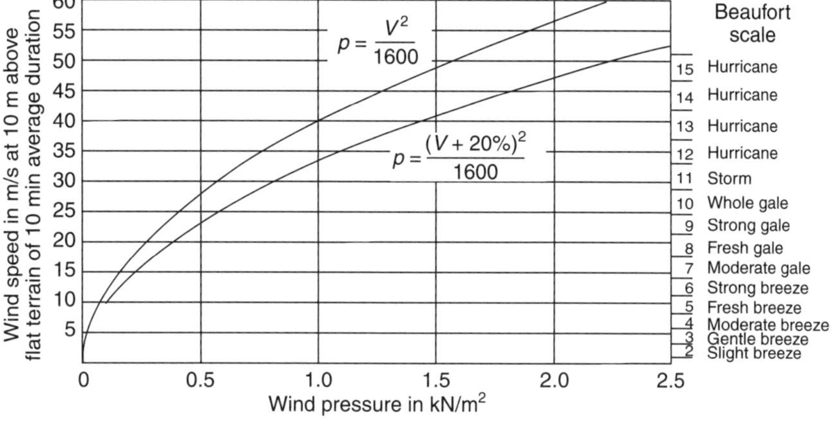

the wind velocity increases above about 25–30 m/s, the ship would normally either leave the berth or take in ballast to reduce its wind area.

It is very important to note that the wind force is proportional to the square of the wind velocity.

2.2.2 Different wind standards and recommendations

Below, different methods and national standards and regulations for calculating the wind forces are shown and compared against each other. It should be noted that these standards and regulations should only be used as a guide to the magnitude of the forces on the ship. If more accurate wind force calculations are needed, these should be established by model testing in a hydraulic institute.

General standard formulas for calculation of the wind forces on a ship moored at a berth structure are as follows:

$$P_w = C_w \times (A_w \times \sin^2 \varphi + B_w \times \cos^2 \varphi) \times \gamma_w \times \frac{V_w^2}{2g}$$

$$P_w = C_w \times (A_w \times \sin^2 \varphi + B_w \times \cos^2 \varphi) \times \frac{V_w^2}{1600}$$

$$P_w = C_w \times (A_w \times \sin^2 \varphi + B_w \times \cos^2 \varphi) \times p$$

where

P_w = wind force in kN
C_w = wind force coefficient
A_w = laterally projected area of the ship above water in m^2
B_w = front area of the ship above water in m^2
φ = angle of the wind direction to the ship's centreline
γ_w = specific gravity of air, 0.01225 kN/m^3 at 20°C
V_w = velocity of the wind in m/s; it is recommended to use a 30 s average wind velocity

g = acceleration of gravity, 9.81 m/s^2
p = wind pressure in kN/m^2

The maximum wind forces in the above equation is when $\varphi = 90°$ (i.e. the wind blows perpendicular to the ship's centreline):

$$P_w = C_w \times A_w \times p$$

The magnitude of C_w depends on the shape of the ship above the water and the orientation of the ship to the wind direction. As average values of C_w for isolated ships, the following are recommended: for wind crosswise to the ship $C_w = 1.3$; for wind dead against the bow $C_w = 0.9$; and for wind dead against the stern $C_w = 0.8$.

The **Spanish standard ROM 0.2-90** (Ministerio de Obras Públicas y Transportes, 1990) recommends that the wind forces or pressure on a ship should be:

Resultant wind force in kN:

$$R_v = \frac{\rho}{2g} \times C_V \times V_W^2 \times (A_t \times \cos^2 \alpha + A_l \sin^2 \alpha)$$

$$= C_v \times (A_t \times \cos^2 \alpha + A_l \sin^2 \alpha) \times \left(\frac{V_w^2}{1600}\right)$$

$$tg\varphi = \frac{A_l}{A_t} \times tg\alpha$$

Lateral or **transverse** wind force in kN:

$$F_T = R_V \times \sin \varphi$$

Longitudinal wind force in kN:

$$F_L = R_V \times \cos \varphi$$

Here

A_l = broadside or lateral projected wind area in m^2
A_t = head-on or transverse projected wind area in m^2
ρ = specific weight of air, 1.225×10^{-3} t/m^3
α = angle between the longitudinal axes of the ship from bow to stern, and the wind direction against the ship
φ = angle between the ship's longitudinal axes from bow to stern, and the resultant wind force R_v on the ship
C_V = shape factor for the ship, which varies between 1.0 and 1.3; in the absence of precise determination from, for example, model studies, the value 1.3 should be used
V_w = design wind velocity in m/s at a height of 10 m with the shortest interval (gust) that will overcome the ship's inertia should be adopted as the basic wind velocity – and a mean velocity corresponding to the following gust values should be adopted:

- 1 min for a ship length longer than or equal to 25 m
- 15 s for ship length less than 25 m

The **British standard BS 6349-1: 2000** (BSI, 2000) recommends that the wind forces or pressure on a ship should be:

Lateral or **transverse** wind force in kN:

$$F_{\mathrm{Twind}} = (C_{\mathrm{TWforward}} + C_{\mathrm{TWaft}}) \times \rho \times A_{\mathrm{L}} \times \frac{V_{\mathrm{W}}^2}{10\,000}$$

Longitudinal wind force in kN:

$$F_{\mathrm{Lwind}} = C_{\mathrm{LW}} \times \rho \times A_{\mathrm{L}} \times \frac{V_{\mathrm{W}}^2}{10\,000}$$

Here

A_{L} = broadside or lateral or transverse projected wind area in m^2
C_{TW} = transverse wind force coefficient forward or aft depending on the angle of wind as shown for the ballasted condition in Table 2.4 and for the loaded condition in Table 2.6
C_{LW} = longitudinal wind force coefficient depending on the angle of wind as shown for the ballasted condition in Table 2.4 and for the loaded condition in Table 2.6
ρ = specific weight of air varying from 1.3096 kg/m^3 at 0°C to 1.1703 kg/m^3 at 30°C
V_{W} = design wind speed in m/s at a height of 10 m above the water level; it is recommended for design of the moorings that a 1 min mean wind speed is used

OCIMF (the **Oil Companies International Marine Forum**) recommends that the wind forces or pressure on an oil tanker should be:

Lateral wind forces:

$$F_{\mathrm{Ywind}} = F_{\mathrm{YAwind}} + F_{\mathrm{YFwind}}$$

Lateral wind force at aft perpendicular:

$$F_{\mathrm{Yawind}} = C_{\mathrm{Yaw}} \times \rho_{\mathrm{w}} \times A_{\mathrm{L}} \times \frac{V_{\mathrm{w}}^2}{2012}$$

Lateral wind force **forward** perpendicular:

$$F_{\mathrm{YFwind}} = C_{\mathrm{Yfw}} \times \rho_{\mathrm{w}} \times A_{\mathrm{L}} \times \frac{V_{\mathrm{w}}^2}{2012}$$

longitudinal wind force:

$$F_{\mathrm{Xwind}} = C_{\mathrm{Xw}} \times \rho_{\mathrm{w}} \times A_{\mathrm{T}} \times \frac{V_{\mathrm{w}}^2}{2012}$$

Here

A_{L} = broadside or lateral projected wind area in m^2

Figure 2.4 OCIMF recommendations for mooring restraints for tankers above 16 000 DWT

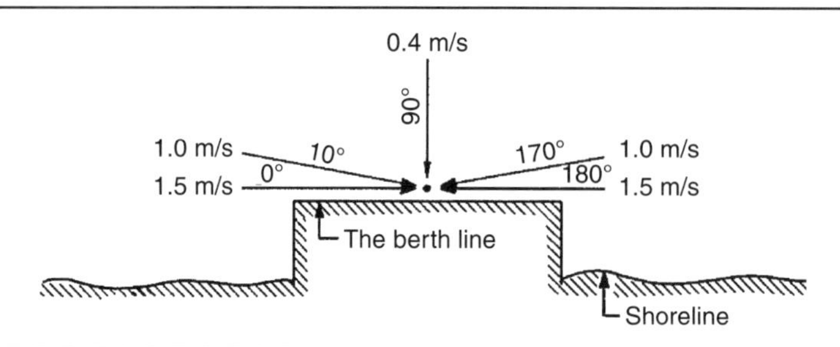

A_T = head-on projected wind area in m^2
C_{Yaw} = transverse wind force coefficient aft depending on the angle of wind as shown for the ballasted condition in Table 2.4 and for the loaded condition in Table 2.6
C_{YFw} = transverse wind force coefficient forward depending on the angle of wind as shown for the ballasted condition in Table 2.4 and for the loaded condition in Table 2.6
C_{Xw} = head-on wind force coefficient depending on the angle of the wind as shown for the ballasted condition in Table 2.4 and for the loaded condition in Table 2.6
ρ_w = specific weight of air, 1.223 kg/m^3 at 20°C
V_W = design wind speed in m/s at a height of 10 m above the water level; it is recommended that a 30 s average wind velocity is used
2012 = conversion factor for the velocity in knots to m/s

OCIMF recommends that, for all tankers above 16 000 DWT (deadweight tonnage) intended for general worldwide trading, the mooring restraint available on board the tanker as permanent equipment should be sufficient to satisfy Figure 2.4.

OCIMF/SIGTTO (Oil Companies International Marine Forum/Society of International Gas Tanker and Terminal Operators) recommends that the wind force or pressure on a gas tanker should be:

Lateral wind forces:

$$F_{Yw} = F_{Yaw} + F_{Yfw}$$

Lateral wind force at **aft** perpendicular:

$$F_{Yaw} = C_{Yaw} \times \rho_w \times A_L \times \frac{V_w^2}{2012}$$

Lateral wind force **forward** perpendicular:

$$F_{Yfw} = C_{Yfw} \times \rho_w \times A_L \times \frac{V_w^2}{2012}$$

Longitudinal wind force:

$$F_{Xw} = C_{Xw} \times \rho_w \times A_T \times \frac{V_w^2}{2012}$$

Here

A_L = broadside or lateral or transverse projected wind area in m^2
A_T = head-on projected wind area in m^2
C_{YAw} = transverse wind force coefficient aft depending on the angle of the wind as shown in Table 2.4 and for the loaded condition in Table 2.6
C_{YFw} = transverse wind force coefficient forward depending on the angle of the wind as shown for the ballasted condition in Table 2.4 and for the loaded condition in Table 2.6; the coefficient will be the same in both the ballasted and loaded conditions since the differences in the gas tanker's loaded condition are not significant due to the relatively small change in draft from the ballasted condition to the fully loaded condition
C_{Xw} = head-on wind force coefficient depending on the angle of wind as shown for the ballasted condition in Table 2.4 and for the loaded condition in Table 2.6; the coefficient will be the same in both the ballasted and loaded conditions since the differences in the gas tanker's loaded condition are not significant due to the relatively small change in draft from the ballasted condition to the fully loaded condition
ρ_w = specific weight of air, 1.248 kg/m^3 at 20°C
V_W = design wind speed in m/s at a height of 10 m above the water level; it is recommended that a 30 s average wind velocity is used
2012 = conversion factor for the velocity in knots to m/s

Comparison between the different wind standards and recommendations for ships in the ballasted condition

For a comparison between the different wind standards and recommendations, the different approximate wind force coefficients on ships in ballast condition are shown in Table 2.4.

Example of wind forces on ships in ballasted conditions

Based on the different wind standards and recommendations formulas and the different wind force coefficients in Table 2.4 with a gust factor recommended by the different standards and recommendations, examples for comparison of the different wind forces in kN with the wind acting longitudinally (0° against the bow) and laterally (90° to the ship's longitudinal axis) for different wind speeds in m/s for ships in the ballasted condition and with a 95% confidence limit (see Chapter 21) are shown in Table 2.5 for the following ship dimensions:

(a) oil tanker, 200 000 DWT
(b) $L_{OA} = 350$ m, $L_{BP} = 346$ m, $B = 56.2$ m
(c) broadside or lateral ballasted projected wind area $A_L = 6930$ m^2
(d) head-on or transverse ballasted projected wind area $A_T = 1730$ m^2.

For a comparison between OCIMF and OCIMF/SIGTTO, the OCIMF/SIGTTO coefficients for a membrane tanker have been used.

As may be seen from the calculations, the wind forces vary considerably between the different standards and recommendations. The above methods for calculation should therefore only be used as a guide to the magnitude of the forces. Since the wind forces can vary considerably for both the type and size of ship, one should therefore, to obtain the most accurate forces for important berth structures, perform wind tests on scale models. This is especially important for wind forces on high-sided ships.

Table 2.4 Wind coefficient in the ballasted condition

Wind coefficients on tankers and container ships in the **ballast** condition

	British standard BS 6349-1: 2000						OCIMF 1997		
Angle from bow to stern	Very large tankers			Large container ships without a deck load			Oil tankers		
	C_{TWf}	C_{TWa}	C_{LW}	C_{TWf}	C_{TWa}	C_{LW}	C_{YFw}	C_{YAw}	C_{Xw}
Forward 0°	0.0	0.0	1.2	0.0	0.0	0.7	0.00	0.00	0.88
15°	0.7	0.3	0.9	0.8	0.4	0.7	0.16	0.06	0.68
30°	1.4	0.9	0.7	1.8	1.0	0.7	0.30	0.16	0.48
45°	2.1	1.4	0.4	2.5	1.7	0.5	0.42	0.30	0.25
60°	2.4	1.9	0.2	2.9	2.2	0.3	0.50	0.38	0.12
75°	2.4	2.3	0.1	3.1	2.6	0.4	0.48	0.46	0.05
90°	2.2	2.6	0.0	2.9	2.9	0.4	0.43	0.52	0.00
105°	1.8	2.8	0.2	2.7	3.0	0.3	0.35	0.55	0.14
120°	1.3	2.7	0.4	2.2	3.1	0.1	0.25	0.54	0.30
135°	0.8	2.5	0.6	1.7	2.9	0.2	0.15	0.48	0.42
150°	0.4	1.8	0.7	1.0	2.0	0.5	0.07	0.36	0.55
165°	0.1	1.0	0.8	0.4	1.2	0.6	0.03	0.20	0.62
Aft 180°	0.0	0.0	0.8	0.0	0.2	0.6	0.00	0.00	0.62

Wind coefficients on tankers and container ships in the **ballast** condition

	OCIMF/SIGTTO 1995					
Angle from bow	LNG membrane tankers			LNG spherical tankers to stern		
	C_{YFw}	C_{YAw}	C_{Xw}	C_{YFw}	C_{YAw}	C_{Xw}
Forward 0°	0.00	0.00	1.02	0.00	0.00	1.02
15°	0.12	0.07	0.95	0.12	0.07	0.95
30°	0.29	0.20	0.80	0.30	0.19	0.80
45°	0.44	0.35	0.56	0.51	0.33	0.56
60°	0.52	0.45	0.30	0.61	0.45	0.30
75°	0.52	0.52	0.10	0.60	0.55	0.10
90°	0.46	0.59	0.02	0.55	0.62	0.02
105°	0.38	0.64	0.15	0.46	0.64	0.15
120°	0.30	0.65	0.39	0.35	0.62	0.48
135°	0.20	0.60	0.65	0.22	0.50	0.83
150°	0.10	0.45	0.79	0.12	0.34	0.97
165°	0.04	0.20	0.89	0.04	0.15	1.02
Aft 180°	0.00	0.00	0.90	0.00	0.00	1.00

LNG: liquefied natural gas

Table 2.5 Comparison of the wind forces on ships in the ballasted condition for the different standards

Forces due to the wind with a gust factor in the **ballast** condition

	General formula		Spanish standard ROM 0.2-90		British standard BS 6349-1: 2000		OCIMF 1997 Oil tankers		OCIMF/SIGTTO 1995 LNG membrane	
Wind speed: m/s	Force: kN		Force: kN		Force: kN		Force: kN		Force: kN	
	0°	90°	0°	90°	0°	90°	0°	90°	0°	90°
10	140	810	186	745	144	576	133	576	158	650
15	315	1824	418	1675	324	1296	300	1297	355	1462
20	560	3243	744	2979	567	2304	533	2305	630	2600

Comparison between the different wind standards and recommendations for ships in the loaded condition

For a comparison between the different wind standards and recommendations, the different approximate wind force coefficients on ships in the loaded condition are shown in Table 2.6. The wind force coefficients for OCIMF/SIGTTO are applicable to draft conditions ranging from ballasted to fully loaded.

Example of wind forces on ships in the loaded condition

Based on the different wind standards and recommendations formulas and the different wind force coefficients in Table 2.6 with a gust factor recommended by the different standards and recommendations, examples for comparison of the different wind forces in kN with the wind acting longitudinally (0° against the bow) and laterally (90° to the ship's longitudinal axis) for different wind speeds in m/s for ships in the loaded condition and with a 95% confidence limit (see Chapter 21) are shown in Table 2.7 for the following ship dimensions:

(a) oil tanker, 200 000 DWT
(b) $L_{OA} = 350$ m, $L_{BP} = 346$ m, $B = 56.2$ m
(c) broadside or lateral fully loaded projected wind area $A_L = 4300$ m^2
(d) head-on or transverse fully loaded projected wind area $A_T = 1210$ m^2.

For an approximate comparison between OCIMF and OCIMF/SIGTTO, the OCIMF/SIGTTO coefficients for a membrane tanker have been used.

As may be seen from the examples above, there are significant differences in the wind forces between the various wind standards and recommendations. Therefore, in the design of, for example, mooring structures, one must be very careful which standard to use.

2.2.3 Drag coefficient

The drag coefficient in the different recommendations is empirical and thus its value is estimated from experimentation. Since its value depends mainly on the vessel's geometry, it is difficult to determine whether one recommendation is better than the others. On the one hand, the ROM formulation provides an expression for the projected areas, seeming to reduce the uncertainty introduced by the

Table 2.6 Wind coefficients on ships in the loaded condition

Wind coefficients on tankers and container ships in the **loaded** condition

Angle from bow to stern	British standard BS 6349-1: 2000						OCIMF 1997		
	Very large tankers			Large container ships without a deck load			Oil tankers		
	C_{TWf}	C_{TWa}	C_{LW}	C_{TWf}	C_{TWa}	C_{LW}	C_{YFw}	C_{YAw}	C_{Xw}
Forward 0°	0.0	0.0	1.8	0.0	0.0	0.6	0.0	0.0	0.98
15°	0.2	0.3	1.7	0.6	0.4	0.6	0.04	0.06	0.88
30°	0.5	0.8	1.4	1.6	0.9	0.6	0.1	0.16	0.72
45°	0.9	1.4	1.1	2.4	1.6	0.5	0.2	0.32	0.5
60°	1.1	1.7	0.7	2.8	2.1	0.3	0.24	0.38	0.34
75°	1.2	2.0	0.3	2.8	2.5	0.4	0.25	0.42	0.15
90°	1.1	2.1	0.0	2.6	2.7	0.4	0.24	0.46	0.04
105°	1.0	2.2	0.3	2.3	2.8	0.3	0.21	0.49	0.19
120°	0.8	2.3	0.6	1.9	2.8	0.1	0.17	0.49	0.3
135°	0.6	2.1	0.8	1.5	2.6	0.5	0.12	0.47	0.4
150°	0.4	1.7	1.2	0.9	1.8	0.8	0.07	0.36	0.62
165°	0.2	1.0	1.4	0.3	0.8	0.5	0.04	0.2	0.75
Aft 180°	0.0	0.0	1.4	0.0	0.0	0.5	0.0	0.0	0.75

Wind coefficients on tankers and container ships in the **loaded** condition

Angle from bow	OCIMF/SIGTTO 1995					
	LNG membrane tankers			LNG spherical tankers to stern		
	C_{YFw}	C_{YAw}	C_{Xw}	C_{YFw}	C_{YAw}	C_{Xw}
Forward 0°	0.00	0.00	1.02	0.00	0.00	1.02
15°	0.12	0.07	0.95	0.12	0.07	0.95
30°	0.29	0.20	0.80	0.30	0.19	0.80
45°	0.44	0.35	0.56	0.51	0.33	0.56
60°	0.52	0.45	0.30	0.61	0.45	0.30
75°	0.52	0.52	0.10	0.60	0.55	0.10
90°	0.46	0.59	0.02	0.55	0.62	0.02
105°	0.38	0.64	0.15	0.46	0.64	0.15
120°	0.30	0.65	0.39	.35	0.62	0.48
135°	0.20	0.60	0.65	0.22	0.50	0.83
150°	0.10	0.45	0.79	0.12	0.34	0.97
165°	0.04	0.20	0.89	0.04	0.15	1.02
Aft 180°	0.00	0.00	0.90	0.00	0.00	1.00

Table 2.7 Comparison of wind forces on ships in the loaded condition for the different standards

Forces due to the wind with a gust factor in the **loaded** condition

	General formula		Spanish standard ROM 0.2-90		British standard BS 6349-1: 2000		OCIMF 1997 Oil tankers		OCIMF/SIGTTO 1995 LNG membrane	
Wind speed: m/s	Force: kN		Force: kN		Force: kN		Force: kN		Force: kN	
	0°	90°	0°	90°	0°	90°	0°	90°	0°	90°
10	98	503	130	462	89	357	93	358	110	403
15	221	1132	293	1040	201	804	210	805	248	907
20	392	2012	520	1848	357	1430	373	1430	441	1613

air–structure interaction. On the other hand, the BS and OCIMF formulations seem to provide more accurate estimations of the drag coefficients, depending on the wind angle of attack relative to the vessel.

2.2.4 Wind area and wind loads

The area of ship above the water, part projected on a plane perpendicular to the wind direction, varies greatly, not only due to the different sizes of ships but also because of the different types of ships, and it also depends on whether the ship is in a ballasted condition or not. As an example, a modern general cargo ship of 30 000 t displacement, fully loaded, has a wind area of about 15 m^2/linear m of ship, while the same ship in a ballast condition has an area of about 20 m^2/lin m of ship. Large passenger ships will have wind areas of about 35 m^2/lin m or more.

How high the design wind velocity should be assumed to be will depend on the location of the berth structure, but it is generally not justified to use a higher value than 40 m/s for the gust factor when calculating the mooring forces without closer investigation (i.e. Beaufort number 13, which corresponds to a pressure of about 1.5 kN/m^2). Therefore, assuming too small a wind velocity could be critical, keeping in mind that in the wind loading formula the velocity occurs to the second power. Table 2.8 provides guidelines for wind loads for design purposes.

Table 2.8 Wind loads for design purposes

Ship displacement upper limit: tonnes	Wind load: kN/m of ship
2 000	10
5 000	10
10 000	15
20 000	20
30 000	20
50 000	25
100 000	30

For **piers** where ships can berth on both sides, the total wind load acting on the pier should be approximately the wind load on the largest ship plus 50% of the wind load on the ship on the other side of the pier.

If the wind blows at an angle to the ship, there will be a transverse and a longitudinal load component plus a possible moment on the berth structure. The pressure on a ship of length L_S caused by wind and/or current must be transmitted to the berth structure over the berth length L_Q. The load or pressure on the berth structure should therefore equal the pressure on the ship's side $\times$ L_S/L_Q.

It should be noted that if a ship protrudes outside the end of a berth structure, the contact length L_Q could be very small. The wind and/or the current will also try to turn the ship around the corner of the berth structure.

2.3. Waves

Waves are traditionally, and for practical reasons, classified into the following different types:

(a) **Wind waves** or locally generated waves. These are generated by winds that are acting on the sea surface bordering on the port site.
(b) **Swell** or **ocean waves**. These are normally also wind-generated waves, but are created in the deep ocean at some distance from the port site, and the wind that created them may be too distant to be felt in the port or may have stopped blowing or changed its direction by the time the waves reach the port.
(c) **Seiching** or **long waves**. Waves of this type have very long periods – typically from 30 s up to the tidal period 12 h 24 min – and are mostly found in enclosed or semi-enclosed basins, such as artificial port basins, bays or fjords.
(d) **Waves from passing ships**. Ship waves may be a significant problem in certain ports, especially since they are generated by a moving source and may appear in areas where large waves would not be expected. Ship waves may also be very complex.
(e) **Tsunamis** and waves created by large, sudden impacts, such as earthquakes, volcanoes or landslides that end up in the ocean.
(f) **Breaking waves**. These types of waves' impacts are short waves creating high-pressure impulses to the vertical structure, and can exert considerable forces of the order of between approximately 150 and 600 kN/m^2.

Waves are also classified according to the ratio of the water depth d in which they occur to the wave length L:

(a) **Deep-water waves**, for which $d/L < 0.5$.
(b) **Intermediate-water waves**, for which $0.04 < d/L < 0.5$.
(c) **Shallow-water waves**, for which $d/L = 0.04$.
(d) **Breaking waves** are those which, for example, fall forward since the forward velocity of the crest particles exceeds the velocity of the propagation of the wave itself. In deep water this normally occurs when $L < 7H$, and in shallow water when d is approximately equal to $1.25H$ and H is wave height. The still-water depth, where the wave breaking commences, is called the breaking depth.

Wind-generated waves are defined by their height, length and period. The height, length and period are dependent on the fetch (the distance the wind blows over the sea in generating the waves) and the velocity, duration and direction of the wind. The wave characteristics for deep-water waves are

Figure 2.5 Wave characteristics in deep water

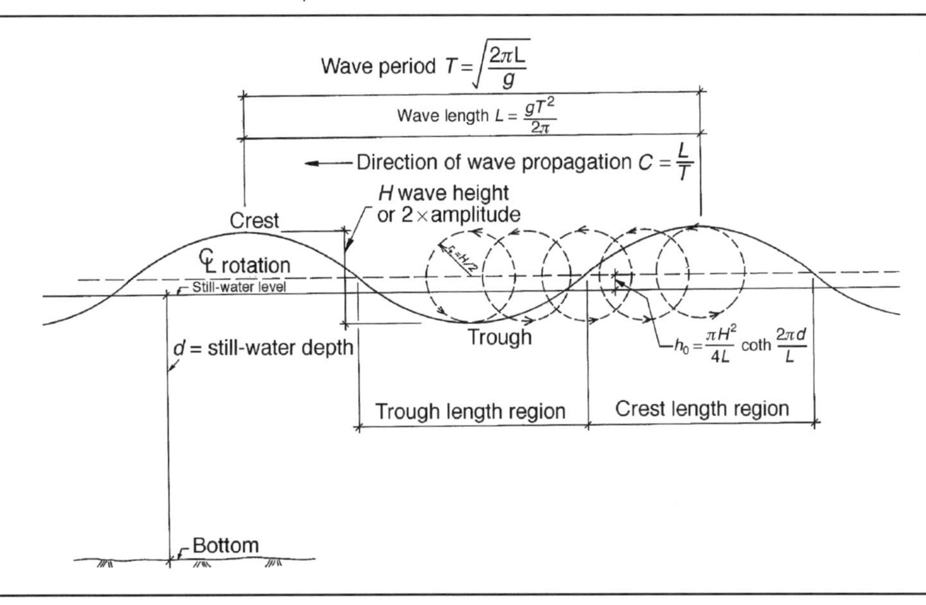

shown in Figure 2.5. The wave period is the time between successive crests passing a given point. The wave steepness is defined as the wave height divided by its length. As the waves propagate in water, it is only the waveform and part of the energy of the waves that move forward.

The **wave heights** H may be defined as follows:

H_m = arithmetical mean value of all recorded wave heights during a period of observation $= 0.6H_s$
H_s = significant wave height, which is the arithmetical mean value of the highest one-third of the waves for a stated interval
$H_{1/n}$ = average value of the $1/n$ highest waves in a series of waves, usually of length 15–20 min – commonly used values of n are 3 (significant wave height), 10, 100
$H_{1/10}$ = arithmetical mean value of the height of the highest 10% $= 1.27H_s$
$H_{1/100}$ = arithmetical mean value of the height of the highest 1% $= 1.67H_s$
H_{max} = maximum wave height $= 1.87H_s$ or rounded to $= 2H_s$ when a high risk of danger is present, or if storms of long duration are to be considered.

The variables in wind wave height computations are:

V_{10} = wind speed at 10 m above sea level, usually taken as a 10 min mean, which is representative of the entire fetch and the entire duration of the situation
F = fetch length
t = duration of the wind.

Accurate calculations of the wave heights at the end of a fetch require a detailed knowledge of the fetch and the wind field. An earlier wave prediction monogram for determining the effective fetch distance is shown as an example in Figure 2.6. It consists of constructing 15 radials from the berth site at intervals

Figure 2.6 Calculation of the effective fetch for an irregular shoreline

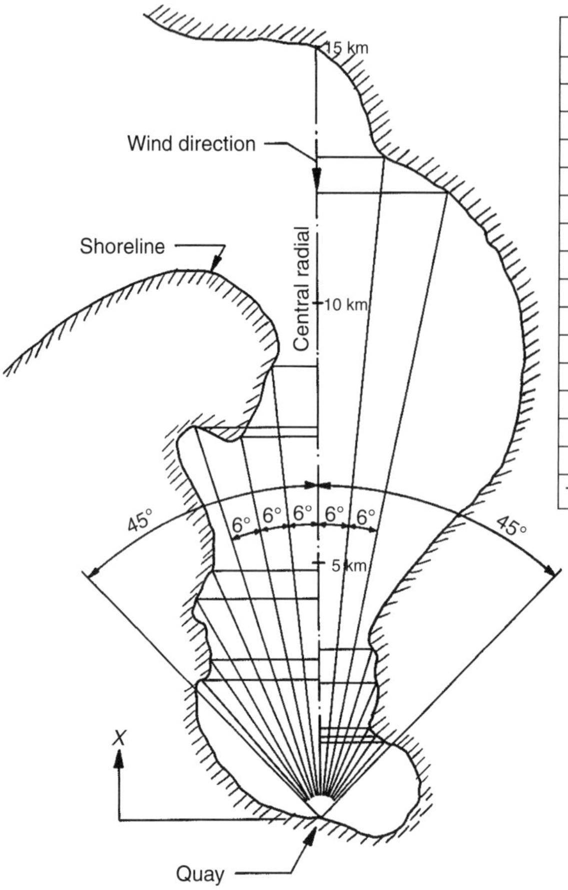

α	$\cos\alpha$	X	$X\cos\alpha$
42°	.743	2.7	2.0
36°	.809	2.9	2.3
30°	.866	4.2	3.6
24°	.914	4.8	4.4
18°	.951	7.6	7.3
12°	.978	7.4	7.2
6°	.995	9.8	9.8
0°	1.000	15.0	15.0
6°	.995	12.7	12.6
12°	.978	12.0	11.7
18°	.951	3.3	3.2
24°	.914	2.6	2.4
30°	.866	1.7	1.5
36°	.809	1.5	1.3
42°	.743	1.5	1.1
Total 13.512			85.4

$$F_{eff} = \frac{\Sigma X\cos\alpha}{\Sigma\cos\alpha}$$

$$\frac{85.4}{13.512} = 6.3\ \text{km}$$

of 6°, limited by an angle of 45° on either side of the wind direction. These radials are extended from the berth site until they first intersect the shoreline, as shown in the figure.

The length component of each radial in the direction parallel to the wind direction is measured and multiplied by the cosine of the angle. The resulting values for each radial are added together and divided by the sum of the cosines of all the individual angles.

Investigations by Resio and Vincent in 1979 (personal communication) suggested that wave conditions in the fetch area were actually relatively insensitive to the width of a fetch. It is thus recommended that the fetch width is not used to estimate an effective fetch in nomograms. Instead, it is recommended that the straight line fetch should be used to define the fetch length for applications.

Where the fetch region is rectangular with a relatively uniform width, Figure 2.7 may be used to obtain the effective fetch length. As shown from these examples, fetches limited by landform will be significantly lower than fetches over more open waters, and will result in lower wave generation.

Figure 2.7 Effective fetch length for a rectangular fetch region

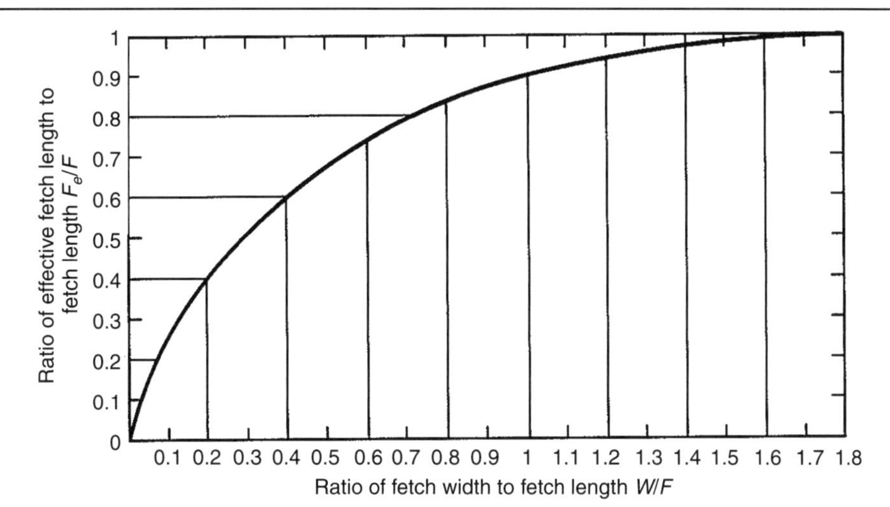

Generally, waves are generated in the transfer of energy from the air moving over the sea surface in the following ways:

(a) The first variations in the sea or water level are created due to the reactions of the water level to the small pressure difference in the moving air. These will increase by the pressure difference exerted by the moving wind on the front and back of the wave.
(b) Tangential stress between the water and the air, which are moving at different speeds relative to each other.

It is clear that the wave characteristics will be a function of the wind velocity, since both the pressure difference and the tangential stress are functions of the wind velocity. Waves generated by wind of a certain velocity will have only attained the characteristics typical for this wind velocity after the wind has blown for a certain time. Generally, the waves will at first increase rapidly in size and then grow at a decreasing rate the longer the wind lasts.

The significant wave height, period and the necessary minimum wind duration for constant wind velocity are shown as a function of the fetch for deep-water conditions in Figure 2.8.

The significant wave height is shown both as a function of wind duration for constant wind velocity and for unlimited and limited fetches in deep water in Figure 2.9.

The following cases are examples of the practical use of these curves:

(a) Case A: from Figure 2.8 with an effective fetch length of 20 km and a constant wind velocity of 20 m/s, the significant wave height is 1.90 m, the significant period is 5.35 s and the necessary wind duration to develop the significant wave height for that particular fetch length is 1.80 h.
(b) Case B: from Figure 2.9 with a wind duration of 5.5 h for a constant wind velocity of 20 m/s in an unlimited fetch area, the significant wave height is 3.80 m.

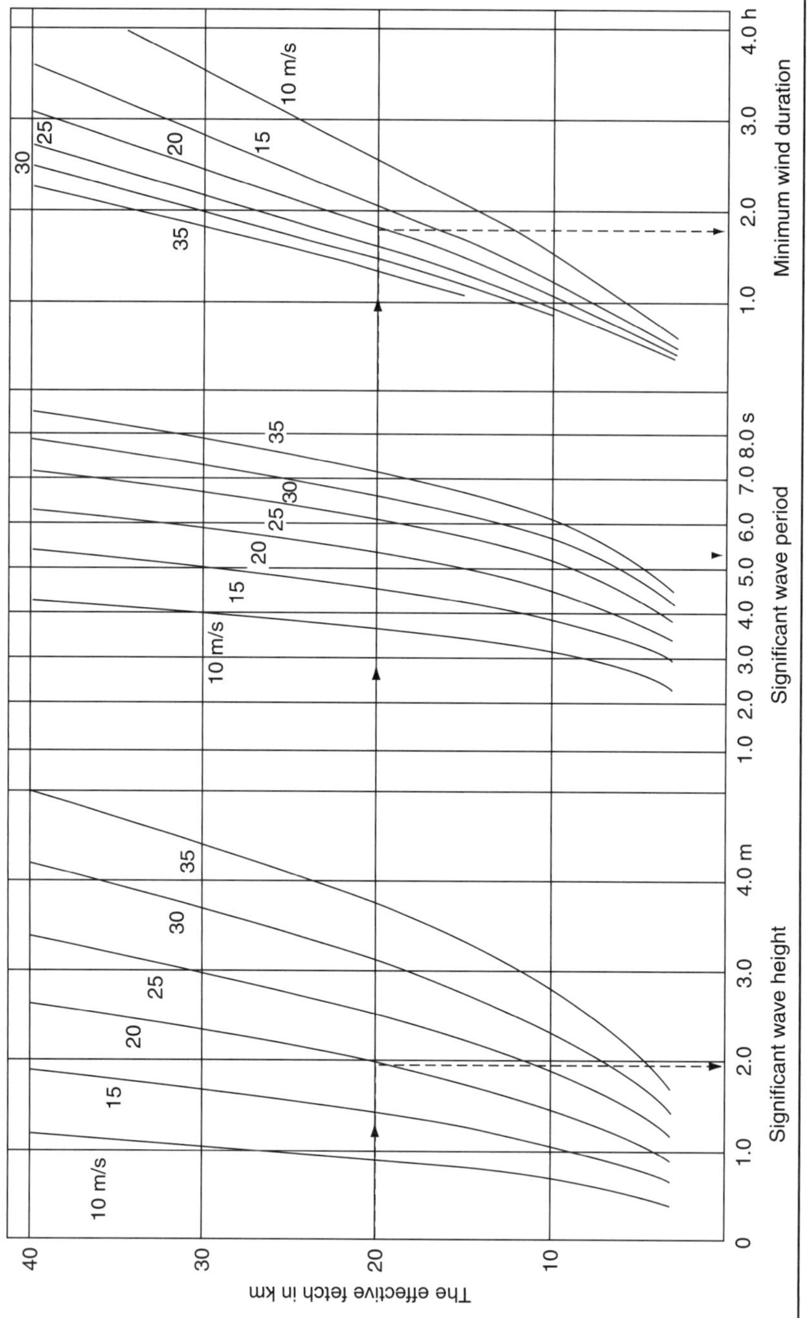

Figure 2.8 Significant wave length, period and duration for constant wind velocity as a function of the fetch for deep-water conditions

The effective fetch in km

Significant wave height

Significant wave period

Minimum wind duration

Figure 2.9 Significant wave height as a function of wind duration for constant wind velocity and for unlimited and limited fetches in deep-water conditions

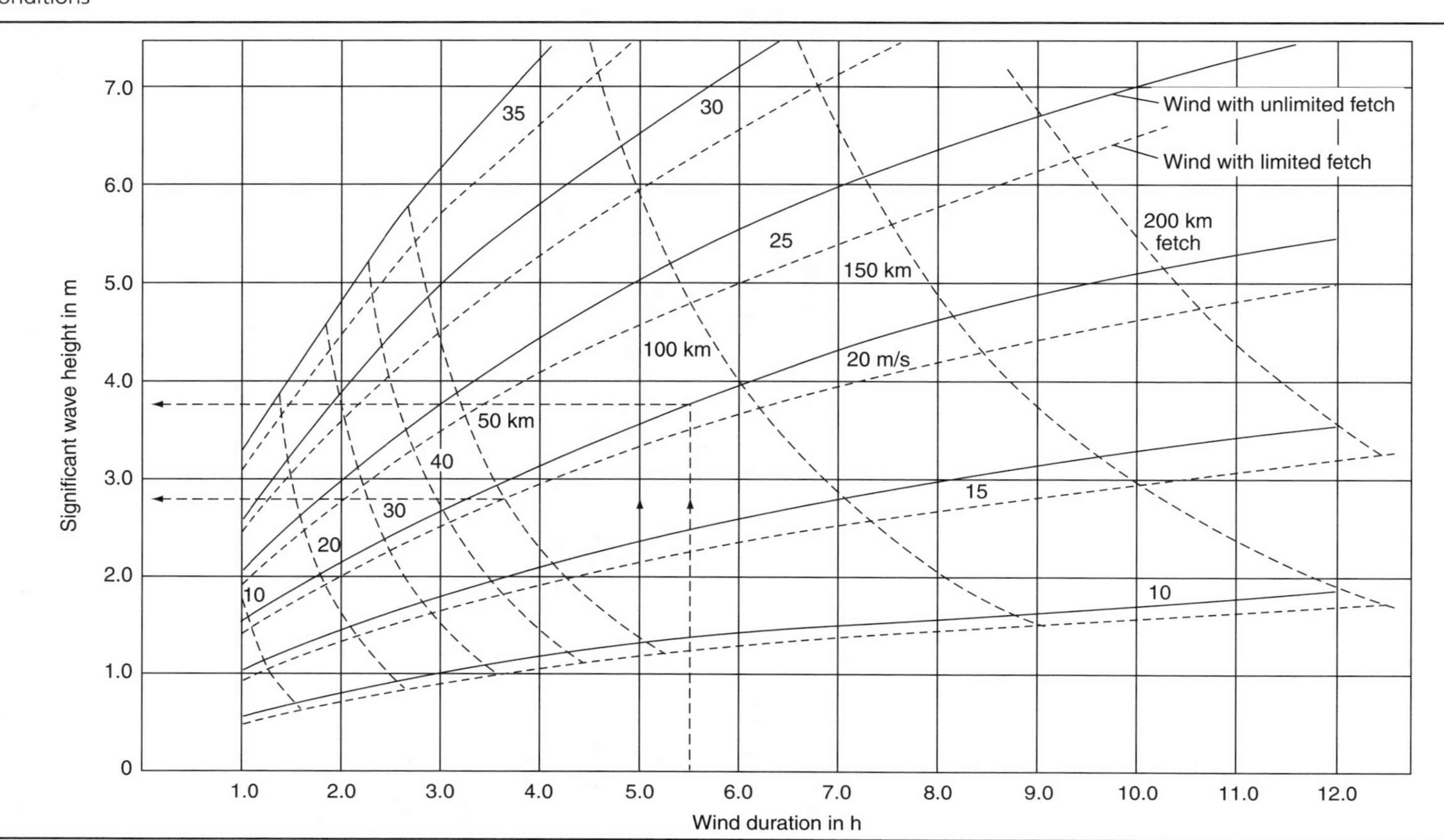

(c) Case C: from Figure 2.9 with a wind duration of 5.0 h for a constant wind velocity of 20 m/s with a limited fetch of 50 km, the significant wave height is 2.80 m.
(d) Case D: the effective fetch for an irregular shoreline as shown in Figure 2.6 is 6.3 km. With a wind velocity of 25 m/s and an unlimited wind duration, the significant wave height for deep water is 1.60 m. If the average constant water depth along the central radial is 10 m and the bottom friction factor is assumed to be 0.01, the significant wave height from Figure 2.9 is found to be $H_s = 0.022 \times V^2/g = 1.40$ m.

In the figures the significant wave height is in metres, the wave period in seconds, the wind speed in m/s, the fetch in metres and the wind duration in seconds or hours. For practical wave predictions it is usually satisfactory to regard the wind speed as reasonably constant if the variations do not exceed 2.5 m/s (5 knots) from the mean.

The water depth will have an effect on the wave generation. For a given wind velocity and fetch conditions, the wave heights will be smaller and the wave periods shorter if the generation takes place at intermediate- and shallow-water depths rather than at a deep-water depth. The significant wave height is shown as a function of the fetch and the wind velocity with unlimited wind duration for intermediate and shallow-water depths in Figure 2.10, with an assumed bottom friction factor equal to 0.01.

2.3.1 Waves near ports

Where it is necessary to carry out instrumental wave recording, it is advisable to install the recording system as early as possible to enable the recording programme to be as long as possible. A minimum recording time should be 1 year, to obtain reasonably reliable data, because a shorter duration is unlikely to yield a representative set.

In most cases, the waves that constitute the design wave condition in a port are a combination of ocean waves and locally generated wind waves. Where the port is situated in sheltered waters such as a bay or in a fjord, the distinction between the two types of waves and the reason for treating them separately is quite obvious. On open coastlines, however, the two types may become inseparable, and one may choose to consider ocean waves only.

A flowchart illustrating a suggested method for calculating wave height is shown in Figure 2.11.

Local wind waves and ocean waves are traditionally calculated separately and combined by adding the energy components of the two sea states:

$$H_{s,i} = (H_{s,w}^2 + H_{s,o}^2)^{1/2}$$

where the subscript i denotes combined inshore waves, w wind waves and o ocean waves.

If one of the wave types is totally dominant over the other, one may choose to ignore the contribution from the lesser component.

2.3.2 Breaking waves

Wave breaking occurs when the wave crest travels faster than the rest of the waveform and becomes separated from it. From the port designer's point of view, wave breaking induced by a limited water depth is the most relevant type of breaking.

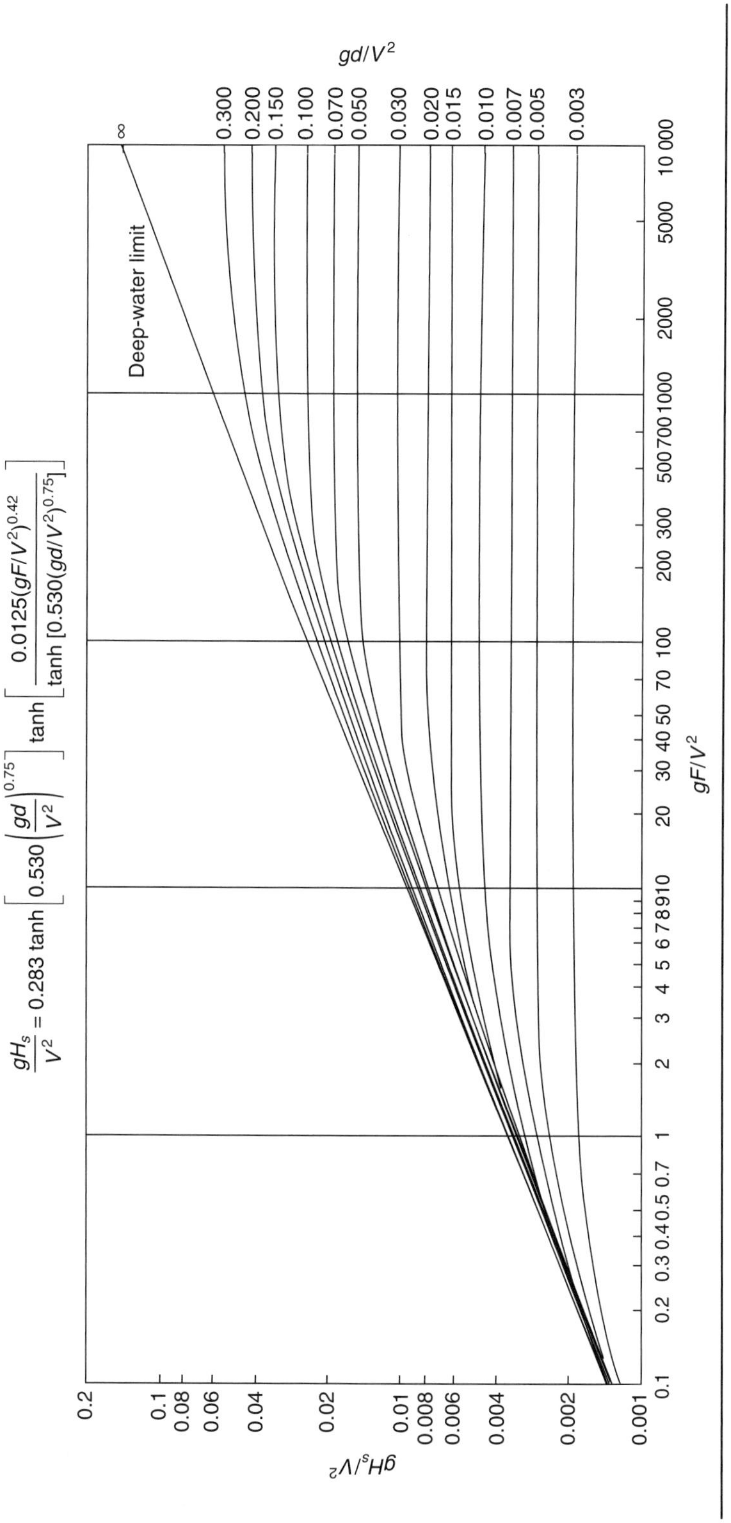

Figure 2.10 Significant wave height as a function of the fetch and unlimited wind duration for intermediate and shallow water of constant depth

$$\frac{gH_s}{V^2} = 0.283 \tanh\left[0.530\left(\frac{gd}{V^2}\right)^{0.75}\right] \tanh\left[\frac{0.0125(gF/V^2)^{0.42}}{\tanh\left[0.530(gd/V^2)^{0.75}\right]}\right]$$

Figure 2.11 Procedure for calculating wave heights

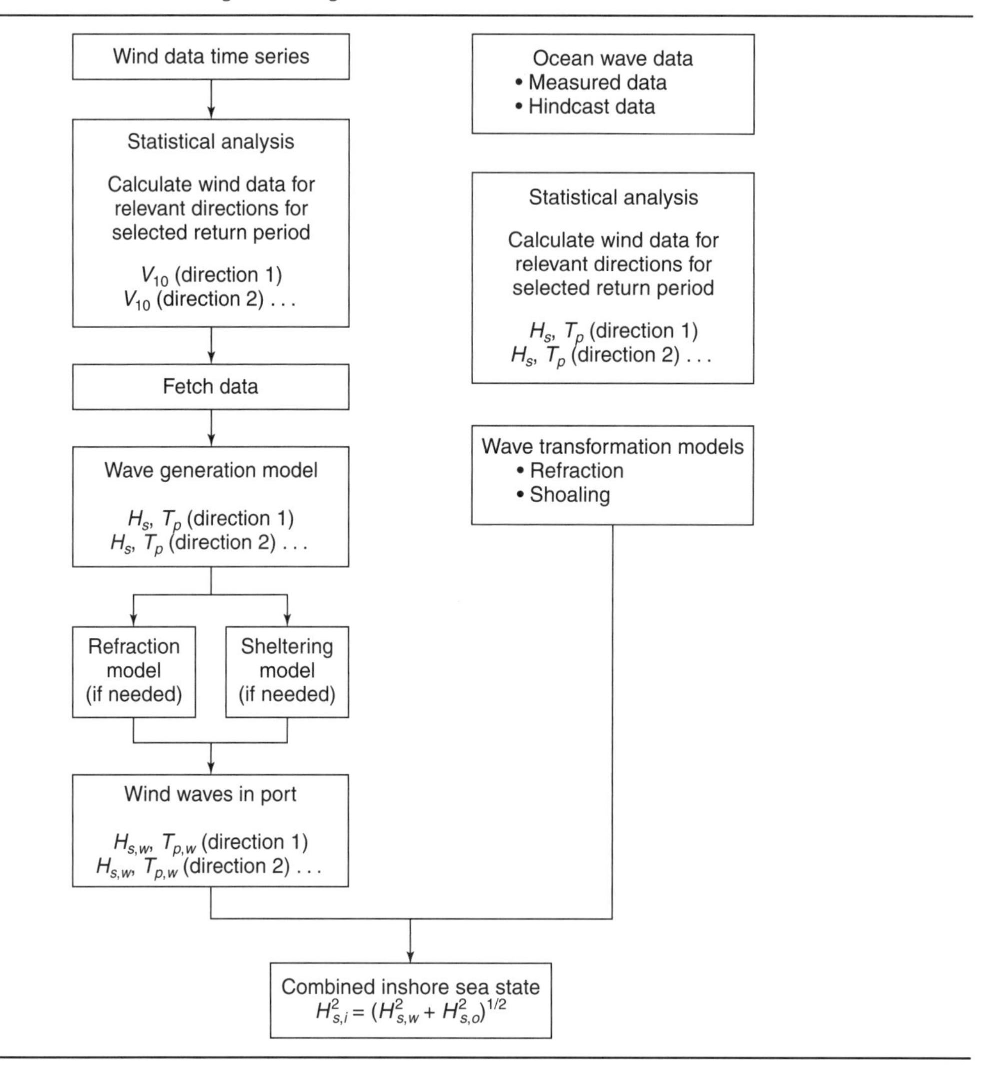

A first approximation of the breaker height can be obtained from the simple expression

$$d_b = 1.28H_b$$

where d_b is the depth at breaking and H_b is the individual wave height at breaking.

2.3.3 Wave action

Since most berth structures are sheltered against sizable waves, the static calculations of berth structures normally do not explicitly deal with forces and reactions due to wave action. Such forces are assumed to be taken care of by the very fact that the structure is also designed for impact and mooring forces. For breakwaters, and similar structures heavily exposed to waves, the wave actions must of course be studied very closely in each case and be performable by model tests in a hydraulic institute.

The characteristics of the waves in a port or berth area and their effect on the berthing structure are influenced by the following factors:

(a) Bathymetry (bottom topography) in the vicinity of the berth structure itself.
(b) Waves can be reflected from the near-shore slopes.
(c) Refraction of waves can or will take place as the waves enter shallow water at an inclined angle.
(d) Wave shoaling will influence the design wave heights as the waves enter shallow water.
(e) The wave regime in the harbour area will be due to swell and/or wind-generated waves.

Structures, which are very resistant to overload stresses, may be designed for a wave height that is lower than H_{max}. This increase in risk for, for example, flooding or destruction can be justified by the lower cost of construction, as long as it does not decrease the safety of personnel. Therefore, one should always evaluate both the construction costs and the capitalised maintenance costs. If it does not decrease the safety of personnel, it can, in some cases, be cheaper, for example, to apply a shore protection that requires regular maintenance work instead of constructing an expensive and maintenance-free shore protection.

Figures 2.12 and 2.13 show wave actions against a breakwater at the beginning of a storm where the top of the breakwater is too low.

Figures 2.14, 2.15 and 2.16 show the collapsed breakwater after the storm. The collapse of the breakwater was due to two main reasons, namely that the top of the breakwater was too low, and that the protection stone layer at the front and at the back of the breakwater had stones that were too small, as can be seen from the photograph.

Figure 2.17 shows the final repair of the collapsed breakwater.

Figure 2.12 Wave action along a breakwater

Figure 2.13 Wave action along a breakwater

Figure 2.14 The washout behind the breakwater

Figure 2.15 The breakwater after the storm

Figure 2.16 Detail of the collapsed breakwater

Figure 2.17 The final repair of the collapsed breakwater

2.3.4 Design wave

The design wave may be chosen by the following means:

(a) Selecting a design wave return period (R_p). In this context, the design return period is defined as the average time lapse between two consecutive events where the wave height is equal to or greater than a given significant wave height (the design wave height). This method is often chosen for its simplicity, and is preferred for smaller structures such as breakwaters, piers, small ports, etc. The design return period should be in the range 50–100 years for simple technical structures.
When selecting the return period, attention must be paid to the consequences of encountering an exceedance of the design wave height. Low return periods (in some cases as low as 25 years) are chosen when the consequences of an exceedance are minor or easily repaired. Where the economic consequences are great, or human life and health may be threatened, the normal return period is a minimum of 100 years. One may also specify a differentiated set of design return periods for different types of functions and types of structures in a port.

(b) Specifying a required technical **lifetime** of the structure; it should be noted that the design life is not necessarily the same as the return period of the design. If a lifetime or design working life n of a structure is specified, then it is assumed that n number of years must pass from the start of the project without the structure experiencing a situation (i.e. significant wave height) that exceeds its design load. As such, the designer cannot give a guarantee, and one is therefore forced to introduce a probability P of encountering such a situation over the projected lifetime.

The relationship is given as

$$T = \frac{1}{1 - \sqrt[n]{1 - \frac{P}{100}}}$$

where

T = design return period
n = design lifetime
P = probability (in per cent) of encountering a significant wave height greater than the design wave height during the lifetime n

The relationship between the design working life, return period and probability of wave heights exceeding the normal average is shown in Figure 2.18.

A port structure is required to have a design lifetime of 50 years. If the associated probability of exceedance of the design wave height level is set equal to 10%, then the required design return

Figure 2.18 The relationship between design working life, return period and probability of wave

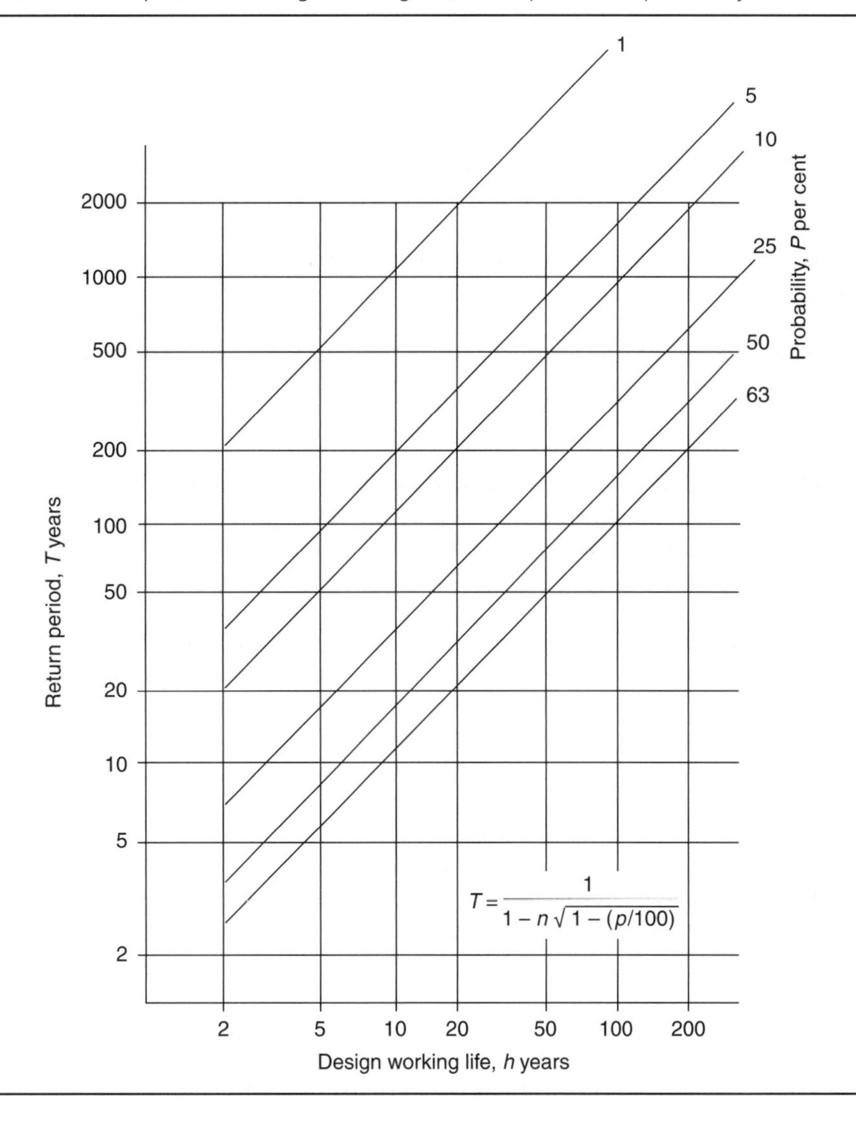

Table 2.9 Design wave height

Type of structure	H_{des}/H_s
Erosion protection	1.0–1.4
Rubble-mound breakwater	1.0–1.5
Concrete breakwater	1.6–1.8
Berth structures	1.8–2.0
Structure with high safety requirements	2.0

period is approximately 500 years. By setting the design return period equal to the design life (e.g. $T = n = 50$ years), Figure 2.18 shows that there is a 63% chance of exceeding the design wave height over the next 50 years. Likewise, by setting the design return period equal to twice the lifetime (100 years), the probability of exceedance during the lifetime is approximately 40%. Therefore, it is necessary to evaluate the consequences and probabilities of damage against the costs of reducing or avoiding these risks.

The return period of the wave should be at least 100 years, but not less than the design life of the structure. If the consequences from failure of the structure are so grave as to be unacceptable at other than very low probabilities, the structure should be able to withstand design conditions with return periods of 1000 years or more.

The design wave height H_{des}, which should be chosen for the design, may, depending on the severity of the allowable risk, be as in Table 2.9.

2.3.5 Wave forces

There is no standard procedure for calculating wave loads on a moored ship. Wave forces and mooring system reactions arise from a complex interaction of the water particle kinematics, ship motion and mooring system response. Determinations of added mass, damping and higher-order non-linear effects add to the difficulty.

Wave forces on moored ships manifest themselves in two components. The first is a **linear oscillatory** force at the frequency of the waves. This force can be found by integrating the varying water pressures around the submerged portion of the hull (Figure 2.19). Diffraction theory must be employed in order to do this, as the presence of the ship modifies the incident wave train. The oscillating ship scatters waves, which propagate away from the hull and contribute to the damping of the oscillations. The hydrodynamic coefficients of added mass and damping are different for each of the six degrees of freedom, and are dependent upon the frequency of the oscillation. In shallow water, the water depth-to-draught ratio or the vessel underkeel clearance has a significant effect on these coefficients. Buoyant restoring forces may result in heaving and/or pitching oscillations at some natural period of the system. The restoring forces of mooring lines and fenders are non-linear and may result in resonant horizontal motions.

The second type of wave force is **non-linear** in nature and a result of the irregular sea state. This force, known as the **drift force**, is primarily a consequence of wave grouping and set-down effects. The momentum flux and radiation stress of progressive groups of higher and then lower waves result in

Figure 2.19 Schematic illustration of wave force on fixed moored vessel in beam sea

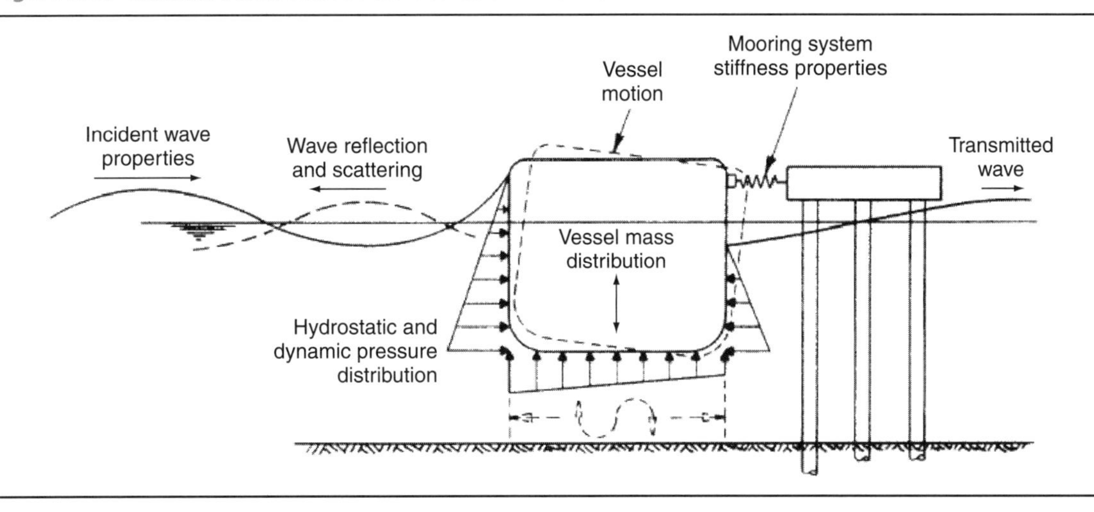

a net force in the direction of wave propagation. The drift force is a steady force in regular waves and a slow-varying/low-frequency force in irregular waves. Because damping is relatively low at a lower frequency and because the usual range of periods of 20–100 s is also within the range of natural periods of moored ships, the slow-varying drift force may cause overstressing of mooring lines and large fender forces. Analytical and computational and physical modelling techniques can be used to approach the general moored-ship problem.

Important conclusions to the problem of waves are that:

(a) mooring line forces are induced almost entirely by waves of longer than 20 s, with major contributions occurring in the period range 40–200 s
(b) rolling is excited primarily by swells in the 13–19 s range
(c) pitching is small and insensitive to the wave period, whereas heave motions tend to equal the wave height at long periods and are smaller with short-period waves.

For wave forces that can act on a ship, the **Spanish standard ROM 0.2-90** (Ministerio de Obras Públicas y Transportes, 1990) recommends that the wave forces or pressure on a ship should be:

Lateral or **transverse** wave force in kN:

$$F_{\mathrm{Twave}} = C_{\mathrm{fw}} \times C_{\mathrm{dw}} \times \gamma_{\mathrm{w}} \times H_{\mathrm{s}}^2 \times D' \times \sin^2 \alpha \times 10$$

Longitudinal wave force in kN:

$$F_{\mathrm{Lwave}} = C_{\mathrm{fw}} \times C_{\mathrm{dw}} \times \gamma_{\mathrm{w}} \times H_{\mathrm{s}}^2 \times D' \times \cos \alpha \times 10$$

where

$$D' = L_{\mathrm{bp}} \times \sin \alpha + B \times \cos \alpha$$

Here

C_{fw} = waterplane coefficient depending on the wave longitude L_w at the location and the ship's draught D; if $(2\pi/L_w) \times D > 1.4$, then $C_{fw} = 0.064$, and if $(2\pi/L_w) \times D < 0.2$, then $C_{fw} = 0.0$

C_{dw} = depth coefficient depending on the wave longitude L_w and the water depth at the location; if $(4\pi/L_w) \times h > 6.0$, then $C_{dw} = 1.0$, and if $(4\pi/L_w) \times h = 0.0$, then C_{dw} is 2.0

γ_w = specific gravity of water (seawater, 1.034 t/m^3; freshwater, 1.00 t/m^3)

H_s = design significant wave height

α = angle between the longitudinal axis of the ship, considered from bow to stern, and the direction of the wave

D' = projection of the ship length in the direction of the incident waves

L_{bp} = length between perpendiculars

B = beam of the ship

D = draught of the ship

h = water depth at the location

2.3.6 Wave loads by breaking waves against vertical walls

Breaking waves against vertical walls can create considerable forces of the order of about 150–600 kN/m^2. Therefore, in the design of berth structures one should try to prevent breaking waves against the berth, and for quay structures the loading platform should be higher than the crests of the waves.

2.4. Current

The magnitude and direction of the tidal current and the wind-generated current must be evaluated to establish any influence on the berthing and unberthing operations.

Current can arise in a port basin due to wind transporting water masses, differences in temperature and salt contents, tidal effects, water flow from river estuaries, etc. At some Norwegian harbours, situated at the mouth of a river, the currents are known to have reached a velocity of 3 m/s or 6 knots.

When designing new berth structures, it is always important to ensure that the berth front is directed as parallel as possible to the prevailing current. Since the direction of the current can vary, it is also necessary to investigate over a long time period the magnitude of the current perpendicular to the direction of the berth front. Should such a component reach a value of about 0.5 m/s perpendicular to, for example, an open pier, the berthing operation would be very difficult.

Even if currents do not usually set up loads of vital importance to, for example, a finished berth structure, they can still be of importance during the construction of a steel sheet pile berth structure. For instance, the normal driving of piles is almost impossible if the current velocity is higher than 1.5 m/s. Divers will be unable to work properly if the current velocity is higher than 0.5 m/s.

The **tidal current** must be measured at various depths in the harbour area. The tidal current is referred to as a flood current on a rising tide and an ebb current on a falling tide.

The magnitude of the **wind-generated current** will, in the open sea, be approximately 1–2% of the wind speed at 10 m above the water level.

The **berthing structure** and the **mooring equipment** for oil and gas tankers should generally be in line with OCIMF, and at least capable of resisting loads due to any one of the following current conditions

Figure 2.20 The effect of the underkeel clearance on the current force

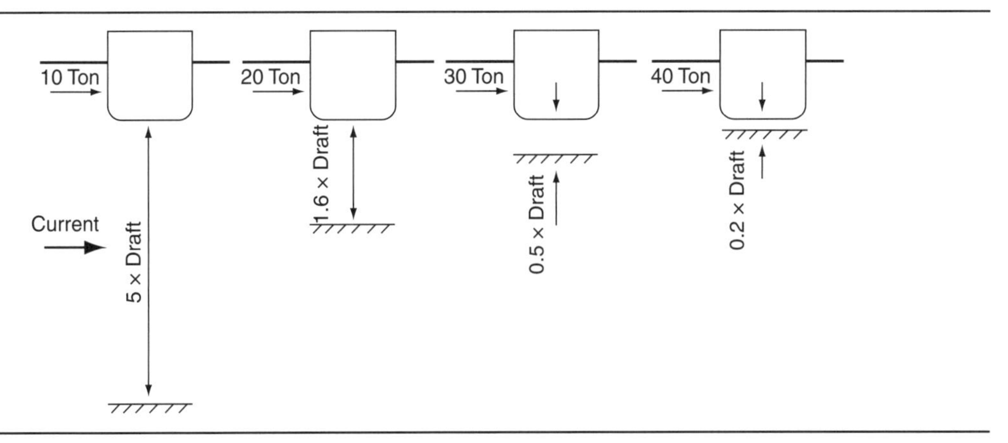

acting simultaneously with the design wind from any of the following directions: 1.5 m/s or 3 knots at 0° and 180° (current parallel to the berth), 1.0 m/s or 2 knots at 10° and 170°, and for pier structure 0.4 m/s or 0.75 knots from the direction of the maximum beam current loading, as shown in Figure 2.20.

2.4.1 Current forces

The current forces acting on a ship may vary considerably with both the type and size of the ship, and should therefore best be established by testing. The longitudinal current forces are especially very scale-dependent.

In general, the current forces on a ship due to velocity and direction follow a pattern similar to that for wind forces. When evaluating the mooring arrangement, the current forces must be added to the wind forces.

The current forces are complicated compared with the wind forces due to the significant effect of clearance beneath the keel. The effect of the underkeel clearance on the current force is shown in Figure 2.20 by the increase in the current force due to a reduced underkeel clearance. Therefore, one should always try to orientate the berth front parallel to the main current direction. Even a small current angle off the ship's longitudinal axis can create a transverse force, which must be ascertained during the evaluation of the mooring system.

2.4.2 Different current standards and recommendations

Below, different methods and national standards and regulations for the calculation of the current forces are shown and compared against each other. It should be noted that these standards and regulations should only be used as a guide to the magnitude of the forces on the ship. If more accurate current force calculations are needed, these should be established by model testing in a hydraulic institute.

A **general standard formula** for the calculation of the current forces on a ship is as follows.

The current pressure or force on a moored ship would be in kN:

$$P_c = C_C \times \gamma_w \times A_c \times \left(\frac{V_c^2}{2g}\right)$$

where C_C is the factor for calculation of the transverse and longitudinal forces. The magnitude of C_C depends to a large extent on the shape of the ship and the water depth along the front of the berth structure.

Factors for the calculation of the **transverse** current force are:

(a) for a deep-water depth, $C_C = 1.0–1.5$
(b) water depth/design ship draught = 2, $C_C = 2.0$
(c) water depth/design ship draught = 1.5, $C_C = 3.0$
(d) water depth/design ship draught = 1.1, $C_C = 5.0$
(e) water depth/design ship draught nearly = 1, $C_C = 6.0$.

Factors for the calculation of the **longitudinal** current force vary from 0.2 for deep water to 0.6 for a water depth to design ship draught ratio of nearly 1.

γ_w = specific gravity of seawater, 10.34 kN/m^3
A_c = area of the ship's underwater part projected onto a plane perpendicular to the direction of the current
V_c = velocity of the current in m/s
g = acceleration of gravity, 9.81 m/s^2

The **Spanish standard ROM 0.2-90** (Ministerio de Obras Públicas y Transportes, 1990) recommends that the current forces or pressure on a ship should be:

Lateral or transverse current force in kN:

$$F_{\text{tcurrent}} = F_{\text{TC}} + F'_{\text{TC}}$$

where the lateral or transverse current force in kN on the ship from pressure is

$$F_{\text{TC}} = C_{\text{TC}} \times \gamma_w \times A_{\text{LC}} \times \sin\alpha \times \frac{V_c^2}{2g} \times 10$$

and the lateral or transverse current force in kN on the ship from drag is

$$F'_{\text{TC}} = C_{\text{R}} \times \gamma_w \times A'_{\text{TC}} \times \sin^2\alpha \times \frac{V_{\text{C}}^2}{2g} \times 10$$

where $A_{\text{LC}} = L_{\text{bp}} \times D$ and $A'_{\text{TC}} = (L_{\text{bp}} + 2 \times D) \times B$.

Longitudinal current force in kN:

$$F_{\text{lcurrent}} = F_{\text{LC}} + F'_{\text{LC}}$$

where the longitudinal current force on the ship from pressure is

$$F_{\text{LC}} = \pm\left(C_{\text{LC}} \times \gamma_w \times A_{\text{TC}} \times \frac{V_{\text{C}}^2}{2g} \times 10\right)$$

and the longitudinal current force on the ship from drag is

$$F'_{LC} = C_R \times \gamma_w \times A'_{LC} \times \cos^2 \alpha \times \frac{V_C^2}{2g} \times 10$$

where $A_{TC} = B \times D$ and $A'_{LC} = (B + 2 \times D) \times L_{bp}$

A_{LC} = submerged longitudinal projected area of the ship exposed to the current
A'_{TC} = submerged transverse projected wetted area of the ship
A_{TC} = submerged transverse projected area of the ship exposed to the current
A'_{LC} = submerged longitudinal projected wetted area of the ship
L_{bp} = length between perpendiculars
D = draught of the ship
B = beam of the ship
γ_w = 1.03 t/m^3 for saltwater and 1.00 for freshwater
g = acceleration of gravity, 9.81 m/s^2
α = angle between the longitudinal axis of the ship and the current direction considered from the bow
C_{TC} = factor for the calculation of the transverse current force depending on the water depth/design ship draught as shown in Table 2.10:
water depth/design ship draught = 6, $C_{TC} = 1.0$
water depth/design ship draught = 3, $C_{TC} = 1.7$
water depth/design ship draught = 2, $C_{TC} = 2.7$
water depth/design ship draught = 1.5, $C_{TC} = 3.8$
water depth/design ship draught = 1.1 $C_{TC} = 5.7$
C_{LC} = factor for the calculation of the longitudinal current force due to the geometry of the ship's bow varying between 0.2 and 0.6; for conventional bows (bulb) it is equal to 0.6, as shown in Table 2.10
C_R = friction drag factor, as shown in Table 2.10
for new ships = 0.001
for ships in service = 0.004
V_C = mean velocity of the current at an interval of 1 min at a depth of 50% of the ship draft in m/s
In the absence of defined operating criteria, the following should be adopted as limiting permanence velocities:
cross-current: $0° < \alpha < 180°$, $V_C = 1$ m/s (2 knots)
bow or stern current: $\alpha = 0°$ and $\alpha = 180°$, $V_C = 2.5$ m/s (5 knots)
For mooring calculations under normal operation conditions, loading and unloading, the design velocities should be:
cross-current: $0° < \alpha < 180°$, $V_C = 1$ m/s (2 knots)
bow or stern current: $\alpha = 0°$ and $\alpha = 180°$, $V_C = 1.5$ m/s (3 knots)

The **British standard BS 6349-1: 2000** (BSI, 2000) recommends that the current forces or pressure on a ship should be:

Lateral or transverse current force in kN:

$$F_{Tcurrent} = (C_{TCforward} + C_{TCaft}) \times C_{CT} \times \gamma \times L_{bp} \times d_m \times \frac{V_C^2}{10\,000}$$

Table 2.10 Current coefficients

Spanish standard ROM 0.2-90

Angle from bow to stern	Depth correction $D_w/D_d = 1.1$			Depth correction $D_w/D_d = 1.5$			Depth correction $D_w/D_d = 2.0$		
	C_{TC}	C_{LC}	C_R	C_{TC}	C_{LC}	C_R	C_{TC}	C_{LC}	C_R
0°	5.70	0.60	0.004	3.80	0.60	0.004	2.70	0.60	0.004
15°	5.70	0.60	0.004	3.80	0.60	0.004	2.70	0.60	0.004
30°	5.70	0.60	0.004	3.80	0.60	0.004	2.70	0.60	0.004
45°	5.70	0.60	0.004	3.80	0.60	0.004	2.70	0.60	0.004
60°	5.70	0.60	0.004	3.80	0.60	0.004	2.70	0.60	0.004
75°	5.70	0.60	0.004	3.80	0.60	0.004	2.70	0.60	0.004
90°	5.70	0.60	0.004	3.80	0.60	0.004	2.70	0.60	0.004

British standard BS 6349-1: 2000

(a) Transverse very large tankers

Angle from bow to stern	Depth correction $D_w/D_d = 1.1$			Depth correction $D_w/D_d = 1.5$			Depth correction $D_w/D_d = 2.0$		
	$C_{TCforward}$	C_{TCaft}	C_{CT}	$C_{TCforward}$	C_{TCaft}	C_{CT}	$C_{TCforward}$	C_{TCaft}	C_{CT}
0°	0.00	0.001	0.00	0.00	0.00	5.00	0.00	0.00	2.00
15°	0.50	0.20	8.00	0.50	0.20	4.50	0.50	0.20	1.70
30°	0.80	0.40	6.00	0.80	0.40	3.80	0.80	0.40	1.70
45°	1.20	0.70	5.00	1.20	0.70	3.40	1.20	0.70	1.60
60°	1.40	1.00	4.70	1.40	1.00	3.00	1.40	1.00	1.60
75°	1.50	1.30	4.80	1.50	1.30	2.90	1.50	1.30	1.70
90°	1.40	1.60	4.90	1.40	1.60	2.90	1.40	1.60	1.70
105°	1.20	1.70	4.70	1.20	1.70	2.90	1.20	1.70	1.70
120°	0.90	1.60	4.50	0.90	1.60	3.10	0.90	1.60	1.70
135°	0.60	1.40	4.50	0.60	1.40	3.60	0.60	1.40	1.70
150°	0.30	1.00	5.00	0.30	1.00	4.00	0.30	1.00	1.80
165°	0.10	0.60	6.50	0.10	0.60	4.40	0.10	0.60	2.00
180°	0.00	0.00	9.00	0.00	0.00	5.00	0.00	0.00	2.20

Longitudinal current force in kN:

$$F_{Lcurrent} = C_{LC} \times C_{CL} \times \gamma \times L_{bp} \times d_m \times \frac{V_C^2}{10\,000}$$

where

$C_{TCforward}$ = transverse forward current force coefficient depending on the angle of the current (Table 2.10)

Table 2.10 Current coefficients (Continued)

(b) Longitudinal

Angle from bow to stern	C_{LC}	C_{CL}	C_{LC}	C_{LC}	C_{CL}	C_{LC}
0°	0.40	1.70	0.40	1.40	0.40	1.20
15°	0.20	1.70	0.20	1.40	0.20	1.20
30°	0.20	1.70	0.20	1.40	0.20	1.20
45°	0.40	1.70	0.40	1.40	0.40	1.20
60°	0.20	1.70	0.20	1.40	0.20	1.20
75°	0.20	1.70	0.20	1.40	0.20	1.20
90°	0.00	1.70	0.00	1.40	0.00	1.20
105°	0.10	1.70	0.10	1.40	0.10	1.20
120°	0.20	1.70	0.20	1.40	0.20	1.20
135°	0.20	1.70	0.20	1.40	0.20	1.20
150°	0.10	1.70	0.10	1.40	0.10	1.20
165°	0.40	1.70	0.40	1.40	0.40	1.20
180°	0.50	1.70	0.50	1.40	0.50	1.20

OCIMF 1997

Angle from bow to stern	Depth correction $D_w/D_d = 1.1$ Loaded condition			Depth correction $D_w/D_d = 1.5$ Ballasted condition		
	C_{YFc}	C_{YAc}	C_{Xc}	C_{YFc}	C_{YAc}	C_{Xc}
0°	0.00	0.00	0.04	0.00	0.00	0.04
15°	0.73	0.22	0.04	0.08	0.03	0.05
30°	1.05	0.47	0.05	0.19	0.10	0.05
45°	1.05	0.60	0.09	0.30	0.18	0.03
60°	1.30	0.92	0.11	0.39	0.27	0.01
75°	1.40	1.30	0.07	0.42	0.34	0.00
90°	1.40	1.54	0.00	0.39	0.40	0.01
105°	1.10	1.55	0.06	0.33	0.42	0.01
120°	0.90	1.40	0.09	0.26	0.39	0.02
135°	0.60	1.20	0.09	0.18	0.32	0.04
150°	0.26	0.95	0.04	0.11	0.20	0.06
165°	0.15	0.70	0.01	0.04	0.09	0.06
180°	0.00	0.00	0.04	0.00	0.00	0.06

C_{TCaft} = transverse aft current force coefficient depending on the angle of current (Table 2.10)
C_{CT} = depth correction factor for transverse or lateral current forces (Table 2.10)
C_{LC} = longitudinal current force coefficient (Table 2.10)
C_{CL} = depth correction factor for longitudinal current forces (Table 2.10)
γ = density of water taken as 1025 kg/m^3 for seawater and 1000 kg/m^3 for freshwater
L_{bp} = length between perpendiculars

d_m = mean draught of the ship
V_C = average current velocity in m/s over the mean depth of the ship

OCIMF recommends that the current force or pressure on a ship should be:

Transverse force:

$$F_{Ycurrent} = F_{YFcurrent} + F_{YAcurrent}$$

where the forward lateral or transverse current force on the ship is

$$F_{YFcurrent} = C_{YFc} \times \gamma_w \times D \times L_{bp} \times \frac{V_c^2}{2012}$$

and the aft lateral or transverse current force on the ship is

$$F_{YAcurrent} = C_{YAc} \times \gamma_w \times D \times L_{bp} \times \frac{V_c^2}{2012}$$

Longitudinal current force:

$$F_{XC} = C_{XC} \times \gamma_w \times D \times L_{bp} \times \frac{V_c^2}{2012}$$

Here

C_{YFc} = lateral or transverse forward current force coefficient depending on the angle of the current and the underkeel clearance (Table 2.10)
C_{YAc} = lateral aft current force coefficient depending on the angle of the current and the underkeel clearance (Table 2.10)
C_{Xc} = longitudinal current force coefficient depending on the angle of the current and the underkeel clearance (Table 2.10)
γ_w = density of water, taken as 1025 kg/m^3 for seawater
L_{bp} = length between perpendiculars
D = mean draught of the ship
V_c = average current velocity in m/s acting over the mean depth of the ship
2012 = conversion factor for the velocity in knots/s to m/s

The OCIMF/SIGTTO does not have any recommendations for calculating the current force or pressure on a tanker.

Comparison between the different current standards and recommendations

For a comparison between the different current standards and recommendations, the different current force coefficients due to the ratio of the water depth D_w and the ship's draft D_d and due to the effect of the underkeel clearance on the current forces, are shown, rounded up, in Table 2.10.

Example of current forces on a ship

Based on the above different current standards and recommendations formulas and the different current force coefficients in Table 2.11, an example for comparison of the different current forces

Table 2.11 Comparison of current forces on a ship with different underwater clearances D_w/D_d

Current speed: m/s	General formula Force: kN			Spanish standard ROM 0.2-90 Force: kN			British standard BS 6349-1: 2000 Force: kN			OCIMF 1997 Force: kN		
	0°	90°	180°	0°	90°	180°	0°	90°	180°	0°	90°	180°
$D_w/D_d = 1.1$												
1.0	360	18 600	–	432	21 167	–	492	10 635	615	126	10 572	126
$D_w/D_d = 1.5$												
1.0	240	11 160	–	432	14 126	–	405	6 294	506	–	–	–
$D_w/D_d = 2.0$												
1.0	120	4 460	–	432	10 050	–	347	3 690	434	–	–	–

based on the different standards in kN acting longitudinally (0° against the bow), laterally or transversely (90° to the ship's longitudinal axis) and longitudinally against the stern (180° against the stern, only the British standard and OCIMF give coefficients for this case) for different current speeds in m/s for ships in the loaded condition and with a 95% confidence limit is shown in Table 2.11 for the following ship dimensions:

(a) oil tanker 200 000 DWT
(b) $L_{OA} = 350$ m, $L_{BP} = 346$ m, $B = 56.2$ m, $D_d = 20.4$ m.

As can be seen from the current forces on a ship based on the different standards and recommendations, the current forces vary considerably more between the different current standards and recommendations than do the wind forces between the different wind standards and recommendations. The main reason for the differences seems to be the different conservative attitudes to the design of current forces based on the empirical factors that should include all possible cases in terms of the type, shape and size of different ships, etc. The above methods for calculation should therefore only be used as a guide to the magnitude of the forces. Since the current forces can vary considerably, one should therefore, in order to obtain the most accurate forces for all important berth structures, undertake current testing of scale models in a hydraulic institute.

2.5. Ice forces

2.5.1 Overview

The study of forces acting on berth structures due to the presence of ice in the harbour basin has so far not been given high priority. However, where berths, dolphins, bridge pillars, etc. are exposed to ice during the winter season (drift ice, ice connected with land or ice accumulations), one must take ice actions into consideration. The estimation of ice actions in harbours and other coastal facilities is in general more complex than for offshore structures. The first step is to evaluate local conditions (water level variation, type of structure, ice conditions and properties) and determine which effects/mechanisms to design against. Both vertical and horizontal ice forces can be important. The loads indicated below are meant as guidelines for an assessment of the magnitude of the ice forces.

The magnitude of ice forces depends on the type and form of the structure, the properties of the ice and the hydrodynamic conditions. If ice connected with land prevails, there is little chance of drifting ice hitting the structure, so that vertical ice forces, thermal ice accumulation, tidal jacking and thermal expansion usually govern the ice actions. If drift ice prevails, then forces from drifting ice floes and ice ride-up may give design conditions. Research has shown that the maximum velocity of moving ice under the influence of a steady wind will not be more than 3% of the wind velocity. The tidal amplitude is one of the governing factors, and needs to be known. In general, ice formed in freshwater is stronger and stiffer than ice formed in salt water. The unit weight of both freshwater and sea ice is about 9 kN/m^3. The coefficient of thermal expansion is higher for freshwater ice, but the presence of brine pockets in sea ice makes it smaller for sea ice.

2.5.2 Horizontal forces

Horizontal forces can act on a structure in both the transverse and longitudinal directions. Regarding transverse horizontal forces due to ice drift, four different scenarios can occur (Løset *et al.*, 2006):

1. The ice floe will hit the structure and stop. The momentum of the ice feature governs the outcome (limit momentum scenario).
2. The ice floe will hit the structure and stop. However, in this case the driving force (mostly wind and currents) is the limiting factor. This often happens when the floes are small, or when ridging occurs at a distance from the structure.
3. The driving forces are so high that the ice fails – this usually gives the highest forces. The ice crushes against vertically faced structures, and fails in bending against sloping structures (limit stress scenario).
4. Finally, the ice floe may split if the confinement is low enough. This gives a smaller ice force than the limit stress scenario described above.

As the highest forces come from the limit stress scenario, we consider only that. For a single leg structure (or sheet pile wall), the forces in a limit stress scenario can be calculated by either

$$I_1 = i_1 \times (l_1 + l_2) \times 0.5 \quad \text{or} \quad I_1 = i_1 \times \chi \qquad \text{(reference is made to Figure 2.21)}$$

where i_1 at rivers or berth structures under traffic is 10–20 kN/lin m; i_1 at sounds, fjords and narrow bays is 30 kN/lin m; and i_1 at places heavily exposed to ice is 50–100 kN/lin m. χ is the width of the ice sheet (floe) in metres.

Figure 2.21 Pressure from drift ice

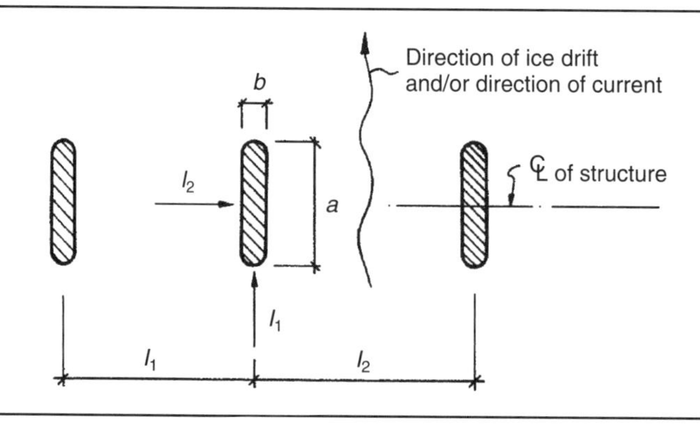

Or one can use the ISO (2010) formulation for an infinite drifting ice floe crushing against a vertical structure:

$$F_H = p_G \times h_i \times b$$

where h_i is the ice thickness and p_G is the global pressure, defined as follows:

$$p_G = C_R \times \left(\frac{h_i}{h^*}\right)^n \left(\frac{b}{h_i}\right)^m$$

where C_R is the strength coefficient (2.8 MPa for the Beaufort Sea and 1.8 MPa for the Baltic), h^* is a reference ice thickness of 1.0 m, m is an empirical coefficient (= 0.16) and n is another empirical coefficient, as follows:

$$\underline{n}\begin{cases} = -0.50 + \dfrac{h_i}{5} & \text{for } h_i < 1.0\,\text{m} \\ = -0.30 & \text{for } h_i \geq 1.0\,\text{m} \end{cases}$$

Experiments have shown that the horizontal force due to ice against a pillar with 45° sloping sides can be reduced to one-third of the force acting against a structure with vertical sides. This is because the ice fails in bending and not crushing. The disadvantages of a sloping water line is that the water line diameter increases so that the wave forces increase, the structure usually becomes more expensive to construct and install, and finally it may be more difficult with ship access.

In harbours the ice floes are often small, so that the limit stress scenario is less probable than for offshore locations. However, there are other effects that can make the horizontal forces even higher on harbour structures. Let us first consider a tidal environment. The ice cover does not have time to adfreeze to the structure, but ice will accumulate, and an ice bustle forms (Løset and Marchenko, 2009). This will increase the effective pile diameter, so that the effective width of the structure (b in the equations above) needs to be increased. In a non-tidal environment the ice cover may adfreeze to the structure. Now there are two effects that may increase the horizontal force compared to the ISO equation above. First, when the ice starts moving horizontally the ice will pull as well as push the structure and, second, the higher thermal conductivity of the structure (in the case of steel piles) gives thicker ice around the structure. Sharapov and Shkhinek (2014) suggest the following modifications:

$$F_H^{\text{Adfreeze}} = F_H \times K^{\text{Adfreeze}}$$

The coefficient K^{Adfreeze} decreases from about 2.9 for low b/h_i, and goes down to between 1.75 and 2 for b/h_i larger than 20.

If the structure consists of several piles (a multi-leg structure), one needs to consider the ratio of the pile spacing to the pile diameter (l_1/b). If this is larger than 5, an individual piles is not affected by the presence of its neighbour (Løset *et al.*, 2006). However, any ice accumulation increases the effective pile diameter, and this needs to be taken into account when evaluating the ice actions.

Finally, the structure may experience longitudinal horizontal ice forces due to thermal expansion. In this case, reference is also made to Figure 2.21:

$$I_2 = i_2 \times a$$

where, if there is open water on the other side of a pillar, i_2 is 100–300 kN/lin m and, if there is firm ice slab on all sides, i_2 is a quarter of the previous value (25–75 kN/lin m).

Alternatively,

$$I_2 = 0.2 \times I_1$$

For berth structures that are being called at frequently and for structures in waters with great tidal variation where the formation of ice is hampered, horizontal ice loading usually causes no problem if the structure has been designed for an ice loading of 10–20 kN/lin m of the berth front.

2.5.3 Vertical forces

In a tidal environment the ice cover does not adfreeze to the structure, but ice bustles may form. Such ice bustles add weight to the structure and increase the effective diameter of the piles. If the water level varies slowly, the ice cover can have time to adfreeze to the structure and give high vertical forces. If the ice cover freezes to a structure with a small deadweight and the water level increases, the pile may be pulled up. Soil may fill in under the pile, so that, when the water level decreases, the pile cannot settle to its original position. This process can be repeated so that piles may slowly be jacked up. As a general guideline, the vertical lifting ice forces on a pile are inversely dependent on the magnitude of the tidal variation: that is, the larger the tidal range, the less will be the lifting forces, owing to the greater difficulty for the ice to freeze to a pile when there is a large tidal variation.

As a guide in the design of structures exposed to ice, Løfquist (personal communication) has suggested the application of the design forces shown in Figures 2.22 and 2.23. The figures are based on ice with a bending strength equal to 2000 kN/m^2. The lifting force/lin m of straight-lined wall and the lifting force/lin m of the circumference of a circular element column are shown. For square-shaped elements, the inscribed circle is used to determine the perimeter length.

In cases where there are severe problems in connection with the formation of ice in the port basin, the installation of one or more compressed-air bubbling plants to keep the water open would probably prove more economical than designing the berth structures for greater loadings. Compressed-air bubbler de-icing is an effective ice suppression and control method. The air bubbles move warmer bottom water upwards, to melt the underside of the surface ice.

Figure 2.22 Lifting forces on a wall

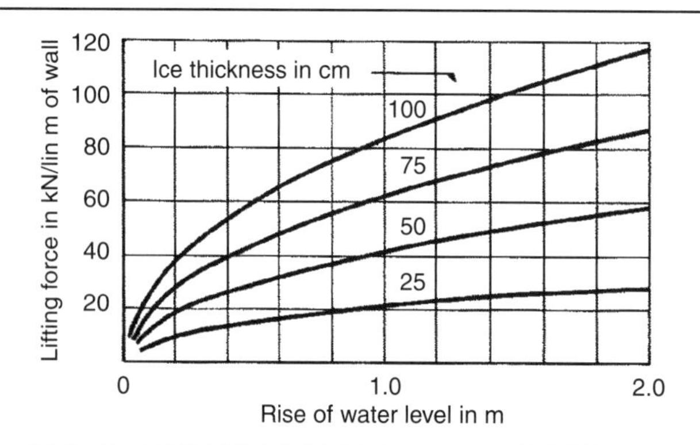

Figure 2.23 Lifting forces on a column

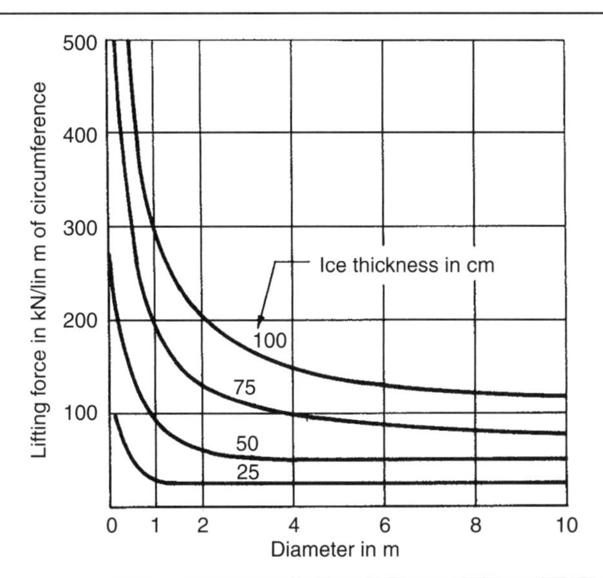

As a rule of thumb, the bearing capacity of good ice in relation to the thickness of the ice is as shown below:

Man on foot	5 cm
Motor cycle	10 cm
Small car	20 cm
Tractor	30 cm
Truck (2.5 t)	40 cm
Aeroplane (9 t)	50 cm

REFERENCES AND FURTHER READING

Bratteland E (1981) *Lecture Notes on Port Planning and Engineering*, Norwegian Institute of Technology. Kluwer Academic, London.

Bruun P (1989) *Port Engineering*, vols 1 and 2. Gulf Publishing, Houston, TX.

BSI (British Standards Institution) (2000) BS 6349: Maritime structures. Code of practice for general criteria. BSI, London.

EAU (Empfehlungen des Arbeitsausschusses für Ufereinfassungen) (2004) *Recommendations of the Committee for Waterfront Structures, Harbours and Waterways*, 8th English edn. Ernst, Berlin.

ISO (International Standardisation Organisation) (2010) ISO 19906: 2010. Petroleum and natural gas industries. Arctic offshore structures. ISO, Geneva.

Løset S and Marchenko A (2009) Field studies and numerical simulations of ice bustles on vertical piles. *Journal of Cold Regions Science and Technology* **58**: 15–28.

Løset S, Shkhinek KN, Gudmestad OT and Høyland KV (2006) *Actions from Ice on Arctic Offshore and Coastal Structures*. LAN, St Petersburg.

Mathiesen M (2002) *Computation of Wind Generated Seas*. SINTEF Fisheries and Aquaculture, Trondheim. Internal memo.

Ministerio de Obras Públicas y Transportes (1990) ROM 0.2-90. Maritime works recommendations. Actions in the design of maritime and harbour works. Ministerio de Obras Públicas y Transportes, Madrid.

Norwegian Petroleum Directorate (1977) *Environmental Loads*. NPD, Oslo.

OCIMF (Oil Companies International Marine Forum) (1997) *Mooring Equipment Guidelines*, 2nd edn. Witherby, London.

OCIMF (2008) *Mooring Equipment Guidelines*, 3rd edn. Witherby, London.

OCIMF/SIGTTO (Society of International Gas Tanker Terminal Operators) (1995) *Prediction of Wind Loads on Large Liquefied Gas Carriers*. Witherby, London.

PIANC (International Navigation Association) (1979) *International Commission for the Reception of Large Ships. Report of Working Group 1*. PIANC, Brussels.

PIANC (2010) *The Safety Aspect Affecting the Berthing Operations of Tankers to Oil and Gas Terminals. Report of Working Group 55*. PIANC, Brussels.

Quinn A De F (1972) *Design and Construction of Ports and Marine Structures*. McGraw-Hill, New York.

Sharapov D and Shkhinek KN (2014) A method to determine the horizontal ice loads on the vertical steel structures with adfreeze to the level ice. Submitted.

Tsinker GP (1997) *Handbook of Port and Harbor Engineering*. Chapman and Hall, London.

US Army Coastal Engineering Research Centre (1984) *Shore Protection Manual*, vols 1 and 2. Department of the Army Corps of Engineers, Washington, DC. See also later editions.

US Army Corps of Engineers (2002) *Coastal Engineering Manual*, Part II. *Coastal Hydrodynamics*. US Army Corps of Engineers, Washington, DC. http://chl.erdc.usace.army.mil/cem (accessed 12/02/2014).

Wortley CA (1984) *Great Lakes Small-craft Harbor and Structure Design for Ice Conditions: An Engineering Manual*. University of Wisconsin, Madison, WI.

Port Designer's Handbook
ISBN 978-0-7277-6004-3

http://dx.doi.org/10.1680/pdh.60043.079

Chapter 3
Channels and harbour basins

3.1. Channels and waterways

3.1.1 General

From a general point of view, channels or waterways can logically be classified into the following four groups:

(a) Group A: main traffic arteries that have satisfactory day and night navigational aids and where given depths are guaranteed.
(b) Group B: as group A but with navigational aids for day navigation only.
(c) Group C: important routes, which may have navigational aids and where depths are checked by regular surveys but are not guaranteed.
(d) Group D: local routes that have no navigational aids and where only estimates of depths are given.

Channels or waterways can be subdivided into unrestricted, semi-restricted and fully restricted channels:

(a) **Unrestricted channels** are channels or waterways in shallow water of width at least 10–15 times the beam of the largest ship using the channel, but without any dredging.
(b) **Semi-restricted channels** are dredged channels in shallow water (Figure 3.1).
(c) **Fully restricted channels** are channels where the entire channel area is dredged (see Figure 3.1). In general, the layout and the alignment of the channels should be such that the channel can be navigated with reasonable safety according to which group the channel is classified into, taking account of tide, current, prevailing wind and wave action.

If possible, the angle between the resultant effect due to the prevailing wind direction and current and the channel axis should be a minimum. The angles of deflection and the number of curves in the channel should also be kept to a minimum.

An example of a fully restricted channel or canal is the Panama Canal, where the maximum dimensions for ships using the canal are: overall length 294 m, width of beam 32.31 m and maximum draft 12 m.

Figure 3.2 shows the dimensions of the new Panama Canal, which is also a fully restricted channel. The new locks will be able to accommodate vessels up to an L_{OA} of 366 m (where L_{OA} = length overall), with beams up to 49 m, a draft up to 15.5 m and a carrying capacity of 12 500 TEUs.

The channels should preferably be located in areas of maximum natural water depth to reduce the cost of initial and maintenance dredging. Areas that are exposed to excessive siltation and littoral drift

Figure 3.1 Semi- and fully restricted channels

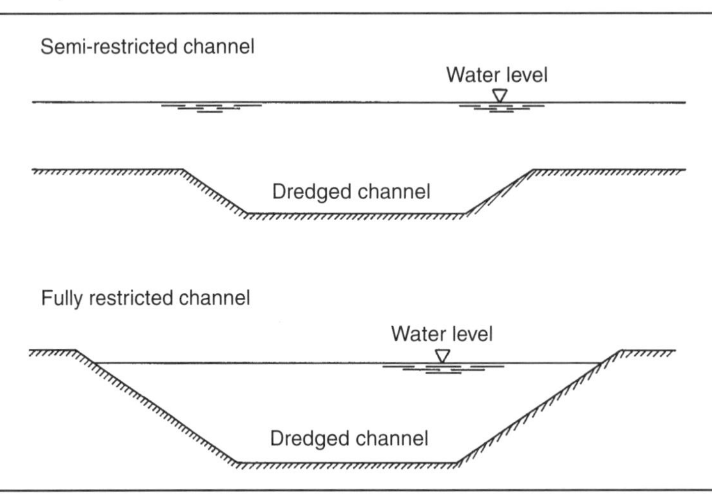

should be avoided if possible. However, to maintain a minimum depth, as shown on navigational charts, maintenance dredging is usually necessary. The volumes to be dredged can vary widely from place to place, depending on the extent of the site, its location and other natural influences such as tides, current and weather conditions.

3.1.2 Straight channel

The minimum width of a straight channel will depend primarily on the size and manoeuvrability of the ships navigating the channel and the effects of wind and current. The channel width is divided into

Figure 3.2 The existing and new channel of the Panama Canal (Source: Panama Canal Authority, February 2011)

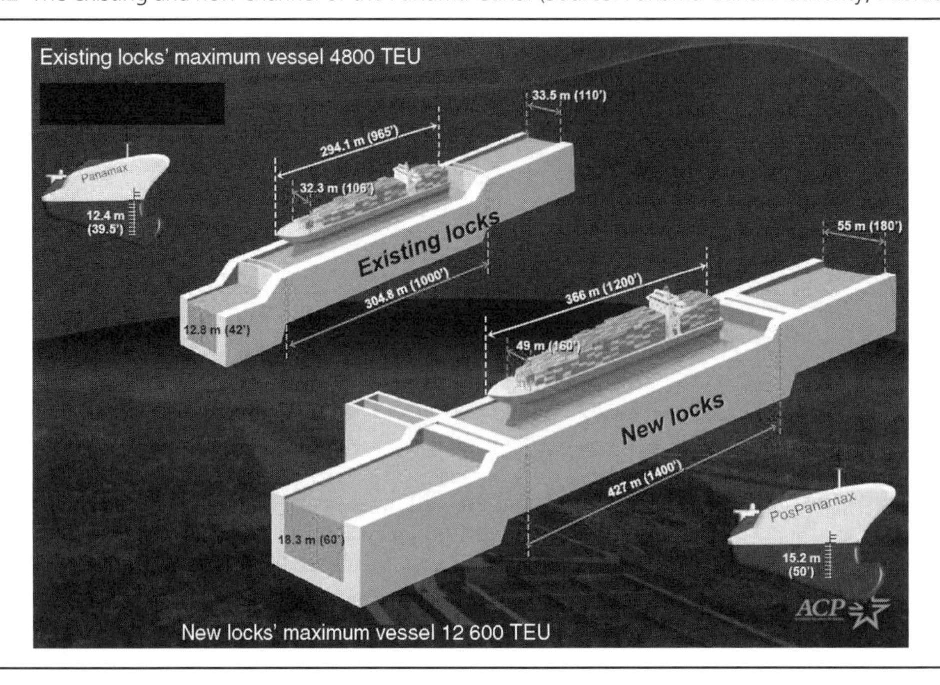

Figure 3.3 Channel width

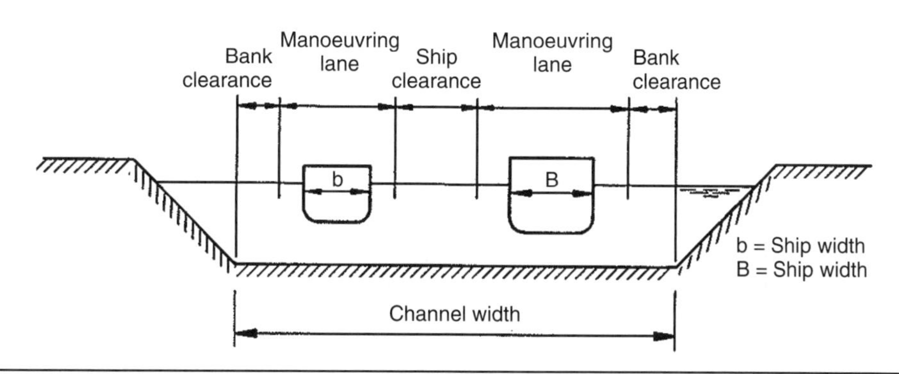

three zones or lanes, as shown in Figure 3.3 for one-way and two-way traffic:

(a) the manoeuvring lane
(b) the bank clearance lane
(c) the ship clearance lane.

The width of a restricted channel should be measured at the bottom of the dredged bed and should be the sum of the lanes.

The width of the **manoeuvring lane** will generally vary from 1.6 to 2.0 times the beam of the largest ship using the channel, depending on wind, current and the manoeuvrability of the ship. The very high superstructures on containerships, car carriers, passenger ships and tankers in ballast present considerable windage area, and may therefore require a greater channel width than their beam would suggest.

Allowance for yaw of the ship must be made if the channel is exposed to cross-current and/or winds. The angle of yaw can be between 5° and 10°. For a large ship, an angle of yaw of 5° can add an extra width, equivalent to half the beam, to the manoeuvring lane.

Ships displaced from the channel centreline towards the banks of the channel will experience a bank suction effect due to the asymmetrical flow of water round the ship, and this will cause a yawing movement. To counteract this effect on the ship an additional **bank clearance** width usually between 1.0 and 2.0 times the beam of the largest ship must be added. A steep-sided channel section produces more bank suction than a channel with a trapezoidal section. Bank suction also increases as the underkeel clearance decreases.

To avoid excessive interaction between two ships travelling past one another, either in the same or in the opposite direction in a two-lane channel, it is necessary to separate the two manoeuvring lanes by a ship clearance lane. To minimise the suction and repulsion forces between the ships, a **clearance lane** of a minimum of 30 m, or the beam of the largest ship, should be provided.

Depending on the sea and wind conditions, the recommended total channel bottom width for **single-lane** channels is 3.6–6 times the beam of the design ship. For oil and gas tankers a minimum bottom width should be 5 times the beam of the design ship. For a **two-lane** channel the total channel width will vary between 6.2 and 9 times the beam of the design ship.

3.1.3 Channels with curves

As a general rule, curves and sharp turns in a channel should be avoided if possible. Where curves are unavoidable, the minimum width of the channel at a curve should be larger than in a straight channel due to the additional manoeuvring width required, because the ship will deviate more from its course in a bend than in a straight section. The definitions of the curve radius and deflection angle are illustrated in Figure 3.4.

In practice, if the deflection angle of the curve is larger than 10° the channel should be widened. It is generally accepted that widening the inside of the curve or bend is the most suitable way to improve safe navigation around a curve. Depending on the manoeuvrability of the ship and the radius of the bend, the width of the manoeuvring lane should be increased from around 2.0 times the beam of the largest ship in a straight channel to around 4.0 times the beam of the largest ship in curved channels.

In the past it was accepted that for ships without tugboat assistance the minimum curve radius should not be less than 3 times the length of the design ship for a deflection angle of the curve up to 25°. Between 25° and 35° the minimum curve radius should be 5 times the length of the design ship. For 35° or more the curve radius should be 10 times the length of the design ship. If the curves must have smaller radii than mentioned above, the channel should be suitably widened. More recent

Figure 3.4 Channel curve

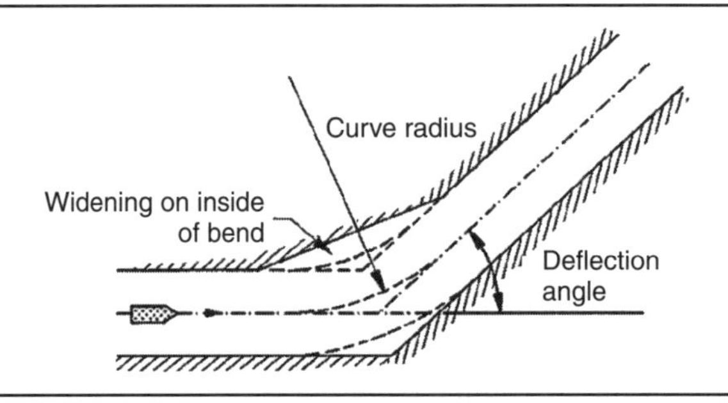

proposals suggest that the minimum curve radius should be in the range 8–10 times the length of the design ship, without being related to the angle of deflection.

If more than one curve is necessary, a straight section equal to at least 5 times the length of the design ship or 1000 m, whichever is greater, should be provided between the two consecutive curves.

3.2. Harbour basin

3.2.1 General

The harbour basin can be defined as the protected water area, which should provide safe and suitable accommodation for ships. Harbours can be classified as natural, semi-natural or artificial. Harbours have different functions, such as commercial (municipal or privately owned) harbours, refuge harbours, military harbours, oil harbours, etc.

Inside the harbour entrance, the areas within the harbour should be allocated different functions, such as berthing or a turning area. If the harbour receives a wide range of ships it should, for economic reasons, be divided into at least two zones, one for the larger and one for the smaller ships. The smaller ships should be located in the inner and shallower part of the harbour. Berths for ships carrying hazardous cargoes such as oil and gas should be located at a safe distance and clearance from other berths. These activities should typically be located in isolated areas at the outer end and on the lee side of the harbour basin.

3.2.2 Entrance

The harbour entrance should, if possible, be located on the lee side of the harbour (Figure 3.5). If it must be located on the windward end of the harbour, adequate overlap of the breakwaters should be provided so that the ship will have passed through the restricted entrance and be free to turn with the wind before it is hit broadside by the waves. Due to this overlap of the breakwaters, the interior of the harbour will be protected from the waves. Accordingly, in order to reduce the wave height within the harbour, and to prevent strong currents, the entrance should be no wider than is necessary to provide safe navigation.

Figure 3.5 The main entrance to a harbour

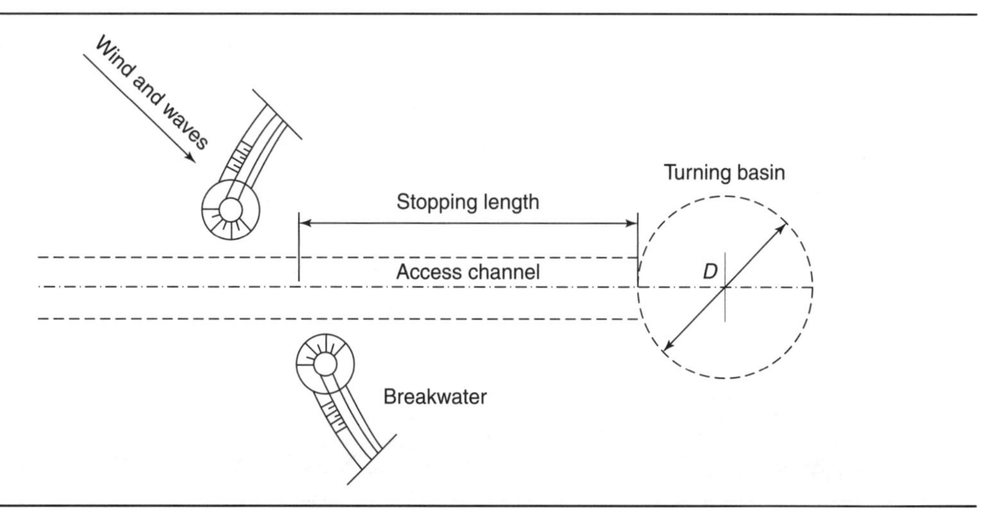

The entrance width measured at the design depth will depend on the degree of wave protection required inside the harbour, the navigational requirements due to the size of ship, density of traffic, depth of water and the current velocity when the tide is coming in or going out. Generally, the width of the harbour entrance should be 0.7–1.0 times the length of the design ship.

The maximum current velocity through the harbour entrance should, if possible, not exceed approximately 1.5 m/s or 3 knots. If the current velocity exceeds this value, the channel cross-section should be adjusted.

3.2.3 Stopping distance

The stopping distance of a ship will depend on factors such as ship speed, the displacement and shape of the hull, and horsepower ratio. As a rough guideline, the following stopping distances are assumed to be sufficient to bring a ship to a complete halt:

(a) ships in ballast, 3–5 times the ship's length
(b) loaded ships (Figure 3.6), 7 to 8 times the ship's length.

In harbours where the entrance is exposed to weather, the stopping distance should typically be measured from the beginning of the protected area to the centre of the turning basin.

3.2.4 Turning area

The turning area or basin should usually be in the central area of the harbour basin. The size of the turning area will be a function of manoeuvrability and of the length of the ship using the area. It will also depend on the time permitted to execute the turning manoeuvre. The area should be protected from waves and strong winds. It should be remembered that ships in ballast have a decreased turning performance.

The following minimum diameters of the turning area are generally accepted. The minimum diameter where the ship turns by going ahead and without use of bow thrusters and/or tugboat assistance should

Figure 3.6 A loaded ore tanker. (Photo by Bergersen DY, Norway)

Figure 3.7 Dredged area around a berth

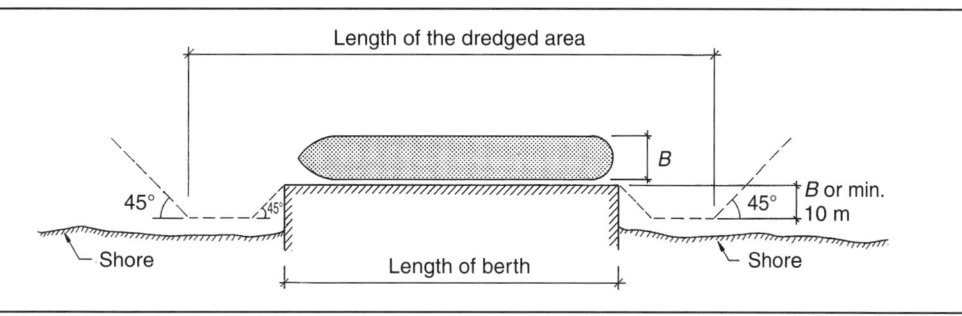

be approximately 4 times the length of the ship. Where the ship has tugboat assistance, the turning diameter could be 2 times the length of the ship. Under very good weather and manoeuvring conditions these diameters might be reduced to 3 and 1.6 times the length, respectively, as a lower limit. With the use of the main propeller and rudder and the bow thrusters, the turning diameter could be 1.5 times the length of the ship.

Where the ship is turned by warping around a dolphin or pier, and usually with tugboat assistance under calm conditions, the turning diameter could be a minimum 1.2 times the length of the ship.

3.2.5 Berthing area

The size of the berthing area and the berth will depend on the dimensions of the largest ship and the number of ships that will use the harbour. The berth layout will be affected by many factors, such as the size of the harbour basin for manoeuvring, satisfactory arrivals and departures of ships to and from the berth, whether or not the ships are equipped with a bow rudder and bow thrusters, the availability of tugboats, and the direction and strength of wind, waves and currents.

If the berthing area in **front of the berth** has to be dredged, the size of the dredging area should be as shown in Figure 3.7. The length of the dredged area should be for ships with tugboat assistance not less than 1.25 times the length of the largest ship to use the berth, and without tugboat assistance not less than 1.5 times that length. The width of a dredged tidal berth should be at least 1.5 times the beam of the largest ship to use the berth.

Where more than one ship has to be accommodated along the berth (Figure 3.8) a clearance length of at least 0.1 times the length of the largest ship should be provided between the adjacent ships. If the harbour basin is subjected to strong winds and tides the clearance should be increased to 0.2 times the length of the largest ship. A minimum distance of 15 m between the ships is commonly adopted.

Figure 3.8 Clearance between ships at berth

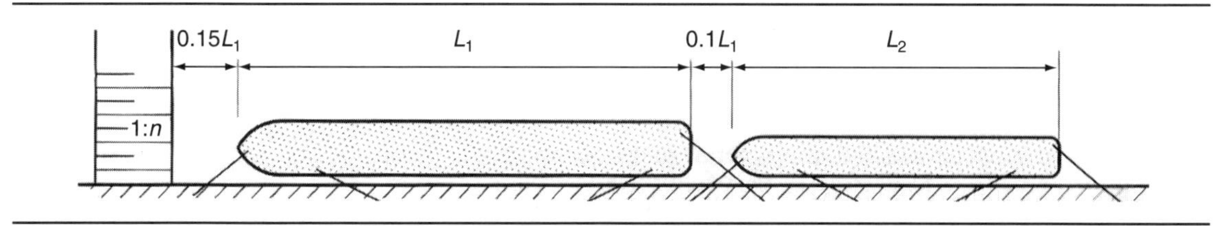

Figure 3.9 Layout of single piers

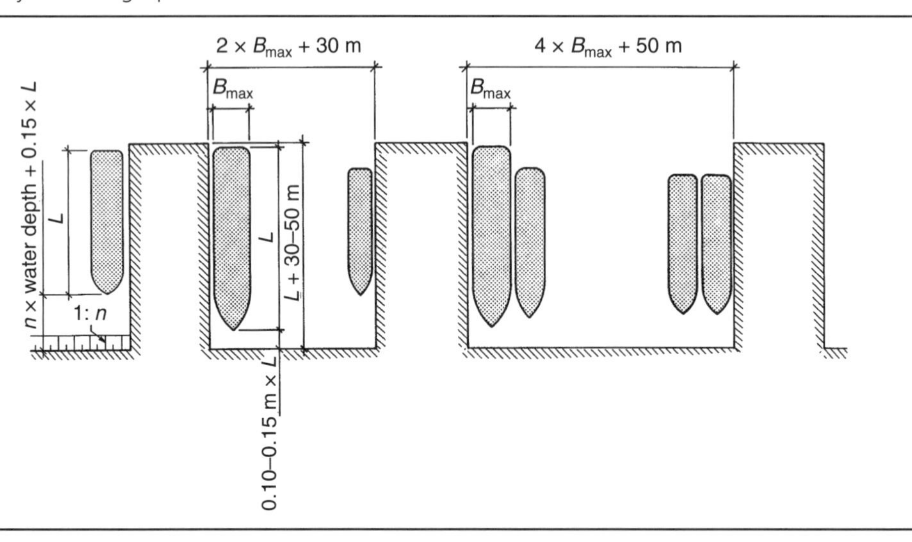

The distance between the bottom of the slope and the distance to the ship should be 0.15 of the ship length. The angle of the slope will depend on the material in the slope and whether the slope is exposed to wave action or erosion due to the ship's propeller. Above the water level the slope can be between 1 : 1.5 and 1 : 2.0.

Berths of the **finger pier type** (Figure 3.9) will provide the greatest amount of berthing space per metre of shorefront. For a single-berth pier, the clear water area between two piers should be 2 times the beam of the largest ship plus 30 m to allow for tugboat assistance. For double-berth finger piers the clear water area between two double-berth piers should be 4 times the beam of the largest ship plus 50 m. The length of the finger pier for a single berth should, if possible, be the length of the ship plus 30–50 m. For very long single-berth piers (Figure 3.10) the clear water area between the two piers should be 2 times the beam of the largest ship plus 50 m.

For harbour basins (Figure 3.11) the width required to permit a ship to swing freely into a berth is 1.5 times the length of the ship for berths at 45°, and 2 times the length of the ship for berths at 90°.

3.2.6 Berthing areas for oil and gas tankers

The layout of berthing structures for **oil and gas** tankers is different from the berth layout for general cargo ships. The major components of an oil and gas berthing structure are shown in Figure 3.12 and include the following elements: mooring structures, breasting structures, a loading platform and an access bridge with a pipeway. For the safety of the tanker and the tugboats it is important that there is enough manoeuvring space provided for the tugboats around the tanker during its berthing and mooring. The berth should, if possible, be oriented so that the predominant wind, wave and current have the least effect on the operation of the berth. The berth should be so oriented that the mooring loads are as small as possible. Usually, this means aligning the berth axis with the direction of the current. Where the currents are weak, it is advisable to locate the berth parallel to the prevailing wind direction. Berths should not be oriented broadside to strong prevailing winds, waves or current.

The loading platform and the breasting structure can be built either as one structure or, preferably, as two separate structures so that the horizontal berthing loads from the tanker against the breasting

Figure 3.10 Layout of long piers

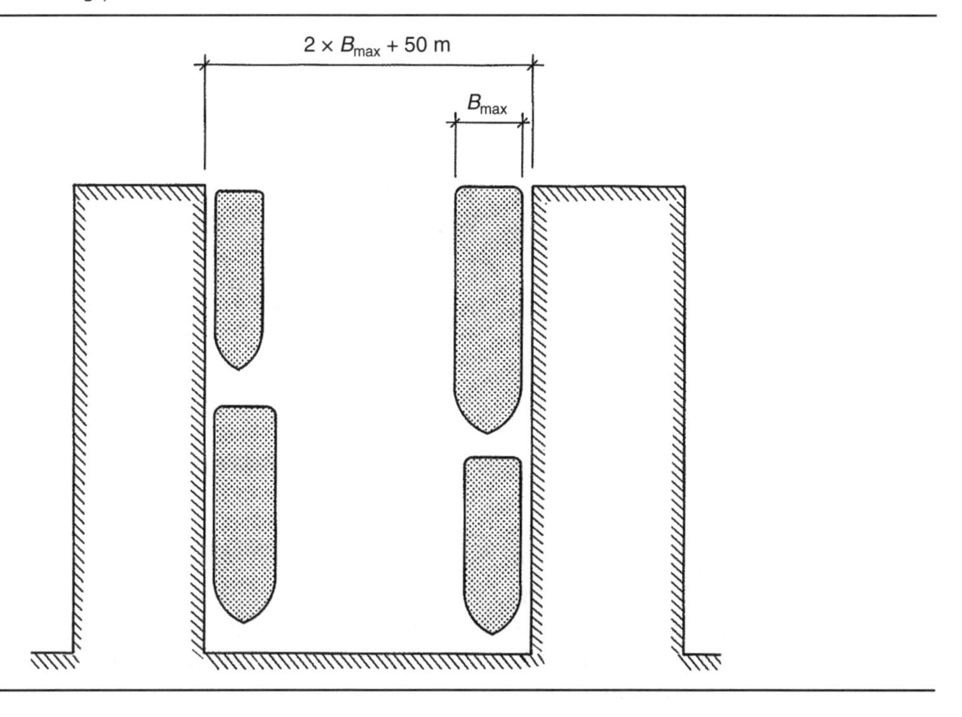

Figure 3.11 Layout of berths

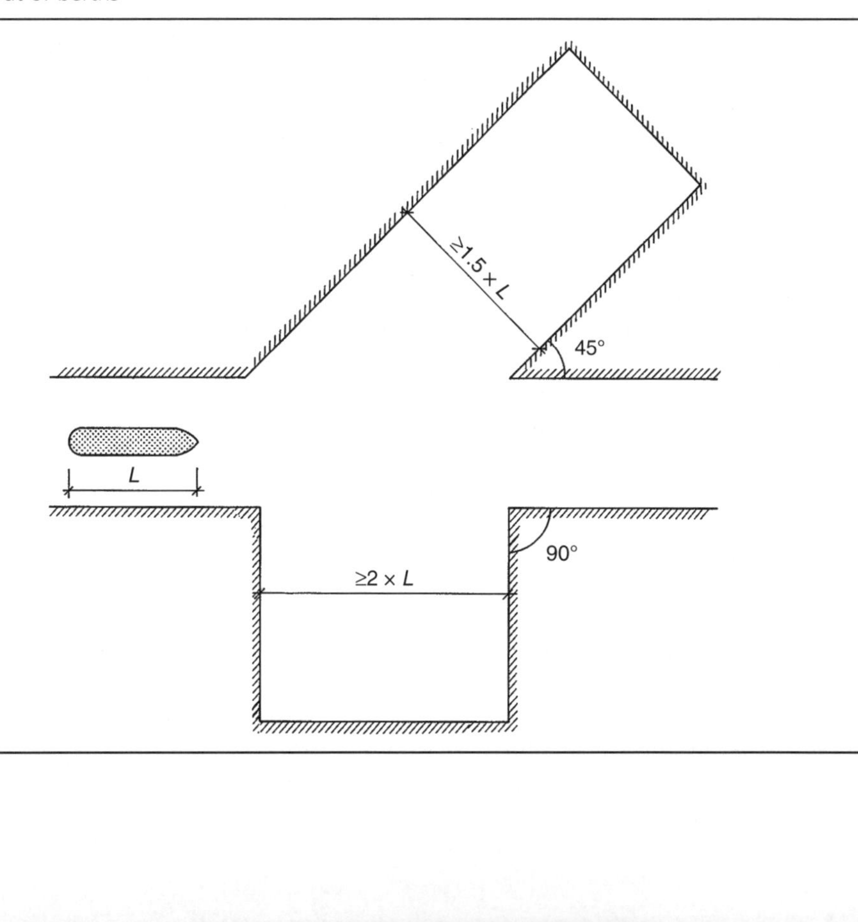

Figure 3.12 Typical berth for tankers

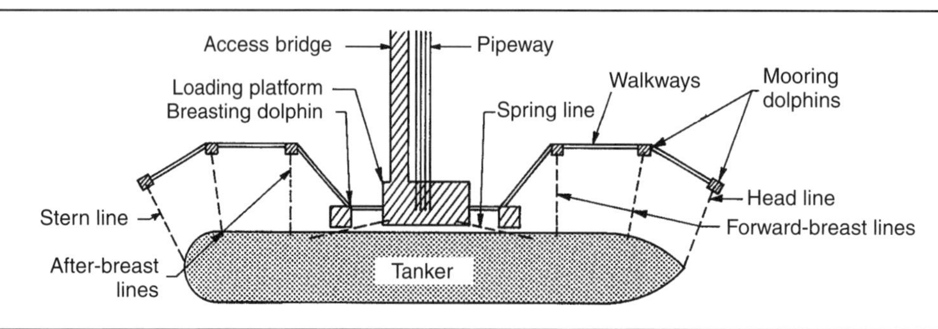

structure are not transferred to the loading platform. The breasting structures should be designed to withstand the berthing impact from the tanker during berthing and the impact from the wind, wave and current forces when moored. The mooring structures should be designed for the mooring and environmental forces.

The loading arms at an oil and gas berth structure are usually at the centre of the loading platform or at the centre between the breasting structures. The manifold position at the tanker should be at about the mid-length of the tanker and in no case more than 3.0 m forward or aft of the mid-length. The presentation flanges should not be more than about 4.5 m from the tanker's side, and the height above the tanker deck should not exceed 2.1 m.

To ensure contact with the parallel sides of the ship, the distance between the breasting structures should be as described in Chapter 12. Walkways should be provided between the mooring structures and the central structures. Although one breasting structure on each side of the loading platform is adequate for the safe berthing of a tanker, it is recommended that for berthing of large tankers two breasting structures are provided on each side in case one of the structures is damaged during berthing.

The **safety distance** between two moored tankers, or a moored tanker and a passing ship, will depend on the overall layout of the harbour, the number of tugboats assisting in the berthing or unberthing operation, the environmental conditions and the safety procedure at the terminal. The distance may also vary from country to country, depending on the safety philosophy in each country.

The manoeuvring area required for the tanker and supporting tugs for safe berthing and unberthing is, in principle, generally as described for the three different cases below.

Case 1. Two tanker berths along the shoreline (Figure 3.13), with tanker 1 loading at berth 1 and tanker 2 berthing with tug assistance to berth 2. The general principle is that tugs with radius RT shall not overlap the safety zone R_1 for tanker 1; if they overlap R_1 the tugs are classified as operating in the safety zone for tanker 1.

The distance d from berth 2 accounts for possible drift along the berth during berthing and mooring, and can be taken as approximately 20 m.

The necessary safety distance varies in the following ranges.

Figure 3.13 Case 1: two tankers and a passing ship

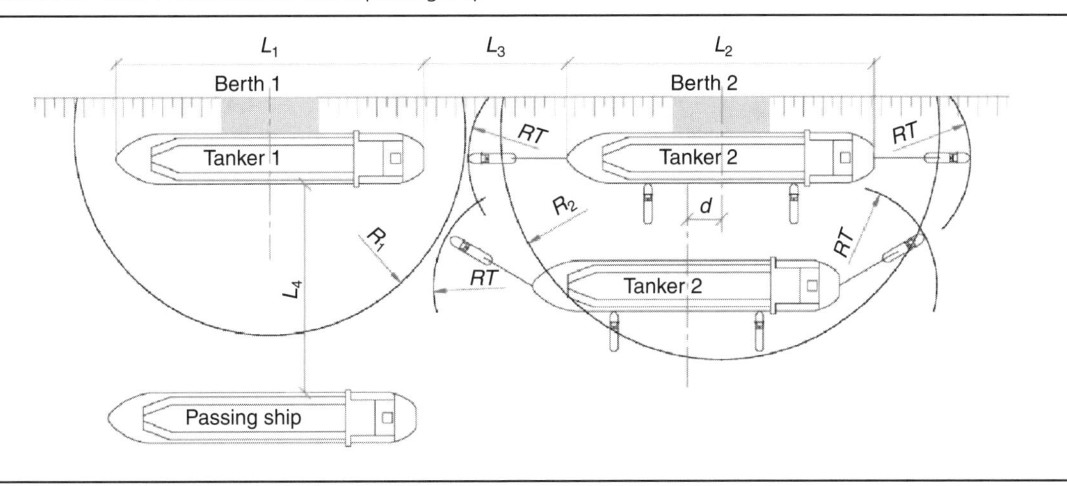

For **oil tankers**:

(a) The distance L_3 (see Figure 3.13) between two moored oil tankers may be from about 50–100 m.
(b) The distance between a moored oil tanker L_4 (see Figure 3.13) and a passing ship may be about 150 m or more. The impact on a moored tanker from, for example, a large passing ship should be subject to study to determine the safe speed and distance in order to minimise the risk of damage to or collision with the moored tanker.

For **gas tankers**:

(a) The distance L_3 (see Figure 3.13) between two moored gas tankers may be from about 80–150 m. For liquefied natural gas (LNG) tankers a minimum clearance of one ship's length between ships or 250–300 m is recommended. The distance required between LNG tankers will, due to safety considerations, also depend on the tugboats' capacity during berthing and unberthing the tankers. Where two LNG tankers are berthed, in order to allow LNG-associated operations to continue safely it is recommended that the safety circle of the berthed LNG ship not be breached unless operations are suspended.
(b) The distance L_4 (see Figure 3.13) between a moored gas tanker and a passing ship may be from 200–250 m. For an LNG tanker the distance shall at least be 300 m. If a distance of 300 m is not possible, safe passing distances must be determined be through studies and risk assessment, taking into account the local conditions and infrastructure.

A large ship passing close to a moored LNG tanker can cause surging along the gas berth, and this can lead to a risk to the mooring lines of the gas tanker.

Case 2. Two tankers moored opposite each other (Figure 3.14), with tanker 1 loading and tanker 2 berthing. The radius *RT* from the tanker 2 tugboats should preferably not overlap R_1; if they overlap R_1 the tugs are classified as operating in the safety zone of tanker 1.

Figure 3.14 Case 2: tankers opposite each other

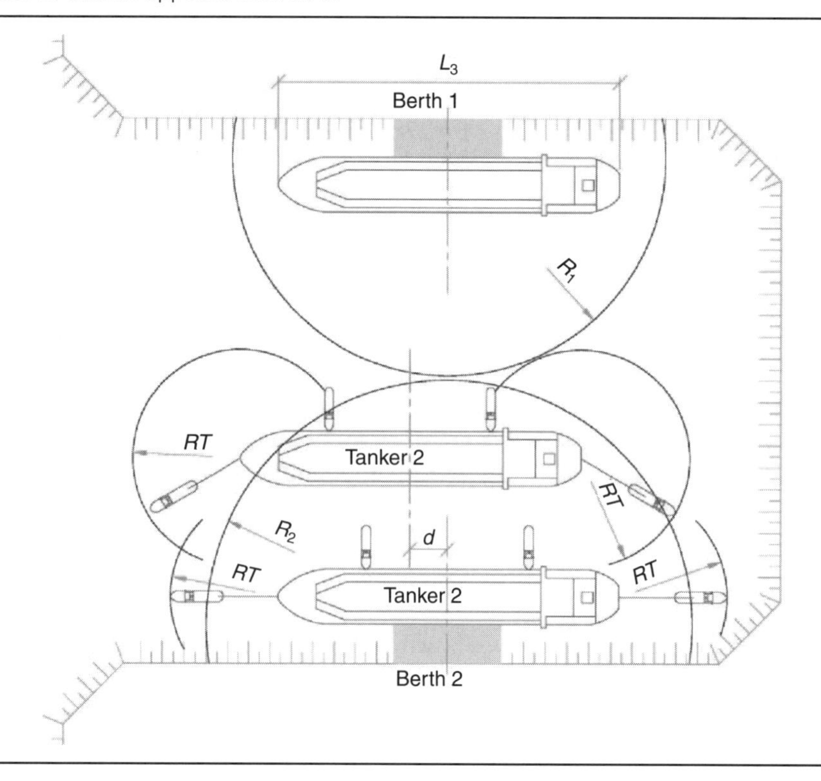

Case 3. Two tankers are moored on each side of a pier as shown in Figure 3.15. Generally the berths should have a layout such that the tankers can berth and unberth without the adjacent safety zone being breached. The berthing tanker 2 and tugs should preferably be able to remain outside of the tanker 1 safety zone during all berthing and unberthing manoeuvres, taking into account the variations in the weather, wind and current; if not, the tugs are classified to operate in the safety zone for tanker 1.

However, reduced berth area may be accepted by adopting a philosophy of interrupting the loading operations when one tanker is berthing or unberthing at an adjacent berth. In this case, two berths may be constructed in a reduced waterfront area, with the operational restriction of having to interrupt loading or deloading operations during movements of the other tanker. This is the terminal's decision, and should be specified in the functional requirements of the terminal. It is generally preferable that tankers located in one berth remain outside the safety zone of an adjacent berth.

A general fairway outside an oil and gas terminal should preferably be outside the turning basin in front of the oil or gas berth, so that the berthing operation will not be disturbed by passing vessels in the fairway. For reasons of safety and risk, it is recommended that, for example, LNG terminals are placed in sheltered locations remote from other port activities so that other ships do not pose a collision risk to a moored LNG tanker.

The **orientation** of oil and gas berths should be chosen to provide the best possible manoeuvring conditions for normal berthing and unberthing, as well as for emergency departure (Figure 3.16). In

Figure 3.15 Case 3: tankers on each side of a pier

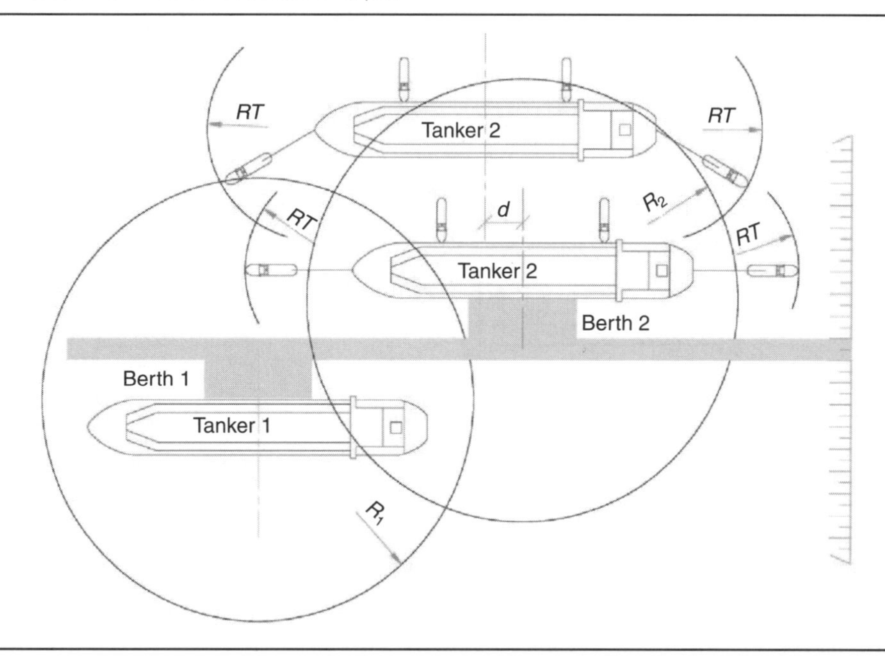

Figure 3.16 Statoil Mongstad Oil Terminal. (Photo courtesy of Øyvind Hagen, Statoil, Mongstad, Norway)

Figure 3.17 General layout of a small craft harbour berthing arrangement

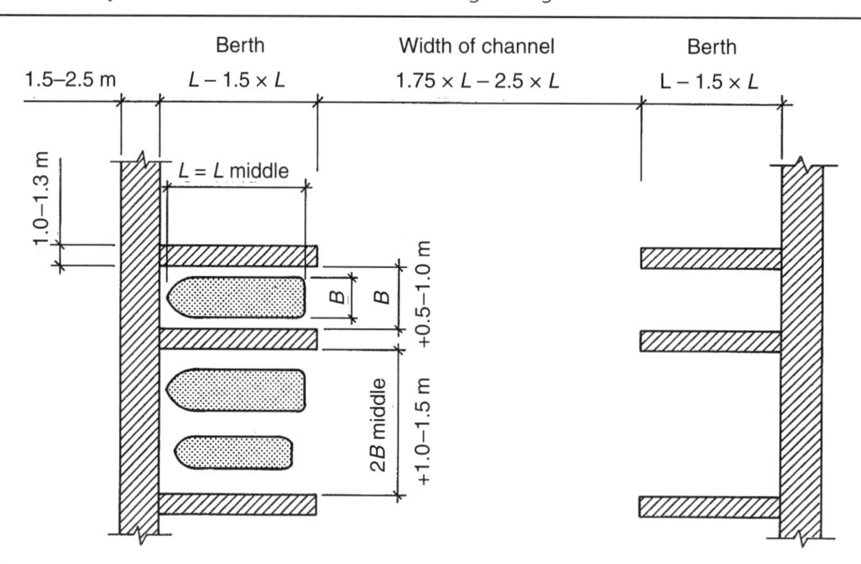

calm weather conditions, tankers should preferably be able to depart without tugboat assistance if possible, although this is not recommended as normal procedure.

At oil terminals, a portable collecting oil barrier should be placed around the oil tanker prior to loading or be able to be placed at very short notice in order to restrain any oil spillage. Equipment for collection and disposal of oil spills must be provided.

3.2.7 Berthing area for small craft

The general area requirements for **small craft** harbour berthing arrangements are shown in Figures 3.17–3.19. The general measurements will vary as shown in the figures, depending on the layout of the harbour. In places with large tidal variations or where the harbour is exposed to wind and/or waves, the maximum figures must be used. The normal figure for the total water area required per boat will vary between 100 and 200 m^2 per boat.

No rules exist for the size of the berthing area for a **fishing port**, but widths of about 100–150 m and lengths of about 200–400 m are common in existing ports. For safety reasons, and depending on the use of the port facilities, it is not desirable to have more than about three or four fishing ships berthing side by side along a berth.

3.3. Anchorage areas

The anchorage area is a place where ships wait for their turn to berth, for more favourable weather conditions or be held back for quarantine inspection or other reasons. Sometimes, special anchorage places are provided for ships carrying dangerous cargo, such as explosives.

The size of water area required for anchorages will, therefore, primarily depend on the number, type and size of ships which require protection and the type of mooring systems available. The selection of the mooring system will depend on the size of ship, the degree of exposure to weather, the degree of restraint required and the quality of the sea bottom material (the anchor holding).

Figure 3.18 Small craft harbour

As a general rule, the harbour should provide anchorage areas for small coastal ships while they are waiting for their turn to be called to berth or for protection in bad weather, while larger ships may be required to anchor or ride out bad weather at open sea if necessary. The anchorage areas should be located in natural protected areas or be protected from waves by breakwaters, and should be located near the main harbour areas but out of the path of the main harbour traffic.

Figure 3.19 Small craft harbour

The water depth at an anchoring area should preferably not exceed approximately 50–60 m due to the length of the anchor chain of the ship. The bottom condition must not be too hard, otherwise the anchor will be dragged along the bottom and not dig into it.

When the ship is anchored, in addition to observing all port traffic and following the port's regulations, the following should be adhered to:

(a) Maintain a 24 h bridge watch by a licensed deck officer monitoring radio contacts.
(b) Make frequent control checks to ensure that the ship is not dragging its anchors.
(c) When the wind exceeds 20 m/s, put the propulsion plant on standby for the possibility of leaving the area.
(d) Provide a 15 minute advance notice to the pilot station before heaving the anchors to get under way.

A ship may be moored either with its own anchors, to a buoy or group of buoys, or by a combination of its own anchors and buoys. Mooring systems can, therefore, be divided into free-swinging systems and multiple-point mooring systems.

When using the **free-swinging mooring** system (Figure 3.20) the ship will swing on its anchor and be located generally parallel to the wind and current. The anchorage area shall, therefore, have a water area exceeding the area of a circle with a radius as shown in Table 3.1 and in accordance with the natural conditions, such as topography, seabed condition and exposure to weather.

The horizontal distance X in Figure 3.20 will usually vary between approximately 6 and 10 times the water depth. The length of the anchor chain can be reduced for a single buoy by adding a deadweight near the buoy (e.g. a concrete block for holding the chain between the buoy and the anchor down to the sea bottom).

In the **multiple-point mooring** system (Figure 3.21) the ship is secured to a minimum of four mooring points, and is thereby held in a more or less fixed position.

Figure 3.20 Free-swinging mooring

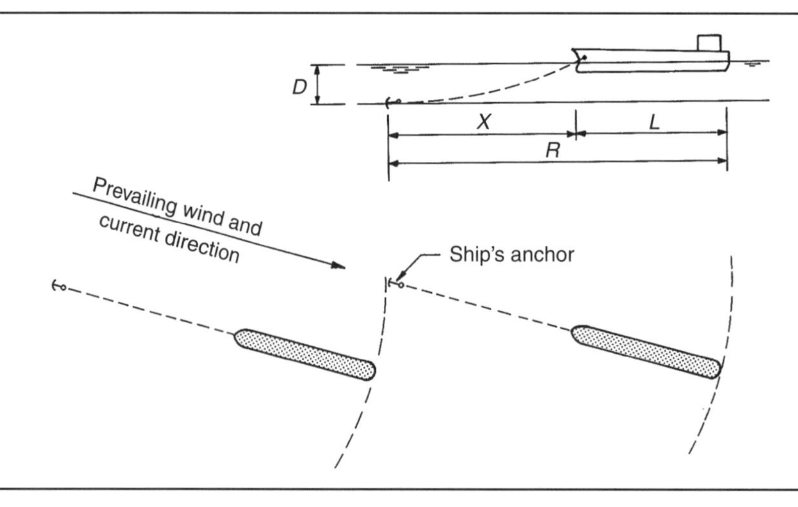

Table 3.1 Approximate radius of anchorage area

Object of anchorage	Seabed soil or wind velocity	Radius
Waiting or cargo handling	Good anchoring	$L + 6D$
	Bad anchoring	$L + 6D + 30$ m
Mooring during	Wind velocity 20 m/s	$L + 3D + 90$ m
	Wind velocity 30 m/s	$L + 4D + 145$ m

To obtain the maximum pullout or anchor resistance, the anchor chain must not be subjected to a pull angle of more than 3° above the horizontal near the anchor. The maximum pullout of the anchor, depending upon the soil condition, is about 7 to 8 times the weight of the anchor. The anchor weight for a 40 000 DWT ship is about 7 t and about 21 t for a 200 000 DWT ship. If the pullout angle is 5° above the horizontal, the maximum pullout of the anchor is reduced by about 25%, and if the angle is about 15° the maximum pullout of the anchor will be reduced by about 50%.

The anchorage area should have enough water area to allow for the possibility of drift of approximately 3 times the water depth when releasing the anchor line. The underkeel clearance of the ship should never be less than approximately 3–4 m at the lowest astronomical tide.

As a rule of thumb, the length of the anchor chain of a ship is approximately 1.5 times the ship's length.

3.4. Area of refuge

If a ship has problems, for example due to machinery difficulties, and needs an area of refuge before it reaches the harbour, it is advisable to adhere to the guidelines of Directive 2002/59/EC of the European Parliament article 20 (OJEC, 2002):

> Member States, having consulted the parties concerned, shall draw up, taking into account relevant guidelines by IMO, plans to accommodate, in the waters under their jurisdiction, ships in distress. Such plans shall contain the necessary arrangements and procedures taking into account operational and

Figure 3.21 Multiple-point mooring

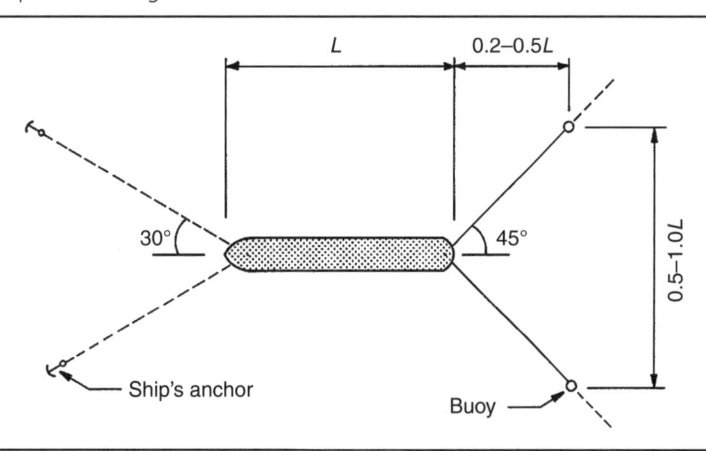

Figure 3.22 Grounding of a battle ship

environmental constraints, to ensure that ships in distress may immediately go to a place of refuge subject to authorisation by the competent authority. Where the Member State considers it necessary and feasible, the plans must contain arrangements for the provision of adequate means and facilities for assistance, salvage and pollution response.

3.5. Grounding areas

In the case of engine trouble or serious damage to ships, and especially for oil tankers, emergency grounding areas along the approach route (AR) between the open sea and the oil and gas harbour should be available. It is very important to have the possibility of grounding an oil tanker, for example in a case where the tanker has been damaged to such an extent that there is a risk of it sinking and thus causing extensive oil pollution. Figure 3.22 shows a battle ship which has been grounded after serious distress.

The consequences of the above potential incidents should be assessed very carefully, and scenarios should be developed for each hazard case, each relevant ship type and each section of the port/channel. For example, the consequence of the grounding of a loaded crude oil tanker will be different from the grounding of a fishing vessel: the risk of environmental pollution is obviously higher with the tanker, whereas the chances of personal injury are higher with the fishing vessel. The risk assessment procedure should identify this, along with the applicable risk control options. For example, the tanker grounding scenario might require extra tugs, traffic control (e.g. 'clear channel' transit) or tidal staging (underkeel limitation), etc., to reduce the risk to acceptable levels, whereas the risk control measures for the fishing vessel grounding would be very different, (e.g. relating to training, certification and inspection procedures).

The water depth at the grounding area should be a little less than the draft of the ship. The sea bottom should preferably be even and soft. The manoeuvring should be as easy as possible, as a damaged ship may have reduced manoeuvrability.

REFERENCES AND FURTHER READING

BSI (British Standards Institution) (1988) BS 6349-2: 1988. Maritime structures. Design of quay walls, jetties and dolphins. BSI, London.

BSI (2000) BS 6349-1: 2000. Maritime structures. Code of practice for general criteria. BSI, London.

Ministerio de Obras Públicas y Transportes (1990) ROM 0.2-90. Maritime works recommendations. Actions in the design of maritime and harbour works. Ministerio de Obras Públicas y Transportes, Madrid.

OJEC (Official Journal of the European Communities) (2002) Directive 2002/59/EC. OJEC, L 208, 5 August.

PIANC (International Navigation Association) (1995) *Port Facilities for Ferries. Practical Guide. Report of Working Group 11*. PIANC, Brussels.

PIANC (1997) *Approach Channels. A Guide for Design. Report of Working Group II-30*. PIANC, Brussels.

PIANC (1997) *Review of Selected Standards for Floating Dock Design*. PIANC, Brussels.

PIANC (2010) *Safety Aspects Affecting the Berthing Operations of Large Tankers to Oil and Gas Terminals. Report of Working Group 116*. PIANC, Brussels.

PIANC (2014) *Design Principles for Container Terminals in Small and Medium Ports. Report of Working Group 135*. PIANC, Brussels.

Port and Harbour Research Institute (1999) *Technical Standards for Port and Harbour Facilities in Japan*. Port and Harbour Research Institute, Ministry of Transport, Tokyo.

Puertos del Estado (2007) ROM 3.1-99. Maritime port configuration design: approach channel & harbor basin. Puertos del Estado, Madrid.

Tsinker GP (1997) *Handbook of Port and Harbor Engineering*. Chapman and Hall, London.

Port Designer's Handbook
ISBN 978-0-7277-6004-3

http://dx.doi.org/10.1680/pdh.60043.099

Chapter 4
Berthing requirements

4.1. Operational conditions

4.1.1 General

In the evaluation of the operational conditions one should always bear in mind that manoeuvring a ship in confined shallow water, in close proximity to other ships such as in a navigational channel or inside a harbour, is entirely different from manoeuvring a ship in deep water on the open sea with infrequent and distant traffic.

The design and evaluation of the operational conditions should be based on the following principles:

(a) common and proven technology from similar harbour projects
(b) internationally acceptable standards and recommendations.

In the planning and evaluation of a proposed harbour site, the collection of information on tide, current, wind, waves, etc., plays an essential role, together with the hydrographical and topographical conditions. These factors are essential both for the safety of the ship during navigation and berthing operations and for cargo-handling operations. Together, all these factors will determine the total operational availability of the harbour or the berth.

Generally, ships are primarily designed for the open ocean, and therefore dramatic changes can occur in a ship's response characteristics in shallow water. Safe manoeuvring and berthing in confined water require an adequate design and layout of the navigational channels and harbour, and a proper understanding of the environmental forces that can act on a ship.

Ships that are difficult to handle in confined water can be roughly divided into two groups:

(a) large, deep-draft ships, such as fully loaded tankers with a small underkeel clearance
(b) ships with very high superstructures, such as container ships, car carriers, passenger ships and tankers in ballast.

The environmental conditions are of prime importance for a marine terminal designed to accommodate ships of the very large crude carrier (VLCC) class. The wind forces on a VLCC tanker in ballast, or current on such a ship in a fully loaded condition, can give rise to considerable forces. Ships with lengths of more than 350–400 m, as is the case for a VLCC tanker, can experience great variations in environmental forces along the ship's length.

For some of these parameters, long-term measurements are needed to establish statistical data as a basis for the preliminary and detailed design of the harbour. These measurements should, if possible, be made for at least 1 year in order to describe the seasonal variations. Measurements of sediment transport should be made under rough weather conditions in order to give reliable information, due to the fact that the processes of interest only occur under extreme current and wave conditions. All these measurements should be as close to the proposed harbour site as possible.

Proper planning of all the field investigations will require a good understanding of the characteristics of the proposed harbour site, and should always be undertaken by an experienced coastal and harbour engineer in order to gain the most valuable and optimal information from the site investigations.

In the planning and evaluation of the approach to a terminal harbour site, the collection of data on the tide, current, wind, waves, fog, etc., plays an essential role, together with the hydrographic and topographical conditions. These factors are essential for the safety of vessels during navigation and berthing operations.

4.1.2 Tide

The tide consists of two components: astronomical effects and meteorological effects.

The astronomical tidal variation can be found in the Admiralty Tide Tables (ATT). The astronomical tidal day is the time of rotation of the Earth with respect to the moon and the planets, and is approximately 24 h and 50 min. The ebb and flood tides are the falling and rising tides, respectively.

The **chart datum** for harbour works is generally the **lowest astronomical tide**, which is the lowest tide level that can be predicted to occur under average meteorological conditions. For the berth structures itself, and for land installations, the chart datum is usually referred to approximately the mean water level.

The following water levels should always be recorded in harbour work:

Highest observed water level	HOWL
Highest astronomical tide	HAT
Mean tidal high-water level	MHW
Mean water level	MW
Mean tidal low-water level	MLW
Lowest astronomical tide (chart datum)	LAT
Lowest observed water level	LOWL

In observing the highest or lowest water level, one should also take into account the changes in atmospheric pressure and the effect of strong winds, either blowing onshore, tending to pile up water against the coast, or blowing water off the coast.

The rise or fall of the water level due to a change in the atmospheric pressure is approximately equal to a 0.9 cm rise or fall of the water level for a 1 mbar fall or rise in the atmospheric pressure. The fall and rise in the atmospheric pressure can, in Norway, give a variation of about maximum ± 50 cm. In combination with other effects, such as strong winds, and intensified by geographical constructions, this effect can be very important.

The rise in sea level due to the greenhouse effect that will occur between the years 2010 and 2060 is assumed to be about 0.25–0.30 m.

4.1.3 Water depth

The water depth in the approach channel and the harbour basin, and in the front of and alongside the berth, should generally be sufficient for safe manoeuvring. The chart datum for tidal areas should be the lowest astronomical tide and, for rivers, the lowest recorded river level.

The water depth should be based on the maximum loaded draft of the maximum design ship, and can be determined from the following factors:

(a) draft of the maximum loaded ship
(b) tidal variations
(c) movement of the ship due to waves
(d) trim due to loading of the ship
(e) squat
(f) atmospheric pressure
(g) character of the bottom
(h) error in dredging
(i) possibility of silting up.

The gross underkeel clearance (Figure 4.1), must be designed to allow for waves, trim, squat, atmospheric pressure, etc., in addition to allowing for a safety margin for unevenness of the bottom.

The squat, or the reduction of the underkeel clearance, is due to the suction effect induced by the higher current velocity between the sea bottom and the ship. This causes a reduction in the water level near the ship, and the ship therefore sinks bodily in the water. The squat increases with the length of the ship, with an increase in the ship speed, and with a reduction in the underkeel clearance and narrowness of a channel. In addition, the water depth is also affected by the water density, and must be greater in freshwater than in seawater. This can be of importance for river or estuary ports.

Ship movements due to waves can be up to two-thirds of the significant wave height for smaller ships. VLCC and large ore carriers, due to their huge size, are only susceptible to waves with a period of more than 10 s. Waves with a shorter period will result in insignificant vertical motions for these ships.

Figure 4.1 Components of depth

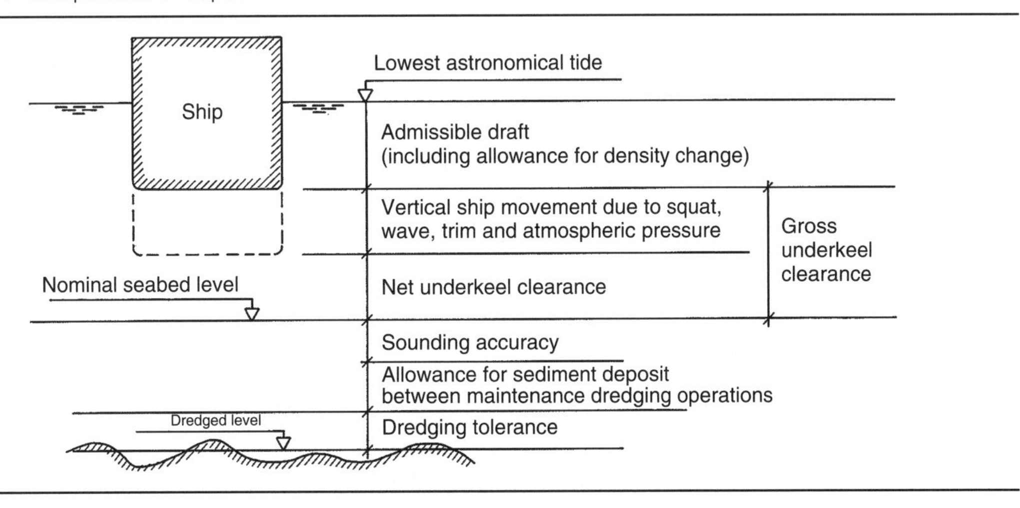

Table 4.1 Allowance for the water depth in front of a gas jetty for a 136 000 m^3 gas tanker

Depth factors due to	Allowance: m
Loaded draft (maximum)	11.3
Approximate LAT	1.6
Atmospheric pressure	0.4
Movements due to waves	0.5
Trim due to loading	1.0
Safety due to dredging	0.5
Character of bottom, rock	1.0
Water depth below MW	**16.3**

Where the bottom is composed of soft materials (sand, etc.) the minimum net underkeel clearance should be 0.5 m, and for a rocky bottom 1.0 m.

Where the bottom of the seabed consists of silt and/or mud, it is usual to define a nautical depth as being from the water surface to the level at which the density of the bottom material is equal to or greater than 1200 kg/m^3, since material layers of a lower density do not significantly impede the passage of a ship.

As an example, the allowance for the water depth in front of a gas jetty below land zero level for a 136 000 m^3 gas tanker is shown in Table 4.1.

The water depth must, during construction dredging, be sufficient to avoid both possible errors in dredging and a yearly excessive cost for maintenance dredging due to the possibility of silting up. The water depth must also take into consideration an increase in draft due to the roll and pitch of a ship when moored, as indicated below:

(a) increase due to roll $= 0.5 \times$ beam $\times \sin \alpha$, where α is the roll angle
(b) increase due to pitch $= 0.5 \times L_{OA} \times \sin \beta$, where β is the pitch angle.

As a rough guide, the gross underkeel clearance above the nominal seabed level for the maximum ship using the seaway, should, as a minimum, be the following:

(a) Open sea areas: for high-ship speeds and exposure to strong swells, the clearance should be approximately 30% of the maximum draft.
(b) Exposed channels: exposed to strong swells, the clearance should be approximately 25% of the maximum draft.
(c) Exposed manoeuvring and berthing areas: exposed to swells, the clearance should be approximately 20% of the maximum draft.
(d) Protected manoeuvring and berthing areas: protected from swells, the clearance should be approximately 15% of the maximum draft.

The nominal seabed level is the level above which no obstacles to navigation exist. For good manoeuvring control, the ship requires a deeper water depth than the absolute minimum requirement from loading of the ship, tidal variations, trim, etc. In a channel it is desirable to have a ratio of the channel depth to the maximum draft of the largest ship of 1.3 for ship speeds under 6 knots and 1.5 for higher speeds.

When erosion protection is needed, it is recommended that the bottom protection be placed 0.75 m below the lowest permitted dredged for berths, which are subjected to maintenance dredging. Therefore the level of the bottom protection requires careful study because the protection installed should not be damaged during maintenance dredging operations. The water depth should therefore take into consideration any maintenance dredging.

At berths where the movement of the largest ships to be accommodated takes place at the higher states of the tide, the underkeel clearance may be achieved by dredging a berth box in front of the berth structure. The berth box should at least have a length of 1.2 times the overall length of the largest ship and a width of 1.5 times the beam of the largest ship that will use the berth.

4.2. Navigation

The total navigation operation, from arrival to departure, can be subdivided into the following operations:

(a) arrival at the outer harbour basin
(b) preparation for berthing, including possible turning of the ship and pre-berthing procedures in the harbour basin
(c) berthing including mooring, etc., to the berth structure
(d) loading and unloading operations while at berth
(e) unberthing from the berth structure
(f) departure from the harbour basin.

Where the ship traffic flow from the open sea to the port is very busy in the approach routes, access channels or harbour area, the total traffic efficiency and safety can be improved by using the recommendations from the **International Association of Lighthouse Authorities** (IALA), the **Automatic Identification System** (AIS), **SafeSeaNet** (for vessel monitoring in the EU) and a **vessel traffic system** (VTS) such as the Norwegian system Norcontrol IT AS or its equivalent. These systems are designed to communicate with each other automatically and exchange critical information about the course, speed and intended route of different ships.

The principle of a typical VTS system is illustrated in Figure 4.2. The radar information is transferred to a traffic-control centre and presented on a bright, high-resolution colour monitor together with detailed chart information. Ships can automatically be acquired and tracked by the VTS system and shown on the monitor as target symbols with their names and vectors indicating the course and speed (in Figure 4.3). Alarm strategies can be set up so that no intervention is required unless an alarm occurs or the operator needs further information. This system can integrate information from several radars, and has proven to be an extremely valuable tool in many ports. All data from traffic movements can be stored and used as references for the port administration, port authorities, coastguards, and search and rescue services.

In the traffic-control centre (Figure 4.4), the information is presented on a bright, high-resolution colour monitor, together with detailed chart information (Figure 4.5).

Figure 4.2 A vessel traffic system (VTS)

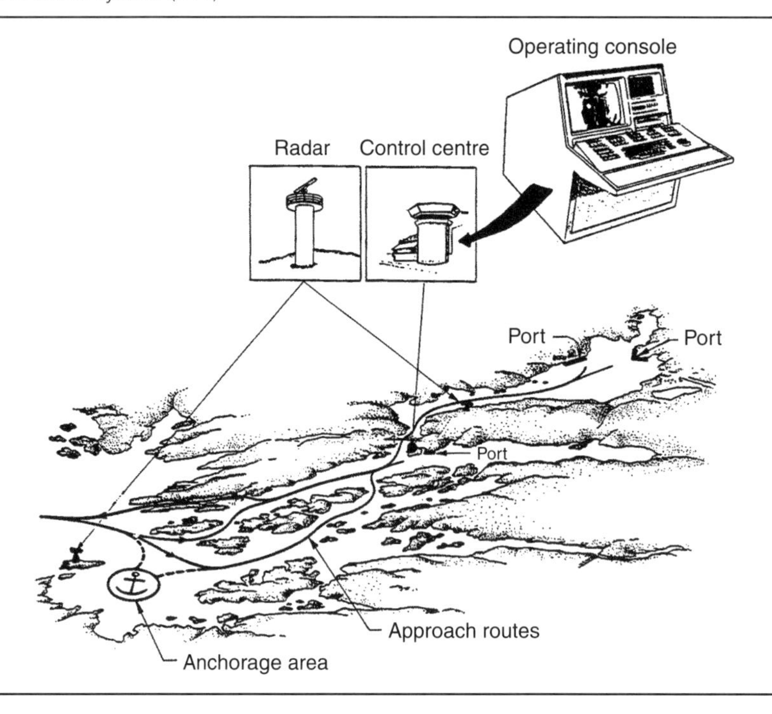

Figure 4.3 The radar information is transferred to the control centre. (Photograph courtesy of Norcontrol, Norway)

Figure 4.4 Inside a traffic control. (Photograph courtesy of Norcontrol, Norway)

Figure 4.5 A monitor showing detailed information. (Photograph courtesy of Norcontrol, Norway)

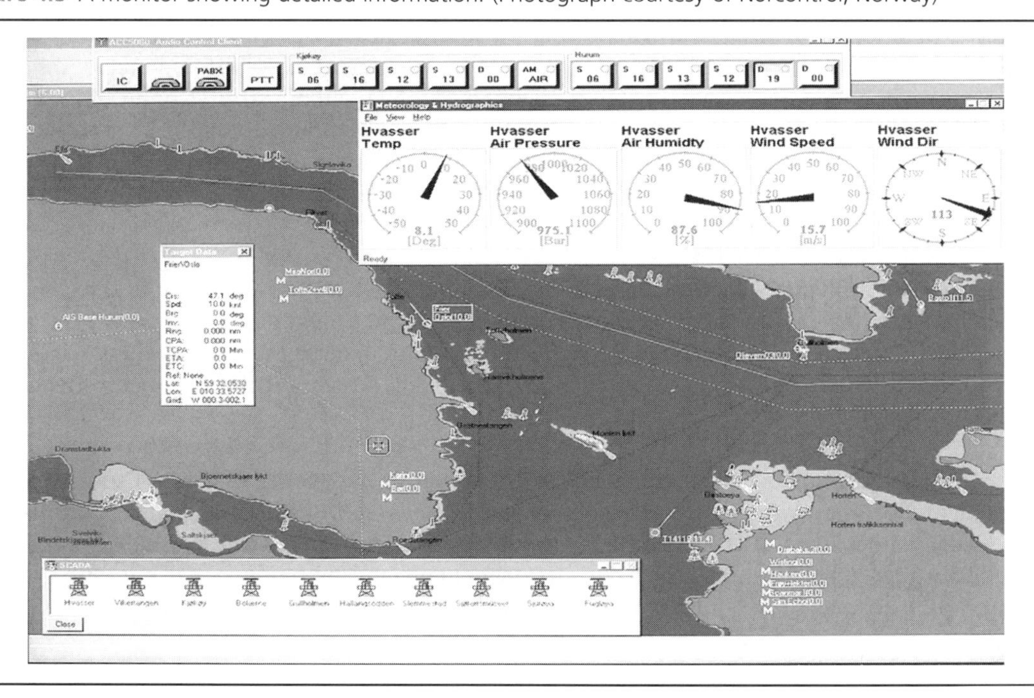

The decision on where the arrival and the berthing operation should start, and where the required number of tugboats should meet and assist the ship to the berth, must be made by the local pilot or the port captain together with the ship's captain, depending on the weather conditions, on whether the ship has bow thrusters or not, etc. Pilotage should be made compulsory for all gas and oil tankers and large ships calling at a harbour. The regulations for the harbour will indicate where the pilot should meet a ship.

For the purpose of manoeuvring, all ships should maintain a reasonable ballasted condition. The ships should carry sufficient ballast so that the propeller remains immersed. At some ports, for example oil terminals, tankers are required to carry a total deadweight of not less than 35% of the summer DWT, including bunkers, freshwater and stores, on arrival.

Manoeuvring during berthing and unberthing of a ship will generally be done in one of the following ways:

(a) by using only the ship's own engine, rudder, bow thrusters and/or the ship's anchor
(b) with the assistance of only one or more tugboats
(c) by using the ship's own anchor, and with the assistance of one or more tugboats
(d) with the use of berth- or land-based winches, and with the assistance of one or more tugboats
(e) with the use of the mooring buoy, and with the assistance of one or more tugboats
(f) a combination of two or more of the above-mentioned systems.

Usually, nearly all berthing and unberthing operations have been done after any one of cases (a), (b) and (c). Case (d), with the use of land-based winches of hauling capacity of about 75 t, can have some technical advantages compared with case (a), (b), (c) or (e). The difference in operation of case (d) is of both a technical and an administrative nature. From a technical point of view, the efficiency, particularly with an offshore wind, will tend to be higher with winches because the responsiveness to action is much faster than for tugboats. The winch system is also safer because the berthing operation is more punctual and faster than for operations with only tugboats. For berthing of larger tankers, the winch system is more economic since it will use fewer tugboats. From an administrative point of view, there may be a shift of responsibility from the tanker's captain to the harbour captain if the mooring ropes are tightened by winches on the berth or land rather than by the tanker's winches. This could cause problems, for instance, with the insurance company if an accident should occur to the berth or the tanker during the berthing operation.

The **nautical chart** in Figure 4.6 shows the approach route to the terminal at the top of the chart. The nautical chart is from the Norwegian Hydrographic Service map number 18. Figure 4.7 shows the manoeuvring of arriving and departing tankers to the terminal under different prevailing wind directions, with the assistance of tugboats or by using the tanker's own anchor together with tugboat assistance.

For emergency evacuation of the tanker in a strong wind without tugboat assistance, for example due to a fire at the loading platform, the tanker may not be able to leave the berth only under its own engine. On the other hand, if the tanker had used its own anchor during the berthing operation, it may be able to leave the berth by using both its own engine and by pulling itself out with the help of the anchor, if the wind is less than approximately 7 Beaufort or between 13.9 and 17.1 m/s.

Figure 4.6 A nautical chart

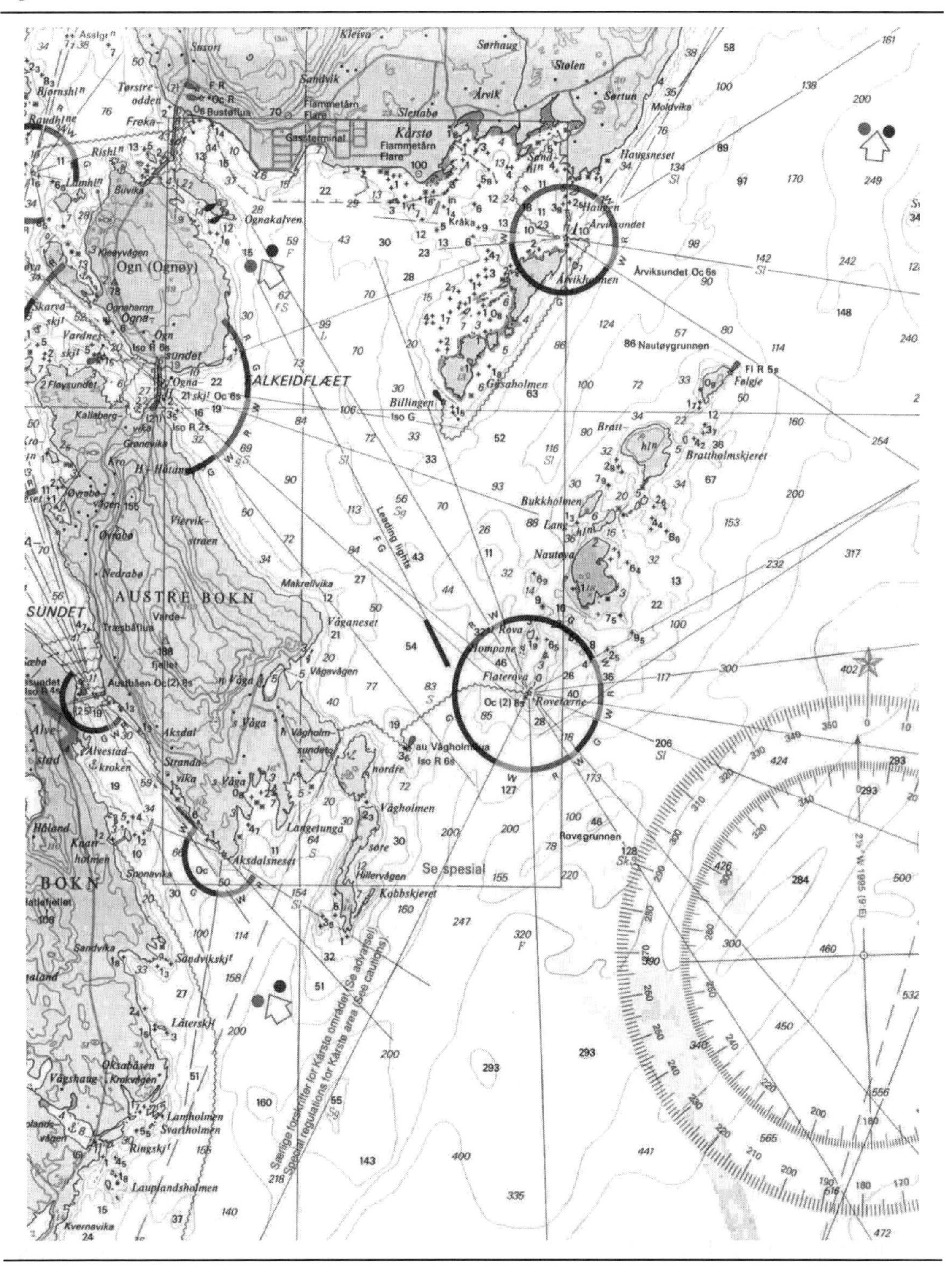

Figure 4.7 Manoeuvring during arrival and departure

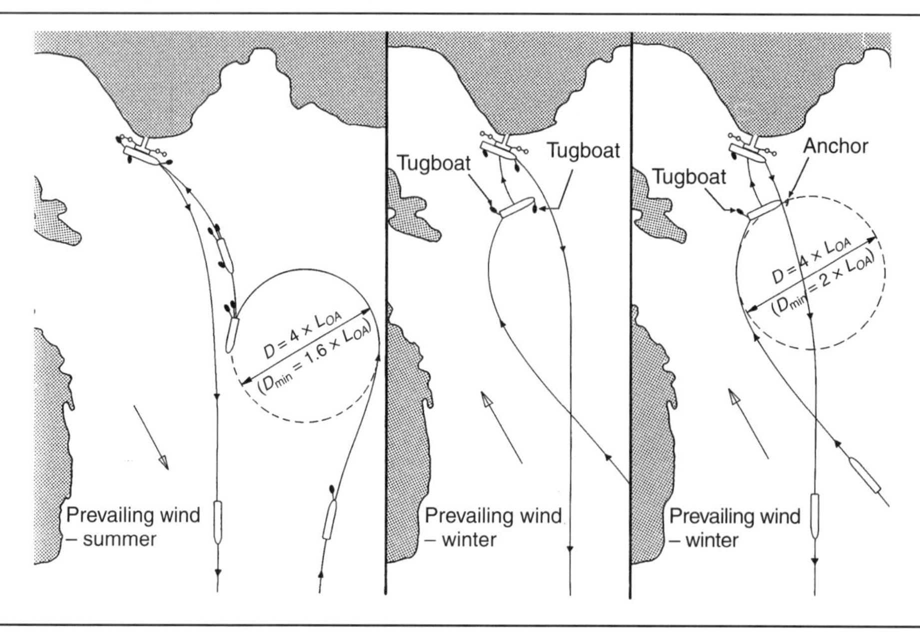

All important berth facilities, for example oil and gas berths, should be equipped with a monitoring berthing or **docking aid system** (DAS) that includes the following data:

(a) wind and sea-current sensors detecting speed and direction
(b) wave heights and tide levels
(c) the ship's approach velocity, distance to the jetty and angle of approach during all stages of the berthing operation
(d) mooring-line loads to the mooring hooks, pulleys or winches
(e) measurement of the ship surge and sway when the oil loading arms are connected.

The purpose of the DAS is to measure and track the speed, distance and angle of the approaching ship over the last 300 m until contact is made with the berthing structure and the ship comes to rest against the breasting fenders. After berthing, the DAS should remain operational, to monitor the position of the ship in terms of transverse drift from the fenders' face or as compression on the fenders.

Particularly in the petrochemical and dry-bulk sectors, most terminals now have a shore-based berthing aid systems and/or a DAS to assist in guiding ships safely to the berth structure. The main berthing aid systems in service are the sonar-in-water and the radar systems, as illustrated in Figure 4.8 and in more detail in Figure 4.9. Practical experience has shown that too much reliance should not be placed on the sonar-in-water system, due to turbulence caused by the ship's propellers, unwanted reflections, maintenance problems in the water, etc. The distance between the two laser sensor units should be at least 40 m.

All information should be permanently recorded as part of the berthing and mooring history. To increase safety during the berthing operation, a display board unit should be provided at the berth structure (Figure 4.10). The display board should be mounted on a fixed, approximately 2 m-high

Figure 4.8 A berthing aid system

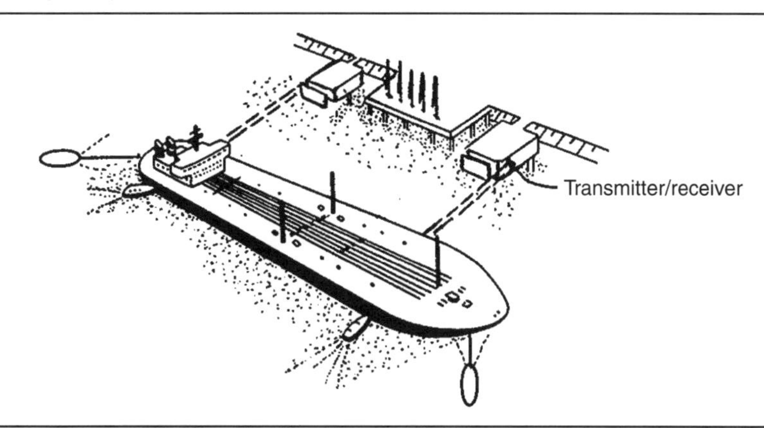

pedestal foundation for measuring both the ship's bow and stern velocity of approach and distance from the berthing line independently. The display board unit should be large enough to allow the ship's captain and the pilot to read information from a distance of approximately 200 m off the berth structure.

All the information from the berthing aid system should be displayed on a visual display unit in the control room showing the berth in plan view, the position of the ship in relation to the berth structure, the actual mooring lines in use and the loads on the mooring lines, the wave heights, the wind speed and directions, and the current speed and directions. A display with the information could be as shown in Figure 4.11 (from Marimatech, Denmark) or an equivalent system.

Figure 4.9 Example of a docking laser system

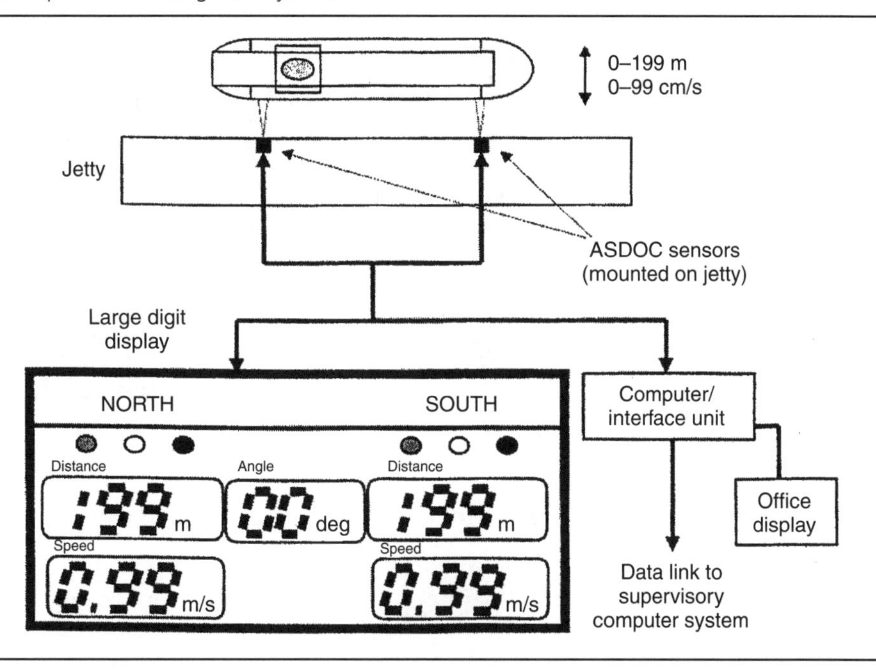

Figure 4.10 A berth with a display board unit

Figure 4.11 A berthing visual display

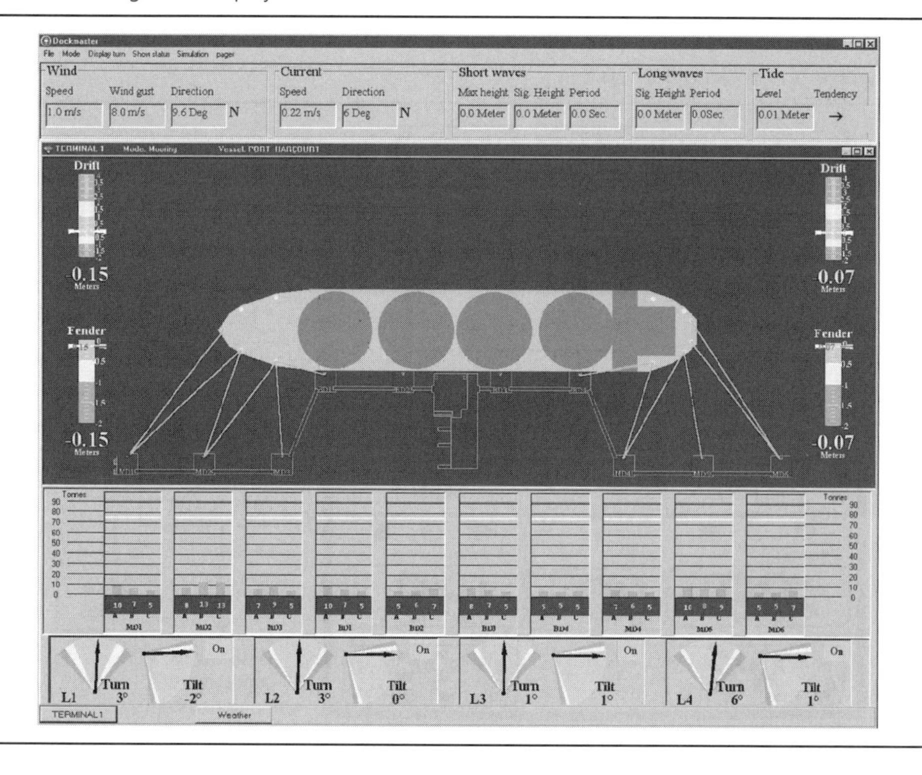

The berthing aid data usually becomes available at a distance of approximately 200 m from the berth. All data should be routed to the terminal control centre, where it can be displayed and recorded. The total navigation operation should be watched by radar, and marine and internal radios should also be provided. The mooring operation should be under the command of the tanker captain assisted by the pilot, tugboat captains and terminal mooring supervisor. The terminal mooring supervisor may interfere in the mooring operation if, for some reason, they find the operation to be hazardous. They may do this by telling the tanker captain to reduce approach speed or, in the worst case, they may deny the tanker mooring permission, stop the berthing operation and instruct the tanker to leave the terminal.

When berthing to an oil or gas berth or terminal, the tanker shall be stopped at a distance of about 100–200 m off the berth, and from this position the tugboats move the tanker transversely at a controlled approach velocity towards the berth. The approach velocity of the ship should gradually be reduced to about 0.05 m/s in the final phase before the ship hits the fender structures. The maximum acceptable approach velocity should be about two-thirds of the design approach velocity for the berth structures. When closing up to the berth, the mooring boats or launches should bring the fore and aft breast mooring lines from the tanker to the mooring points at the berth structure. The tanker's spring mooring lines should then be moored to the berth structure. The tanker's position can now be adjusted by using the fore, aft breast and spring mooring lines, together with the tugboats. The berthing angle between the ship and the berth line during the final stage of the berthing operation should not be more than 3–5°.

In addition to the electronic monitoring systems installed at the berth for measuring the wind, wave, current and tide, it is advisable to install an additional visual wind indicator similar to a windbag of the type that is normally installed at airports and airfields. At a mast on top of the control building, which should be situated in a non-hazardous area, the wind, temperature, humidity, air pressure and visibility should be measured.

The shore-based berthing aid system can play a major role in reducing the risk and damage to the berth and fender structure and/or the ship. It is undeniable that millions of pounds worth of damage is caused annually to berth structures and/or ships around the world through ships approaching too fast, at the wrong angle, etc. From terminals with berthing aid systems, it has also been found that the berthing time over the last 20 m between the ship and the berth structure has been reduced to 30–40%.

4.3. Tugboat assistance

The efficient and safe manoeuvres of large ships in confined waters are largely determined by the degree of controllability of the ship's own rudder, propellers and thrusters and by the assistance of necessary tugboats due to weather conditions. The ship's captain should, before entering restricted areas of a port, also ensure that anchors are ready for letting go prior to entering the pilot operating areas.

From a safety point of view, tugboat assistance should always be used during the berthing and unberthing operations of oil and gas tankers. Particularly during the berthing operation, tugboats should always be used, so that the tanker can approach the berth structure parallel to the fender face of the berth. The tugboats will guarantee a simpler and easier berthing or unberthing procedure, and are also an important safety factor if the ship develops engine or steering trouble. The tugboats may also be used to assist other traffic in the lead (Figure 4.12). It is also strongly recommended that, under normal circumstances, oil and gas tankers should not leave the berth without the assistance of tugboats. However, from an emergency evacuation point of view, it is recommended that they are moored in such a way that unberthing operations will be as easy as possible.

Figure 4.12 Tugboats assisting a large oil tanker from the open sea to the terminal. (Photograph courtesy of Buksér og Berging AS, Norway)

It is particularly important that all oil and gas terminals should be in possession of evidence and give documentation that they can handle the tankers with the operational wind speeds and with the tugboat fleet stationed at the terminal. From a safety point of view, tugboat assistance should always be used during berthing and unberthing operations of oil and gas tankers.

The term 'tugboat escort' refers to the stationing of tugboats in the proximity of the ship as the ship is manoeuvring into port, to provide immediate assistance if a steering and or propulsion failure should develop. The term 'tugboat assist' refers to the situation when the tugboat is applying forces to assist the ship in making turns, reducing speed, berthing, etc.

A ship will generally start to lose steerage if its speed, under its own engine, is less than about 3–4 knots, depending on the type of ship and on whether the ship has bow thrusters. Therefore, when berthing an oil tanker, the tugboats should meet the tanker outside the restricted oil harbour area and before the speed of the tanker is less than about 3–4 knots, and connect up the tow lines, at least one fore and one aft.

The requirements of the tugboat fleet should cover the following services:

(a) Provide necessary assistance during the berthing and unberthing operations to counteract the wind, wave and/or current forces.
(b) Enable the ship to turn in a confined area.
(c) Act as a restraining or anchoring force on a ship moving towards the berth structure.
(d) Act as a standby ship when a gas or oil tanker is moored.
(e) Carry out emergency, fire-fighting and anti-pollution operations. For a ship in an emergency situation, for example due to breakdown of propulsion machinery, steering gear, etc., the tugboat must be able to assist directly.

The steering and manoeuvring capacity of a ship is dependent on the following:

(a) the environmental condition (wind, waves and current)
(b) the mechanical equipment of the ship itself (the main propeller and rudder, bow thrusters, etc., and the available tugboat assistance).

Figure 4.13 Tugboats assisting during the berthing operation. (Photograph courtesy of Buksér og Berging AS, Norway)

Generally, depending on the wind and/or wave direction, the pilot and the ship's captain will normally decide which side should be alongside the berth. However, the ship should, if possible, be moored with the head out, in case an emergency departure is required. The number of tugboats assisting in each berthing and unberthing operation will (see Figure 4.13) depend on the weather conditions at that time.

Figure 4.14 shows the different types of tugs and their capabilities used as harbour tugs. Traditionally, requirements for harbour towage have relied on the skills and experience of masters and pilots. At present, the standards are more often given by port regulations based on the evaluation of relevant risk elements, calculations of environmental forces and practical experience, with due allowance given for the size and type of ship, loading conditions and inherent manoeuvrability. One should always be aware, depending on the tug type, that the tug efficiency factor will be reduced when the significant wave height increases.

Both the operating envelope and the assisting force that can be delivered by a tug depend on a large number of complex factors, including the type of tug, its hull shape and configuration, the number, type and position of the tug's propulsors, the local metocean conditions, the ship's speed and the tug operating mode. As the wave height increases, the effective bollard pull for a tug reduces until, at a particular wave height, the tug will not be effective at all, delivering zero bollard pull. Tugs operating in direct push mode are the most affected, and those operating in indirect assist modes are the least affected.

Figure 4.14 Capability of different types of tugs. (Photograph courtesy of Buksér og Berging AS, Norway)

Conventional tug up to 118 TBP (tons bollard pull) | Azimuth stern drive tug up to 90 TBP

Z-Tractor tug up to 45 TBP | Fin First Voith tractor tug up to 95 TBP

In general, it is accepted that harbour tugs fitted with conventional 'static' winches can operate, albeit at much reduced effectiveness, in waves of up to approximately 1.5–2 m. Beyond this limit, the effectiveness of tugs becomes more uncertain. It is possible to increase the tug's effectiveness, when on lines, by using dynamic winches (sometimes called 'constant tension' or 'render recovery' winches), which act to reduce snatching. Further improvements may be achieved with sea-keeping hull forms and alternative towing point configurations.

Tugboats can generally be divided into two groups according to their size and power:

(a) Harbour tugboats, which operate mainly in sheltered waters. Their horsepower varies roughly between 2000 and 7000 hp, and their size between about 15 and 30 m in length.
(b) Offshore tugboats, which can operate in exposed waters. Their horsepower varies roughly between 10 000 and 20 000 hp or more, and their size between about 30 and 60 m in length.

The necessary tugboat horsepower is approximately 10–12 times the actual tugboat bollard pull in kN; 1 hp is equivalent to approximately 0.75 kW.

The number of tugboats needed for handling the different types of ships is affected by the size and type of ship, the approach route, the exposure and type of the berth structure, the environmental conditions, etc., and on the bollard pull that each tugboat can mobilise.

Therefore, the necessary tugboat capacity in effective bollard pull should be sufficient to overcome the maximum wind, waves and/or current forces generated on the largest ship using a port, under the maximum wind, waves and/or current permitted for harbour manoeuvring and with the ship's main engines and bow thrusters out of action.

SIGTTO recommends that there must always be sufficient tugboat assistance to control the gas tanker in the maximum permitted operating conditions, and this should be specified assuming the tanker's engines are not available. It is recommended to have three, or preferably four, tugboats to assist the tanker, and that they should be able to exert approximately half of the total tugboat power at each end of the gas tanker.

In the evaluation of the total number of tugboats required for manoeuvring of a ship under the berthing and unberthing operations, the following are assumed:

(a) The ship is not equipped with bow thrusters.
(b) The forces acting on the ship can be due to wind (F_{wind}), wave (F_{wave}) and current ($F_{current}$). For the wind forces, a 'gust factor', F_g, of about 1.2 should be applied.
(c) The force required moving a ship against the wind and current is generally assumed to be at least approximately 30% higher than the forces necessary to hold the ship against the forces due to wind and current.

An evaluation of the necessary tugboat bollard pull should at the least be based on the design specification from British standard BS 6349-1 (formulas for wind and current forces), or the OCIMF (Oil Companies International Marine Forum) *Mooring Equipment Guidelines* for the safe mooring of large ships at piers and sea islands (formulas for wind and current forces), SIGTTO (formulas for wind) and/or the Spanish standard ROM 0.2-90 (formulas for wind, wave and current forces) if a model of the specific ship has not been tested in a laboratory.

Due to possible uneven bollard pull when several tugboats are used, and inaccuracy in the method of calculating the bollard pull required to control the ship, it is usually recommended that a tugboat bollard pull factor S_f of between 1.2 and 1.5 is used, depending on the general weather conditions.

The **total required effective tugboat bollard pull** B_P needed to control a ship due to environmental forces can be calculated approximately from the following formula:

$$B_P = S_f \times [(F_{wind} \times F_g) + F_{wave} + F_{current}]$$

It is recommend that there should always be an operational safety factor $S_O \geq 1.2$–1.5 due to the actual total available tugboat capacity (including any bow thrusters) T_C during berthing divided by the total needed effective tugboat bollard pull B_P due to environmental forces during each phase of the berthing and unberthing operations:

$$S_O \geq \frac{T_C}{B_P} \geq 1.2 \text{ to } 1.5$$

It is important to allow safety margins in all berthing/unberthing operations. These can vary from 20% to 50% of the available bollard pull, depending on the length of the tow, if the tug is pushing, the direction of the tow or the push, the swell conditions and the current speed. Careful consideration should be made of the external forces that will influence the effective power of the tug.

When discussing safety during berthing and unberthing, this is very often related to the bollard pull required to overcome the maximum environmental forces of wind, current and waves. We are talking about the tug boat capacity, and not a specified number of tug boats. It should therefore be borne in mind that when considering the number of tug boats from a safety point of view, this differs from the number of tug boats in terms of tug boat capacity to overcome the environmental forces.

In general, the number of tugs to be used can be between two and six, but the norm is to use between three and five. Using too many tugs will be a disadvantage, because the pilot will find it problematic controlling them.

The presence of an oil or gas tanker at a terminal will involve a certain risk to the surroundings. For safety reasons, there should thus always be at least one tugboat on standby in case of possible changes in the weather situation and/or a need for an emergency departure due to fire, etc.

In determining the tugs to be used for berthing and unberthing operations it is essential that the maximum environmental operating criteria are established for the ship and the port/terminal. Essentially, this means the wind speed and direction, the sea, and the swell height, direction and period.

The overall port layout – the channel design, entrance, turning areas, manoeuvring space – will influence the number and power of the tugs to be used. Where space in the port is limited, the length of towing lines may need to be short, and this can limit the effective applied towing power. Where tugs are working in sheltered areas, efficiency in pushing must be considered.

Most of the oil terminals, and especially the gas terminals, around the world require, due to safety reasons, that at least four tugboats are always used during berthing to, and three during unberthing from, the berth.

The physical size and displacement of the ship in loaded and ballast conditions, the longitudinal and end-on windage area, the submerged hull and the underkeel clearance are critical in determining sufficient tug power. The efficiency of the tug design, both its propulsion and hull form, in the prevailing environmental conditions is also a factor.

Different water depths can affect the tug power requirements. The less the depth of water is under the ship then the greater the power required. The addition of current and reduced water depths will also add to the tug power requirement for the same ship.

Using simulation tools, the tugs, the ship and the environmental conditions can be modelled to assess the nominal tug power requirements.

In using simulation tools it is necessary to clearly identify the success criteria beforehand. These can include but are not be limited to the following:

(a) the effective tug power used should not exceed, for example, 75% of the maximum available power on, for example, three out of four tugs, for sustained periods

(b) the manoeuvrability of the vessel is achieved within preset boundaries, and safety distances are maintained.
(c) the ability to tow effectively in maximum sea/swell conditions.

Smaller support vessels involved in ship manoeuvring operations, such as pilot boats, tugs and line boats, are generally more affected by waves than the ships they are assisting. At many ports, it is the performance limitations and safety considerations of the required support vessel operations that govern the limiting wave conditions for safe ship manoeuvring.

In waves, the effective bollard pull of the tugs is reduced due to the following:

(a) The tug is unable to deliver a constant level and angle of thrust due to tug motions and possible line snatching.
(b) The wave-induced motions of the ship and the tug in any particular sea state are generally different, and therefore the tug moves relative to the ship. In long-period waves, the relative motion between the tug and the ship can be less than in short-period waves.
(c) A proportion of the tug's power is required just to keep the tug in position/under control and/or to keep up with the ship's lateral speed.

The efficiency of tug operating in waves can be shown through the use of tug effectiveness curves, as illustrated in Figure 4.15 for ASD (azimuth stern-drive) tugs.

Failure of a tug during a berthing or unberthing operation should be taken into account. In the event of a tug failure, the berthing operation should be able to be either completed or aborted safely with the remaining tugs. Due consideration should be given to reducing berthing/unberthing parameters in the event that operations cannot be safely terminated when a tug fails for whatever reason.

The use of bow/stern thrusters on a ship to reduce the number of tugs for berthing/unberthing should be considered carefully. Providing the reliability and operation is confirmed before arrival, then it could be possible to reduce the number of tugs. However, this gives very little if any safety margin in the event of a shipside failure, and may add unnecessary risk to the operations. On the other hand, if it is assumed that any bow/stern thruster is not included in the determination of required tug power, then in the event of a tug failure the margin of safety is greatly improved if the bow/stern thruster is available for use.

As in all situations, the economic considerations of the number of tugs used for operations should not outweigh the safety considerations, and the master/port should ensure that the safety of the ship and tugs is paramount at all times.

To illustrate and simplify the risk assessment it is possible to establish a matrix where probability is on the horizontal axis and consequence is on the vertical axis (Figure 4.16).

For gas and oil terminals, the probability of losing control over the ship should be at least level 3 or lower. It should not generally be acceptable to use only two tugs, even if the bollard pull is acceptable, because the probability of losing control over the ship is too high if one tug drops out.

For oil and gas tankers, the consequence of losing control over the tanker is always high during the berthing operations. This is the total of risk for damage to the ship itself, to the terminal or to other

Figure 4.15 Tug curves. (Courtesy of HR Wallingford)

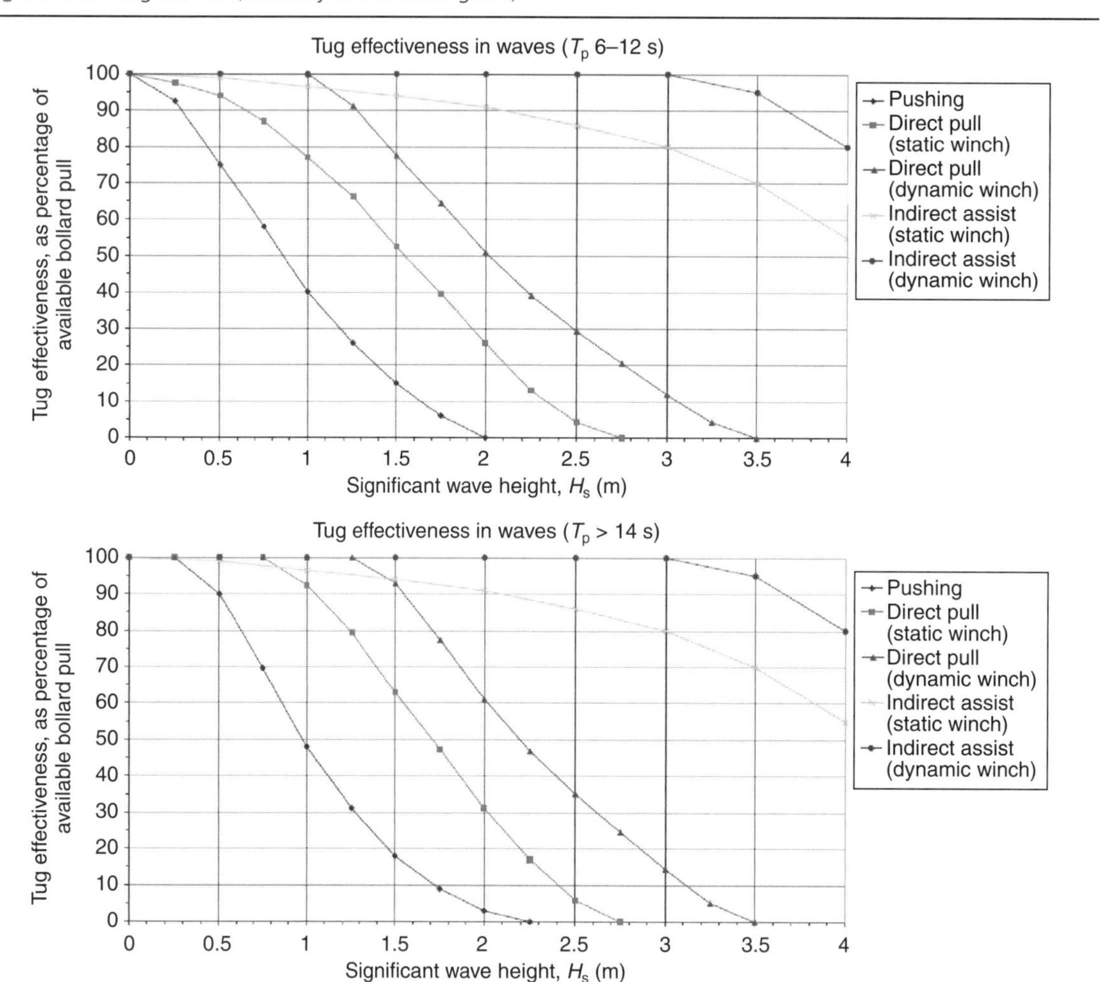

similar ships that are berthed at the terminal. In practice, this means that oil and gas terminals should always be operated at the consequence level of 3 or lower.

Figure 4.16 shows the risk of losing control over the ship if one tug drops out as a product of probability and consequence linked to the numbers of tugs. With the general restrictions in levels in both probability and consequences, the acceptable risk is shown shaded in grey in the table.

This risk assessment must be done before the calculations of the bollard pull from the tugs, and it will overrule the numbers of tugs from these calculations if they are lower than the risk assessment gives.

Therefore, a governmental body should always periodically review and/or modify its rules, regulations, guidelines, etc., to ensure safety at oil and gas terminals.

Figure 4.16 An example of the consequence and probability philosophy

Consequence	Risk = Consequence × Probability					
5 (catastrophic)				Unacceptable	Risk	
4						
3						
2						
1 (very low)	Acceptable	Risk				
	1 (improbable)	2	3	4	5 (very high)	Probability
	5	*4*	*3*	2	1	*Example of number of tugs*

Example of the total forces acting on a ship

The approximate wind and current area dimensions in m^2 for, for example, a 137 000 m^3 spherical LNG gas tanker in ballasted condition, length $L_{OA} = 290$ m, length $L_{BP} = 270$ m, width = 48.1 m and ballasted draft = 7.5 m is used in the example of the total forces acting normal to the tanker in Table 4.2.

The total force acting normal or crosswise to the 137 000 m^3 spherical LNG tanker in ballast condition in open sea (deep water) will, in line with the OCIMF/SIGTTO recommendation (wind forces), the British standard (current forces) and the Spanish standard (wave forces), need a total effective bollard pull B_P (with a wind (10 min mean), a gust factor of 1.2 and an uneven tugboat bollard pull factor $S_f = 1.3$) as follows:

(a) For a wind speed of **13 m/s**, a wave height of 1.0 m and a deep-water current speed of 0.5 m/s, the total bollard pull is

$$B_P = S_f \times [(F_{wind} \times F_g) + F_{wave} + F_{current}] = 1.3 \times (1484 + 178 + 265) = 2505 \text{ kN}$$

(b) For a wind speed of **10 m/s**, a wave height of 0.0 m and a deep-water current speed of 0.0 m/s, the total bollard pull is

$$B_P = S_f \times [(F_{wind} \times F_g) + F_{wave} + F_{current}] = 1.3 \times (878 + 0 + 0) = 1141 \text{ kN}$$

Table 4.2 Total forces acting normal to the tanker

Area of spherical LNG tankers	137 000 m^3
Wind laterally projected area: ballast	8 400
Wind front area: ballast	2 100
Current laterally projected area: ballast	2 100
Current front area: ballast	360
Total current surface area: ballast	13 000

Figure 4.17 Necessary tugboat bollard pull for a crosswise wind and a current acting on an oil tanker in deep water

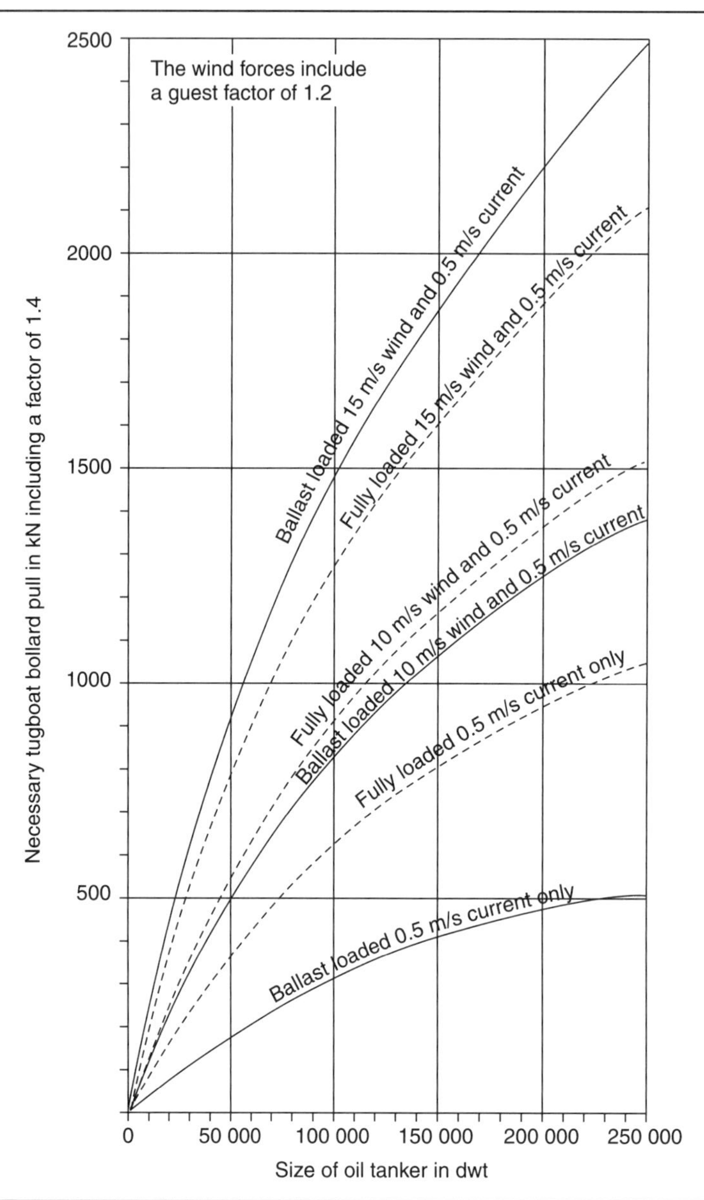

From this example, it should be recommended that at least four tugboats are stationed at the terminal for handling 137 000 m^3 tankers at approximately 13 m/s wind speed, each with approximately 6000 hp and a minimum pulling capacity of 70 t, which would give a total tugboat pulling capacity T_C of 2880 kN. The operational safety factor S_O would then be 1.15.

Figure 4.17 shows the necessary tugboat bollard pull based on British standard BS 6349 (see Chapter 2 for wind and current forces), required to move different sizes of oil tankers in deep water with winds of 10 and 15 m/s, a gust factor of 1.2 and a current of 0.5 m/s, acting crosswise to the oil tanker. For wind

Figure 4.18 Necessary tugboat pull due to a crosswise current acting on a ship as a function of the water depth to ship draft ratio

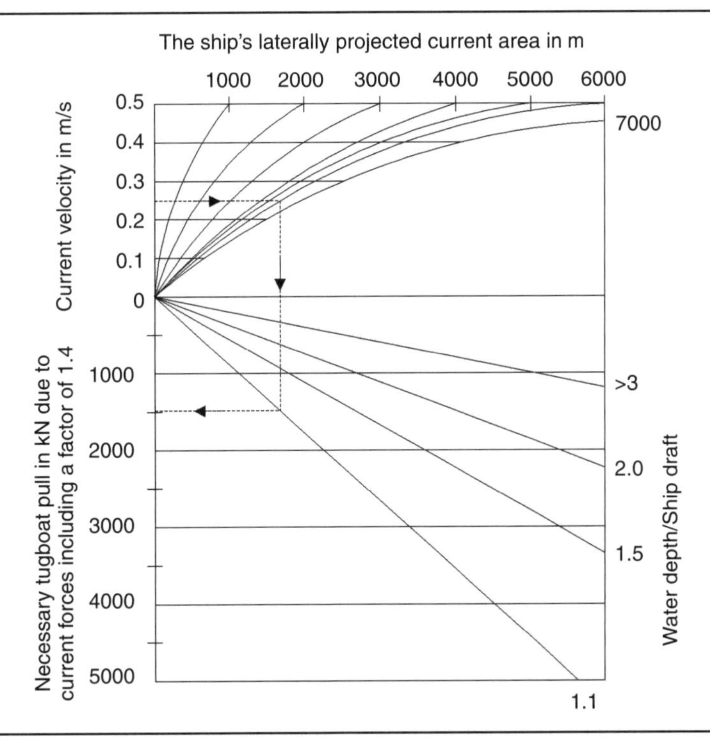

and current areas, see Chapter 21 for ship dimensions. A tugboat bollard pull factor of 1.4 has been used. From the figure it can generally be seen that for a ship in the ballasted condition the wind forces are the dominating forces acting on the ship, while in the loaded condition the current forces are the important forces.

From Figure 4.17 it can also be seen that the necessary tugboat bollard pull in kN is roughly equal to about 0.5–2.5% of the tanker's DWT, including the reserve capacity.

The necessary required tugboat pull in kN due to a crosswise current acting on a ship as a function of the ratio of the water depth to the ship draft, based on the general standard current formula shown in Chapter 2 for current forces, is illustrated in Figure 4.18. A tugboat factor of 1.4 has been used.

Figure 4.19 shows the number of tugboats usually required to assist an oil tanker of 100 000–200 000 DWT while escorting the ship to the harbour basin, turning, berthing, standby during loading or unloading, unberthing and escorting out of the harbour basin in good weather conditions (below Beaufort 5) and with nearly no current. The time or duration of each of these activities will depend on the size of the tugboats, the type and size of the tanker, the layout of the harbour, and the loading or unloading capacity of the piping system.

4.4. Wind and wave restrictions

Generally, the terminal operator and the port captain should have a good understanding of the mooring principle for all the ships using the terminal and the different berths. This includes the

Figure 4.19 Required number of tugboats during different port operations

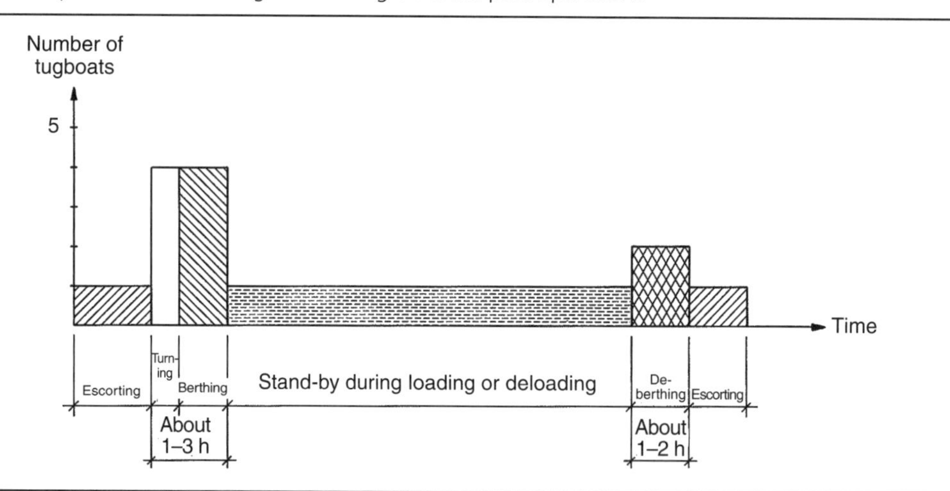

fundamental principles of the design of a mooring system and the loads likely to be expected in the mooring system under different conditions of wind, wave and current, and they should have a clear understanding of the operating limits of the various types of ships and mooring arrangements and systems that may be used at the berth.

The port authority should have the responsibility for implementing the weather policy for the time up until the ship is berthed and for the time when the ship is unmoored to when it is clear outward from the port. During the time at the berth, the terminal will usually implement its weather policy for suspending cargo operations. In the event that any unforecast adverse weather arrives when the ship is alongside, the terminal, the port authority and the ship's master should discuss any safety implications that may arise.

The following limits for wind velocities, as indicated in this chapter, are commonly used in the evaluation of the mooring system, horizontal forces, etc., during berthing and unberthing operations for cruise ships, and for oil and gas tankers (Figures 4.20–4.22).

For port and ship operations it is generally accepted that a wind gust duration shorter than about 1 min will be of secondary importance. Nevertheless, at some gas and oil terminals around the world it has been observed that wind durations down to 20–30 s may affect the tankers in ballast condition.

Therefore, the design wind speed for a moored tanker should be the mean wind speed corresponding to the shortest gust that will affect the tanker at any time, having a return period of at least 100 years and taking account of the height of the tanker and the wind speed/height gradient. In the case of a moored tanker, the gust duration must also be sufficient for the mooring line or fender strains to develop, taking account of the inertia of the tanker.

For oil and gas berths, it is recommended that there is instrumentation for the continuous measurement of the wind velocities in the near vicinity, as indicated in Section 4.2. Guidelines for operating wind limits and mooring arrangements have been developed for all larger terminals to be used by the operators.

Figure 4.20 Oil tankers at Statoil Mongstad Oil Refinery. (Photograph courtesy of Øyvind Hagen, Statoil Mongstad, Norway)

A critical relationship for the manoeuvrability of ships with very high superstructures, such as oil tankers in ballast, containerships and car carriers, appears to be the ratio of the wind speed to the ship speed. At ratios of the wind speed to the ship speed of about 6–7, great difficulty can be expected in controlling these lightly loaded ships or ships with high windage areas. On the other hand, at ratios of about 10, control of even fully loaded ships, such as large fully loaded tankers, will most likely be impossible.

Figure 4.21 Norsk Hydro, Sture Crude Oil Terminal, Norway. (Photograph courtesy of Norsk Hydro, Norway)

Figure 4.22 Snøhvit LNG Terminal, Norway. (Photograph courtesy of Statoil, Norway)

The wind force on a ship will vary with the exposed area of the ship. A beam wind will strike the entire exposed side area of the ship, compared with the relatively small exposed area for the head wind. For a given wind velocity on, for example, a large tanker, the maximum transverse or crosswise wind force is about three to five times as large as the maximum longitudinal wind force.

The values below are based on 10 min average wind velocities (the Beaufort wind scale). The figures will also depend upon the wave and current situation at the berth.

It is very important that all ports and terminals should provide documentation that they can handle ships, taking account of the operational wind speeds, with the tugboat fleet stationed at the terminal.

As a very rough guideline for operation of a berth or terminal, based on experiences around the world, the following operational wind velocities are suggested as limits for cruise ships, ferries, and for oil and gas tankers in ballast condition during the following operations.

4.4.1 Cruise terminals

One should try to minimise cruise ship motions in winds up to 18 m/s or 35 knots to ensure that the gangways remain operationally safe for arriving and departing passengers and cargo transfer. Figure 4.23 shows eight cruise vessels in Port of Bergen, Norway.

4.4.2 Ferry terminals

Due to the large differences between different ferries arising from the hull form and windage area, propulsion system, rudder arrangements, etc., it is difficult to recommend wind limits during berthing and unberthing. For larger ferries, the wind limits for berthing may vary between 15 and 30 m/s, and for unberthing between 12 and 35 m/s, depending on the wind direction.

Figure 4.23 Eight cruise vessels in Port of Bergen, Norway. (Courtesy of Jan. M. Lillebø, Bergen, Norway)

It must be remembered that ferry and cruise terminals are nearly always in locations sheltered from the effects of waves and currents.

4.4.3 Oil and gas terminals

The following can be considered a guide for safe operations, but do not preclude the terminal operator/port authority from carrying out basic due diligence in performing site- and vessel-specific navigation simulations, mooring analyses, risk assessments and obtaining accurate metocean data.

For **oil tankers** and **gas tankers**, the following limits can be recommended, subject to a risk analysis of the local conditions and support infrastructure:

(a) Approximately 10 m/s or 20 knots is the recommended safe wind limit for berthing for vessels with a projected wind area greater than 5000 m^2.
(b) Approximately 15 m/s or 30 knots is the recommended safe wind limit for berthing for vessels with a projected area less than 3000 m^2.
(c) Approximately 15 m/s or 30 knots is the recommended safe wind limit for cargo transfer operations.
(d) Approximately 20 m/s or 40 knots is the recommended safe wind limit for the disconnection of loading arms, subject to confirmation from the manufacturer.
(e) Approximately 22 m/s or 45 knots is the recommended safe mooring wind limit. Vessels should be held out at anchor if higher wind velocities are forecasted. If the vessel is already moored at the berth when such conditions occur unforeseen, extraordinary risk mitigation measures, such as ballasting to reduce the wind area, using additional mooring lines, and using one or more standby tugs (to push the vessel against the berth), etc., are to be taken to reduce the risk of a mooring breakout. Note that it is recommended that loading arms are disconnected before allowing the tug to push the LNG carriers alongside the berth.
(f) At 25 m/s or 50 knots wind velocity, the tanker should, if possible, normally leave the berth for open sea.

(g) At wind speeds above 30 m/s, the line tension can exceed 60–65% minimum breaking load (MBL), and the winch brakes will start to render. The tanker will be in a potentially dangerous situation.

Note: *These limits are applicable for sheltered berths (significant wave height less than 0.6 m). Lower wind limits apply at exposed berths, and can be based on dynamic mooring analyses to provide a higher degree of confidence in the mooring integrity.*

The forces generated by waves that strike moored vessels beam-on or on the bow or the quarter need to be carefully analysed in terms of fender capacity. The period of the wind wave will also have a significant effect on the forces created on the fenders by movement of the ship on the berth.

The analysis of these forces will determine the wave height conditions and the time period that will provide limiting conditions to the operation.

At some large new oil terminals, due to environmental and safety requirements, the acceptable wind limit during berthing is a maximum of 10 m/s for all tanker sizes. The operational limit for the loading arms during loading and unloading is 15 m/s, and the loading arms are disconnected at a wind speed of 20 m/s.

At oil terminals where double moorings are used, where one smaller tanker is moored outside a larger tanker at a jetty, the operation will normally stop, and disconnection between the tankers will take place if the wind exceeds 16 m/s. If prognoses for the wind will exceed 20 m/s, the smaller tanker will depart for open sea from the terminal. Figure 4.24 illustrates the double mooring of two oil tankers.

The rough operational guidelines for oil tankers, as described above, are shown in Figure 4.25.

Figure 4.26 shows an example of the type of information that OCIMF recommends should be valuable to the berth operator for mooring 250 000 DWT tankers. The mooring arrangement with 18 nylon lines shows the maximum freeboard for this type of tanker (or minimum draft) and that the loading arms should be disconnected at a wind speed of approximately 30 knots peak gust. As the draft increases, the permissible wind speed can increase to 50 knots peak gust. The ship motions, in this case, can govern the loading arm limits.

It must also be remembered that gas tankers usually have greater freeboard than oil tankers, as shown in Figure 4.27 for an LNG spherical gas tanker, and hence will be more affected by the wind. The maximum acceptable wind velocities are also usually a little higher for smaller ships.

For LNG tankers, the membrane LNG tankers have smaller dimensions than the spherical LNG tankers of similar capacities, which is of great importance in the determination of wind force effects. At some terminals for approximately 138 000 m^3 of LNG, the wind speed limit for berthing is 14 m/s for flat deck or membrane tankers and 12 m/s for spherical tankers.

The environmental operating limits should always be established for each berth and should be detailed on the Ship/Shore Safety Checklist. The tanker's staff should be advised of any limitations on the tanker movement due to the operating envelopes of the shore equipment such as loading arms, hoses, fender compression limits, etc., and the actions to be taken if these should be reached.

Figure 4.24 Double mooring of two oil tankers. (Courtesy of Dr.techn. Olav Olsen, Norway)

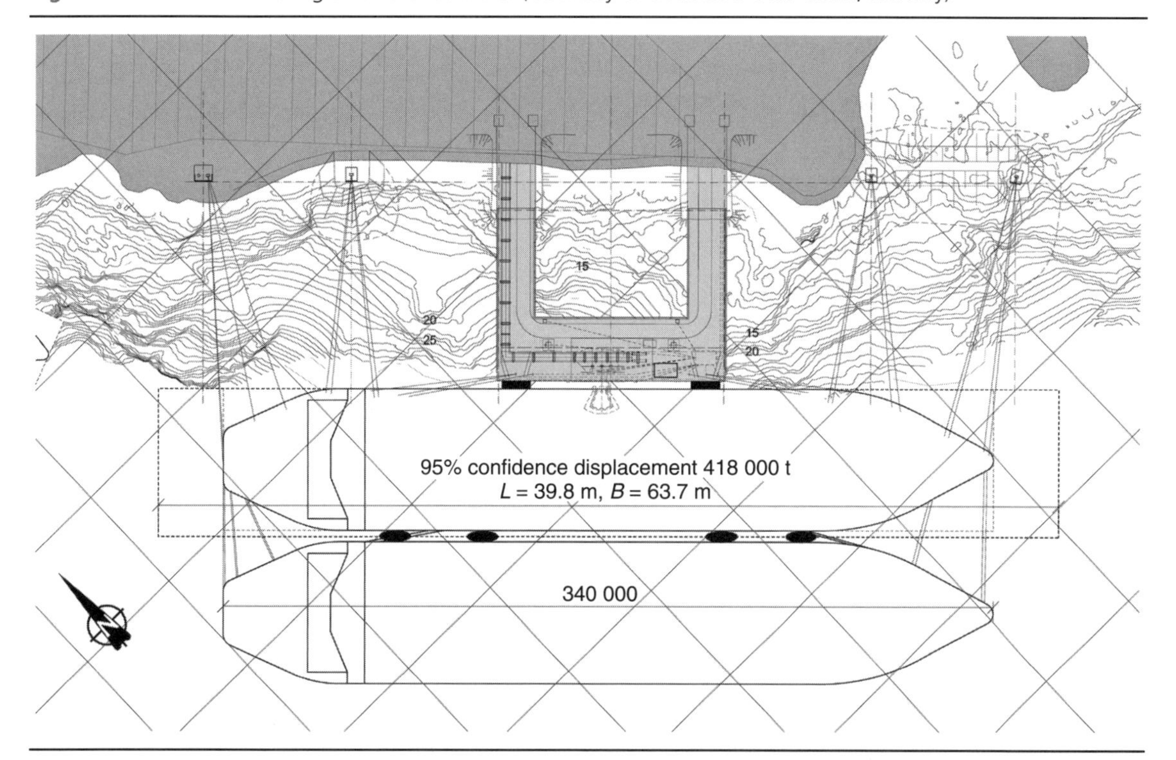

4.4.4 Cargo and container terminals

For comparison, it is generally recommended that equipment such as heavy lifting equipment for cargo and containers, loading towers, etc., will not operate in a wind stronger than about 20 m/s, and that the wind can blow horizontally in all directions. Therefore, for wind above a certain wind speed level, the operation of the cranes, etc., may be impossible. This need not necessarily affect the total loading and unloading operations of the ship, if these operations can be done by trucks, etc., and provided that the wind has not given the ship itself unacceptable ship movements.

4.4.5 Operating limits

Tugboats and mooring boats

It is important that limiting operating conditions for tug operations are determined well in advance of the requirements for tugs. These limiting conditions are included in tenders bids for new or used tugs in ports or terminals. The required environmental conditions in which the tugs are expected to operate must be written into contracts. Tugboats start to lose power efficiency for the control of ships immediately there is any significant wave height, as was noted earlier in Section 4.3.

It should be recognised that, due to wind generated or short periodic waves, tugboats will have operational limits. With significant wave heights of more than 1.0–1.5 m for ordinary tugboats and approximately 1.5 m for tractor tugboats, tugboats start to lose their efficiency for controlling ships.

For modern mooring boats or launches, a wind speed of about 12–15 m/s or a significant wave height of 1.0–1.3 m must be taken as a guideline limit for safe operation. If these limits are exceeded, the

Figure 4.25 Approximate operational guidelines for tankers

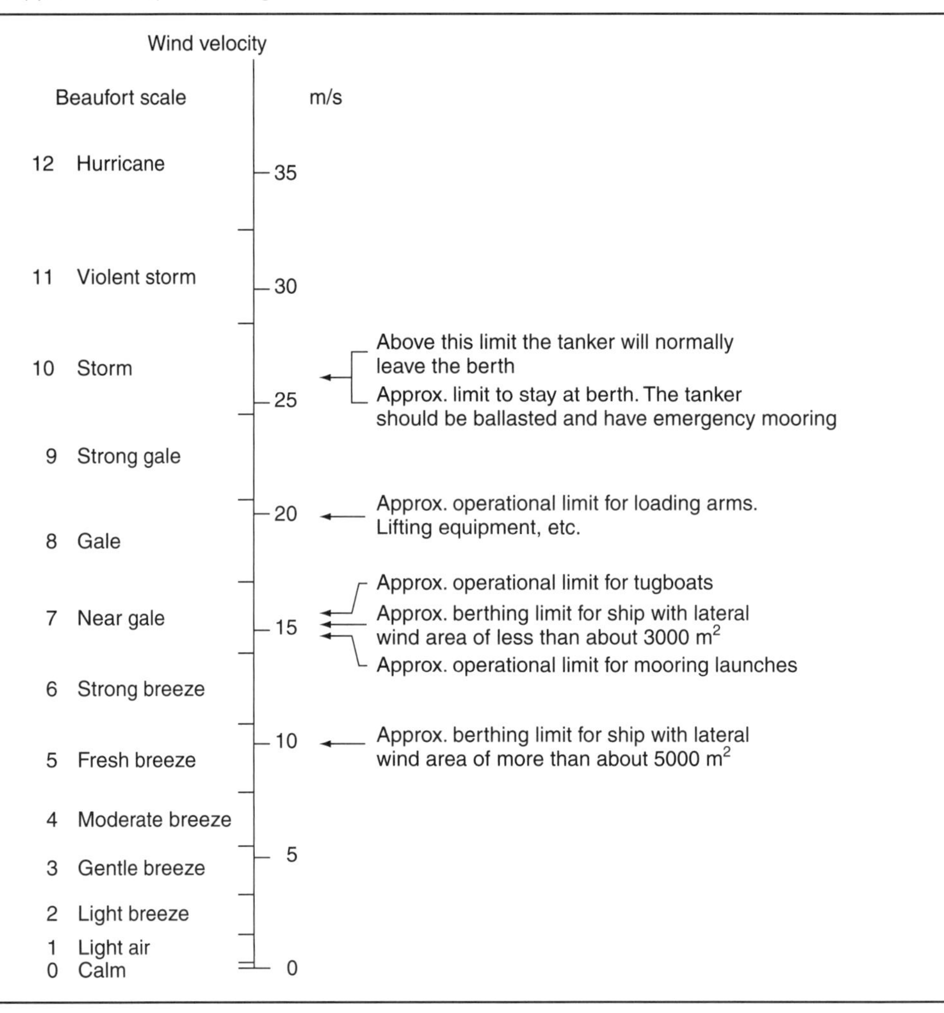

mooring boats will experience difficulty in delivering the mooring lines from the ship to the mooring points at the berth. An assessment of the mooring boat design and operational characteristics against the local conditions and environment may allow deviations from the guideline wave height.

Pilots

The number of pilots used at ports and terminals is normally based on an assessment by the national authority, harbour master and pilots, taking into account the length of pilotage, the complexity, the number of tugs and the berthing manoeuvre. Optional parameters for consideration can be wind, current, sea and swell. An additional consideration should be made with respect to the age profile of the pilot group and the recruitment and development of new pilots. The introduction of new pilots can take up to 2 years to reach full competency.

Where there is more than one pilot on board, then it is most important to ensure that the ship's captain knows who is taking the lead role. Where delegation of duties is made between pilots, this must be clearly communicated.

Figure 4.26 Example of operating wind limits for a 250 000 DWT tanker moored with nylon lines

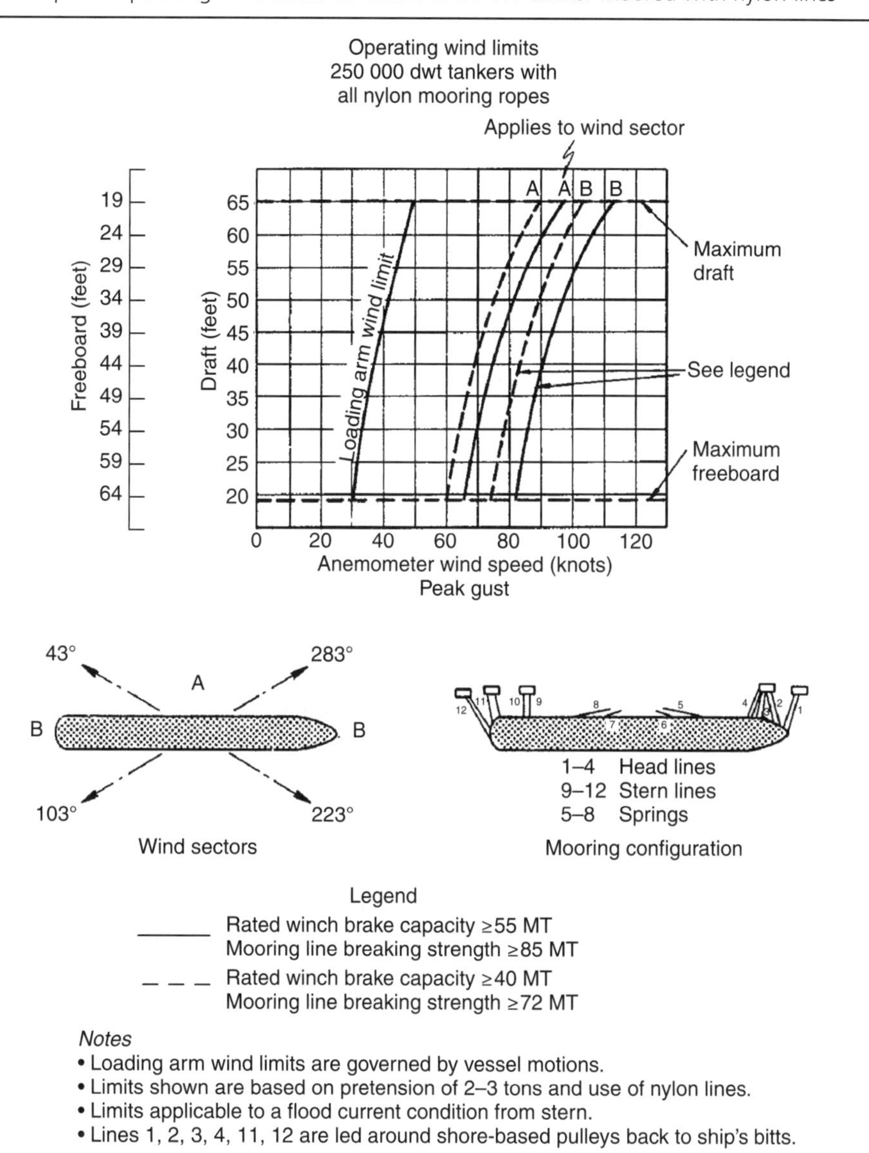

It is recommended that IMO Resolution A.960, *Recommendation on Training and Certification and Operational Procedures for Maritime Pilots* (IMO, 2003), is followed to ensure that competencies are maintained to enhance the reputation of the port.

4.5. Ship movements

It is very difficult to predict the acceptable movements for the different types of ships. It can be done with varying degrees of accuracy and reliability by mathematical models and analytical methods, but the most reliable method of predicting a ship's response under wave action is to build and test a physical model. In this section, some recommendations for acceptable movements are given.

Figure 4.27 A spherical LNG gas tanker. (Photograph courtesy of Høegh, Norway)

Under the impact of current, waves and wind gusts a moored ship is in continuous movement. The magnitude of the movement varies over a wide range, and depends on the magnitude and direction of the waves and wind. Even the best mooring systems will not be able to stop the ship from moving due to waves and wind. The six main components of ship movements are shown in Figure 4.28.

Usually, the overall movement of a ship will be a combination of more than one of the six individual movements shown in Figure 4.28. These six movements can, strictly speaking, be subdivided into two main types of movements or oscillations:

(a) Movements in the horizontal plane: surge, sway and yaw. These movements are related to the forces in the fenders and mooring systems that tend to counteract the movements or displacements of the ship from its equilibrium position. It is the virtual mass of the ship in relation to the fenders and mooring system that governs the natural period of a moored ship.
(b) Movements in the vertical plane: roll, pitch and heave. These are the natural movements of a free-floating ship, and are almost unaffected by the fenders and mooring systems.

Figure 4.28 Types of ship's movements

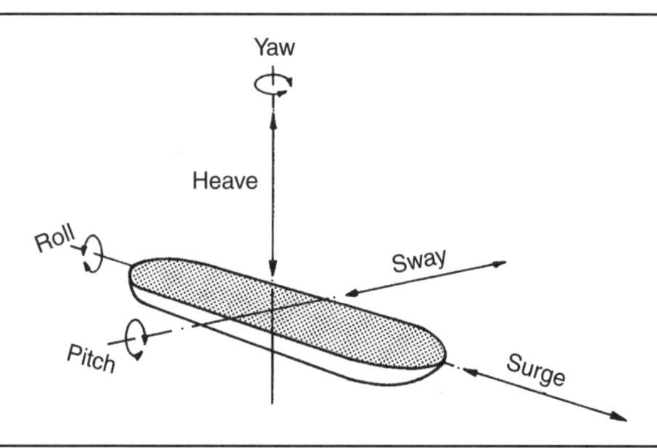

Figure 4.29 Wave directions

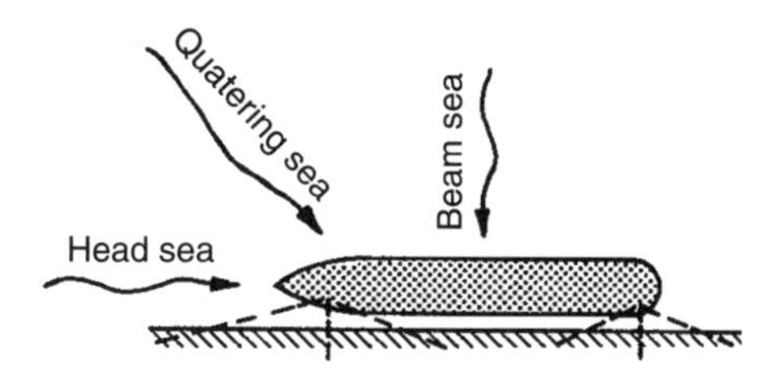

The ship at berth can be exposed to the wave directions shown in Figure 4.29, and the combination of long and short waves shown in Figures 4.30 and 4.31. Waves with short periods will typically only affect small ships, while long-period waves will generally affect all ships.

The wave system is mainly responsible for the unacceptable ship movements and forces in mooring systems. In harbours, for fishing boats or small ships, the shorter periodic waves (less than about 6–8 s) normally determine the berthing and acceptable movement conditions for the ship. For larger ships, it is the longer periodic waves (above 20 s) with a wavelength of about 5000–8000 m and a wave slope of 1 in 2000 to about 1 in 3000 that can cause serious movements and forces in the mooring systems. The reason for this is the risk of the resonance of the long periodic waves having periods of the same magnitude as the natural periods of large moored ships. The ship movement can therefore increase significantly if the periods of the driving forces are in the same range as the natural periods of the moored ship.

Figure 4.30 Combinations of waves

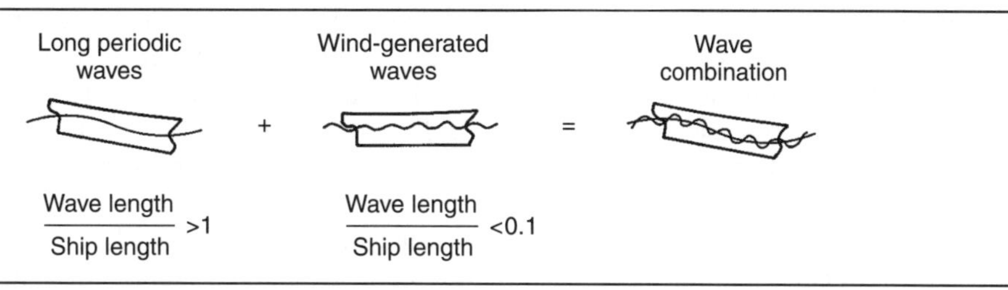

Figure 4.31 Waves affecting different sizes of ships

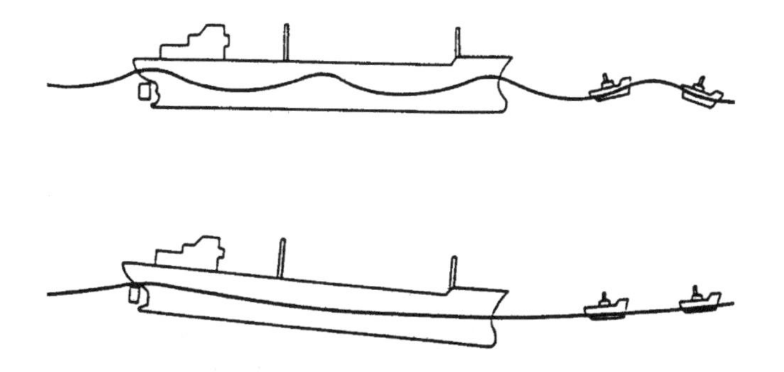

The physics of a moored ship is a complex system because the stiffness characteristics are not symmetrical since the fenders are usually stiffer than the moorings, and the deflection characteristics of the fenders and mooring lines are not linear. In addition to the moored ship itself, the wind and wave forces can be time-varying and irregular. For a traditional mooring system, a typical natural period for a large ship would be 1 min or more. Therefore, for the purpose of estimation of the port operation rate based on acceptable ship movements, it is necessary to know the amount of ship movements against the expected wind, wave and current conditions for a ship moored to any kind of mooring system. For the estimation of the ship movements, a numerical simulation method or a model test will be the best solution.

Wind forces normal to a moored ship may either press the ship against the fenders or push it away from the fenders. These situations are important to the free movements of a ship at berth because the ship movements are generally larger when the ship is not in contact with the fender system.

For a ship moored at a berth with a strong current, either parallel or at an angle to the ship, the current may create long periodic ship movements. These periodic movements will depend on the stiffness of the fender and mooring system, the inertia of the ship around the centre of gravity of the virtual mass of the ship, the moment of the respective forces around the centre of gravity and the current velocity. These movements occur when the moored ship is dynamically unstable to lateral displacements and when the current velocity exceeds a velocity of around 1 m/s.

For ships moored at an offshore terminal, such as a single-buoy mooring, the movements in the vertical plane, roll, pitch and heave caused by ocean waves will be the critical movements for the terminal's operations.

One of the main criteria for the evaluation of a good harbour is whether the ship has sufficiently calm conditions when moored at the berth. It is therefore important to establish realistic criteria for acceptable wave heights and movements at the berth. The acceptable wave heights and movements of a ship at a berth will depend on the elastic properties of the fenders and the mooring system of the ship, the type of ship, the methods used for loading and unloading, the orientation of the berth with respect to the current and wave directions, the wave period and the ship's natural period of oscillation.

The combination of all the physical factors that affect ship movements when at berth are difficult to handle mathematically. Another complicating factor is the human factor, for example the captain's experience, evaluation of a given critical situation, willingness to take risks, etc. Therefore, the most reliable method of predicting a ship's response to the wave action is to build a physical model of the harbour and the berth. In a physical model, the many interactions can automatically be built into the model. For example, current flowing against the waves tends to make the waves steeper and sometimes break, whereas current flowing with the waves tends to reduce the wave height.

As a general guide, the maximum significant wave heights H_s (in metres) for a head sea (Table 4.3) have generally been assumed as acceptable for ships at berth. The table shows that the generally acceptable wave height increases with increasing ship size. The acceptable wave height is highest for a head sea and lowest for a beam sea. The figures can generally be accepted for wave periods up to 10 s. For longer wave periods the figures must be reduced.

In an article in *PIANC Bulletin*, No. 56, Velsink (1986) of the Netherlands published (Table 4.4) maximum significant wave heights H_s for different wave directions before loading and unloading

Table 4.3 Maximum significant wave heights

Ship at berth	H_s at berth
Marinas	0.15
Fishing boats	0.40
General cargo (<30 000 DWT)	0.70
Bulk cargo (<30 000 DWT)	0.80
Bulk cargo (30 000–100 000 DWT)	0.80–1.50
Oil tankers (<30 000 DWT)	1.00
Oil tankers (30 000–150 000 DWT)	1.00–1.70
Passenger ships	0.70

operations have to be stopped. The values refer to the heights of residual deep-water waves with periods in the range of about 7–12 s.

Locally wind-generated waves inside a harbour will generally have shorter periods, and will therefore have relatively little effect on a moored ship. Waves with very long periods, for example seiches, can have disastrous effects at much lower wave heights than indicated in Table 4.4.

The figures given in Tables 4.3 and 4.4 express the wave criteria in terms of maximum acceptable wave heights at berth. A more realistic criterion would, in most cases, be an expression of the maximum tolerable movement of the ship itself relative to the berth that the mooring system and the cargo handling equipment can tolerate.

The figures for the maximum acceptable significant wave heights must, therefore, always be checked against the ship movement that can be accepted by the loading and unloading systems. A movement of up to ±2.00 m along the berth is usually acceptable for oil tankers if allowed for in the design of the loading system, but, for the same type of movement along the berth (surge), only ±0.20 m can be accepted for gas tankers. Among the most sensitive berths are those for gas tankers, lo/lo container ships and ro/ro ships, due to the safety requirements for the loading and unloading operations.

Table 4.4 Maximum significant wave heights for different wave directions

	Limiting wave height H_S: m	
Type of ship	0° (head-on or stern-on)	45–90°
General cargo	1.0	0.8
Container, ro/ro ship	0.5	
Dry bulk 30 000–100 000 DWT loading	1.5	1.0
Dry bulk 30 000–100 000 DWT unloading	1.0	0.8–1.0
Tankers 30 000 DWT	1.5	
Tankers 30 000–200 000 DWT	1.5–2.5	1.0–1.2
Tankers >200 000 DWT	2.5–3.0	1.0–1.5

Figure 4.32 The relative importance between the different ship movements

	Movement in: Horizontal plan			Vertical plan		
	Surge	Sway	Yaw	Roll	Pitch	Heave
VLCC	••	••	••	•	•	•
Cool bulk	•••	••	••	•	•	•
Ore bulk	•••	••	••	•	•	•
Grain bulk	•••	••	••	•	•	•
Supply	•••	••	••	••	••	••
General cargo	•••	••	••	••	••	••
LPG	•••	•••	•••	••	••	••
LNG	•••	•••	•••	••	••	••
Lo/lo	•••	•••	•••	•••	••	••
Ro/ro	•••	•••	•••	•••	•••	••

• = Less important • • • = Most important

Figure 4.32 shows the relative importance of the linear and angular movements of the ship for the safety and efficiency of operations at berth. As shown, the most dangerous movements are the movements in the horizontal plane (surge, sway and yaw), which could break the ship loose from the berth. From a safety point of view, these movements should therefore be minimised, and the possibility of the build-up of a resonant system must be avoided. Some of these movements are more dangerous to safety than others, due to the different types of ships: for example, pitch is dangerous to the operation of ro/ro ships, while it barely affects the operation of a VLCC ship.

The ship behaviour at berth should always be considered from an operational point of view for this reason. For all ro/ro operations, where trucks are moving over ramps in either the bow or the stern, any movement beyond a couple of centimetres can be of danger to the safety of the operations. It may, therefore, generally be concluded that horizontal movements – surge, sway and yaw – are almost equally dangerous to safety, but that surge is usually the most damaging to operational efficiency.

Therefore, in order to improve the safety and the efficiency of the operations at berth, the mooring and fender system must be designed so that the ship can lie as still as possible while at berth, to avoid the build-up of inertia in the system. The fender system should not only be as energy-absorbing and non-recoiling (high hysteresis effect) as possible but, together with the mooring system, provide a damping effect on all movements, and especially by friction between the fender and the ship against surge movements. The ideal system is the one which, by proper tension in pertinent mooring lines, makes the ship rest against the fenders and limits all movements to the smallest possible. This system requires very strong and safe breast lines. The line material may be steel or synthetic, but if fenders are of a recoiling type, ropes have to be as rigid as possible. That means combinations of synthetic ropes and non-recoiling fenders are ideal.

The term '**acceptable ship movement**' can be more precisely specified in accordance with the following three main aspects:

(a) Safety limits. If the ship's movement exceeds a certain value, this will or can result in damage to the ship, moorings, port installations, etc. This limit is usually specified by an upper limit in the mooring system: mooring lines or mooring winches. The degree of ship movement allowed will be a function of the stiffness in the mooring system due to the fact that a soft system will allow more movement than a stiff system before the mooring loads reach their safety limit.

(b) Operational limits of ship movement. There is a limit to ship movement beyond which the operations of loading and unloading of the cargo can no longer be efficiently or safely performed. The amount of ship movement allowed will depend upon the type of ship, for example an oil tanker, general cargo ship or container ship. Generally, the larger the ship the less it will respond in horizontal movements – surge, sway and yaw – to the primary wave system. The amount of vertical movement limit will, for upward movement, be a minimum when the ship is in ballast at high tide, while the limit to downward movement will be a minimum when the ship is loaded at low tide.
In the case of container ships, sudden horizontal movements of 1.0 m can significantly lower the rate of loading and unloading. For oil tankers, practical limiting movements would be about $\pm$2.0 m for surge and 1.0 m for sway off the fender. For loading and unloading of gas tankers, the low temperature of the gas can cause ice to form by condensation at the joints of the loading arms, and also, because gas is considered to be more dangerous than oil, the movements considered acceptable are lower than those for oil tankers. In the case of ro/ro ships, there are quite stringent limits on the amount of movement allowed.

(c) Different opinions of the harbour authorities and ship operators. These opinions may be expected to vary considerably, but it appears that surge, yaw and roll are considered to be the most dangerous movements in head, quartering and beam seas, respectively. This is perhaps due to the fact that hydrodynamic damping is less for these three movements compared with sway, pitch and heave. It is generally felt that for ships above 40 000 t displacement, the surge should not exceed 1.0 m and the yaw and roll movements should not exceed 0.5 m. For smaller ships, it appears that slightly larger movements of up to 1.0 m are considered acceptable for surge and yaw, while roll movements of up to 0.7 m are thought tolerable. These figures take no account of the speed of the ship movement, so that an oscillatory movement of 1.0 m completed in 10 s will appear far more alarming than the same movement completed in, say, 1 min. The acceptable amount of long-period surge, sway and yaw movements of large ships may therefore be in excess of the figures given above.

In general, the acceptable movements of the ship due to safety-limit criteria will exceed the values due to the limiting degree of ship movement criteria, which, according to the very little published information, again will exceed the values given by the opinions of the harbour authorities and the ship operator's criteria.

PIANC Working Group 24 (PIANC, 1995) gives recommendations, reproduced here as Table 4.5, which more or less agree with the research carried out by a joint Nordic group for the different types of ships.

Table 4.5 Recommended motion criteria for safe working conditions

Ship type	Cargo-handling equipment	Surge: m	Sway: m	Heave: m	Yaw: °	Pitch: °	Roll: °
Fishing vessels	Elevator crane	0.15	0.15	–	–	–	–
	Lift on/lift off	1.0	1.0	0.4	3	3	3
	Suction pump	2.0	1.0	–	–	–	–
Freighters, coasters	Ship's gear	1.0	1.2	0.6	1	1	2
	Quarry cranes	1.0	1.2	0.8	2	1	3
Ferries, ro/ro	Side ramp[a]	0.6	0.6	0.6	1	1	2
	Dew/storm ramp	0.8	0.6	0.8	1	1	4
	Linkspan	0.4	0.6	0.8	3	2	4
	Rail ramp	0.1	0.1	0.4	–	1	1
General cargo	–	2.0	1.5	1.0	3	2	5
Container vessels	100% efficiency	1.0	0.6	0.8	1	1	3
	50% efficiency	2.0	1.2	1.2	1.5	2	6
Bulk carriers	Cranes	2.0	1.0	1.0	2	2	6
	Elevator/bucket-wheel	1.0	0.5	1.0	2	2	2
	Conveyor belt	5.0	2.5	–	3	–	–
Oil tankers	Loading arms	3.03	3.0	–	–	–	–
Gas tankers	Loading arms	2.0	2.0	–	2	2	2

[a]Ramps equipped with rollers
Motions refer to peak–peak values (except for sway: zero peak)
For exposed locations, 5.0 m (regular loading arms allow large movements)

The purpose of Table 4.5 is to quantify and optimise how often different types of ship will be able to use the berth safely. This will be an important input to the evaluation of the operational availability of the berth.

D'Hondt (1999) provides very interesting comments on the PIANC report for acceptable movements for container ships during loading and unloading due to the small tolerances in the cell guide location in a container ship. Based on the cell tolerances, he advises that the maximum container ship movements to allow unimpeded cargo handling should be as follows:

(a) Pitching: 0.4° with respect to the horizontal plane.
(b) Rolling: 0.24° with respect to the horizontal plane.
(c) Combined pitching and rolling: 0.45° with respect to the horizontal plane.
(d) Heaving: a maximum amplitude of 20 cm with respect to the point of rest, and a maximum speed of 7.5 cm/s in the most affected cell fore or aft. Most movement calculations relate to the centre of gravity.

Ranges for maximum allowable sudden movements for different types of ships larger than 200 m at berth during loading operations for wave periods between 60 and 120 s are shown in Table 4.6.

Table 4.6 Ranges for maximum allowable sudden movements

Type of ship	Surge: m	Sway: m	Heave: m	Yaw: m
Oil tanker	±2.0	+0.5	+0.5	1.0
Ore bulk (crane unloading)	±1.5	+1.0	±0.5	0.5
LNG tanker	±0.2	+0.1	±0.1	0.5
Container	±0.5	+0.3	±0.3	0.5
Ro/ro (side)	±0.3	+0.2	±0.1	0.2
Ro/ro (bow or stern)	±0.1	0.0	±0.1	0.2

Investigations show that the acceptable significant wave heights increase for increasing ship size and that the acceptable ship movement relative to the berth decreases for increasing ship size (Figure 4.33).

A joint Nordic group involving Denmark, Finland, the Faroe Islands, Iceland, Sweden and Norway published the report *Criteria for Ship Movements in Harbours* (Jensen *et al.*, 1990). The purpose of this project was to determine criteria for acceptable movements of moored ships in a harbour under working and safe mooring conditions. The project primarily concentrated on the assessment of criteria for fishing boats, coasters, container ships and ferries.

For the working conditions during loading and unloading operations, the Nordic group has suggested that the maximum ship movements for working conditions should not be higher than shown in Table 4.7. The figures assume that the occurrence frequency of critical ship movements for fishing boats, coasters and container ships should be less than 1 week/year (2% of the time), and for ferries less than 3 h/year (0.3% of the time).

For safe mooring conditions only, the Nordic group suggested that the maximum ship movements should not be higher than shown in Table 4.8. The figures assume that the ships are reasonably well moored and that the berth structures are well equipped with fenders. For the berth to be acceptable, the frequency of these movements should be less than 3 h/year (0.3% of the time).

Figure 4.33 Acceptable wave heights and ship movements at berth

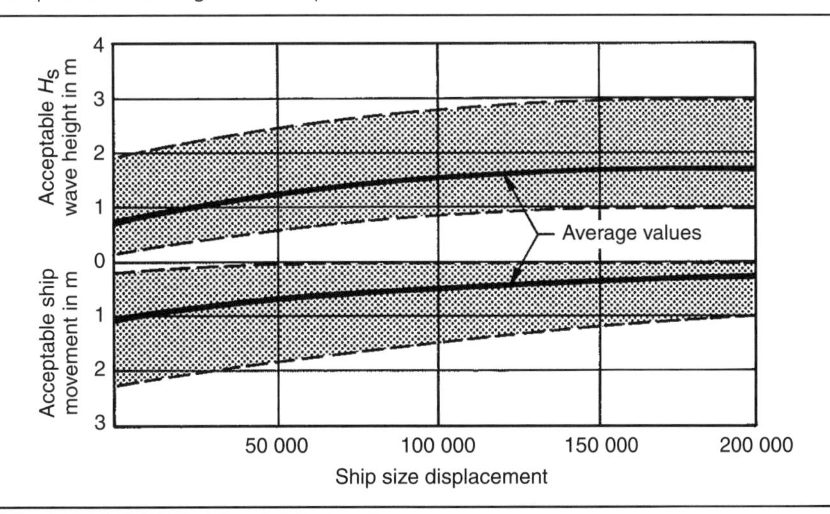

Table 4.7 Limiting criteria for ship movements under working conditions

Type of ship	Surge: m	Sway: m	Heave: m	Yaw: °	Pitch: °	Roll: °
Fishing boats						
L_{OA} = 25–60 m						
Lo/lo	±0.75	+1.5	±0.3	3–5	4	3–5
Elevator crane	±0.08	+0.15	–	–	–	1.5
Suction pump	±1.5	–	–	–	–	–
Coasters						
L_{OA} = 60–150 m						
Ship crane	±1.0	+1.5	±0.5	1–3	1–2	2–3
Berth crane	±1.0	+1.5	±0.6	2–4	1–2	3–5
Container ships						
L_{OA} = 100–200 m						
90–100% efficiency	±0.5	+0.8	±0.45	0.5	1.5	3
50% efficiency	±1.0	+2.0	±0.6	±0.6	2.5	6
Ferries						
L_{OA} = 100–150 m	–	+0.8	±0.5	1	1	–

The transhipment operations for gas tankers, containers and ro/ro cargo are very sensitive to ship movements, and can lead to considerable downtime, and therefore the location of the terminal should be chosen with especial care.

In the case when the ship movements, in the captain's judgement, exceed the safety limit for the ship to stay at the berth, it is important that the ship has time to leave the berth and the possibility of doing so. If the assistance of tugboats is required in the unberthing manoeuvre of a ship, the operational conditions for the tugboats must not prevent them from doing their job satisfactorily. This could be the case when the operational limit for the safe operation of the tugboats is reached before the safety limit, or the limiting degree of ship movements, is reached.

Table 4.8 Limiting criteria for ship movements under safe mooring conditions

Type of ship	Surge: m	Sway: m	Heave: m	Yaw: °	Pitch: °	Roll: °
Fishing boats						
L_{OA} = 25–60 m	±0.75	+2.0	±0.5	6	4	8
Coasters						
L_{OA} = 60–120 m	±1.0	+2.0	±0.75	3–5	2–3	6
Velocity						
Size of ship	m/s		m/s	°/s		°/s
About 1000 DWT	0.6	–	0.6	2.0	–	2.0
About 2000 DWT	0.4	–	0.4	1.5	–	1.5
About 5000 DWT	0.3	–	0.3	1.0	–	1.0

4.6. Passing ships

It is important to determine the impact of the speed of passing ships on a moored ship. Excessive surge of a ship on the berth can cause damage to fenders, loading arms and moorings.

Studies can be carried out to determine these limiting conditions. As guidance, it is generally accepted that the speed of the passing ship, if 50 m off, should not be above 2 m/s and, if 100 m off, should not be more than 2.5 m/s.

Where possible, these studies should be confirmed by carrying out trials once the terminal is commissioned.

4.7. Visibility

The types of weather condition that can cause bad visibility are fog, heavy rain and snow. Fog is defined by some standards as a weather condition in which the visibility is less than 1000 m. The combination of heavy snow or rain together with strong wind is considered more difficult for berthing operations than fog, which usually appears in calm weather when the ship is easier to handle.

In general, visibility between 500 and 1000 m can be acceptable for the manoeuvring and berthing process inside a harbour. If the visibility is less than 2000 m, the ship's velocity should be reduced to 6 knots or less for ship sizes above 10 000 DWT. For visibility less than 1000 m, it is advisable for safety reasons that all larger ships and oil and gas tankers should have tugboat assistance in restricted areas such as main traffic channels, inner harbours, oil terminals, etc.

Regulations in some of the larger ports state that no ship should start manoeuvring within the port area if the visibility is less than three times the ship's overall length. For tugboats with a tow, the ship's length includes the length of the entire tow.

As a general rule, most oil and gas terminals will close for arrival and berthing or unberthing and departure of tankers if the visibility is less than 1000–2000 m.

4.8. Port regulations

Generally, it is the ship's captain who is responsible for the manoeuvring of the ship to and from the berth and the mooring of the ship at berth. The port captain and staff are responsible for the berth structure itself, the berth and mooring equipment, and that the ship's captain follows all the port regulations. If the ship's captain does not comply with the port safety requirements or the port regulations, the port captain has the right to stop all operations and order the ship off the berth for appropriate actions to be taken by the ship's owners and charterer concerned.

4.9. Availability of a berth

Nearly all the items discussed in this chapter will together lead to the total availability or, conversely, the downtime of the berth, which can be subdivided into the following two cases:

(a) **Navigational availability**: this describes the percentage of time that the ship is able to call at the harbour or berth safely from the open sea or ocean.
(b) **Operational availability**: this describes the percentage of operational time during which the ship can operate by loading and unloading at the berth.

Table 4.9 Yearly estimated availability

Non-availability due to	Frequency of estimated percentage downtime
Navigation	
Ice problems	–
Excessive current	–
Wind above 10 m/s, which will stop the manoeuvring of the tanker	4.5
Waves above 1.5 m, which will stop tugboat assistance	0.2
Swell and long-period waves	–
Visibility less than 1000 m	0.2
Tugboat non-availability	0.05
Operation	
Stop in loading operation due to wind above 20 m/s	0.2
Excessive ship movements at berth	0.1
Maintenance on berth structure	0.5
Average estimated percentage downtime	5.75

The availability should not only give the overall availability of the berth per year but also the availability of the berth for each month. Dependent on the type of berth, the total yearly navigational and operational average availability should not be less than about 90–95% due to the extra cost of waiting time for the ships to call at the port for example, or the **downtime** should not be more than about 5–10%.

The designer should always try to evaluate the downtime of a berth due to the navigational and operational availability. The acceptable figures for the downtime for a berth would vary according to the type of cargo handled. For an oil and gas berth, an approximate limit of 10% downtime on an annual basis is the norm. The yearly preliminary estimated availability is illustrated in Table 4.9 for a possible harbour location for oil tankers of 150 000–300 000 DWT.

In general, the downtime calculation of a berth should involve the determination of critical wind, wave and current conditions that could cause unacceptable ship motions or mooring line loads, and/or an inability to operate (e.g. because of the berth crane equipment).

These critical conditions would define the boundary between the acceptable operations and the availability or downtime of a berth.

The estimated accuracy for this type of availability or downtime evaluation is about ±1–3%. It has not been taken into consideration in the example in Table 4.9 that some of the non-availability factors can act together and therefore reduce the sum of the yearly estimated percentage of downtime obstacles.

The yearly berth availability should generally be approximately 95%, and not less than approximately 85% for any particular month of the year. However, this will depend on the type of traffic and the importance of the cargo traffic.

REFERENCES AND FURTHER READING

Bratteland E (1981) *Lecture Notes on Port Planning and Engineering*. Norwegian Institute of Technology, Tapir, Trondheim.

Bruun P (1983) Mooring and fendering rational principles in design. *Proceedings of the 8th International Harbour Congress*, Antwerp.

Bruun P (1990) *Port Engineering*, vols 1 and 2. Gulf Publishing, Houston, TX.

BSI (British Standards Institution) (1988) BS 6349-2: 1988. Maritime structures. Design of quay walls, jetties and dolphins. BSI, London.

BSI (1994) BS 6349-4: 1994. Code of practice for design of fendering and mooring systems. BSI, London.

BSI (2000) BS 6349-1: 2000. Maritime structures. Code of practice for general criteria. BSI, London.

Danish Hydraulic Institute (1983) *Danish Hydraulics*, No. 4. Danish Hydraulic Institute, Copenhagen.

D'Hondt IrE (1999) Port and terminal construction. Design rules and practical experience. *Proceedings of the 12th International Harbour Congress*, Antwerp.

EAU (Empfehlungen des Arbeitsausschusses für Ufereinfassungen) (2004) *Recommendations of the Committee for Waterfront Structures, Harbours and Waterways*, 8th English edn. Ernst, Berlin.

Happonen K, Sassi J, Rytkonen J and Nissinen H (2002) Defining wind limits in ports and approaching fairways. *30th PIANC-AIPCN Congress*, Sydney.

Harbour and Marine Engineering (2002) *Marine Mooring and Docking Systems. Design Manual.* Harbour and Marine Engineering, Melbourne.

IMO (International Maritime Organisation) (2003) *Resolution A.960. Recommendation on Training and Certification and Operational Procedures for Marine Pilots*. IMO, London.

Jensen OJ, Viggosson G, Thomsen J, Bjørdal S and Lundgren J (1990) Criteria for ship movements in harbours. *Proceedings of the International Conference on Coastal Engineering*, Venice.

Marine Board (1985) An assessment of the issues. In *Dredging Coastal Ports*. National Academy Press, Washington, DC.

Ministerio de Obras Públicas y Transportes (1990) ROM 0.2-90. Maritime works recommendations. Actions in the design of maritime and harbour works. Ministerio de Obras Públicas y Transportes, Madrid.

NATO Advanced Study Institutes (1987) *Advances in Berthing and Mooring of Ships and Offshore Structures*. NATO Advanced Study Institutes, Trondheim.

OCIMF (Oil Companies International Marine Forum) (1997) *Mooring Equipment Guidelines*, 2nd edn. Witherby, London.

OCIMF (2008) *Mooring Equipment Guidelines*, 3rd edn. Witherby, London.

OCIMF/SIGTTO (Society of International Gas Tanker Terminal Operators) (1995) *Prediction of Wind Loads on Large Liquefied Gas Carriers*. Witherby, London.

PIANC (International Navigation Association) (1985) *Underkeel Clearance for Large Ships in Maritime Fairways with Hard Bottom*. PIANC, Brussels.

PIANC (1995) *Criteria for Movements of Moored Ships in Harbours. Report of Working Group No. 24*. PIANC, Brussels.

PIANC (2002) *Mooring Systems for Recreational Craft. Report of Working Group 10*. PIANC, Brussels.

PIANC (2010) *The Safety Aspect Affecting the Berthing Operations of Tankers to Oil and Gas Terminals. Working Group 55*. PIANC, Brussels.

Port and Harbour Research Institute (1999) *Technical Standards for Port and Harbour Facilities in Japan*. Port and Harbour Research Institute, Ministry of Transport, Tokyo.

SIGTTO (Society of International Gas Tanker and Terminal Operators) (1997) *Site Selection and Design for LNG Ports and Jetties*, Information Paper No. 14. Witherby, London.
Velsink H (1987) *PIANC Bulletin*, No. 56. PIANC, Brussels.

Port Designer's Handbook
ISBN 978-0-7277-6004-3

http://dx.doi.org/10.1680/pdh.60043.143

Chapter 5
Impact from ships

5.1. General

In this chapter the berthing forces that can arise between a berth structure and a berthing ship will be discussed. The berthing forces transmitted to the structure will consist of impact loads normal to, and frictional loads parallel to, the berthing face.

While the vertical loads on a berth from deadweight, live load, crane loads, etc. can be determined very accurately, it can be very difficult to evaluate the horizontal loads caused by the impact of ships. The size and velocity of ships when berthing, the manoeuvring, direction and strength of current, wind and waves at the berth are factors that often escape an exact quantification, and therefore tend to complicate the correct calculation of the impact forces of ships.

The following design criteria should be used in the calculation of the berthing and mooring energies and in the selection of the fender system to be used:

(a) the design codes and regulations
(b) the desired design life of the berth structure and the safety factors to be used
(c) the design berthing ship and the ship's allowable hull pressure
(d) the design berthing velocity under both normal and abnormal conditions.

For normal berthing procedures, the berthing energy and the impact forces from the berthing ship against the berthing structure can be estimated using one of the following:

(a) the theoretical method
(b) the empirical method
(c) the statistical method.

The ship's berthing energy is proportional to the virtual mass of the ship and the square of the approach velocity, and is reduced according to the rotation of eccentric berthing when the ship hits the berth structure at a distance from its centre of gravity.

5.2. The theoretical or kinetic method

The theoretical method is based on the general basic kinetic energy equation due to the impact of a ship on a berthing structure:

$$E = 0.5 \times M_v \times V^2 = 0.5 \times (M_d + M_h) \times V^2$$

where

E = kinetic energy in kNm
M_v = virtual mass in tons, which equals the displacement of the ship M_d plus the hydrodynamic or added mass moving with the ship M_h
M_d = the mass of the design ship (the displacement in tonnes) should, in line with the PIANC (International Navigation Association) *Guidelines for the Design of Fenders Systems* (PIANC, 2002), be based on the 95% confidence level. For guidance on the displacement for various types of ship see Chapter 21 on ship dimensions. In most cases the displacement of the fully loaded ship should be used for the fender design
M_h = when the ship is moving through water it is also necessary to consider the a movement of a volume of water around the ship, which is entrained with it. When the ship comes to a stop this additional mass of water will continue to move and press the ship against the berth. This additional mass of water is also known as the 'added mass' or the 'hydrodynamic mass'
V = the velocity in m/s of the ship normal to the berth line. In line with the PIANC (2002) fender guidelines, the 50% confidence level should be used.

For all berth structure design, the displacement for a fully loaded ship should be used, except where the berth will be used exclusively for the export of cargo, in which case the displacement of the ship and the draft may be reduced to the actual value for the ship when berthing, but not less than the ballast displacement.

Of the total kinetic energy of the ship, the fender system must absorb:

$$E_f = C \times (0.5 \times M_d \times V^2)$$

where the adjusting factor or berthing coefficient is $C = C_H \times C_E \times C_C \times C_S$.

Hydrodynamic mass factor C_H

The hydrodynamic or added mass factor C_H allows for the movement of the water around the ship to be taken into account in the calculation of the total berthing energy of the ship by increasing the mass of the system.

$$C_H = \frac{M_d + M_h \times C_{HR}}{M_d} = 1 + \frac{M_h \times C_{HR}}{M_d}$$

where

C_H = hydrodynamic or added mass factor
M_d = displacement of the ship
M_h = hydrodynamic or added mass
C_{HR} = reduction factor due to the ship moving at an angle to its longitudinal axis. In principle, the reduction factor C_{HR} for the hydrodynamic mass of a ship moving normal to the berth line in open water will be 1.0, but for a ship moving along its longitudinal axis in open water it can be assumed to be about 0.1

Over the years, different formulas for the hydrodynamic or added mass factor have been suggested, as shown in Table 5.1, where

Table 5.1 Formulas for hydrodynamic or added mass factor

Author	Year	Type of test and comments	Formula for C_H
Stelson (PIANC, 2002)	1955	Model test	$1+\frac{1/4\times\pi\times\rho\times D^2\times L}{M_d}=1+\frac{\pi\times D}{4\times C_b\times B}$
Grim (PIANC, 2002)	1955	Model test	$1.3+1.8\times\frac{D}{B}$
Saurin (PIANC, 2002)	1963	Full-scale observation and model test	1.3 (mean value) 1.8 (safe value)
Vasco Costa (PIANC, 2002)	1964	Model test	$1.0+2.0\times\frac{D}{B}$
Giraudet (PIANC, 2002)	1966	Model test	$1.2+0.12\times\frac{D}{H-D}$
Rupert (PIANC, 2002)	1976	Full-scale observations	$0.9+1.5\times\frac{D}{B}$
Ueda (PIANC, 2002)	1981	Full-scale observations	$1+\frac{1/2\times\pi\times\rho\times D^2\times L}{M_d}=1+\frac{\pi\times D}{2\times C_b\times B}$

ρ = specific gravity of seawater (10.3 kN/m^3)
D = draft of ship
B = width of ship
L = length of ship
H = water depth
M_d = displacement of ship
C_B = block coefficient $=\frac{M_d}{\rho\times L\times B\times D}$

The displacement is the product of the length between the perpendicular L_{BP} times the draft D times the width B times the block coefficient C_B. The PIANC (2002) fender guidelines recommend the following block coefficient C_B be adopted if other data are lacking:

Container ship	0.6–0.8
General cargo ship and bulk carriers	0.72–0.85
Tankers	0.85
Ferries	0.55–0.65
Roll-on/roll-off ship	0.7–0.8

Professor Vasco Costa of Portugal assumes that the ship moves sideways to, for example, a berth, or rotates about its centre of gravity. The Vasco Costa formula is valid if the keel clearance is more than $0.1D$ and the ship velocity is more than 0.08 m/s. If the ship moves along its longitudinal axis, Vasco Costa assumes that $C_H = 1$.

The formulas of Professor Vasco Costa and Professor Shigeru Ueda of Japan are nowadays the most used formulas for calculating the hydrodynamic mass factor.

The exact value of the hydrodynamic mass is very difficult to determine. Investigations and research have shown that the hydrodynamic mass will vary with the shape of the ship, the underkeel clearance, the ship velocity and the water depth. The hydrodynamic mass usually varies between about 25% and 100% of the displacement of the ship. Generally, it is recommended that for a water depth of 1.5 times the draft of a ship or more, C_H be taken as 1.5. When the water depth is only 1.1 times the draft of the ship, C_H is taken as 1.8.

Eccentricity effect C_E

The eccentricity factor C_E is due to the consideration of the energy dissipation which arises from the rotational motion after berthing around the contact point at either the bow or at the stern.

$$C_E = \frac{i^2 + r^2 \times \cos^2 \phi}{i^2 + r^2}$$

where

I = the ship's radius of inertia, generally between 0.2L and 0.25L
r = the distance of point of contact from the centre of mass
ϕ = the angle as shown in Figure 5.1

Figure 5.1 shows C_E as a function of the angle ϕ and the ratio r/L when $i = 0.2L$. If ϕ is 90°, the equation will be:

$$\frac{1}{1 + (r^2/i^2)}$$

Figure 5.1 The eccentricity effect C_E as a function of ϕ and r/L

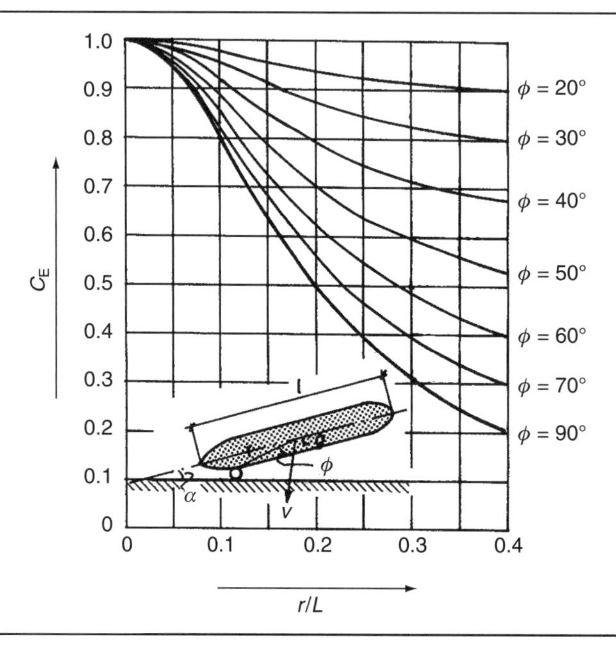

Figure 5.1 indicates that the way in which a ship comes alongside is a very important factor. This should, therefore, be studied as closely as possible. The value of $\phi = 90°$ may give values that are too favourable.

Normally, either the ship will come alongside under its own power and at an angle to the berth line, or the ship will have almost come to a stop outside the berth structure and be manoeuvred carefully towards the structure with the help of current, wind and/or waves. When the ship is berthing at an angle, it will usually make contact only with one fender.

Figure 5.1 further indicates that the value of C_E also depends on which part of the ship makes contact with the berth. Usually, the berthing angle or angle of approach, which is the angle between the berth line and the ship, will be about 1–5° if the ship is berthing with tugboat assistance.

If the manoeuvring into berth is done without tugboat assistance, the berthing angle will usually be less than about 10–15°. Then the distance r between the centre of gravity (c.g.) of the ship and the point of impact is about 0.25–$0.35L$. If the angle ϕ approaches 90° there will be a minimum amount of impact energy on the berth structure. For a continuous fender system, C_E is generally taken as 0.5–0.6, and for berth structures with, for example, individual breasting dolphins, C_E is taken as 0.7–0.8.

If the ship comes alongside parallel to the berth front, i.e. $\alpha = 0°$, the ratio r/L also approaches 0, and the impact energy will be at a maximum. On the other hand, the part or length of the ship hitting the structure is now far greater, implying that the energy to be absorbed per linear metre of berth structure will be less than in the above case.

Therefore, if one assumes the most favourable values of α, ϕ and r/L, the impact energy would theoretically be very moderate. In practice, however, manoeuvring will deviate from the ideal, and it is advisable to choose realistic values of these parameters.

Water cushion effect C_C

The water cushion effect, C_C, ranges between 0.8 and 1.0 at solid and open berths, respectively. If the bottom is steeply sloping under the berth, the resistance from the water will increase when the ship comes near the berth front. This is particularly true at solid berths (e.g. steel sheet pile structures, where the water between the ship and the berth has to be squeezed aside before the ship can touch the berth structure). For open berth structures, where there is usually an easy way out for the water between the berth and the ship, the water cushion effect will be very small.

Softening effect C_S

This factor is determined by the ratio between the elasticity and/or the flexibility of the ship's hull and that of the fender system or berth structure. Therefore, part of the berthing kinetic energy will be absorbed by elastic deformation of the ship's hull and/or flexibility of the berth structure. For a small ship C_S is generally taken to be 1.0. For hard fenders and larger ships (e.g. large tankers or flexible wood piers) C_S is 0.9–1.0.

BS 6349-1: 1994, 'Code of practice for maritime structures', states that a hard fender system can be considered as one where the deflection of the fenders under impact from ships for which the fenders are designed is less than 0.15 m.

Figure 5.2 The fender energy E_f with a berthing coefficient $C = 1.0$

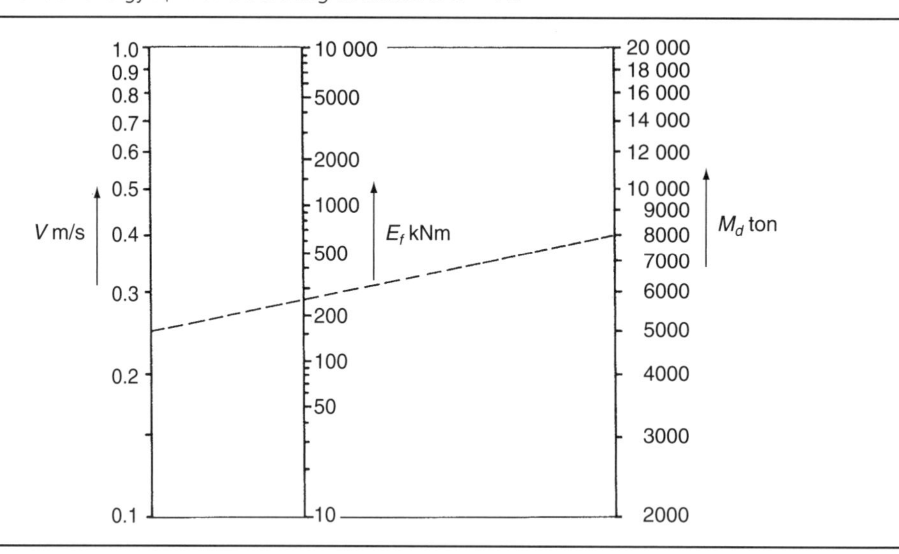

Approach velocity v

It appears from the above that sophisticated calculations to establish the magnitude of the adjusting factor or the berthing coefficient C are not justified if the values of the velocity v are only approximate. The approach berthing velocity v is the most influential variable in the calculation of the berthing energy. The approach velocity is defined as the ship's speed at the initial berthing contact, measured perpendicular to the berth line. In line with the PIANC (2002) fender guidelines, the mean velocity value v should be taken to be equivalent to the 50% confidence level.

Although determining the correct value of the approach velocity is very difficult, as it appears to the second power in the energy formula it is important to try to find as accurate a value as possible. This is illustrated in Figure 5.2 for a berthing coefficient $C = 1.0$.

The actual berthing approach velocity will be influenced by a large number of factors, such as:

(a) The experience of the crew working during the berthing operation.
(b) The influence of the wind, waves and current around the berth structure.
(c) Whether the navigation approach to the berth is easy or difficult, and whether the approach channel is equipped with navigational aid systems.
(d) Whether the ship is equipped with bow thrusters and has good manoeuvrability, or whether tugboat assistance is necessary.
(e) The type of ship (e.g. container ship, tanker, general cargo ship), the windage area of the ship, etc.

Normally, smaller ships have higher velocities when hitting the berth structure than do larger ships. Figure 5.3 indicates some velocities for small and medium-sized ships berthing without tugboat assistance and under various weather and manoeuvring conditions. The ship velocities in Figure 5.3 and Table 5.2 are taken at the 50% confidence level. In the event of the berthing manoeuvre taking place without tugboat assistance, the figures given below for tugboat-assisted berthing should be increased considerably.

Figure 5.3 The velocity of a ship coming alongside without tugboat assistance

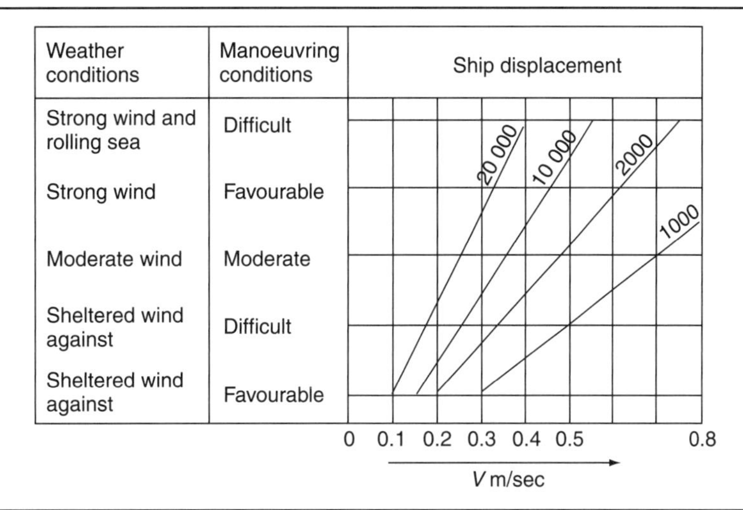

For reasons of safety and to reduce the probability of damage to the fender systems, PIANC recommends that when designing fender systems for larger ships the berthing velocities with use of tugboat assistance should not be less than:

Very favourable conditions	10 cm/s
In most cases	15 cm/s
Very unfavourable conditions with cross-current and/or much wind	25 cm/s

In 1977, Brolsma *et al.* (PIANC, 2002) recommended the berthing velocities (mean values) with tugboat assistance shown in Figure 5.4, differentiated according to the following five navigation conditions and the size of the ship in displacement:

(a) good berthing conditions, sheltered
(b) difficult berthing conditions, sheltered
(c) easy berthing conditions, exposed
(d) good berthing conditions, exposed (this figure is considered to be too high)
(e) navigation conditions difficult, exposed (this figure is considered to be too high).

Table 5.2 Ship velocity during berthing with tugboat assistance

Ship displacement (*t*)	Velocity: m/s		
	Favourable conditions	Moderate conditions	Unfavourable conditions
Under 10 000	0.20–0.16	0.45–0.30	0.60–0.40
10 000–50 000	0.12–0.08	0.30–0.15	0.45–0.22
50 000–100 000	0.08	0.15	0.20
Over 100 000	0.08	0.15	0.20

Figure 5.4 Design berthing velocity (mean value) according to the ship displacement with tugboat assistance

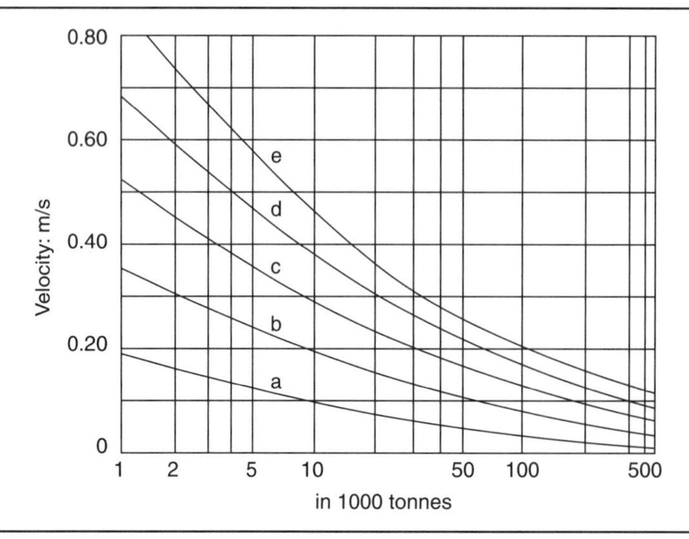

The Japanese National Section of the PIANC has collected information on the relationship between a ship's displacement and the approach velocity for large cargo ships and tankers (Figure 5.5). The approach of the ship was made in such a manner that it was stopped parallel with and about 10–20 m off the berth structure and then gradually pushed to the berth structure under the full control of several tugboats.

When designing the fenders for the ramps for ferries and roll-on/roll-off ships, the berthing bow or stern velocity for these ships berthing under their own power will, depending on the stopping distance which the actual ship will use in relation to the length of the berth, generally vary between about 0.4 and 1.0 m/s.

Figure 5.5 The relationship between berthing speed and ship displacement

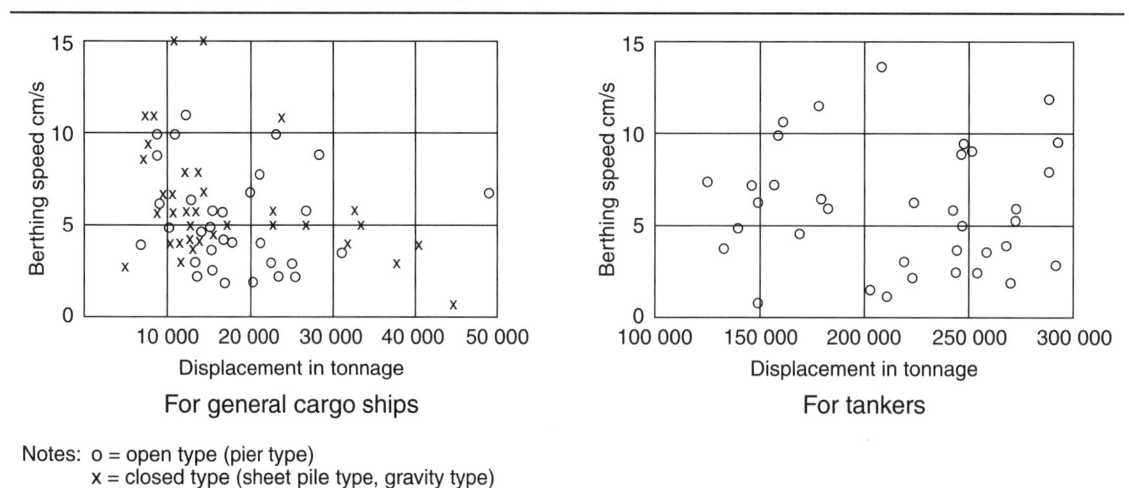

Due to the rapid turnaround times and high engine power of most ferries, the berthing velocities are generally higher than for other ships. It is therefore recommended that the berthing velocity in the direction of approach for the fender design should be:

(a) For fenders at a corner of the berth structure or the outer or end breasting dolphin: 2.0–3.0 m/s.
(b) For fenders along the berth structure: 0.5–1.0 m/s.

5.3. The empirical method

In the theoretical method, the velocity of approach is the most significant and difficult to determine element in the evaluation of the berthing energy imparted to the fender. Therefore the following empirical formula (Girgrah, 1977) for the maximum impact energy (in kNm) to be absorbed by the fender, based solely on a ship's displacement, may be considered:

$$E_{\mathrm{f}} = \frac{10M_{\mathrm{d}}}{120 + \sqrt{M_{\mathrm{d}}}}$$

where M_{d} is the displacement tonnage of the berthing ship. A factor of 0.5 may be applied in cases where the impact would be either shared between two fenders or accompanied by rotation of the ship.

5.4. The statistical method

The statistical design method is based on measurements of the impact energies actually absorbed by the fenders during berthing. As the method is based on impact energies actually observed at existing berth sites, it automatically includes the effect of the berthing velocity, hydrodynamic mass, eccentricity, etc.

In Figure 5.6 the impact energy during berthing operations in normally protected harbours is shown as a function of the displacement of the ship. One curve shows the measurements of the energy in Rotterdam. The other two curves show the impact fender energies recommended by the British Code of Practice and the Norwegian Standard for berth structures.

Figure 5.6 The impact energy during a ship berthing to a berth structure

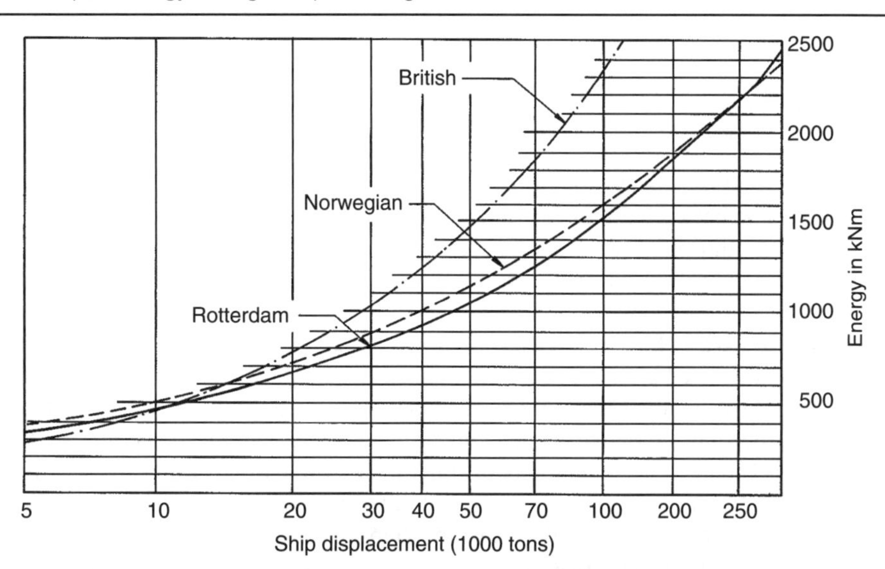

The Norwegian Standard also mentions that for harbours exposed to strong winds and current, or with particularly difficult manoeuvring conditions, the impact energy given in Figure 5.6 should be increased by up to 50%. For structures in the open sea the impact energy should be increased by up to 100%.

5.5. Abnormal impacts

An abnormal impact occurs when the normal calculated energy to be absorbed on impact is exceeded. The fender systems have to cater for the normal impacts of the design ship and be capable of catering for a reasonably abnormal impact. The reasons for abnormal impacts can be bad manoeuvring, mishandling, exceptional winds or currents, or a combination of these factors. The abnormal impact factor should enable reasonable abnormal impacts to be absorbed by the fender system without damage.

The abnormal impact factor should, according to the PIANC (2002) fender guidelines, take into account the following:

(a) Berths with a high frequency of berthing will have a higher probability of abnormal impact and therefore a higher factor.
(b) The effect a failure of the fender system would have on berth operations.
(c) Berths that have been designed for very low approach velocities would be more likely to incur abnormal impact than would berths designed for higher approach velocities.
(d) The vulnerability and importance of the berth structure supporting the fender system. If the time and costs involved in repairs are likely to be disproportionately large, a higher abnormal impact factor should be used.
(e) Where a wide range of ships use the berth and the largest ships use the berth only occasionally, the factor of abnormal impact may be reduced.

Care should be taken not to increase the factor for abnormal berthing to such an extent that the required fender capacity, and consequently the fender reaction forces, becomes prohibitive for small ships using the berth. It should be noted that a factor of safety of 2.0 for the berthing energy provides for only a 40% increase in ship speed, as the berthing energy is related to the square of the speed.

Suggested fender safety factors for abnormal impact to be applied to the designed fender energy are as shown in Table 5.3.

For the chain system used for suspension of the fenders, and for fender panels and fender walls, normal impact factors should be between 3 and 5. For abnormal impacts when the chain loads may be higher, the abnormal factor should be at least 2. The highest factored load from either normal or abnormal impacts should be less than or equal to the minimum breaking load of the chain.

5.6. Absorption of fender forces

Once the energy to be absorbed by the fender system has been established, a fender that will transmit an acceptable horizontal force against the front of the berth structure should be selected. This horizontal force will depend on the characteristics of the fender. It should be taken into account that the ship will also have to resist this force. Generally, it is desirable to have these horizontal forces, or reaction force and corresponding reaction pressure, as low as possible, to avoid damage to the side of the ship and to minimise the construction cost of the berth. It is an often-discussed question how great a part of the fender will be actively resisting the impact.

Table 5.3 Fender safety factors for abnormal impact

Type of ship	Factor of abnormal impact	General comments
Small ships	1.5–2.0	Depends on operation
General cargo	1.75–2.0	Depends on ship size
Roll-on/roll-off	2.0–3.0	Stern berthing
Ferries	2.0–3.0	Depends on berth exposure and operation
Tankers and bulk	1.3–2.0	Smallest ships
	1.3–1.5	Largest ships
Container ships	2.0	Smallest ships
	1.5	Largest ships

If the ship comes alongside under its own power, which is the most usual way, it will form an angle with the berth line (Figure 5.7). Therefore, it is generally accepted design practice that each fender unit in a system should have sufficient energy-absorbing capacity to absorb the largest impact load. Each fender unit must be capable of absorbing the full impact energy or load, as ships almost always contact only one fender on the first impact.

If the ship has been assisted by tugboats and is berthed parallel to the structure, or is manoeuvred parallel to the structure with the help of wind or current, the length of the area of contact between the ship and the berth structure will still be smaller than the length of the ship, i.e. $L_{sf} < L_s$ (Figures 5.8 and 5.9). This fact is of great importance with regard to the choice of the fender type, the spacing of the fenders and the horizontal force acting on the structure and the ship. For instance, the length of the contact area for some container ships can be as small as about 20% of the ship's length, while it can be up to 70% for a traditional general cargo ship.

To ensure that the front of the berth structure is satisfactorily safe under normal operations, the German *Recommendations of the Committee for Waterfront Structures, Harbours and Waterways* (EAU, 2004) recommend that the berth front is designed for a horizontal point load equal to the bollard load. This point load should be allowed to act anywhere at the berth front without the

Figure 5.7 A ship coming alongside a berth under its own power

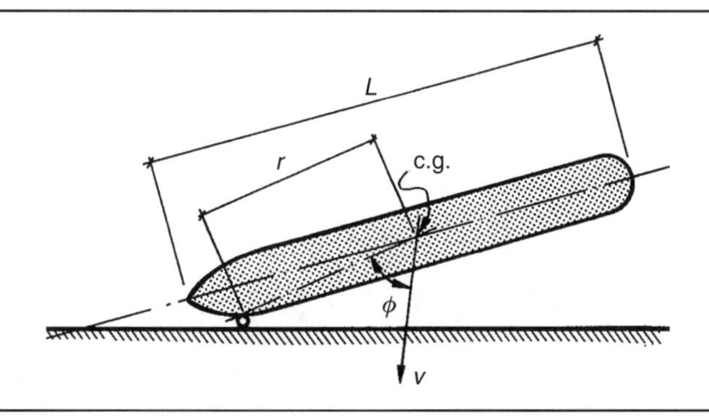

Figure 5.8 The length of contact L_{sf} is smaller than the length of the ship L_s

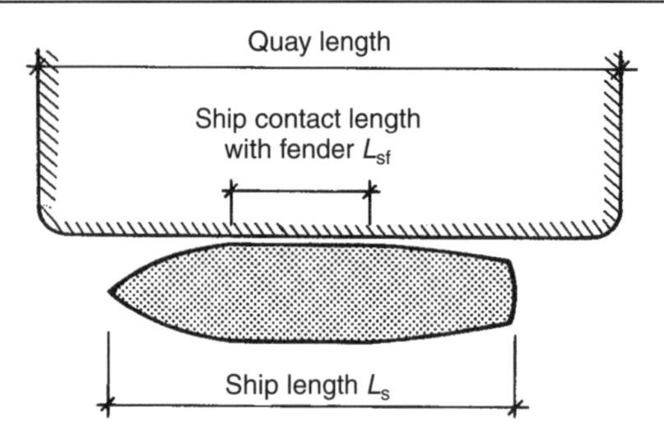

allowable stresses being exceeded, and its contact area should be limited to 0.25 m^2. Table 5.4 gives point loads and the corresponding increased loadings for various ship displacements.

When a ship comes alongside at an angle to the berth line, longitudinal friction forces parallel to the berthing face will be transmitted by way of the fenders to the berth structure. The ship will skid along the structure after the impact while the fenders are still in a compressed state. The front of the berth structure must take up this friction force $F = \mu \times P$, where μ is the friction coefficient between the ship and the fender, and P is the impact force.

The friction coefficients μ are: for steel to steel 0.25, for steel to polyethylene 0.2, for timber to steel 0.4–0.6 and for rubber to steel 0.6–0.7. The friction force F is usually acting simultaneously with the impact force perpendicular to the front of the structure. The longitudinal friction forces can be reduced by the provision of low-friction contact surface materials.

Figure 5.9 The berth contact area of a very large container ship

Table 5.4 Point loads and line loads for various ship displacements

Displacement (*t*)	Point load: kN per 0.25 m^2	Line load: kN per linear metre of berth
2 000	100	15
5 000	200	15
10 000	300	20
20 000	500	25
30 000	600	30
50 000	800	35
100 000	1000	40

Table 5.5 Up-and-down directed loads

Ship displacement in tons up to	Vertical up/down directed load: kN per linear metre of berth
2 000	10
5 000	15
10 000	20
20 000	20
30 000	25
50 000	25
100 000	30

The horizontal forces due to long-period wave action along the berth front are dependent on the length of the long-period wave in relation to the size of the moored ship. On a wave slope of 1 : 2000, the parallel berthing forces for a 300 000 t ship are about 1500 kN. With a friction coefficient between the fenders and the ship's hull of 0.7, the fenders and the berth structure must take up a horizontal force along the berth front of about $1500 \times 0.7 = 1050$ kN. If the ship is moored by forced or tension mooring to reduce the surge movement, the mooring forces normal to the berth front will be $1500/0.7 = 2150$ kN, which is equivalent to four winches of about 550 kN capacity each.

5.7. Ship 'hanging' on the fenders

When a ship is moored at a berth structure it can 'hang' itself on the fenders due to tidal variations, or friction between the ship's hull and the fender, or it can chafe on the fenders during loading and unloading. The front of the structure should, therefore, in cases where such effects are possible, be designed for up-and-down directed loads, as suggested in Table 5.5.

Where large energy-absorbing rubber fenders or similar protruding fenders are used, such vertical loads must be estimated in each case.

REFERENCES AND FURTHER READING

BSI (British Standards Institution) (1994) BS 6349-4: 1994. Maritime structures. Code of practice for design of fendering and mooring systems. BSI, London.

Dubruell J (2002) Fender design for complex nautical situations. *Proceedings of the PIANC 30th International Congress*, Sydney.

EAU (Empfehlungen des Arbeitsausschusses für Ufereinfassungen) (2004) *Recommendations of the Committee for Waterfront Structures, Harbours and Waterways*, 8th English edn. Ernst, Berlin.

Girgrah M (1977) Practical aspects of dock fender design. *Proceedings of the International Navigation Congress*, PIANC, Leningrad, pp. 5–13.

Japanese National Section of the PIANC (1980) *Design of Fender System*. Bureau of Ports and Harbours, Ministry of Transport, Tokyo.

Ministerio de Obras Públicas y Transportes (1990) ROM 0.2-90. Maritime works recommendations. Actions in the design of maritime and harbour works. Ministerio de Obras Públicas y Transportes, Madrid.

PIANC (International Navigation Association) (2002) *Guidelines for the Design of Fender System. Report of Working Group 33*. PIANC, Brussels.

PIANC (2013) *Collection of Berthing Velocities and Resulting Design Recommendations by PIANC Working Group MarCom 145*. PIANC, Brussels.

Port and Harbour Research Institute (1999) *Technical Standards for Port and Harbour Facilities in Japan*. Port and Harbour Research Institute, Ministry of Transport, Tokyo.

Ueda S (2002) Reliability design method of fender for berthing ship. *Proceedings of the PIANC 30th International Congress*, Sydney.

Port Designer's Handbook
ISBN 978-0-7277-6004-3

http://dx.doi.org/10.1680/pdh.60043.157

Chapter 6
Design considerations

6.1. General

A coastal construction might be exposed to a high number of various loads that it has to withstand throughout the whole design life. During design all these loads need to be noted by the designer, and the maximum expected intensity (characteristic value) of each load should be set. Figure 6.1 shows a typical variation of loads on a berth structure.

Figure 6.1 shows in some detail the various types of force which normally might occur on a berth structure. In this chapter the characteristic loads acting on the berth structure and the loads from the land side will be considered in detail. The characteristic loads from the sea side are mainly given in Chapters 2 and 5.

Norwegian design practice for open berth structures recommends that, if possible, all vertical loads on the berth structure are taken by the piles or the columns, and that all horizontal loads are taken by the friction slab behind.

Loads can be divided into **normal loads** and **extreme loads**, where normal loads refer to any loads that may reasonably be expected to occur during the design life of the structure and under normal operating conditions. Typical normal loads are the self-weight and operating live loads. Extreme loads are any loads that *may* occur, but are not expected to occur during the design life of the structure. Typical extreme loads are uncontrolled impacts from vessels, and seismic activity.

The self-weight from the construction and equipment is expected to occur with more or less full intensity continuously during the design life of the construction. Normally, the self-weight can easily be calculated, and the self-weight should be implemented in all combinations of loads with full intensity.

Live loads are both general live loads on the construction and natural loads that are expected to occur with an intensity close to estimated values *at least once* during the design life of the construction.

Normally, **live loads** are set by the designer in accordance with the construction owner, based on input for equipment dealers, values from experience, and given values in standards and guidelines. Different live loads should be combined together with self-loads, but in situations with several live loads occurring together, it is unlikely that the live loads will occur with full intensity simultaneously, and probability reduction factors might be used. Normally, combinations of live loads are covered in standards and guidelines. National standards are considered the straightforward way to provide documentation in order to satisfy the demands of legislation. Guidelines are only informative.

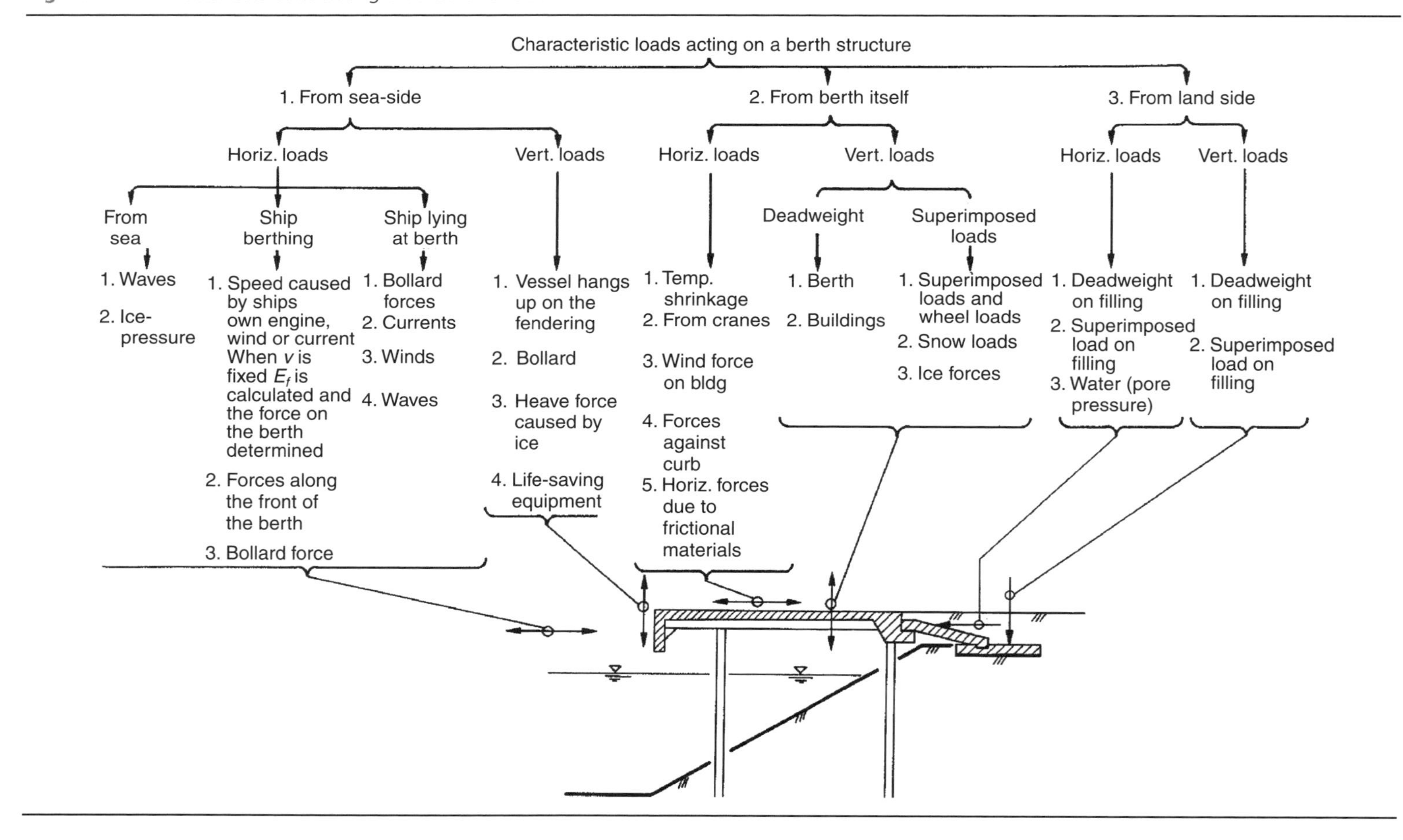

Figure 6.1 Characteristic loads acting on a berth structure

Extreme loads are typical responses from the construction due to, for example, seismic activity or the uncontrolled impact from vessels. These are not expected to occur, but might do so during the design life of the construction. Normally, extreme loads and combinations of extreme loads are covered in standards and guidelines.

Different countries have different legislation, and all constructions have to be in accordance with the legislation in the country in which it is built. Many countries have their own standards for marine and costal construction, but some do not. The designer has to find out whether or not there are national standards or guidelines that have to be used for design. If such national standards exist, they have to be used, and if national guidelines exist, it is recommended to use them. If no national legislation, standard or guideline exists, the designer is not required to use specific design manuals, but it is highly recommended to use internationally accepted standards and guidelines instead of trying to create a specific guideline for the project. It is also very important to inform the construction owner of the chosen design criteria.

Standards and guidelines for the design of constructions have developed from quite simple combinations with only self-weight and one live load to more modern legislation with combinations of loads due to probability studies, failure mechanisms and calculations. For modern standards, it is also common to take into account the intensity between the self-load and live loads and the intensity of the highest live load (dominating live load) when setting the load factors. If the self-weight is quite high compared with live loads, the self-weight might be the dominating load, and all live loads will be non-dominating, with an intensity reduction in the specific combination.

In modern standards, load combinations are divided into several limit states due to different failure mechanisms. The bearing capacity, the durability, the operability, the fatigue and the construction's ability to withstand accidental loading have to be checked in different limit states, which are the following:

(a) the **ultimate limit state (ULS)** is related to (the risk of) local and global failure or large inelastic displacements or strains characteristic of a failure
(b) the **accidental limit state (ALS)** is related to (the risk of) global failure of the structure in accidental situations such as earthquake, and it is assumed that local parts of the structure will be beyond capacity but no progressive collapse will occur
(c) the **fatigue limit state (FLS)** is related to (the risk of) failure caused by repeated loading
(d) the **serviceability limit state (SLS)** is related to operability and/or durability.

In the different limit states, there are three main categories of characteristic loads or forces acting on a berth structure:

(a) characteristic loads from the sea side
(b) characteristic loads on the berth structure
(c) characteristic loads from the land side.

The characteristic load may be a permanent load, a variable load (normally of a return period of 50 years), a fatigue load or an accidental load.

Limit states are defined loading situations that the construction has to withstand during the design life of the structure. The limit states implement loading factors and simultaneously factors on loads in different combinations of the loads. Each limit state has a function, and the construction has to contain sufficient internal strength in all design limit states in order to fulfil satisfactorily its intended functions.

The function of the load factors is to provide *sufficient safety* regarding the probability of each load exceeding the characteristic value during the design life of the structure.

The function of the concurrency factor is to reduce the total loading of the structure due to the probability of several live loads occurring simultaneously at a characteristic value. The total load at the structure is kept above a minimum statistical level specified in legislation as the lowest probability of failure allowed, and at the same time the structure is not exposed to a total load of higher intensity than necessary. There might be a need for calculations in several combinations in each limit state in order to find the most unfavourable combination that should be used as the basis for design.

Some loads, both self-weight and live loads, might give dynamic effects, both singular and regular, that have to be implemented in design. The load coefficient does not include the dynamic allowances, and singular dynamic effects can to a certain point be implemented with a dynamic factor, while regular dynamic effects might need to be controlled due to the dynamic response and the dynamic amplification. For most berth constructions, regular dynamic effects can be neglected due to a quite massive and high dynamic absorbing construction compared with most regular loads.

6.2. Design life

The design life of a structure can be described in many ways. One common description of the design life is the assumed operational time for the structure including planned maintenance and without major repair.

The design life of a structure can be described as:

(a) a relevant limit state
(b) a given number of years
(c) a reliability level in a limit state that is not passed during a given period.

Figure 6.2 gives a description of the operational design life for a concrete structure. Cracking of the concrete cover will probably occur at the end of the design life period, with subsequent concrete cover failure.

The design life can be described in similar ways for materials other than concrete, although the visibility of the damage in most cases will be higher in concrete. A steel structure will end its design life when corrosion has decreased the thickness beyond the critical level, and a timber structure will end its design life in a similar way due to decay, sea worms or other effects that weaken the structure.

A general design rule is that ordinary berth structures in commercial ports should have a design life of 50 years or higher. For berths serving special industries, container traffic, oil traffic, etc., a period of not more than 30 years is often more relevant. Modern specialised cargo handling is more subject to rapid development, which may also lead to the berth facilities becoming outdated relatively early (e.g. because of the influence of container handling technology on the berth structure). For shore protection works and breakwaters, a design life of 100 years, and for flood protection works a design life of more than 100 years, will normally be appropriate.

Figure 6.2 Design life period for a concrete construction

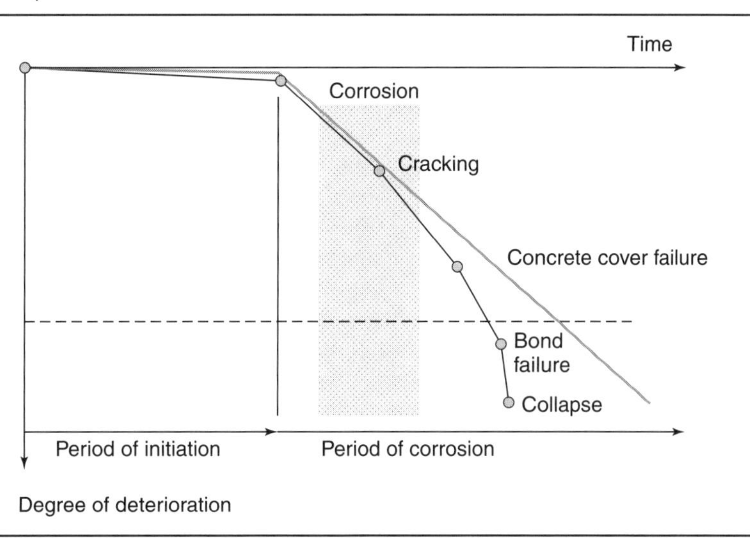

There is a well-known joke that the design life of a structure should at least be as long as the remaining engineering career of the designer.

Note that quite often the minimum allowed design life is given in national standards.

The total design and service life of a berth structure, or indeed any type of structure, will depend on all the following factors:

(a) the design engineer's experiences in the design work
(b) the design standards and recommendations used
(c) the correct materials being used for the environment at the site
(d) the construction of the berth structure, and the contractor's ability to do the job properly
(e) management of the operation and maintenance, including proper periodic inspections of the berth structures
(f) repair of any damage.

The determination of the design life should consider the following:

(a) Assessment of the factors that influence the security of the structure. These may include fatigue loading, corrosion, marine growth and a reduction in the soil strength. Therefore, the acceptable probability of failure or the acceptable degree of damage during the lifetime of the structure should be decided at an early stage of the design.
(b) Evaluation of the probabilities that particular limit states will occur during the design life.
(c) Appraisal of economic feasibility, and the necessity to allow for developments and related matters. Therefore, the costs of, say, repair work should be estimated and included in the evaluation of the economic feasibility of the project.

Temporary berth constructions or berth constructions of a short design life might be designed with material and elements without major concern for durability. For instance, a berth slab consisting of

Table 6.1 Average maintenance costs

Part of the port structure and type of equipment	Average economical design life: years	Annual average maintenance costs: % of new cost or replacement value
Breakwater	100	2
Reinforced open berth structure	50–100	1–2
Steel sheet pile berth structure	50	1–2
Rubber fenders	10	1
Concrete aprons and roads	20	1–2
Asphalt surfacing	10	2
Container gantry cranes	20	5
Mobile container cranes	15	10
Fork-lift and reach stackers	10	10
Straddle carriers	5–10	10–15
Road tractors	10	10
Warehouses and sheds	40	5

slender prefabricated prestressed concrete elements may represent an economically good solution if the useful design life of the berth is estimated to be only approximately 10–15 years. Based on experience, such types of slab should not be chosen for berths with a design life beyond 50 years, due to the fact that it is practically impossible to repair prestressed concrete elements.

The most important feature of maintenances costs is regular inspections and reporting. A routine maintenance system is recommended in order to monitor the structure and have a long service life rather than unknown deterioration during the operation period. A good maintenance system will discover deviations, permitting a simple repair at an early stage instead of a major and costly repair later.

Typical values for economical design life and maintenance are given in Table 6.1.

6.3. Standards, guidelines and design codes

Structural design calculations are generally performed according to the partial safety factor method, by applying partial safety factors on both the loads and the capacity. As a general check, the capacity including reduction factors should give a higher capacity than the actual loading combination including load factors. A consistent set of standards should be used for the whole structure, as design formulas, loads and material factors are all calibrated to give the intended reliability. When using a non-national standard, note that the total security given in legislation might differ from one country to another, or in some countries might not exist at all.

A characteristic value for a load is calculated, given in standards, found in guidelines or set according to experience. The characteristic value should equal the highest value expected during the design life or the highest statistical value expected during the period given in a standard. There will always be a certain probability of higher values occurring, but this is dealt with in the specific load combinations in a statistical way.

Within the territorial zone for the EU and in the EEA area, the Eurocodes are the default standards for all constructions. These standards are common to the whole zone, and default values for different factors are given, but each country has its National Annex, which can override the default values. In this area, there are also other design codes such as the Spanish **ROM 3.1-99** (Recommendations for the Design of Maritime Configurations of Ports, Approach Channels and Harbour Basins) or the German **EAU 2012** (*Recommendations from the Committee for Waterfront Structures: Harbours and Waterways*), which coexist together with the Eurocodes. Other countries might have other coexisting codes. In the case of inconsistency with the Eurocodes, these coexisting codes should be perceived as guidelines, and do not override the Eurocodes. Similar situations might be present in other countries and regions as well.

As a general recommendation, national standards should be used if present, and internationally recognised standards and guidelines should be used as supplements in areas that the standards do not cover. If no national standard exists, national legislation should be studied, and an internationally recognised standard should be used for design, with the possible correction of safety factors to take account of national safety legislation.

6.4. Load combinations and limit states

Structures might be exposed to a large number of combinations of loads that may also be in the same limit state.

The following subsections will focus on safety and factors given in the Eurocodes. As stated in Section 6.1, there are four major safety limit states (ULS, ALS, FLS and SLS) that a construction has to be designed to. Most modern standards have a similar way of handling safety.

6.4.1 Ultimate limit state (ULS)

In the ULS, the total loading affecting the stability of the whole construction, the capacity of each element and the construction as a complete element, the capacity during erection and geotechnical safety are calculated. There are three major cases that have to be calculated, and both load factors might have different values in different sets:

- Set A – global stability and equilibrium of the whole construction:

$$E_{\mathrm{d}} = \gamma_{\mathrm{G},j,\mathrm{sup}} G_{\mathrm{k},j,\mathrm{sup}} + \gamma_{\mathrm{G},j,\mathrm{inf}} G_{\mathrm{k},j,\mathrm{inf}} + \gamma_{\mathrm{P}} P + \gamma_{\mathrm{Q},1} Q_{\mathrm{k},1} + \gamma_{\mathrm{Q},i} \psi_{0,\iota} Q_{\mathrm{k},i}$$

- Set B – structural capacity of elements or large inelastic deformations:

$$E_{\mathrm{d}} = \gamma_{\mathrm{G},j,\mathrm{sup}} G_{\mathrm{k},j,\mathrm{sup}} + \gamma_{\mathrm{G},j,\mathrm{inf}} G_{\mathrm{k},j,\mathrm{inf}} + \gamma_{\mathrm{P}} P + \gamma_{\mathrm{Q},1} \psi_{0,\mathrm{i}} Q_{\mathrm{k},1} + \gamma_{\mathrm{Q},i} \psi_{0,\iota} Q_{\mathrm{k},i}$$

$$E_{\mathrm{d}} = \xi \gamma_{\mathrm{G},j,\mathrm{sup}} G_{\mathrm{k},j,\mathrm{sup}} + \gamma_{\mathrm{G},j,\mathrm{inf}} G_{\mathrm{k},j,\mathrm{inf}} + \gamma_{\mathrm{P}} \mathrm{P} + \gamma_{\mathrm{Q},1} Q_{\mathrm{k},1} + \gamma_{\mathrm{Q},i} \psi_{0,\iota} Q_{\mathrm{k},1}$$

- Set C – geotechnical bearing capacity or large deformations on foundations:

$$E_{\mathrm{d}} = \gamma_{\mathrm{G},j,\mathrm{sup}} G_{\mathrm{k},j,\mathrm{sup}} + \gamma_{\mathrm{G},j,\mathrm{inf}} G_{\mathrm{k},j,\mathrm{inf}} + \gamma_{\mathrm{P}} P + \gamma_{\mathrm{Q},1} Q_{\mathrm{k},1} + \gamma_{\mathrm{Q},i} \psi_{0,\iota} Q_{\mathrm{k},i}$$

where

γ = different load factors that might have different values in different formulas and different sets
ψ = concurrency factor
G = self-load (both favourable (inf) and unfavourable (sup))

P = prestressed forces
$Q_{k,i}$ = different characteristic live loads 1, 2, 3, etc., denoted by i

Satisfactory capacity is proven by calculating a higher resistance than the total load.

6.4.2 Accidental limit state (ALS)

In the ALS, the total stability of the whole construction exposed to accidental forces is checked. Accidental forces are forces that are not expected to occur, but might occur during the design life of the construction. Accidental forces might be seismic activity, which quite often is given with a return period of a few hundred up to several thousand years, uncontrolled impacts from ships or collision from other objects. Damage to the construction is accepted, but no progressive collapse. The capacity of bearing of the construction after collapse of singular units or elements is checked.

The constructions should be able to be repaired without total demolition and rebuilding.

(a) Accidental loading:

$$E_d = G_{k,j,sup} + G_{k,j,inf} + \gamma_P P + A_d + \psi_{1,1} Q_{k,1} \text{ (or } \psi_{2,1} Q_{k,1}) + \psi_{2,\iota} Q_{k,i}$$

(b) Seismic loading:

$$E_d = G_{k,j,sup} + G_{k,j,inf} + \gamma_P P + A_{Ed} \, (=\gamma_1 A_{Ek}) + \psi_{2,\iota} Q_{k,i}$$

where

A_d = design value of the accidental load
A_{Ek} = characteristic seismic load

6.4.3 Fatigue limit state (FLS)

Normally, the FLS is not a limit state that needs to be considered for coastal constructions. Fatigue will occur with frequent high loads repeated over a long time. During the lifetime of harbour constructions, the number of repeated load cycles will normally be significantly beyond the critical level of fatigue. Special care needs to be taken for constructions where loads from waves are dominating loads and the element reaction from wave loads varies from tension to compression. This issue is quite unique, and specialists should be consulted.

6.4.4 Serviceability limit state (SLS)

In the SLS, the design value of the loading affecting the durability and operability of the construction is calculated. The load combinations are divided into different sets according to the probability of occurring:

Characteristic combination:

$$E = G_{k,j,sup} + G_{k,j,inf} + P + Q_{k,1} + \psi_{0,\iota} Q_{k,i}$$

Frequently occurring combination:

$$E = G_{k,j,sup} + G_{k,j,inf} + P + \psi_{1,1} Q_{k,1} + \psi_{2,\iota} Q_{k,i}$$

Quasi-permanent combination:

$$E = G_{k,j,\mathrm{sup}} + G_{k,j,\mathrm{inf}} + P + \psi_{2,1}Q_{k,1} + \psi_{2,i}Q_{k,i}$$

The correct combination for the verification of the durability and/or operability is given in various material standards. Typical operability limitations are deflections and vibrations, while a durability restriction might be the maximum crack width for concrete.

6.5. Load and concurrency factors

Load factors and concurrency factors are normally given in standards together with load combinations. Note that the factors may vary from general default values in different National Annexes and from one standard to another and between different countries. Examples of such factors are given in Tables 6.2 and 6.3.

As a general recommendation, the entire set of standards for the country in which you are designing constructions should be acquired in order to obtain an overview of the design codes. In design codes applying to several countries, the National Annex should be used, if present.

6.6. Material factors and material strength

In order to provide sufficient safety for the capacity of a structure, all construction materials are given a design strength with a very low probability of encountering values beneath this value. Normally, the material strength is linked to strength classes, and specimens of different materials are tested in order to verify the material properties. Procedures and testing methods are given in material standards and design codes.

Typical material factors are:

Concrete	$\gamma_c = 1.5$
Reinforcement	$\gamma_s = 1.15$
Structural steel	$\gamma_m = 1.05$
Timber	$\gamma_m = 1.15$

The design strength of the material is found by dividing the characteristic strength by the material factor. For example, the concrete compression strength is determined as $f_{cd} = \alpha_{cc} f_{ck}/\gamma_c$, where α_{cc} is a factor taking account of the long-term effects on the compressive strength of concrete resulting from the duration of the load and the way it is applied.

6.7. Characteristic loads from the sea side

Most common loads from the sea side are shown in Figure 6.1, and calculation of these forces is described in Chapter 5. In addition to these loads, there might also be a possibility of the berth structure being exposed to slamming and/or uplifting forces from wave action. Further information can be found in British standard BS 6349-1 (BS, 2000) for piers, jetties and related structures exposed to waves.

Reaction loads from fenders and mooring load bollards can be very high. Normally, the reaction from fenders is a horizontal respond force, which has to be transmitted to the foundations. Special care needs to be taken to parallel body fenders, which in addition to a direct reaction force will also apply a pair of tension/compression loads to implement eccentricity between the load from the vessel and the rubber respond unit in the fender.

Table 6.2 Example of concurrency factors from Eurocode EN 1990

Action	Symbol		ψ_0	ψ_1	ψ_2
Traffic loads (see EN 1991-2, Table 4.4)	grla (LM1 + pedestrian or cycle-track loads)[a]	TS	0.75	0.75	0
		UDL	0.40	0.40	0
		Pedestrian + cycle-track loads[b]	0.40	0.40	0
	grlb (Single axle)		0	0.75	0
	gr2 (Horizontal forces)		0	0	0
	gr3 (Pedestrian loads)		0	0	0
	gr4 (LM4 – Crowd loading))		0	0.75	0
	gr5 (LM3 – Special vehicles))		0	0	0
Wind forces	F_{Wk}				
	■ Persistent design situations		0.6	0.2	0
	■ Execution		0.8	–	0
	F_W^*		1,0	–	–
Thermal actions	T_k		0.6[c]	0.6	0.5
Snow loads	$Q_{Sn,k}$ (during execution)		0.8	–	–
Construction loads	Q_c		1.0	–	1.0

[a] The recommended values of ψ_0, ψ_1 and ψ_2 for grla and grlb are given for road traffic corresponding to the adjusting factors α_{Qi}, α_{qi}, α_{qr} and α_Q equal to 1. Those relating to the UDL correspond to common traffic scenarios, in which a rare accumulation of lorries can occur. Other values may be envisaged for other classes of routes, or of expected traffic, related to the choice of the corresponding α factors. For example, a value of ψ_2 other than zero may be envisaged for the UDL system of LM1 only, for bridges supporting severe continuous traffic. See also EN 1998

[b] The combination value of the pedestrian and cycle-track load, mentioned in Table 4.4a of EN 1991-2, is a 'reduced' value. ψ_0 and ψ_1 factors are applicable to this value

[c] The recommended ψ_3 value for thermal actions may in most cases be reduced to 0 for ultimate limit states EQU, STR and GEO. See also the design Eurocodes

Note 1. When the National Annex refers to the infrequent combination of actions for some serviceability limit states of concrete bridges, the National Annex may define the values of $\psi_{1,infq}$. The recommended values of $\psi_{1,infq}$ are:

- 0.80 for gr1a (LM1), gr1b (LM2), gr3 (pedestrian loads), gr4 (LM4, crowd loading) and T (thermal actions)
- 0.60 for F_m in persistent design situations
- 1.00 in other cases (i.e. the characteristic value is used as the infrequent value)

Note 2. The characteristic values of wind actions and snow loads during execution are defined in EN 1991-1-6. Where relevant, representative values of water forces (F_{wa}) may be defined in the National Annex or for the individual project

Note 3. This table gives concurrency factors for bridges. For example, in Norway the National Annex contains more suitable concurrency factors than the general part of the Eurocode, and includes forces from the wind, waves and current that are present on coastal constructions

Respond from bollards might be horizontal or oblique. The horizontal respond force has to be taken to the foundations, and the vertical part has to checked regarding self-weight. If the vertical part of the respond force is higher than the local self-weight, the berth has to be anchored to avoid uplift.

Table 6.3 Example of load factors and load combination from Eurocode EN 1990

Persistent and transient	Permanent actions		Leading variable action[a]	Accompanying variable actions	
	Unfavourable	Favourable		Main (if any)	Others
Equation 6.10	$\gamma_{G,j,sup}G_{k,j,sup}$	$\gamma_{G,j,inf}G_{k,j,inf}$	$\gamma_{k,i}Q_{k,i}$		$\gamma_{Q,i}\psi_{0,i}Q_{k,i}$

[a]Variable actions are those considered in Table A1.1 of EN 1990.

Note 1. The γ values may be set by the National Annex. The recommended set of values for γ are:

$\gamma_{G,j,sup} = 1.10$
$\gamma_{G,j,inf} = 0.90$
$\gamma_{Q,i} = 1.50$ where unfavourable (0 where favourable)
$\gamma_{Q,i} = 1.50$ where unfavourable (0 where favourable)

Note 2. In cases where the verification of static equilibrium also involves the resistance of structural members, as an alternative to two separate verifications based on Tables A1.2(A) and A1.2(B) of EN 1990, a combined verification, based on Table A1.2(A), may be adopted, if allowed by the National Annex, with the following set of recommended values. The recommended values may be altered by the National Annex

$\gamma_{G,j,sup} = 1.35$
$\gamma_{G,j,inf} = 1.15$
$\gamma_{Q,i} = 1.50$ where unfavourable (0 where favourable)
$\gamma_{Q,i} = 1.50$ where unfavourable (0 where favourable)

provided that applying $\gamma_{G,j,inf} = 1.00$ both to the favourable part and to the unfavourable part of permanent actions does not give a more unfavourable effect

Note 3. This table gives load factors for bridges in ULS set A. For example, in Norway the National Annex contains more suitable load factors than the general part of the Eurocode, and includes forces from the wind, waves and current that are present on costal constructions

6.8. Vertical loads on berth structures

6.8.1 Live loads and wheel loads

The live-load capacity of a berth construction should be set in cooperation between the designer and the owner. All the expected uses of the berth should be noted, and also all the expected loads. As a general rule, berths with special functions such as oil jetties can be designed for a more specified live load closer to actual live loads than is the case for multi-purpose berths.

It is difficult to lay down guidelines for live loads on aprons as a function of the ship's size. The loads on the apron deck are determined by the type of traffic utilising the berth, and not so much by the size of the ships. Special berths such as oil piers accommodate ships of hundreds of thousands of tonnes displacement, but have live loadings of, say, 10 kN/m^2. On the other hand, berths accommodating supply ships for the offshore oil industry of only, say, 2000 t displacement must be designed for a live load of 50–200 kN/m^2. Berths for heavy industry should be designed for a live load of 40–100 kN/m^2. In fishing harbours, the berth structures should be designed for a live load of at least 15 kN/m^2. The loads are therefore essentially dependent on the type of cargo, on the handling equipment, local practices, etc., so that uniformity can only be achieved to a limited extent.

Table 6.4 Typical live loads

Type of traffic and cargo	Loading: kN/m^2
General loading	
Light traffic or small cars	≥5
Heavy traffic or trucks	≥10
General cargo	≥20
Palletised general cargo	20–30
Multi-purpose facility	≥50
Offshore feeder bases	50–200
Heavy vehicles, heavy crane, crawler crane, etc., that operate from the berth front and 3 m inboard	≥60
Heavy vehicles, heavy crane, crawler crane, etc., that operate from 3 m behind the berth front and further inwards	40–100
Container loading:	
Empty and stacked four high	15
Full and stacked two high	35
Full and stacked four high	55
General ro/ro loads	30–50
Special berths	
Oil jetties	≥20
Fish industry	≥15
Timber berths	30

As a general guideline, Table 6.4 lists recommended live loads for the apron and the terminal area.

In the case of a very exposed open berth structure, the possibility of uplifting of the deck structure due to waves passing under should be considered.

Table 6.4 gives typical values for different types of berth constructions, and may be used as the first guideline for predesign.

Most public berths (multi-purpose berths), accommodating ocean-going dry-cargo ships, should be designed for container loads. Twenty-foot containers stacked two high imply a load of 25–35 kN/m^2, depending on the cargo they are loaded with. The sizes of a 20 ft and 40 ft container are, respectively, 6.06 m × 2.44 m × 2.44 m and 12.12 m × 2.44 m × 2.44 m. The empty weight of a 20 ft container is in the range 19–22 kN, and the maximum total weight permitted by the ISO (the international container standard) is 240 kN. For a 40 ft container, the empty weight ranges between 28 and 36 kN, with a maximum total weight of 305 kN. Aprons and ramps for container traffic should be designed for a live load of 40 kN/m^2 or higher.

Useful loads in transit sheds and warehouses depend to a great extent on the height to which palletised cargo can be stacked with fork-lift trucks. Design loads might vary between 20 and 50 kN/m^2 (or more) over the whole floor area, depending on the type of cargo. To prevent overloading of the

Figure 6.3 A rubber-tyre gantry

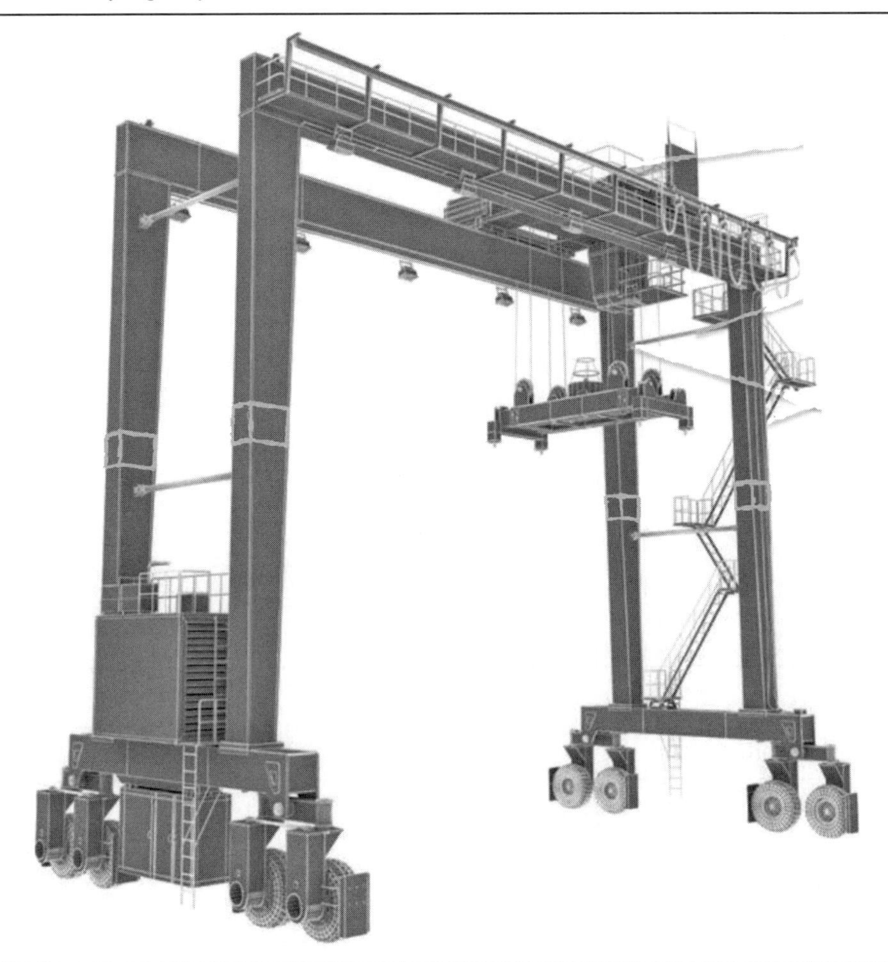

berth structure, the allowable load should be marked in clear letters and numbers on a signboard at the apron.

Wheel loads from trailers, fork-lift trucks, mobile cranes, container cranes and other cranes on rails, railways, etc. should be evaluated in each case, because there is, in the market nowadays, a spectrum of types and makes of such equipment with individual loading specifications. Fork-lift trucks for handling 40 ft containers can have axle loads of up to 1200 kN. A fork-lift truck will give a wheel load slightly higher than the maximum wheel loads transmitted to the pavement during take-off by a Boeing 747.

A rubber-tyre gantry (RTG), like the one shown in Figure 6.3, has a lower wheel load than a fork-lift truck.

Where a ship-to-shore gantry (Figure 6.4) or mobile harbour cranes (Figure 6.5) operate in the area behind the berth line, then provision should be made for the outrigger reactions and bearing pressures

Figure 6.4 Example of a ship-to-shore gantry crane

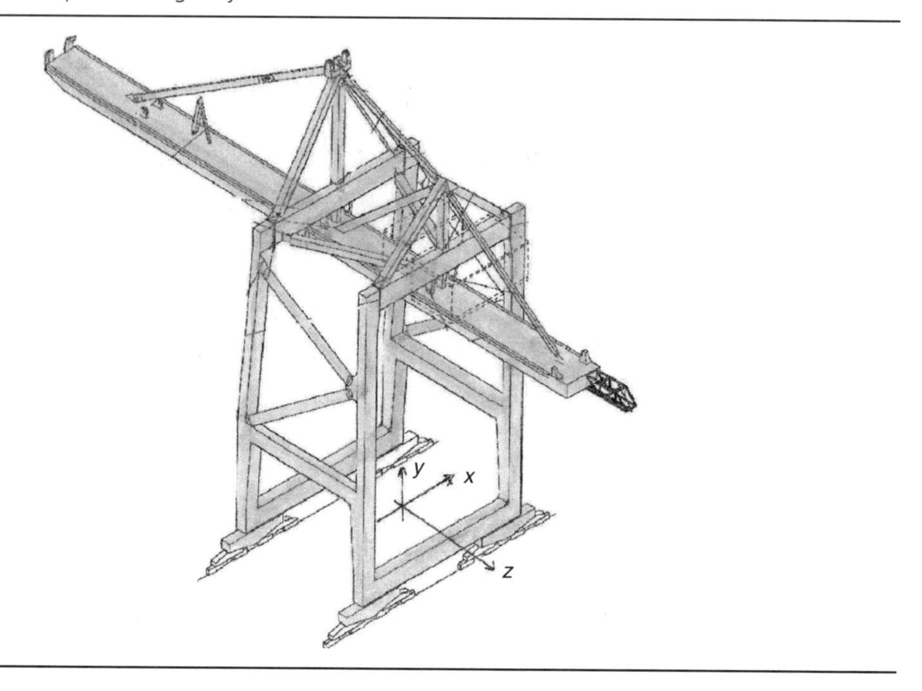

Figure 6.5 A mobile harbour crane

that may be imposed by the maximum size of the crane anticipated. The outrigger reactions are largely dependent on the crane lifting capacity and the radius of the jib length.

Typically, dedicated beams support the rails of gantry cranes, but mobile harbour cranes might be randomly placed with, normally, significantly higher pad loads than general live loads on the top slab of the berth. Support loads from mobile harbour cranes (even distributed on pads) are normally the dominant live load for the top slab, if present.

If no information on mobile cranes or other equipment can be obtained, the berth structure or the apron should be designed for a concentrated point load of at least 700 kN on an area of 1.0 m × 1.0 m in an arbitrary location.

Dynamic effects from moving equipment or loads should be implemented if present. As a conservative first recommendation, wheel loads for railways might be increased by 10%, and for fork-lift trucks and cranes by 20%, due to dynamic impacts. Both the berth apron and the whole container yard should be designed in a homogeneous way, and must be able to carry the heaviest combination of wheel or static loads for all handling equipment that may be present in the areas (container cranes, trailers, fork-lifts, straddle carriers, etc.). In Chapter 13, on container terminal equipment, the different types of container handling equipment are described.

To reduce the effect from a concentrated point load acting directly on a concrete deck slab, or to increase the loading area from a concentrated point load, a layer of sand and asphalt can be placed on top of the concrete slab (Figure 6.6).

Berth structures that have a direct road connection to the public highway network should at least be designed for loads in accordance with Highways Department regulations. A concentrated load of 150 kN on an area of 0.2 m × 0.5 m randomly placed or a live load with an intensity of 20 kN/m^2 or higher should be specified.

Useful loads in transit sheds and warehouses depend to a great extent on the height to which palletised cargo can be stacked with fork-lift trucks. Design loads vary between 20 and 50 kN/m^2 (or more) over the whole floor area, depending on the type of cargo.

To prevent overloading of the berth structure, the allowable load should be marked in clear letters and numbers on a signboard at the apron.

Figure 6.6 Load distribution area

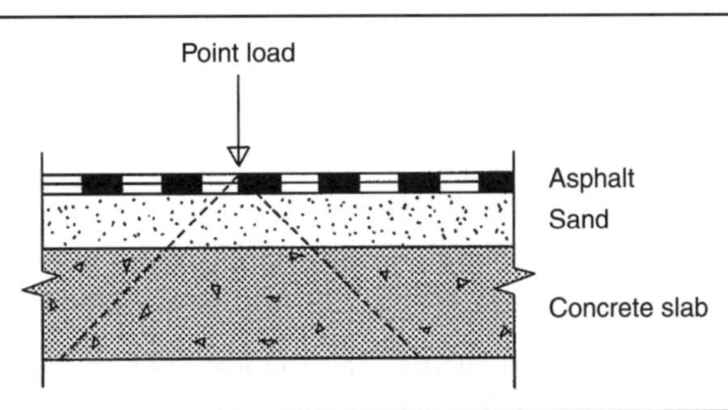

6.8.2 Temperature and shrinkage forces

In the design of, for example, berth decks of reinforced concrete, allowances must be made for temperature and shrinkage forces in the transverse direction as well as in the longitudinal direction of the deck. Guidance can be found in design standards.

6.8.3 Seismic loads

In many countries, an obligatory check regarding seismic or earthquake loads on the berth structure is introduced even in areas of low seismographic disturbance. The seismic loads will act at the centre of gravity of the structure as a horizontal and/or a vertical force. Combinations of load directions are given in standards and guidelines. In principle, a design coefficient is multiplied by the weight of the structure. The design seismic coefficient will usually be in the range 0.05–0.25.

The actual seismic load due to an earthquake will depend on the magnitude of the earthquake, the type of structure, the type of equipment and the soil conditions in the area. Generally, unless the berth structure is of a massive or gravity type, the seismic effect on the design will usually be small in regions with low seismic activity, but can be a major force in areas of high seismic activity. This applies to both the transverse and longitudinal directions of the berth structure.

The seismic performance requirements for a particular berth structure should be based on national standards if present or else in accordance with international standards and guidelines based on acceptable risk procedures. The requirements for a berth structure should be based on the importance of the berth structure, the acceptable levels of risk to life, safety, the port operations, etc.

6.9. Horizontal loads on the berth

The berth might be exposed to a high number of horizontal loads. Compared with vertical loads, the horizontal loads on top of the berth are relatively small, but can be of high relevance, especially regarding local fixings. Typical horizontal loads are wind on equipment and buildings, and braking loads from cranes and trucks.

As a first recommendation, a load in the direction perpendicular to the rails is about 1/10 of the vertical wheel load. For rubber-tyre-mounted equipment, a factor of 1/10 is also applied. In order to prevent vehicles from rolling over the berth line into the water, a front curb should be installed along the berth front. This curb should be designed for a horizontal point load of 15–25 kN, depending on the type of traffic, and should be about 0.20 m high.

Storing of frictional material will imply a horizontal load component being transmitted to the apron deck in addition to the vertical load. This component, acting as a tensile force in the top of the deck, is equal to the maximum static friction of the stored material.

6.10. Characteristic loads from the land side

The weight of the fillings and loads on fillings may cause horizontal loading to the berth structure.

Differences in water pressure might also cause horizontal loading, especially on steel sheet berths, which have a dense front towards the sea. For instance, the magnitude of such forces in a blockage of the drainage system behind the berth structure can be significant, and must be evaluated for each separate case.

The weight of the fill behind the berth structure and the live load on top of the fill will serve as a stabilising load, for example on berth anchoring friction plates.

Table 6.5 Example design loads

	Design load
Ship size, maximum	137 000 m^3
Ship size, maximum fully loaded displacement	100 000 t
Ship size, minimum	20 000 m^3
Approach, ship velocity normal to jetty front	0.15 m/s
Approach, ship velocity parallel to jetty front	0.02 m/s
Ship angle of approach to jetty front fully loaded	5°
Hull pressure between the fenders and the ship, maximum	0.20 MPa
Friction coefficient between the ship hull and fender, horizontal and vertical	0.2
Quick release hook with capstan, minimum capacity of each hook	1 000 kN
Loading platform and access road:	
Point load at any point on the loading platform and access road on an area 1 m × 1 m	700 kN
Vertical live load, general	20 kN/m^2
Pipeline:	
Vertical live load	2.5 kN/m^2
Walkways:	
Vertical live load	3.0 kN/m^2
Horizontal load top handrail	0.8 kN/m
Mooring dolphins:	
Vertical live load	3.0 kN/m^2

A further discussion of forces acting from the land side is outside the scope of this book, but both EAU 2012 and ROM 3.1-99 give useful recommendations.

6.11. Summary of loads acting from the sea side

In addition to static and dynamic conditions involved when design loads on the berth structure are being established, the human factor during, for example, manoeuvring the ship to the berth, also needs be taken into account. More conservative design load values than those that are strictly necessary according to detailed calculations are recommended.

As an example, if the contractors carrying out a tender are allowed to give alternative designs for a berthing structure for gas carriers, the minimum design loads shown in Table 6.5 should be given.

REFERENCES AND FURTHER READING

BSI (British Standards Institution) (2000) BS 6349-1: 2000. Maritime structures. Code of practice for general criteria. BSI, London.

BSI (1990) BS EN 1990. Eurocode. Basis of structural design. BSI, London.

CUR (Centre for Civil Engineering Research and Codes) (2005) *Handbook of Quay Walls*. CUR, Gouda.

EAU (Empfehlungen des Arbeitsausschusses für Ufereinfassungen) (2004) *Recommendations of the Committee for Waterfront Structures, Harbours and Waterways*, 8th English edn. Ernst, Berlin.

McConnel K, Allsop W and Cruickshank I (2004) *Piers, Jetties and Related Structures Exposed to Waves. Guidelines for Hydraulic Loadings*. Thomas Telford, London.

Ministerio de Obras Públicas y Transportes (1990) ROM 0.2-90. Maritime works recommendations. Actions in the design of maritime and harbour works. Ministerio de Obras Públicas y Transportes, Madrid.
Ministerio de Obras Públicas y Transportes (2007) ROM 3.1-99. Recommendations for the design of the maritime configuration of ports, approach channels and harbour basins. English version. Puertos del Estado, Madrid.
PIANC (International Navigation Association) (1998) *Life Cycle Management of Port Structures. General Principles. Report of Working Group 31*. PIANC, Brussels.
PIANC (2001) *Seismic Design Guidelines for Port Structures. Report of Working Group 34*. PIANC, Brussels.
Port and Harbour Research Institute (1999) *Technical Standards for Port and Harbour Facilities in Japan*. Port and Harbour Research Institute, Ministry of Transport, Tokyo.

Port Designer's Handbook
ISBN 978-0-7277-6004-3

http://dx.doi.org/10.1680/pdh.60043.175

Chapter 7
Safety considerations

7.1. General

The safety measures that have to be considered in a harbour project relate to the safety of the specification, the design, the construction, the personnel and the operation. Safety-related factors are important in the work of the consulting engineer, and all the above aspects should therefore be given the highest priority during his or her consulting work.

7.2. Specification safety

On a harbour project, securing satisfactory implementation of safety should be considered at the specification or start-up phase. In this phase all the engineering standards, design codes, governmental laws and regulations have to be defined and listed as the project engineering specifications.

The safety routines for all the work to be performed by the consulting engineer should be implemented through the quality assurance and the control system for the project, and through the project coordination and engineering procedures. The project quality assurance manual and the project coordination and engineering procedures should give detailed regulations for review and approval, both internal for the project team, and in relation to client and the external interfaces.

7.3. Design safety

The design of berth structures should be based on common and proven design methods and technology. In order to determine the different load conditions that may occur over the lifetime of a berth structure, it is convenient to distinguish between the following three conditions.

(a) **Operational condition**: the design condition, which takes into account the normal design loads based on the national standards or recommendations for berthing structures. In this case, for example, the fender system should be able to absorb the normal design berthing energy related to the ship approach velocity without damage occurring to either the berth structure or the ship itself. For reasons of safety, the maximum acceptable approach or berthing velocity should not be more than about two-thirds of the design approach velocity for the berth structure. The design berthing energy should not be higher than the fender manufacturer's recommended absorbed or rated energy of the fender.
(b) **Accidental condition**: for example, the 'engines out' condition that may affect a berthing or unberthing ship. In this case, the berthing energy may be higher than in the operational condition. Damage to the fenders may be allowed or expected to occur in this condition, but the concrete breasting structure itself should be constructed so as not to collapse under such an impact. Under accidental conditions, the concrete structure should be able to resist a horizontal force due to a 20–25% higher berthing energy than the design berthing energy in the operational condition, without a total collapse of the berth structure (see Section 5.5 in

Chapter 5 on abnormal impacts). This is because it is very difficult to define limiting values for the exact value of a ship's approach velocity. The velocity will depend on the wind, waves, current and the number of tugboats assisting during the berthing operation.
The more difficult it is to estimate the characteristic loads against a structure, or if the consequences of a collapse of a structure will be very serious, the more important it is that the structure has as high as possible a reserve capacity against collapse.

(c) **Catastrophic condition**: this condition covers the situation where a large unfamiliar ship (a large cargo or tourist ship which would not normally use the berth) impacts, for example, an oil berth construction at speed, possibly causing total collapse of the structure. It is uneconomical and often impossible to construct a berth structure that will resist such an impact. Therefore, decreasing the probability of such accidents occurring by changing the sailing routes or imposing restrictions on other ship traffic is often the only possible course of action.
The layout of an oil jetty (Figure 7.1) is based on the principle that the loading arms and other items of functional equipment are placed on a separate **loading platform** that is free from normal horizontal berthing forces from the oil tankers.

The horizontal berthing or impact forces, normal to the berthing face, are taken up by separate breasting structures, or breasting dolphins, as shown in Figure 7.1. This design philosophy of separating the breasting dolphins from the loading platform reduces the possibility of damage to the loading platform during the berthing operation from normal operational or accidental berthing forces.

Different construction solutions for the cross-sections of the **breasting dolphins** are shown in Figure 7.2. Unlike breasting dolphins that have a large reinforced concrete caisson filled with sand or rock fill (Case B), breasting dolphins founded on raking piles with prestressed rock anchors (Case A) do not

Figure 7.1 The layout of an oil jetty

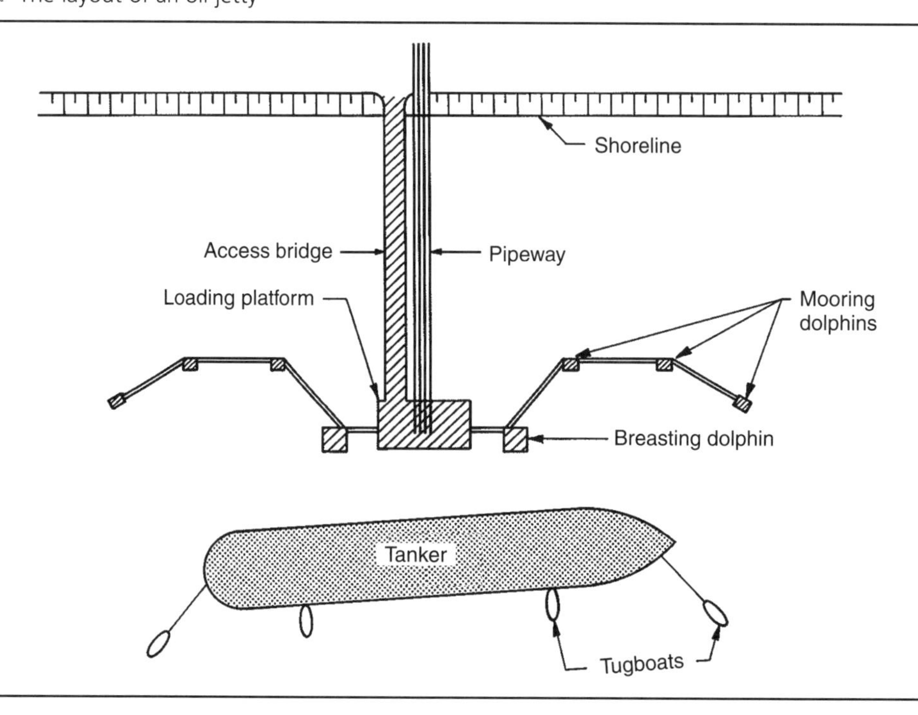

Figure 7.2 Breasting dolphins of different construction

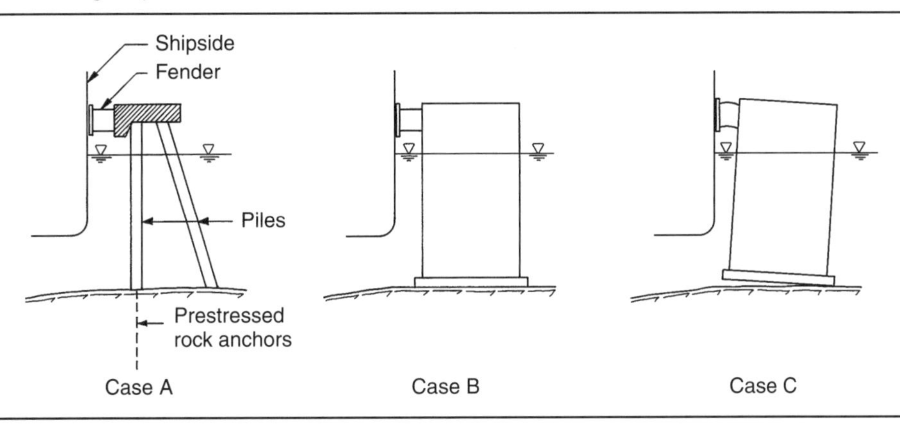

have any internal stability due to deadweight, and make use of rock anchors to take the horizontal forces arising from a berthing tanker. If a breasting dolphin founded on raking piles with rock anchors is overloaded, due to an accidental condition, it may collapse beyond repair, or may be unusable for a long time while under repair. On the other hand, if the caisson construction is overloaded, the caisson should be so designed that the overloading will only tilt it as shown in Case C, or push it a certain distance along the sea bottom. The repair work on the caisson could, therefore, be just a matter of adding some extra material to re-establish the original fender line, and the interruption of the operation of the terminal due to repair work will only be for a very short period.

If a breasting dolphin founded on raking piles with prestressed rock anchors has to be used, an additional increase in safety can be achieved by constructing either an energy-absorbing concrete or steel overloading **collapsible unit** between the fender and the dolphin head (Figure 7.3), or by designing

Figure 7.3 Concrete overloading collapsible unit

Figure 7.4 Berthing to an oil quay

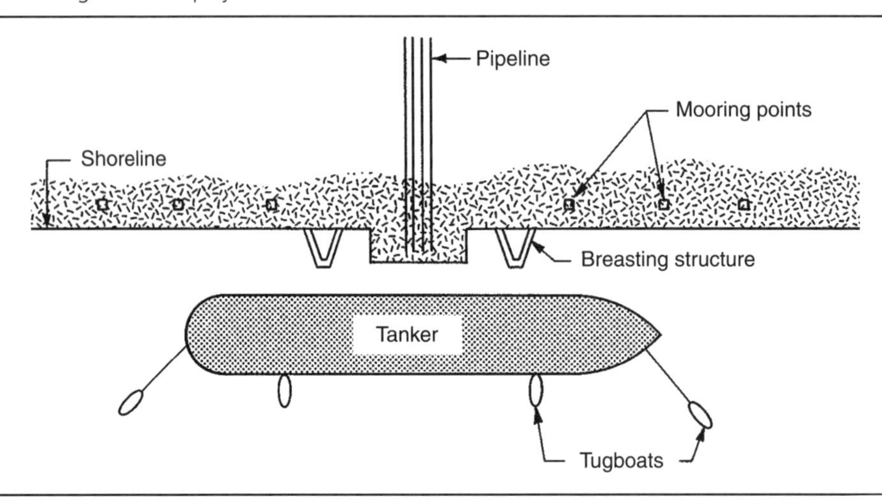

the breasting dolphin itself for a reaction force 2–3 times higher than the fender reaction forces. If few tugboats are used and the weather conditions at the site are generally rough, a higher reaction force should be used in the design of the breasting dolphin itself.

The design philosophy for **mooring dolphins** shown in Figure 7.1 assumes that the safety against overloading from the mooring hawsers can be taken care of by the anchor bolts for the mooring equipment in the concrete top slab of the mooring dolphin. The anchor bolts are only designed for a maximum horizontal force, which will not affect the stability of the mooring dolphin itself. As shown in Figure 7.1, there are three mooring dolphins on each side of the loading platform. If an accidental condition should destroy one of the mooring dolphins, the other two can, in an emergency situation, be used alone while the destroyed dolphin is being repaired.

In situations where the sea bottom slopes steeply under the berth structure up to the water level so that the distance from the berth line to shoreline is small (Figure 7.4), the tanker has to squeeze aside the water between the itself and the shoreline during the berthing operation. This squeezing effect will increase the stopping effect of the tanker compared with berthing in a totally open berth structure (see Figure 7.1). This effect is called the 'water cushion effect', and can reduce the berthing energy by up to about 10–20%. To get this water cushion effect, the shoreline must be parallel to the berth line, and the parallel length of the shoreline must be at least as long as the length of the berthing tanker. To get the maximum reduction in the berthing energy due to the water cushion, the distance between the shoreline and the berth line should not be more than about 2 m. When the distance between the berth line and the shoreline is about 20 m (see Figure 7.4), the reduction in the berthing energy is, according to the experience of ship captains, only about 5%.

In the case of **oil spill** or pollution from a moored tanker in front of a berth structure as in Figure 7.4, it will be easier to control the pollution where there is this type of structure than it would around a totally open jetty structure (see Figure 7.1). In the case of oil pollution and/or fire, a structure such as the one shown in Figure 7.4 will also provide easier and safer escape routes for personnel and easier and safer vehicular access.

7.4. Construction safety

Safety considerations with regard to the construction of berth structures should, if possible, always be based on common and proven construction methods and technology. The safe construction solutions should be easy to implement, and the structure should be adapted to the construction equipment that is available.

When constructing harbour structures sophisticated solutions to issues should be avoided. Experience has often shown that the amount of maintenance required for harbour structures is generally proportional to the amount of sophisticated solutions selected at the construction phase.

7.5. Personnel safety

Safety considerations associated with the personnel working at, for example, petroleum jetties are characterised by the following:

(a) Designed for maximum fire prevention. This means that piping should be routed and designed to avoid failure from predictable causes. Known sources of ignition should be shielded. There should be effective control against ignition from static current.
(b) Provided with effective fire protection. Valves should be available to quickly stop the flow of hydrocarbons through piping that is susceptible to leakage. Fire-fighting facilities should be simple to operate and easily maintained. All fire-fighting equipment should be focused on piping elements that are susceptible to leakage or vulnerable to damage, etc.
(c) Provided with effective emergency evacuation routes for jetties. Emergency and escape routes for personnel should be clearly marked. For the safety of personnel the routes should be constructed so that it should not be necessary to jump into the water for rapid evacuation. Therefore, with regard to emergency evacuation from the jetty platform, escape routes should be designed taking into account the overall design of the jetty (e.g. all outer mooring dolphins should be designed with facilities for the mooring of lifeboats).

In addition, as far as the operation of the jetties permits, handrails should be provided in all places where there is a risk of falling into the sea, and safety ladders should form part of the design of all jetty platforms, dolphins, etc. All rescue and access ladders, bollards, curbs, walkways between mooring dolphins, etc. should be painted in orange or red self-illuminating paint.

7.6. Operational safety

For operations related to the environmental conditions during berthing and unberthing, it is very important that information on tide, current, wind, waves, etc., which together with the hydrographical and topographical conditions plays an essential role, has been collected. These factors are essential both for the safety of the ship during navigation and berthing operations and for the safety of cargo-handling operations, and should be carefully considered in the planning and evaluation of proposed harbour sites, and the layout of berths, breakwaters and other structures.

7.7. Total safety

The total safety of a harbour will depend on the combined safety inherent in the specification, design and construction of the harbour, and the personnel and operational safety. To increase total safety it is necessary to decrease the risk of accidents occurring. The risk is equal to the probability times the consequence, and so to reduce the risk one has to reduce the probability and/or the consequences of an accident.

Figure 7.5 The relative likelihood of an accident occurring during each phase of the total navigation operation of an oil tanker

	Possibility for accident to		
	Berth structure	Personnel	Environmental
Arriving outer-harbour basin			● ● ●
Turning in harbour basin	●		● ● ●
Berthing operation	● ● ●	● ●	● ● ●
Mooring	● ● ●	●	●
Loading/unloading	●	● ● ●	● ●
At berth due to bad weather	● ● ●	●	● ●
Deberthing operation	● ●		● ● ●
Departure from harbour			● ● ●

● = small possibility ● ● ● = high possibility

The berthing manoeuvre of an especially large ship is always a slightly hazardous operation, and the berthing structures may be damaged, resulting in consequential losses in the operation of the facility or the terminal. Therefore, these events deserve the attention of the designer, as their probability and consequences may increase the total risk of the harbour operation. Figure 7.5 shows the relative possibility of an accident occurring during each phase of the total navigation operation, from arrival to departure, for the example of an oil tanker and potential oil spillage.

REFERENCES AND FURTHER READING

BSI (British Standards Institution) (2000) BS 6349-1: 2000. Maritime structures. Code of practice for general criteria. BSI, London.

Bruun P (1989) *Port Engineering*, vols 1 and 2. Gulf Publishing, Houston, TX.

EAU (Empfehlungen des Arbeitsausschusses für Ufereinfassungen) (2004) *Recommendations of the Committee for Waterfront Structures, Harbours and Waterways*, 8th English edn. Ernst, Berlin.

Ministerio de Obras Públicas y Transportes (1990) ROM 0.2-90. Maritime works recommendations. Actions in the design of maritime and harbour works. Ministerio de Obras Públicas y Transportes, Madrid.

Port and Harbour Research Institute (1999) *Technical Standards for Port and Harbour Facilities in Japan*. Port and Harbour Research Institute, Ministry of Transport, Tokyo.

Port Designer's Handbook
ISBN 978-0-7277-6004-3

http://dx.doi.org/10.1680/pdh.60043.181

Chapter 8
Types of berth structures

8.1. General

The purpose of a berth structure is mainly to provide a vertical front where ships can berth safely. The berth fronts are constructed according to one of the following two main principles (Figure 8.1):

(a) **Solid berth structure**: the fill is extended right out to the berth front, where a vertical front wall is constructed to resist the horizontal load from the fill and a possible live load on the apron. The solid berth structures can be divided into three main groups, depending on the principle on which the front wall of the structure is constructed in order to obtain sufficient stability:
- Gravity-wall structure: the front wall of the structure with its own deadweight and bottom friction will be able or self-sufficient to resist the loadings from backfill, useful load and other horizontal and vertical loads acting on the berth wall structure itself.
- Sheet pile structure: the front wall is not adequate to resist any horizontal loads acting on the structure, and must, therefore, be anchored to an anchoring plate, wall or rock behind the berth.
- Structure with a relieving platform: this is a type of sheet pile wall with a relieving platform or a slab behind the sheet pile to reduce the horizontal forces against the sheet pile structure (the principle is shown in Figure 8.6).

(b) **Open berth structure**: from the top of a dredged or filled slope and out to the berth front a load-bearing slab is constructed on columns or lamella walls.

A diagrammatic classification of berth structures according to the type and the construction method is shown in Figure 8.2.

It is difficult, however, to formulate precise guidelines for the choice of berth type in each individual case. With a view to choosing the technically and economically most favourable type, the factors mentioned below should be considered.

Berth structures should be designed and constructed to safely resist the vertical loads caused by live loads, trucks, cranes, etc., as well as the horizontal loads from ship impacts, wind, fill behind the structure, etc.

In general, solid berth structures are considered more resistant to loadings than open berth structures, both vertically and horizontally. Since the deadweight of a solid berth structure constitutes a greater part of the total structure weight than the deadweight of an open berth structure, the former is less sensitive to overloading. On the other hand, the safety factor applied for solid structures is normally lower than for open structures.

Figure 8.1 The two main principles of berth structures

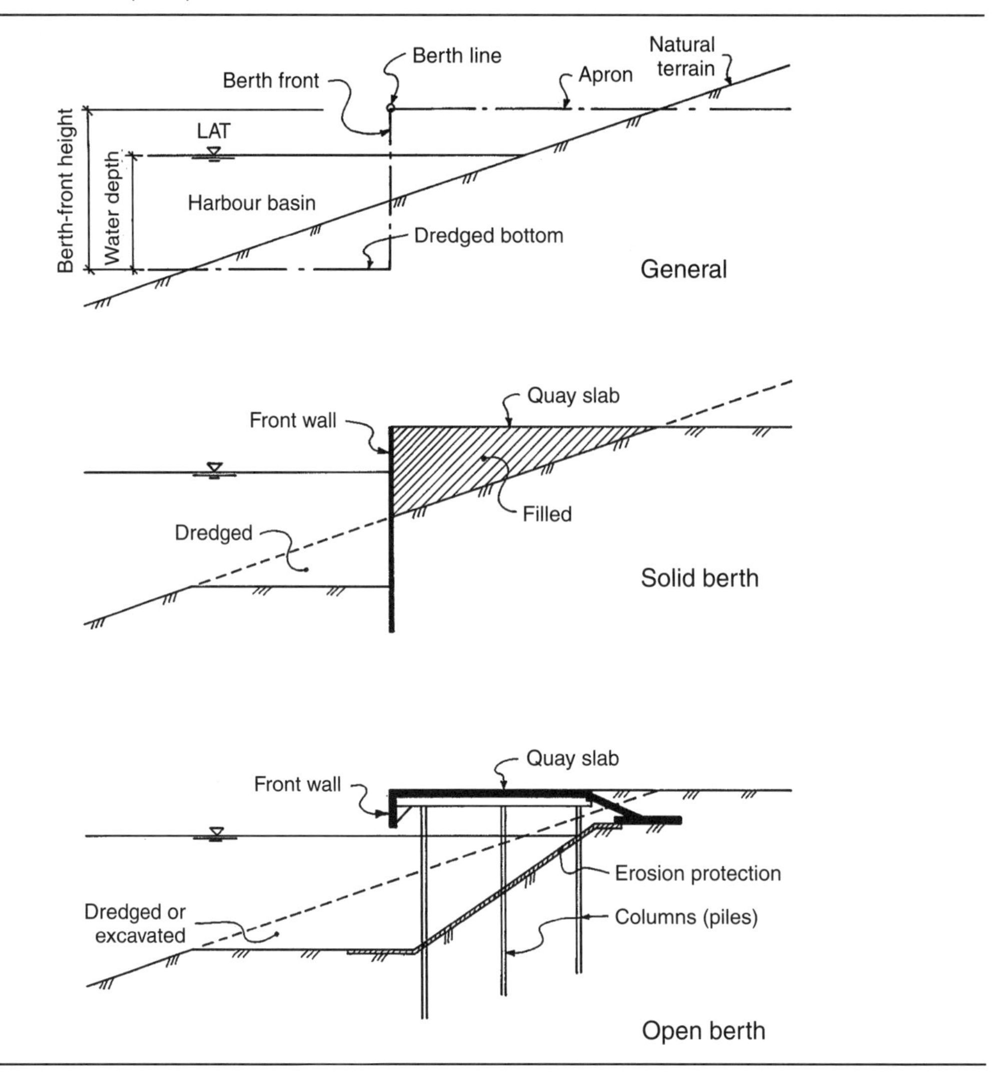

Figure 8.2 Types of berth structures

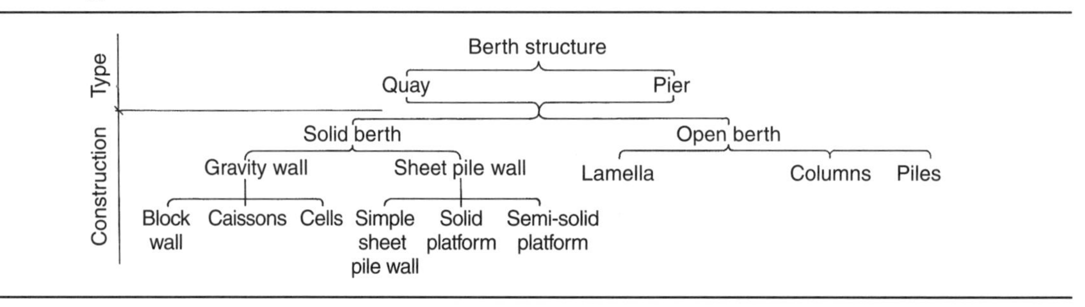

Figure 8.3 The wave height compared with the berth height. (Photograph courtesy of Trondheim Port Authority)

For instance, in an open-column berth for ocean-going ships the deadweight of beams and slab is about 15 kN/m^2 berth deck area, while the live load is normally 40 kN/m^2.

Such a berth of length 50 m and width 15 m weighs only about 1200 t, but will have to resist the impacts from ships of, say, 30 000 t displacement or more.

Solid structures are usually more resistant to impact than open structures; that is, the resistance to impact from ships decreases with increasing slenderness of the structure. For instance, a block wall wharf is far less vulnerable than a pier built as an open berth on piles. An exception to this rule is the open berth on wooden piles, where the whole structure is flexible and, when ships come alongside, yields sufficiently to absorb a substantial part of the impact energy.

8.1.1 Top elevation of the berth slab

The top elevation of berth structure should be determined by the following factors:

(a) the elevation of the terminal area behind the berth apron
(b) the highest observed water level and the tidal level
(c) the wind-raised water level in the harbour basin
(d) the wave action in the harbour basin
(e) the type of ship using the berth
(f) the harbour installations and the cargo operation.

Generally, for a cargo berth structure within an impounded dock, the top elevation of the berth slab and apron should be at least 1.5 m above the working water level. For berth structure directly connected to the open sea, the top elevation of the berth slab should be 0.5–1.0 m above the highest observed crest of waves in the port, depending on the type of cargo handled at the berth. The required height of the berth slab is illustrated very clearly in Figure 8.3, when a storm hit the berth structure. It is therefore always advisable to try to evaluate the highest wave height in the port.

8.2. Vertical loads

As shown in Figure 8.4, the vertical load on solid structures including live loads, crane loads, etc., will also cause a horizontal load on the front in addition to the load from the fill.

Figure 8.4 Loading on a solid berth

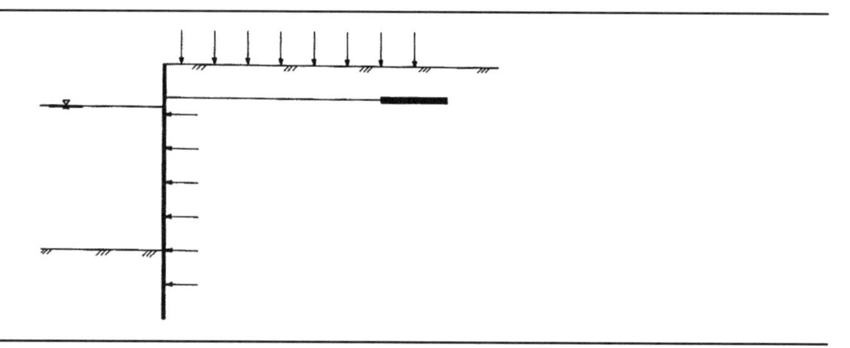

If the height of the berth front and/or the live load is large, the influence of the above horizontal load can be reduced or eliminated by building the structure as a solid or semi-solid platform berth (Figure 8.5).

A very common Dutch relieving platform structure is shown in Figure 8.6.

In open structures, all vertical loads are transmitted by way of the columns or lamella walls to rock, or to a load-resistant subsoil stratum (Figure 8.7).

In a slab/beam structure, the vertical loads are taken up by a system similar to a system of beams on elastic supports. The importance of this effect depends on the elasticity and slenderness of the columns, and the properties of the seabed material or the depth to rock. Since the distribution of the loads is determined by the rigidity of the slab and beam system in relation to the spacing of the columns, it can sometimes be recommended that the beams should be made more rigid, enabling a distribution of the loading to take place via a greater number of columns.

The columns are considered immovably connected with, and partly fixed to, the beams. The degree of fixation depends on the torsion resistance of the beams. At the bottom of the columns there are various degrees of fixation, depending on the thickness and properties of the seabed material above rock. To assure full fixation is hardly justified in any case. If the seabed layer is only 0–3 m thick, it must be

Figure 8.5 Loading on a solid or semi-solid platform

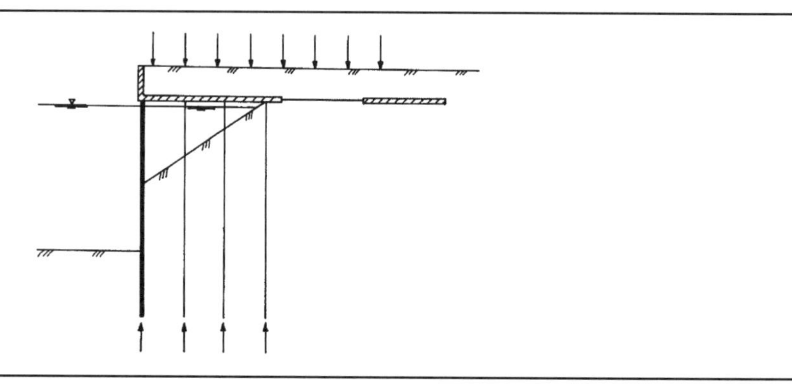

Figure 8.6 A common Dutch relieving platform structure

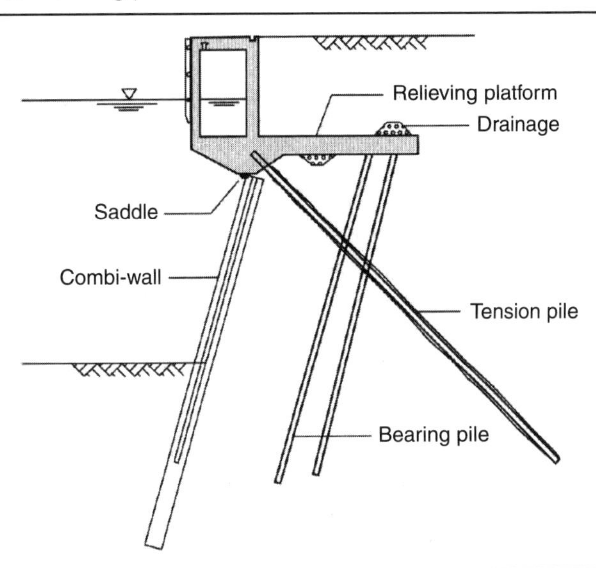

assumed that there is a joint at the rock. If the layer is of substantial thickness, it is usually correct to assume a joint located 3–5 m below the sea bottom, depending on the properties of the seabed material.

8.3. Horizontal loads

The bearing or absorption of the horizontal loads can take place at three levels:

(a) at the berth deck level
(b) between the deck and sea bottom levels
(c) at the bottom level.

In any case, the bearing of the loads should be arranged as simply and clearly as possible.

Figure 8.7 Loading on an open berth structure

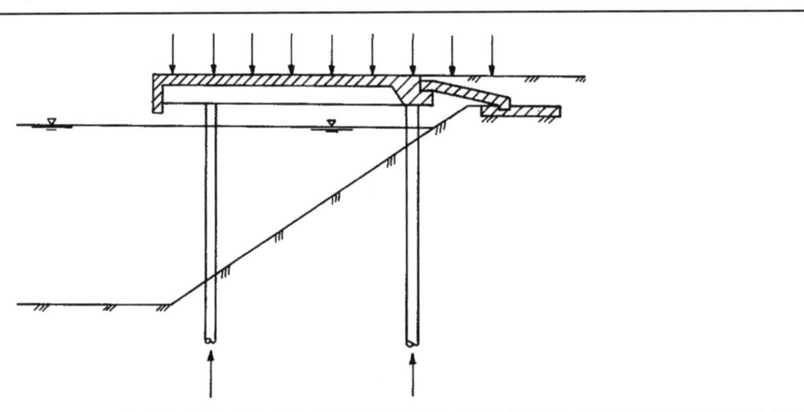

Figure 8.8 Bracing and anchoring at the berth deck level

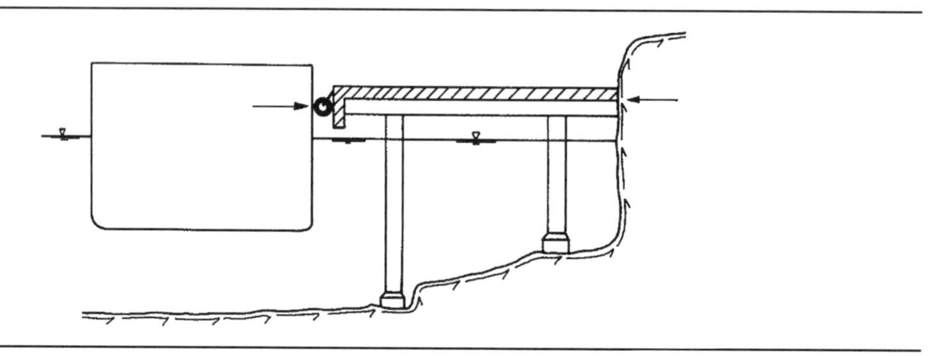

8.3.1 At the berth deck level

The simplest way of taking up horizontal loads at the berth deck level is to brace and anchor the berth deck (Figure 8.8) to, if possible, rock behind the berth.

However, open structures with high and slender columns or piles are sometimes difficult to anchor at the berth deck level. Horizontal loads must then be taken up by lamella walls or by anchoring the deck to the ground behind the berth structure.

The horizontal loads from ship impacts should not be transmitted by way of the columns or piles to the bottom level, assuming rigid frame conditions between the deck and columns. The reason is that even if the moments occurring at the column tops could be resisted in theory, there is a great danger of cracks developing in the column tops (i.e. in perhaps the most vulnerable part of the structure). A structural design rule therefore states that columns should not transfer horizontal loads from ship impacts.

In an isolated column berth, the horizontal load bearing can be arranged as shown in Figure 8.9, the berth deck (slab plus beams) is considered here as a rigid plate transmitting the ship impacts to the supports of A and B. From there, all horizontal loads are transmitted by way of rods 1, 2 and 3 to the immovable shore rock anchors at C, D and E. The anchor bolts in the rock must be protected against corrosion, and they must have a length sufficient to provide good anchorage in the rock. A compressive force in the rods would normally not cause any problem, even if the rock were cracked.

The principle behind this type of anchoring is that one of the supports (A) is made immovable in both the longitudinal and transverse directions of the berth deck, while the other support (B) is designed to allow movements in the longitudinal direction caused by temperature changes and shrinkage.

The berth deck itself will normally be able to resist the horizontal bending moments occurring between A and B due to ship impacts, although some additional reinforcement along the longer sides of the slab may prove necessary.

The support A can be connected to the anchors C and D by two tie rods, or these rods can be replaced by a trafficable bridge between A–F and C–D, or only rod 2 can be replaced by a bridge, keeping rod 1 as it is. The angle α should be wide, due to the considerable longitudinal forces that can act in the berth deck.

Figure 8.9 An isolated column berth

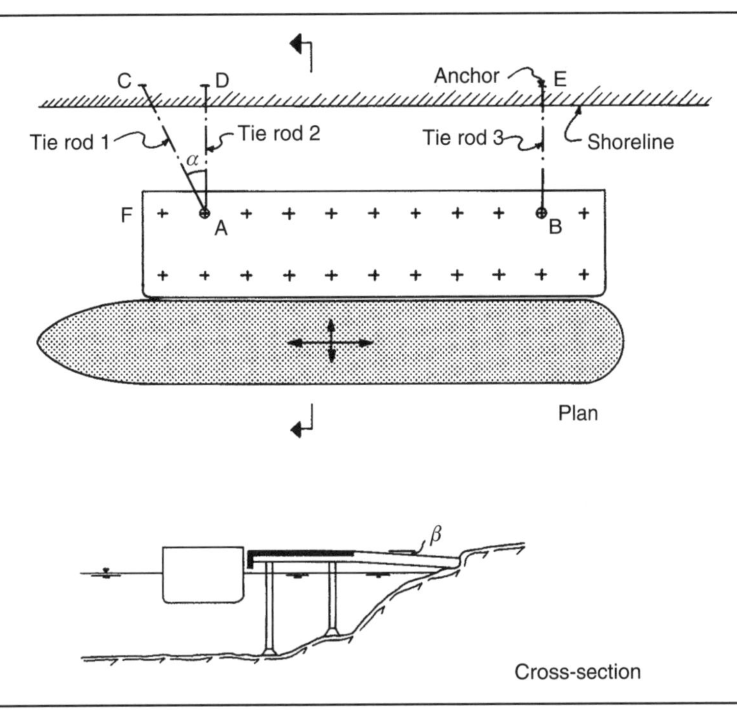

Since the tie rods transmit substantial forces, the connection between the rods and the berth deck should be arranged in such a way that the rods, the berth beam axis and the column axis meet at one point. This is to avoid secondary moments being set up. As shown in Figure 8.9, angle β should be as small as possible so that the vertical component to be transmitted by the column also becomes small.

The tie rods can be very long, and they are often supported by separate columns. They must be designed to carry dead load plus some live load and axial loads, and are often shaped as T beams.

8.3.2 Between the slab and sea bottom levels

As mentioned above, the anchors should take up the horizontal forces directly without any forces being transmitted to the columns or piles. Figure 8.10 shows a type of anchorage that is not acceptable because a passive soil pressure has to be established in front of the retaining wall before it can absorb the horizontal forces. In other words, the wall must have been subjected to a certain movement first.

In an approximately rigid structure, such as the berth on columns shown in Figure 8.10, the greater part of the horizontal force will be transmitted to the columns and possibly cause some damage to them. The most rigid element of a structure tends to attract the forces acting on the structure. The anchorage should therefore be as shown in Figure 8.11, consisting of a friction plate and a concrete tie rod, so that the force is transmitted directly to the anchorage.

The anchoring system must be able to take up horizontal forces acting both parallel to and perpendicular to the berth front, and must also be designed with a view to the deformations in the longitudinal

Figure 8.10 A steel tie rod and retaining wall anchoring

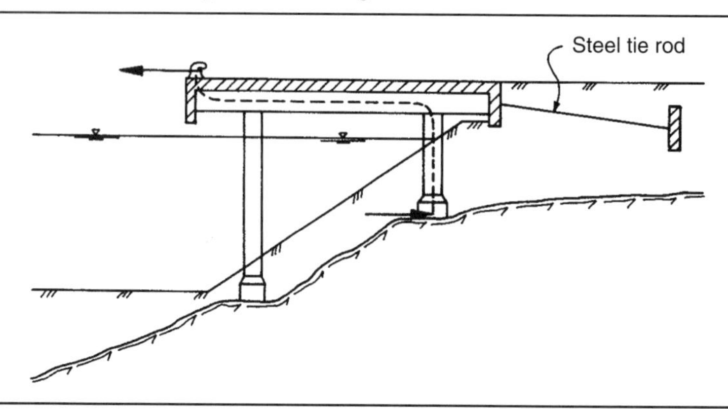

direction of the berth due to shrinkage and changes of temperature. As shown in Figure 8.9, the angle β should be as small as possible, and the axes of the tie rod, the beam and the column should intersect at one point.

8.3.3 At the bottom level

If the berth structure cannot be anchored at the deck level or between the deck and sea bottom levels, the forces must be transmitted to the bottom level and be taken up by one of the following means:

(a) batter columns or piles
(b) lamella walls
(c) cells.

One of these means will normally be applied for the bracing of a pier head. Problems are involved in this connection due to the width of the structure being relatively small, so that it can be difficult to get a stabilising moment sufficient to resist the overturning moment. In particular, open structures have a small stabilising deadweight.

In open structures, where bracing is provided by batter piles, the stabilising weight of the deck alone is seldom sufficient to take up the horizontal load. Figure 8.12 shows an open pier structure where ships

Figure 8.11 A concrete tie rod and friction plate anchoring

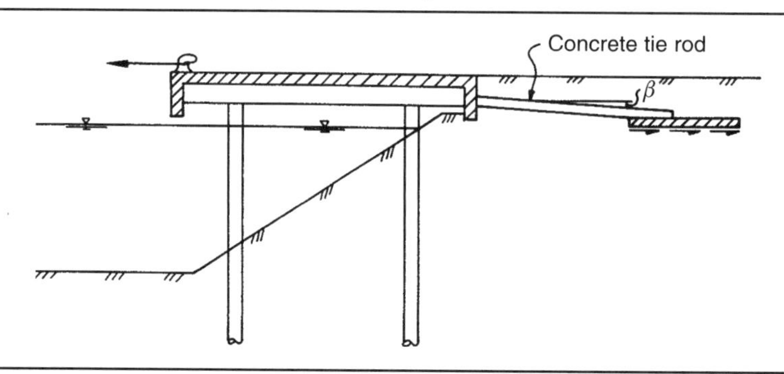

Figure 8.12 An open pier structure

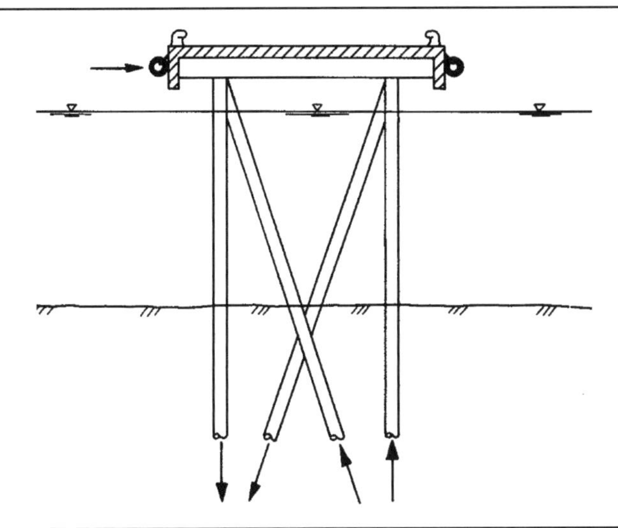

can berth at both sides. The horizontal load is taken up by help of the deadweight of the structure and adherence between the piles and soil.

Where sufficient adherence cannot be mobilised, the deadweight can be increased (Figure 8.13) by adding a layer of sand on top of the slab, or by making the concrete slab thicker.

Where the thickness of the seabed material above rock is too small to provide adequate adherence capacity, a sheet piling cell can provide an alternative for taking up horizontal loads. At the head of an important berth, construction of two cells should be considered for safety reasons.

If the sea bottom is rock, a lamella wall at the end of the open structure can be a good alternative (Figure 8.14). The stabilising moment from the deadweight of the wall and deck should be greater

Figure 8.13 Increased deadweight

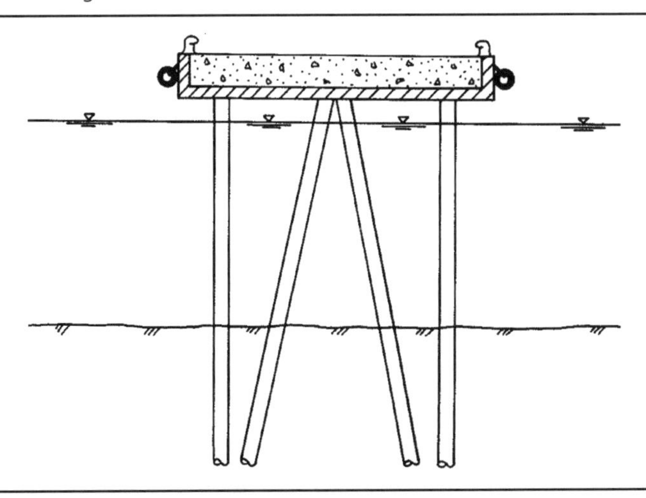

Figure 8.14 Anchoring through a lamella wall

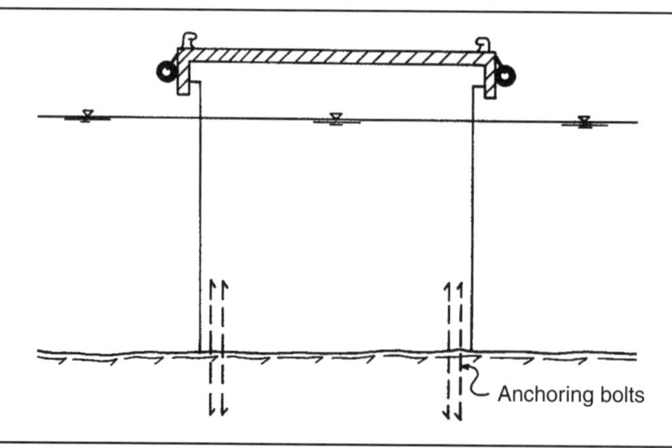

than the overturning moment from the horizontal load. Rock anchor bolts or post-tensioned anchor cables bored into the rock should give the necessary additional safety against overturning. The bolts or rods should be designed to allow for possible corrosion. Such bolts should have a factor of safety of at least 2 based on their net cross-sectional dimensions after corrosion, and the amount of rock that will be involved. In order to simplify the pouring of concrete under water, the wall can be designed as a frame wall (Figure 8.15).

Where the horizontal load is taken up by one of the above methods, the most economical scenario is to involve as few load transfer points as possible, and to design the columns for axial loads only (Figure 8.16).

If extremely severe impacts may occur once or twice during the lifetime of a structure, this should be taken into consideration. For instance, a dolphin could be designed to take up normal loads, while its foundation could be designed in such a way that it permits the dolphin to skid along the foundation under an extreme load. The cheapest solution would probably be to construct the dolphin in this way, and, in that case, to jack it back into its original position afterwards, if any skidding occurs.

Figure 8.15 A frame wall

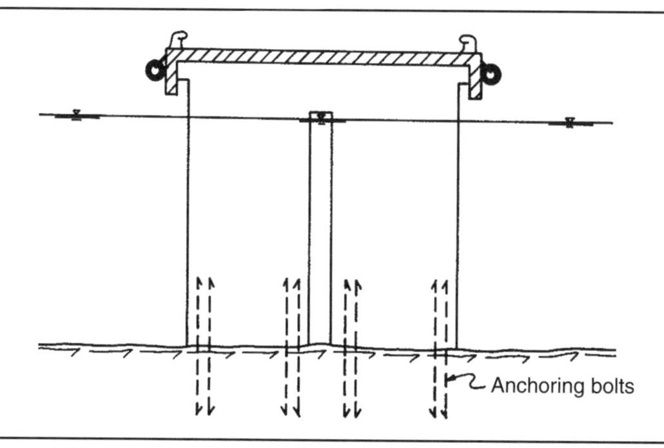

Figure 8.16 There should be as few load transfer points as possible

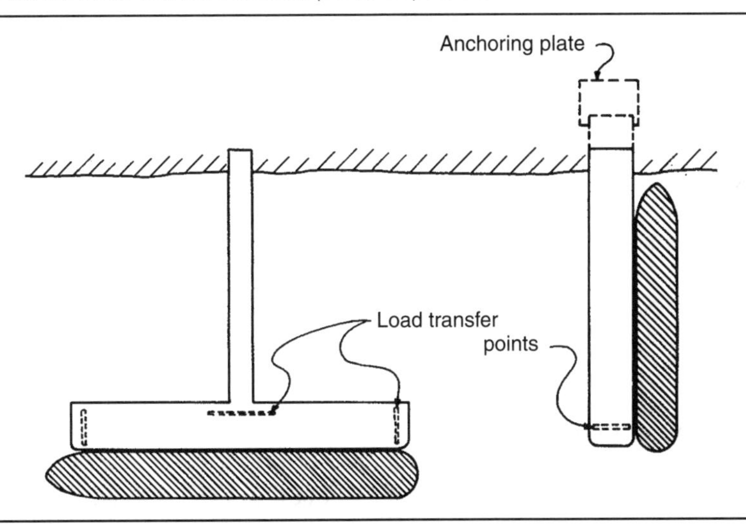

To construct the whole structure to resist possible extreme loads once or twice in the lifetime of the berth structure would in many cases be prohibitively expensive.

8.4. Factors affecting the choice of structures

8.4.1 Soil conditions

The fact that soil conditions can vary considerably from one site to another has led to the development of a wide spectrum of berth structure types. If the soil is loose and has a low bearing capacity, it would serve no useful purpose to consider constructing a solid block wall-type structure. In such a case, it would be better to consider an open type founded on piles driven down to rock or another sufficiently firm stratum.

In other words, reliable and complete soil investigations must be carried out at the site of the new berth structure. The soil engineer should normally be consulted about the type of foundation to be chosen.

8.4.2 Underwater work

Avoidance of construction work that must be carried out under water is an important goal in modern berth design. Emphasis is placed on the application of construction methods that allow as much work as possible to be carried out from a position above the water, thus keeping the amount of diving to a minimum. Sheet pile structures and structures founded on steel pipe piles are ideal in this respect.

One reason why diving work should be eliminated is that only one or two divers undertake important structural elements, making close supervision of their work difficult to achieve. The capacity for work under water is often limited compared with the situation above water; this applies equally to visibility. The work that has to be carried out by divers, if any, should therefore be of a simple nature. When underwater work is unavoidable, the structural engineer should consult an experienced diver in advance regarding the working methods to be used.

8.4.3 Wave action

Open berth structures are normally more favourable than solid ones in respect of the reflection of incoming waves against the berth front. At an open structure the waves will be damped to a great extent against the rough rubble-covered slope. At the vertical wall of a solid berth any wave reflections and other disturbances can have harmful effects, particularly if the shape of the harbour basin is such that it supports wave reflection.

8.4.4 Design experience

Particular qualifications are required from those who are entrusted with the responsibility for the design and construction of marine structures. Most of the construction work takes place in connection with water, and therefore working techniques, plant and machinery are, to a considerable extent, different from those applied to construction on land.

Documented relevant experience for engineers and other personnel in the maritime sector is therefore a requirement. In the Norwegian Concrete Association's *Guidelines for the Design and Construction of Concrete Structures in Marine Environments*, 2003) this requirement is emphasised.

8.4.5 Construction equipment

When designing a berth structure, thought should be given to which types of construction plant and machinery can reasonably be procured for the site in question. One should also keep in mind that a number of contractors ought to be able to procure the necessary equipment so that true competition is secured.

It is sometimes argued that various contractors own different types of equipment and therefore tend to practise construction methods deviating from contractual specifications. However, a closer study of the methods used by the most experienced contractors shows that their methods are very much the same. This fact should be considered by structural engineers, and could possibly lead to a certain degree of standardisation in this sector.

The heavy equipment used in foundation works does not, on the other hand, easily lend itself to standardisation. This equipment usually involves high transportation and installation costs. Therefore, alternative foundation methods are sometimes considered with a view to the utilisation of equipment that is found locally. The magnitude of the foundation works in each particular project is also a factor to consider in this respect.

8.4.6 Materials

A berth structure can be constructed out of timber, steel or concrete or a combination of these materials. The general choice of construction materials to be adopted will be dependent on the purpose of the structure and the economic considerations. The durability under marine environmental conditions is of particular importance for marine structures. The aggressive action from seawater requires especial attention.

The likelihood of actually getting the specified materials delivered to the site is a matter that must be carefully investigated. A modification of the structural system can be the result of such investigation.

Figure 8.17 shows the cross-section of a pier where the horizontal load P is resisted by stabilising lamella walls founded on 500 mm steel piles driven to rock. It could have been cheaper to use 800 mm piles extended up to the berth slab, and to increase the slab thickness to obtain sufficient weight G. However, in this case it was not possible to obtain 800 mm pipes from any source in time, and 500 mm pipes and lamella walls had to be used instead.

Figure 8.17 Cross-section of a pier: design is dependent on the availability of materials

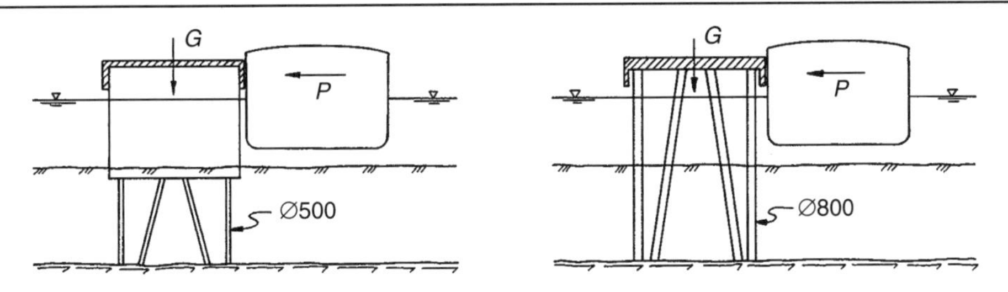

8.4.7 Construction time

If a new ferry berth is to be built in the same place as an old one that is still in operation, or a new berth structure is going to be erected close to an existing berth, the operation of which would be hampered during the construction period, a time-saving construction method should be emphasised even if its construction cost is higher.

Among the solid berth structures, the types called 'cells', 'simple steel pile wall' and 'solid platform' require a construction time of 3–5 lin m of berth front per week. For a lamella berth structure, the performance is about 15 m^2 of slab per week, and for column berths 30–60 m^2 of structure slab per week. In special circumstances, and/or when using an advanced formwork system, a production of 150 m^2 of slab per week may be achieved in a column or pile structure project. These figures are based on one working team working for 8 h/day.

8.4.8 Future extensions

Usually, provisions should be made for possible future extensions of the berth in one or more directions. There can be a need for an increase in the water depth in front of existing berth structures, to provide greater depths along the front due to the increase in the size of ships over recent years. This can create, and has created, problems for the stability of the front of the berth structure. The following methods illustrate some solutions to obtaining greater depths in the front while utilising as much of the existing berth structure as possible (Figure 8.18):

(a) Large floating fenders between the ship and the berth structure. The solution is cheap from a construction point of view, but there can be problems with the reach of cranes, etc.
(b) Additional anchorage of the existing sheet pile structure with new grout anchors or a reduction in the live load on the quay loading area, or replacing the existing filling behind the quay front by lighter materials.
(c) Sheet piling in front of the existing berth structure.
(d) A new berth structure in front of the existing berth structure. This type of solution can offer more possibilities than cases (a), (b) and (c), because it provides additional area to the apron and can be designed for higher live loads.

It is very important to remember that, due to many of the newer ship types with very strong bow or side thrusters, erosion in front of the wall structures shown in Figure 8.18 can present a serious problem to the stability of the berth structure because of the increase in the water depth. Therefore, many of these berth structures have to be erosion protected, as described in Chapter 15.

Figure 8.18 Methods for increasing the water depth in front of a berth structure

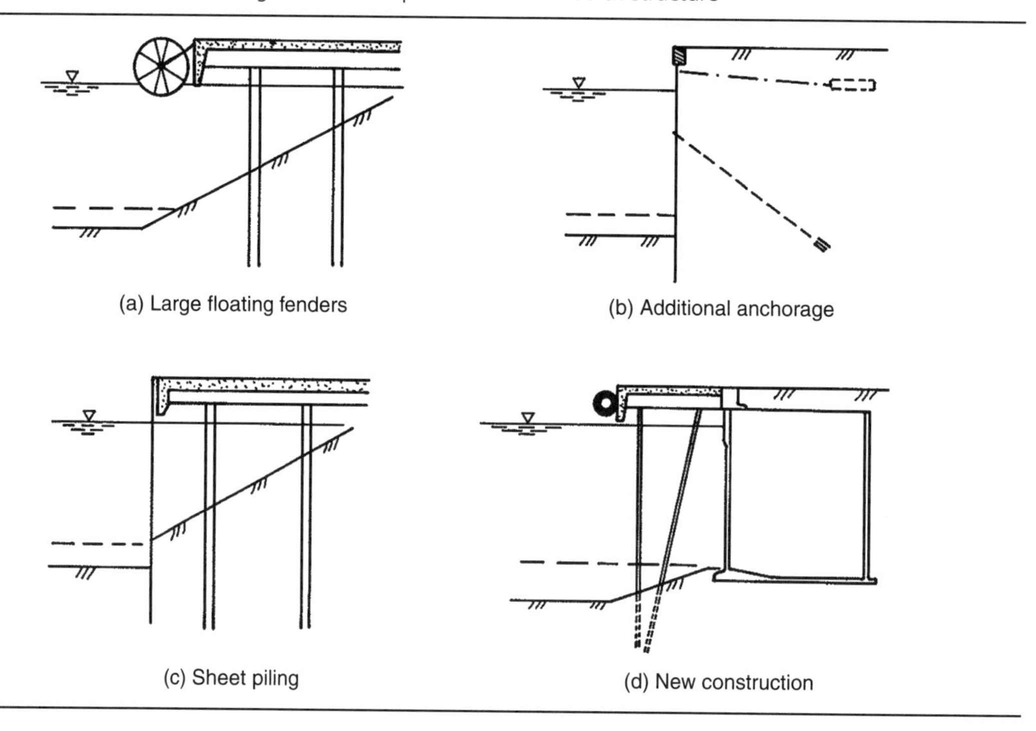

8.4.9 Expansion joints

The distance between the joints along the berth structure will depend upon the type of berth or harbour structure, the soil condition and temperature variations. Expansion joints must be provided in order to accommodate the movements arising from temperature changes, shrinkage and some yielding of the foundations. Steel and/or reinforced concrete structures should be designed to prevent temperature and shrinkage cracks.

The expansion joints in the sections should be keyed for the mutual transfer of shear forces, and should be so designed that changes in the length of the sections are not hindered. The expansion joints should, especially for solid berth structures, be covered to prevent the backfill from being washed out.

If the berth slab exceeds a certain length, expansion joints must be provided at certain intervals perpendicular to the quay front. Usually, their spacing can vary from 30–40 m up to about 60–100 m, but this depends very much on the system used for the lengthwise anchoring of the berth. If a lamella wall at the middle of the berth stretch anchors the berth, it is quite possible to make the berth 100–200 m long without providing any expansion joint. Horizontal forces acting perpendicular to the joints have to be absorbed by some kind of indentation.

8.4.10 Construction costs

Construction unit prices tend to vary from one part of a country to another, and will also depend on the competition existing among contractors at any particular time.

Generally, it can be said that open berth structures are relatively cheaper than solid structures the greater the water depth at the front.

8.5. Norwegian and international berth construction

Norwegian berth construction methods differ in several ways from international practice. This is partly because in Norway the authorities permit the construction of slender load-bearing concrete column structures poured under water. An experienced contractor is able to construct, in this way, columns up to about 30 m in length and about 90–100 cm in diameter. Tests carried out on such structures have shown that they are fully intact 60 years after pouring.

This method of design and construction of berths founded on slender concrete elements poured under water represents pioneer work developed in Norway by engineers over the last three generations. In the international market, much heavier berth structures built for similar purposes under similar conditions are encountered. Foreign port contractors who have designed berths in Norway have often been faced with the Norwegian contractors' alternative designs of the above type, which are faster and cheaper to build.

In Norway, on average during the last 15 years, approximately 20% of all berth structures have been built as solid-berth structures and approximately 80% have been built as open berth structures. The reason for these figures is mainly due to geotechnical and geological conditions. Berth structures founded on steel pipe piles that are reinforced and filled with concrete after driving are representative of a type that has been successfully developed in Norway.

REFERENCES AND FURTHER READING

Bruun P (1989) *Port Engineering*, vols 1 and 2. Gulf Publishing, Houston, TX.

BSI (British Standards Institution) (1988) BS 6349-2: 1988. Maritime structures. Design of quay walls, jetties and dolphins. BSI, London.

BSI (2000) BS 6349-1: 2000. Maritime structures. Code of practice for general criteria. BSI, London.

CUR (Centre for Civil Engineering Research and Codes) (2005) *Handbook of Quay Walls*. CUR, Gouda.

EAU (2004) *Recommendations of the Committee for Waterfront Structures, Harbours and Waterways*, 8th English edition. Ernst, Berlin.

Ministerio de Obras Públicas y Transportes (1990) ROM 0.2-90. Maritime works recommendations. Actions in the design of maritime and harbour works. Ministerio de Obras Públicas y Transportes, Madrid.

Norwegian Concrete Association (2003) *Guidelines for the Design and Construction of Concrete Structures in Marine Environments*. Norwegian Concrete Association, Oslo, Publication No. 5 (in Norwegian).

PIANC (International Navigation Association) (1978) *Standardisation of Roll On/Roll Off Ships and Berths*. PIANC, Brussels.

PIANC (1990) *The Damage Inflicted by Ships with Bulbous Bows on Underwater Structures*. PIANC, Brussels.

Port and Harbour Research Institute (1999) *Technical Standards for Port and Harbour Facilities in Japan*. Port and Harbour Research Institute, Ministry of Transport, Tokyo.

Quinn A De F (1972) *Design and Construction of Ports and Marine Structures*. McGraw-Hill, New York.

Tsinker GP (1997) *Handbook of Port and Harbor Engineering*. Chapman and Hall, London.

Port Designer's Handbook
ISBN 978-0-7277-6004-3

http://dx.doi.org/10.1680/pdh.60043.197

Chapter 9
Gravity-wall structures

9.1. General

Gravity-wall structures can be subdivided into the following three groups depending on the type of structural design:

(a) block wall berths
(b) caisson berths
(c) cell berths.

The gravity berth wall structure can generally only be used where the seabed is good and the risk of settlement is low.

9.2. Block wall berths

Block wall berths belong to the oldest type of berth structures. They consist of large blocks placed one upon another in a masonry wall pattern (Figure 9.1). Such berths, built on firm ground using blocks of good-quality natural stone or concrete, are structures having a long life, and require only modest amounts of maintenance. Due to the present high costs of mining natural stone blocks, only concrete blocks can be considered economical for projects nowadays.

Many of the newer ship types have strong bow thrusters, and the erosion of the pitching in front of the wall structure (see Figure 9.1) has been a serious problem for many of the older block wall structures. Therefore, many of block wall berth structures have had to be erosion protected, as described in Chapter 15.

As a great deal of the construction work on block wall berths has to be carried out under water by divers, the construction costs are usually very high. Therefore, nowadays only special local conditions would justify the use of this type of structure. Such conditions could involve, for instance, a long berth to be founded on very firm ground, and, where cheap unskilled labour could be engaged, casting a sufficient number of concrete blocks before the start of the actual construction (thus minimising any idle time of the skilled labour). In order to minimise the extent of underwater work, the blocks should be of equal size, insofar as is possible, and, after casting the blocks, each course should first be arranged and marked onshore in order to facilitate its final placing in the water.

To ensure the stability of individual blocks, the blocks should be sufficiently large that the maximum capacity of the block-handling equipment (cranes, etc.) is fully utilised.

The size of concrete blocks is also determined by the available casting equipment and storage space for the blocks. The weight of the blocks may vary from about 150 to 2000 kN. Blocks made of natural

Figure 9.1 A block wall berth

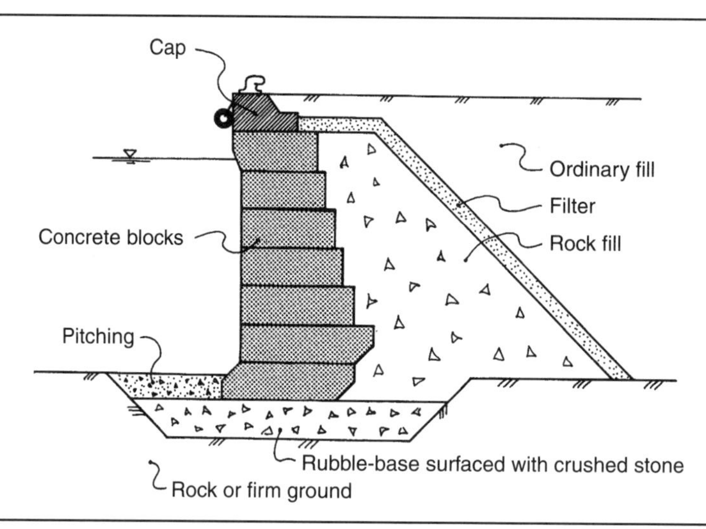

stone normally have weights in the range 150–500 kN, and the choice of block size will depend on the distance to the quarry and the transport equipment available.

Due to their great deadweight, block wall berths should be used only on very firm ground, in order to avoid settlement. The structure can cause great stresses at the outer edge of the bottom course of the wall, which should therefore be laid on a rubble-base that is surface levelled with crushed stone. The ideal foundation is achieved where the wall can be laid directly on rock levelled with an in situ concrete footing.

Figure 9.2 shows a method of improving the ground condition by dredging away the loose layer above the rock and then replacing it with a filling of sand or gravel. To reduce the settlement in the sand, a vibro-compaction method is used. Vibro-compaction is an in situ method of compacting loose cohesionless granular soils, such as sand and gravel. The process is based on the principle that granular soils below maximum density are compacted under the influence of vibration motion.

Another method of achieving improvement and reinforcement of soft cohesive soils is by the installation of stone columns to carry structural loads (Figure 9.3). The stone columns will improve the bearing capacity of the foundation and reduce the overall settlement, as they are stiffer than the soil they have replaced. This method is called vibro-replacement, and can generally be done using the same basic vibratory equipment as for the vibro-compaction method.

A modified method of constructing block wall berths has been developed. As an alternative to concrete blocks, reinforced concrete retaining wall elements have been developed. These L-elements, or L-blocks, are constructed in the same way as the concrete caissons on the shore, but are transferred and installed at the berth site by cranes. The berth wall is made by installing the L-elements side by side in position on a prepared gravel and/or rubble-base at the sea bottom. The elements are shown in Figure 9.4. Elements without ribs are constructed with a maximum height of about 7 m, while those with ribs have been constructed up to a height of about 20 m. The length of the elements has varied in the range 3–12 m, depending on the capacity of the mobile or floating cranes. Mobile cranes can

Figure 9.2 Improving the ground condition by vibro-compaction

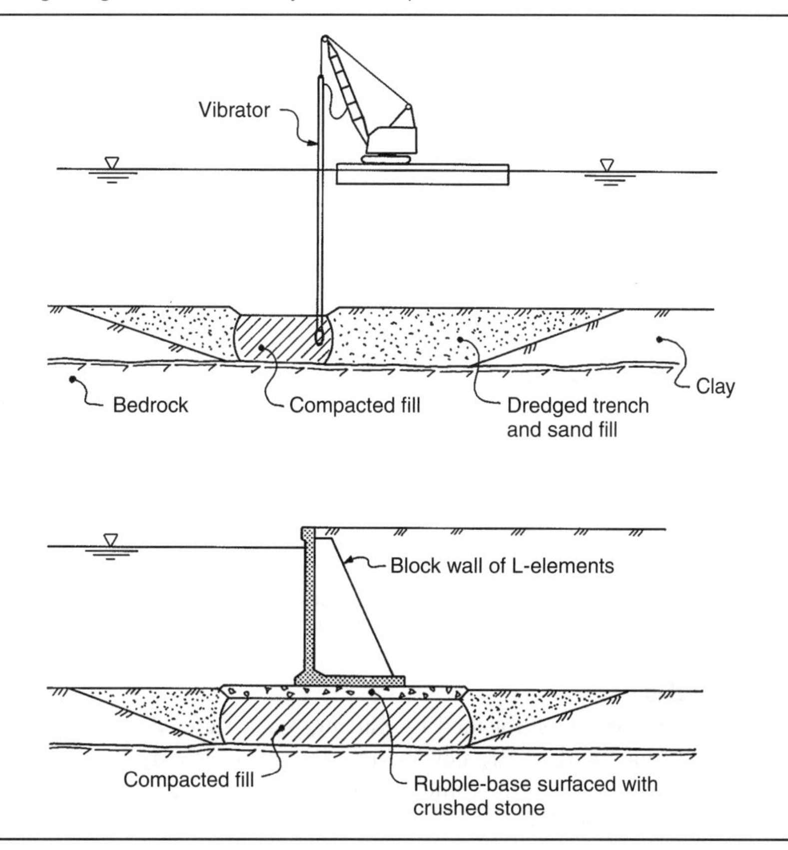

have a lifting capacity of up to 100 t but, due to the lack of availability of suitable equipment, a more practical limit is about 30 t. Floating port cranes can have a lifting capacity of about 200 t, but special heavy floating cranes exist that have a lifting capacity of about 800 t.

An article by Lauri Pitkala of Finland, printed in *PIANC Bulletin*, No. 54 (PIANC, 1986), describes in detail the construction of berth structures composed of L-elements. The typical building order of a berth structure composed of L-elements is shown in Figure 9.5. The usual tolerances for element installation are a deviation in the x, y and z directions of 50 mm, inclinations of 1 : 400 and an angular misalignment of 0.5°.

The fill placed at the back of the wall should have as great a frictional angle as possible in order to reduce the lateral earth pressure. Nearest to the wall the fill should consist of stone or crushed rock, while finer or mixed fill can be used further back. Between the coarser and the finer masses there must be a filter to prevent the penetration of the finer material into the rock fill. Above the low-water level, all the fill should be compacted.

When the placing of the blocks or elements has been completed, the wall should be left for a while to settle before an in situ reinforced concrete cap (capping beam) is placed on top of it. This cap will keep

Figure 9.3 Improving the ground condition by vibro-replacement

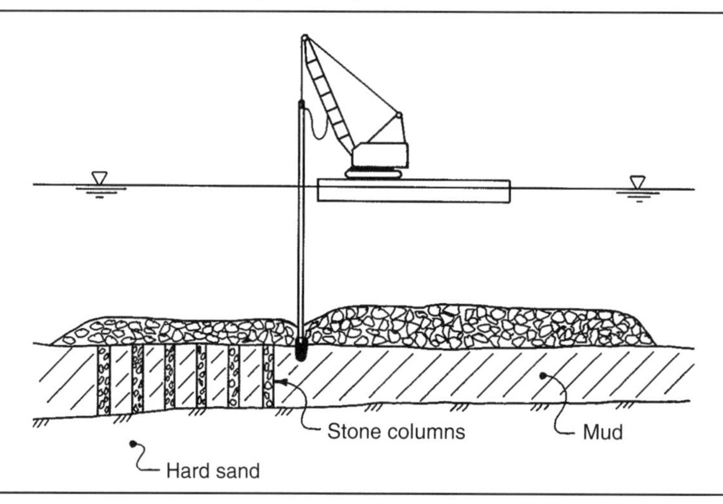

the blocks in the top course in place, and will also provide a base for the installation of bollards, a quay-front kerb and other equipment.

9.3. Caisson berths

In caisson quays the berth front is established by placing precast concrete caissons in a row, corresponding to the planned alignment of the new berth. The caissons may be differently shaped and designed, depending on the site conditions and the available construction equipment. Rectangular caissons are the most usual (Figure 9.6).

The caissons are usually placed on a firm base of gravel and/or rubble that has been well compacted and accurately levelled. It is very important that. before placing the caissons, most of the settlement has reached a minimum, particularly any uneven settlement. If the site is exposed to waves and currents, the base and the caissons should be designed in such a way that the time required for launching, towing and placing the caissons is as short as possible. After the caissons have been put in place they are filled with a suitable material, and a reinforced concrete cap is placed on the top, as is done on block wall berths.

Figure 9.4 An L-element

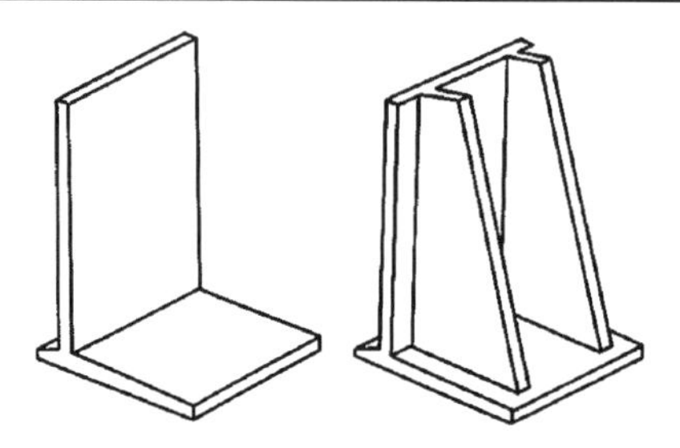

Figure 9.5 Construction of an L-element wall

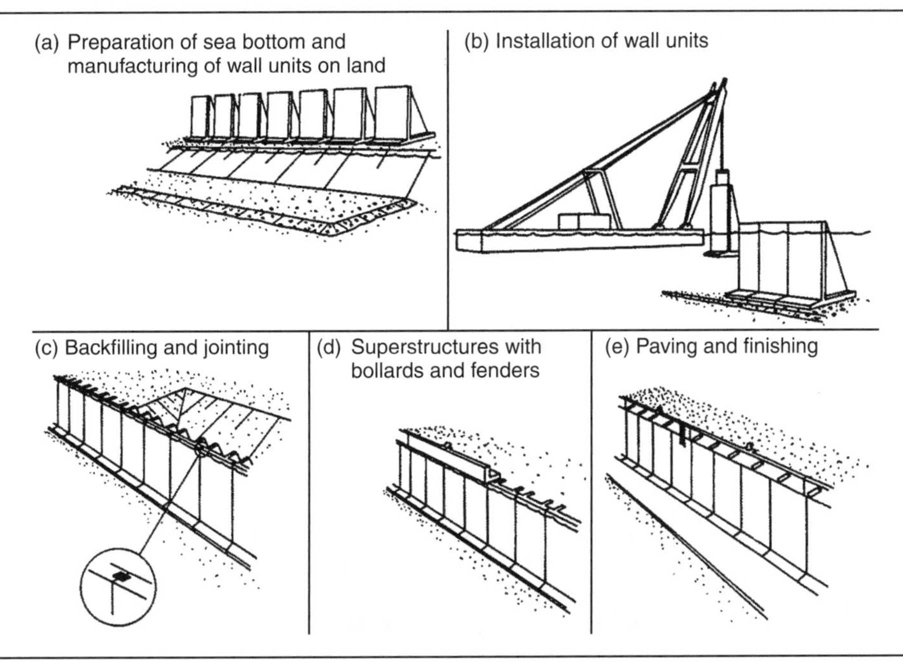

As the caissons can have different shapes and construction layout, as described above, with suitable design underwater work can be reduced to a minimum. It is both very economical and convenient if the caissons can be made on an existing slipway or in a dry dock, from which they can easily be launched. For economic reasons, the caissons should also preferably be made in a considerable number so that the production can be arranged in a rational way with multi-use of the formwork units.

For convenience of constructing, launching, towing, placing, etc. the caissons, experience has shown that, for economic reasons, the caisson dimensions should usually not be greater than about 30 m long, 25 m wide and 20 m high, but the concrete elements can be produced in a dry dock in lengths

Figure 9.6 A caisson berth

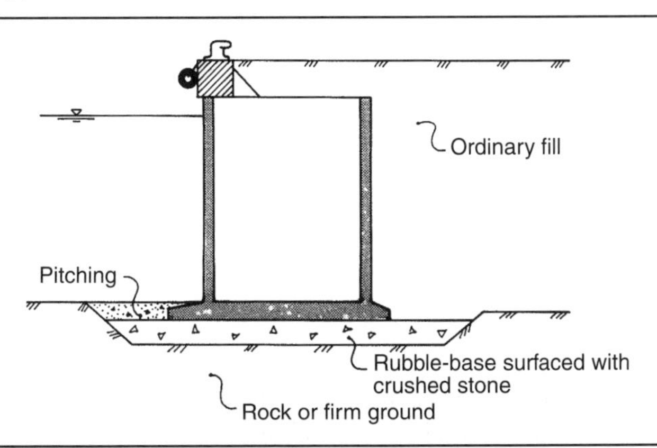

Figure 9.7 Lifting a caisson element during the construction of a double roll-on/roll-off quay. (Photo courtesy of Birken & Co., Norway)

exceeding 100 m. The caissons should be designed such that they are suitable for all stages during construction and service.

The caissons are usually made ashore and then launched, towed out and sunk in position, or lifted into position, on a prepared gravel and/or rubble base (as shown in Figures 9.2 and 9.3) before placing the elements on the seabed. Figures 9.7–9.9 show a crane vessel lifting one prefabricated caisson element weighing 560 tonne.

It is easier to reduce the stresses at the outer edge of the caisson foot in caisson berths than it is to reduce the stresses in block wall berths. Increasing the width of the caisson, or providing it with two or three chambers of which only the rear chambers are filled (Figure 9.10), can reduce the stresses. The caissons must be designed to also resist the loads and stresses that occur during their production, launching, towing, placing and filling.

All joints between the caissons must be sealed if the caissons are used to retain materials behind them and/or prevent waves or current from passing through the gaps between them. The joints should be designed for placing tolerances and uneven settlements. The placing tolerances should be ± 150 mm in sheltered water.

9.4. Cell berths

In recent years, sheet pile cell berths have become one of the most used types of gravity-wall berth. One of the main reasons for this is the increasing ratio of the cost of labour (divers, etc.) to the cost of

Figure 9.8 The installation of the final caisson element under the roll-on/roll-off ramp. (Photo courtesy of Birken & Co., Norway)

Figure 9.9 The completed double roll-on/roll-off quay. (Photo courtesy of Birken & Co., Norway)

Figure 9.10 A caisson berth with three chambers

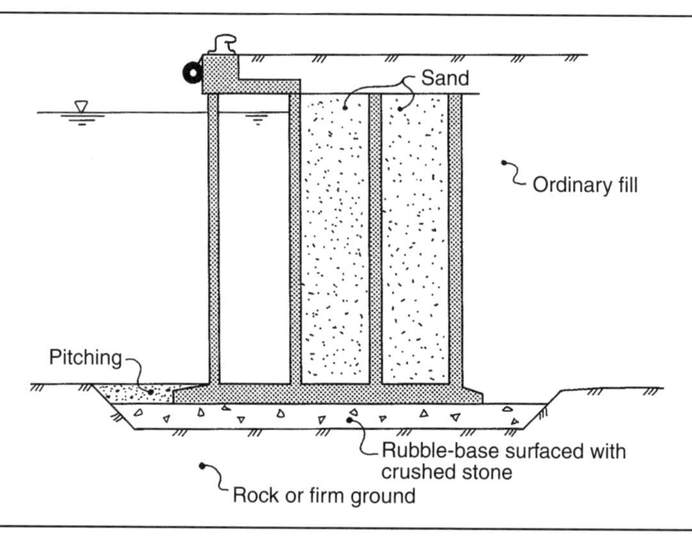

materials, as compared with the construction of block wall and caisson berths. Various geometric configurations of sheet pile cell berths are used, but the usual designs are shown in Figure 9.11.

Circular main cells connected by arched cells are the most used form of construction. The circular cells have the advantage that each cell can be individually constructed and filled, and the cells are therefore independently stable. In the diaphragm cells or straight web cells design (see Figure 9.11), sheet pile arches are connected by straight sheet pile walls.

Circular steel sheet pile cells of large diameter are one of the most used berth structures in the ice-affected waters in the arctic, as this type of berth structure can resist large horizontal forces.

A cell berth consists of a row of cells filled with sand, gravel, etc., connected as shown in Figure 9.11. The diameter of the main cells normally varies from 10 to 20 m. The circular and arched cells are formed of flat steel sheet piles of width 50 cm, web thickness 9.5–12.7 mm and weight 128–154 kg/m^2.

Figure 9.11 A sheet pile berth

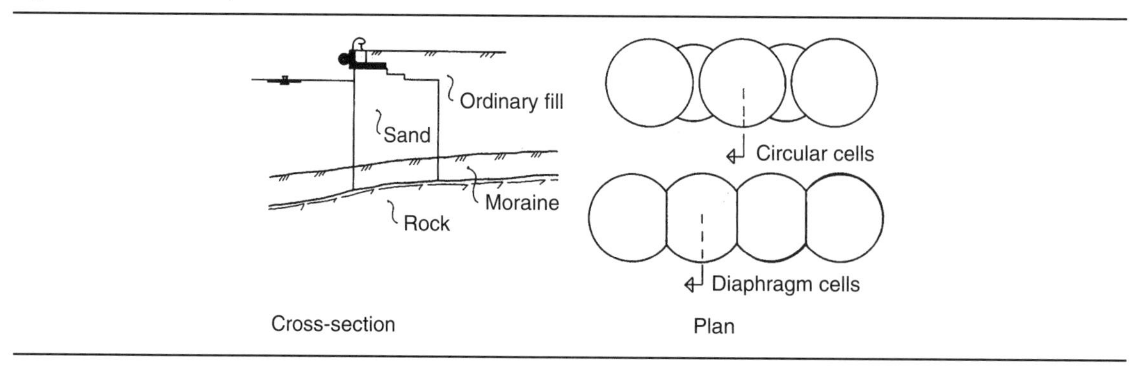

Figure 9.12 Straight web sections produced by Arcelor, Luxembourg

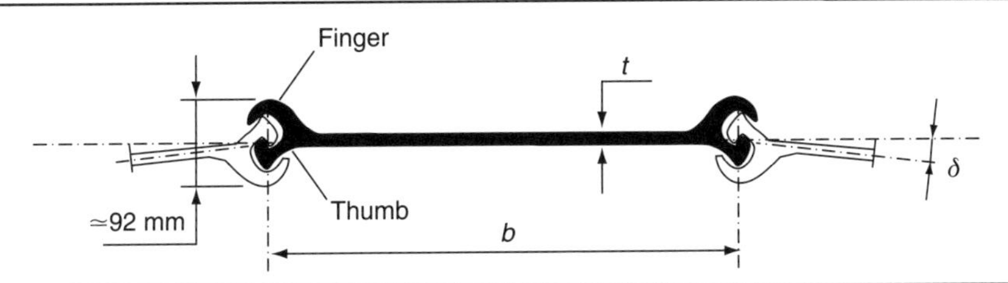

Section	Nominal width* b (mm)	Web thickness t (mm)	Deviation angle δ°	Perimeter of a single pile (cm)	Steel section of a single pile (cm^2)	Mass per m of a single pile (kg/m)	Mass per m^2 of wall (kg/m^2)	Moment of inertia of a single pile (cm^4)	Section modulus of a single pile (cm^3)	Coating area*** (m^2/m)
AS 500-9.5	500	9.5	4.5**	139	81.6	64.0	128	170	37	0.58
AS 500-11.0	500	11.0	4.5**	139	90.0	70.6	141	186	49	0.58
AS 500-12.0	500	12.0	4.5**	139	94.6	74.3	149	196	51	0.58
AS 500-12.5	500	12.5	4.5**	139	97.2	76.3	153	201	51	0.58
AS 500-12.7	500	12.7	4.5**	139	98.2	77.1	154	204	51	0.58

Note: all straight web sections interlock with each other.

* The effective width to be taken into account for design purposes (lay-out) is 503 mm for all AS 500 sheet piles

** Max. deviation angle 4.0° for pile length >20m

*** One side, excluding inside of interlocks

The following interlock strengths can be achieved for an S 355 GP steel grade:

Sheet pile	F_{max}: (kN/m)
AS 500-9.5	3000
AS 500-11.0	3500
AS 500-12.0	5000
AS 500-12.5	5500
AS 500-12.7	5500

For verification of the strength of piles, both yielding of the web and failure of the interlock should be considered.

Figure 9.12 shows straight web sections from Arcelor, Luxembourg. Sheet pile cells have the advantage that they can be designed as stable gravity walls without external anchoring. The nature of the cell fill material must be carefully specified and controlled. Experience has shown that the permeability of sheet pile cells interlocking under tension is low, so the need for drainage through the cell constructions should be carefully investigated.

The sheet piles are joined by lock arrangements acting in tension to retain the fill inside the cell. The locks have a guaranteed interlock tensile strength of 2000–6000 kN/lin m, depending on the type of steel and sections. The interlocks between the flat steel sheet piles may have either one-point or three-point contacts, depending on the manufacturer of the sheet pile. The sheet pile profiles can resist an angular deviation of about 10° in the locks. Tables indicating cell diameters, distances to neighbouring cells, dimensions of intermediate arches, etc. are published by the manufacturers of these profiles. Examples of circular cell constructions from Hoech Stahl AG, Germany, and Arcelor-Mittal Commercial, Luxembourg, are shown in Figures 9.13 and 9.14, respectively. Figure 9.15 gives calculations for the equivalent widths and ratios of the different circular cells and diaphragm cells shown.

Figure 9.13 Flat steel sheet piles produced by Hoech Stahl AG, Germany

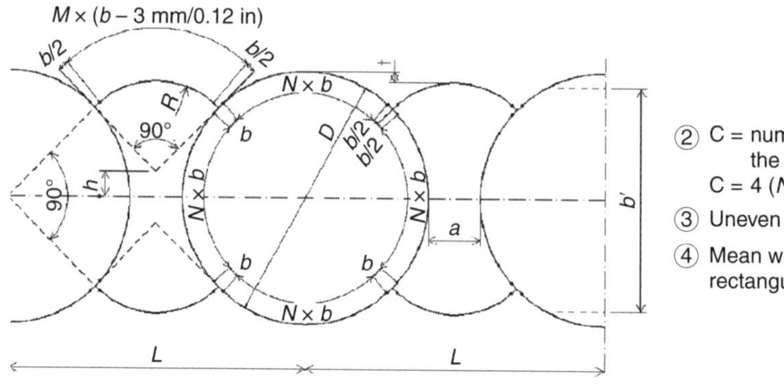

Circular cell walls				Diaphragm walls					Analytical width	Diam. of guide table
C ②	N ③	D	L	M	R	h	t	a	b ④	d
no.	no.	ft m	ft m	no.	ft m	ft m	ft m	ft m	ft m	ft m
				Sections FL 511, FL 512, FL 512.7						
56	13	29.24	35.45	9	10.44	2.95	1.22	6.20	25.38	28.67
		8.91	10.80		3.18	0.90	0.37	1.89	7.89	8.74
64	15	33.42	36.40	9	10.44	4.43	1.84	4.98	29.40	32.82
		10.19	11.70		3.18	1.35	0.56	1.52	8.96	10.00
72	17	37.60	41.35	9	10.44	5.91	2.45	3.76	32.97	36.98
		11.46	12.60		3.18	1.80	0.75	1.15	10.05	11.27
80	19	41.77	47.26	11	12.53	5.91	2.45	5.49	36.70	41.13
		12.73	14.41		3.82	1.80	0.75	1.67	11.19	12.54
88	21	45.95	50.21	11	12.53	7.38	3.06	4.26	40.28	45.28
		14.01	15.31		3.82	2.25	0.93	1.30	12.28	13.80
96	23	50.13	56.12	13	14.62	7.38	3.06	5.99	44.01	49.43
		15.28	17.11		4.46	2.25	0.93	1.83	13.41	15.07
104	25	54.31	59.08	13	14.62	8.86	3.67	4.77	47.59	53.59
		16.55	18.81		4.46	2.70	1.12	1.45	14.50	16.33
112	27	58.48	64.98	15	16.71	8.86	3.67	6.50	51.32	57.74
		17.83	19.81		5.09	2.70	1.12	1.98	15.64	17.60
120	29	62.66	67.94	15	16.71	10.34	4.28	5.28	54.90	61.89
		19.10	20.71		5.09	3.15	1.31	1.61	16.73	18.86
128	31	66.84	70.89	15	16.71	11.82	4.89	4.05	58.50	66.04
		20.37	21.61		5.09	3.60	1.49	1.24	17.83	20.13
136	33	71.01	76.80	17	18.80	11.82	4.89	5.78	62.21	70.20
		21.65	23.41		5.73	3.60	1.49	1.76	18.96	21.40

The choice of cell diameter depends on the water depth and the loads, as well as the subsurface conditions. The sheet pile area or the sheet pile weight per linear metre of berth front is virtually independent of the cell diameter. The unit price of the fill in the cells is, therefore, often the decisive factor in the construction cost.

Figure 9.14 The geometry of circular cells produced by ArcelorMittal, Luxemburg

Once the equivalent width has been determined, the geometry of the cells can be defined. This can be done with the help of tables or with computer programs.

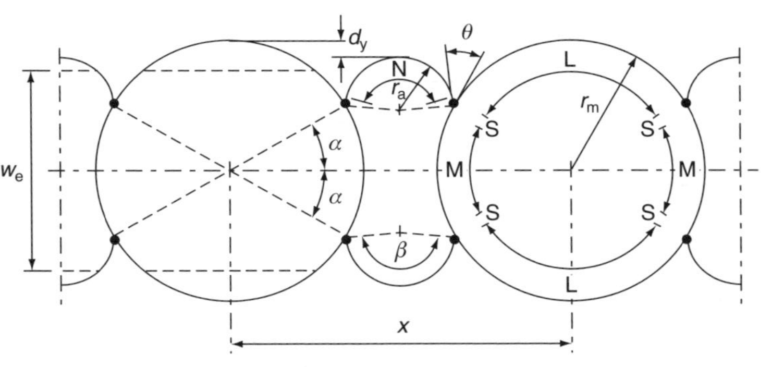

Junction piles with angles θ between 30° and 45°, as well as θ = 90°, are available on request.

Standard solution

θ = 35°

b/2

b/2 b/2

r_m = radius of the main cell
r_a = radius of the connecting arcs
θ = angle between the main cell and the connecting arc
x = system length
d_y = positive or negative offset between the connecting arcs and the tangent planes of the main cells
w_e = equivalent width

The table below shows a short selection of circular cells with 2 arcs and standard junction piles with θ = 35°.

Number of piles per						Geometrical values						Interlock deviation		Design values	
Cell				Arc	System							Cell	Arc	2 arcs	
pcs.	L pcs.	M pcs.	S pcs.	N pcs.	pcs.	$d = 2r_m$ m	r_a m	x m	d_y m	α °	β °	δ_m °	δ_a °	w m	R_a
100	33	15	4	25	150	16.01	4.47	22.92	0.16	28.80	167.60	3.60	6.45	13.69	3.34
104	35	15	4	27	158	16.65	4.88	24.42	0.20	27.69	165.38	3.46	5.91	14.14	3.30
108	37	15	4	27	162	17.29	4.94	25.23	0.54	26.67	163.33	3.33	5.83	14.41	3.27
112	37	17	4	27	166	17.93	4.81	25.25	0.33	28.93	167.86	3.21	6.00	15.25	3.35
116	37	19	4	27	170	18.57	4.69	25.27	0.13	31.03	172.07	3.10	6.15	16.08	3.42
120	39	19	4	29	178	19.21	5.08	26.77	0.16	30.00	170.00	3.00	5.67	16.54	3.38
124	41	19	4	29	182	19.85	5.14	27.59	0.50	29.03	168.06	2.90	5.60	16.82	3.35
128	43	19	4	31	190	20.49	5.55	29.09	0.53	28.13	166.25	2.81	5.20	17.27	3.32
132	43	21	4	31	194	21.13	5.42	29.11	0.33	30.00	170.00	2.73	5.31	18.10	3.39
136	45	21	4	33	202	21.77	5.82	30.61	0.36	29.12	168.24	2.65	4.95	18.56	3.35
140	45	23	4	33	206	22.42	5.71	30.62	0.17	30.86	171.71	2.57	5.05	19.39	3.42
144	47	23	4	33	210	23.06	5.76	31.45	0.50	30.00	170.00	2.50	5.00	19.67	3.39
148	47	25	4	35	218	23.70	5.99	32.13	0.00	31.62	173.24	2.43	4.81	20.67	3.44
152	49	25	4	35	222	24.34	6.05	32.97	0.34	30.79	171.58	2.37	4.77	20.95	3.42

Figure 9.15 The geometry of the different cells produced by ArcelorMittal, Luxembourg

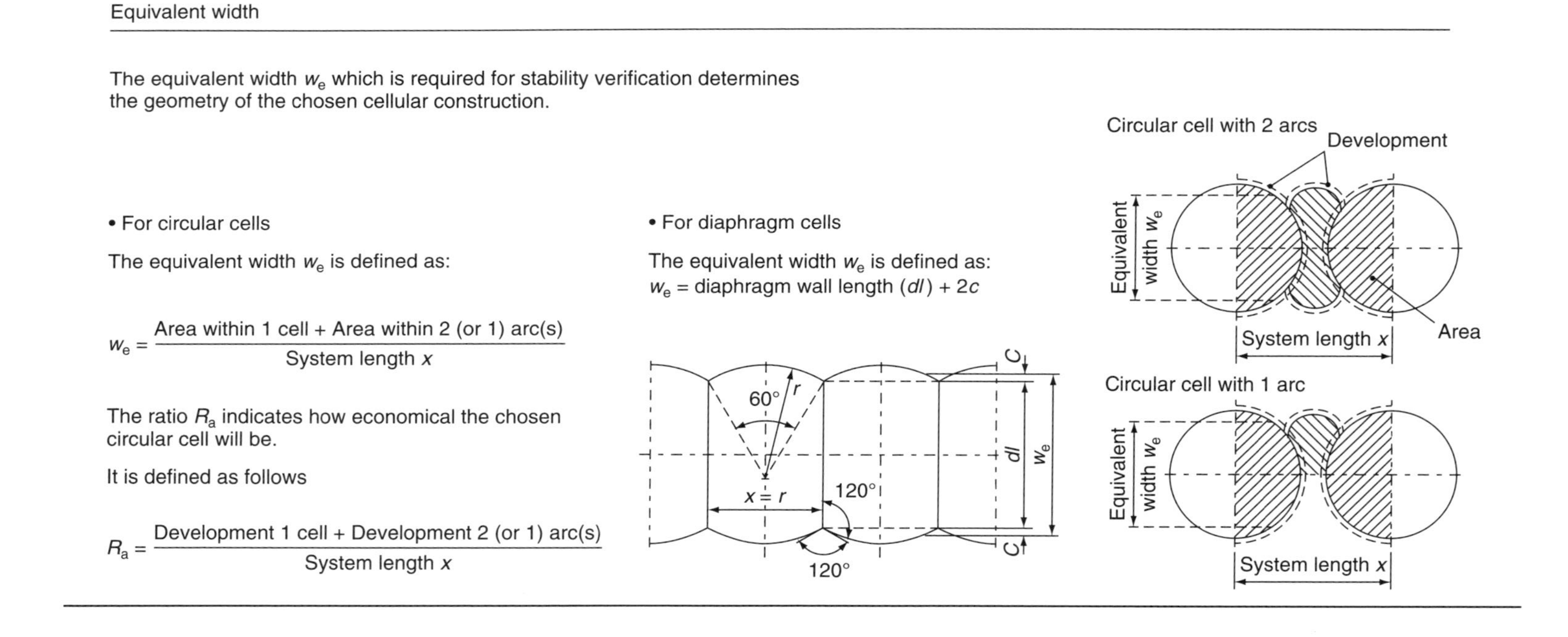

Figure 9.16 The stair–step method for circular sheet piling

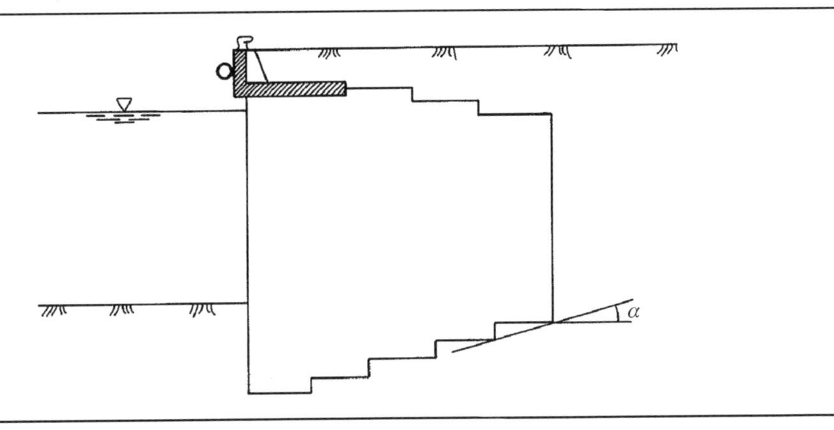

The depth to which the sheet piles are to be driven may be reduced by using the stair–step method in order to save steel (Figure 9.16). This method can be used provided the backfill material is of high quality and the angle α is less than 15°.

The design of cell structures is based on the conventional theory of structures and empirical formulas, and slightly different calculation methods are described in the literature. As illustrated in Figure 9.17, the cell structure must be designed to resist the following modes of failure or general bearing capacity:

(a) Tilting due to external loading of the cell structure. Normally this will give the minimum cell diameter.
(b) Tilting due to failure in vertical or horizontal shear within the fill material.
(c) Tension failure in the sheet pile locks. Normally this will give the maximum cell diameter.
(d) Horizontal sliding of the cell structure.
(e) Tilting due to rotational failure on a curved rupture surface at or near the base of the cell structure.
(f) General shear failure of the soil beneath the cell structure.
(g) Settlement of the cell structure.

If the sheet pile cell structures have been driven into soil below the dredged or sea bottom level, the resistance due to tilting, sliding, rotational failure, etc. of the structure will be provided by the passive resistance of the soil below the sea bottom and the pullout resistance of the sheet piles on the landward side.

The cylindrical cell type of berth is the easiest to build because each cell is stable by itself when filled with sand, gravel, etc. The finished cell can then be used as a working platform for further work. The arched cells between the cylinders must be filled after the filling of the cylinders. The cellular structure is able to resist horizontal forces without the assistance of any anchoring. The cells can, therefore, be built independently of the backfill. They are also suitable for use in supporting structures such as dolphins and head sections in piers.

For the construction of a cell berth consisting of, for instance, three circular main cells and two intermediate arched cells, as shown in Figure 9.8, the procedure will be as follows:

Figure 9.17 Modes of cell failure

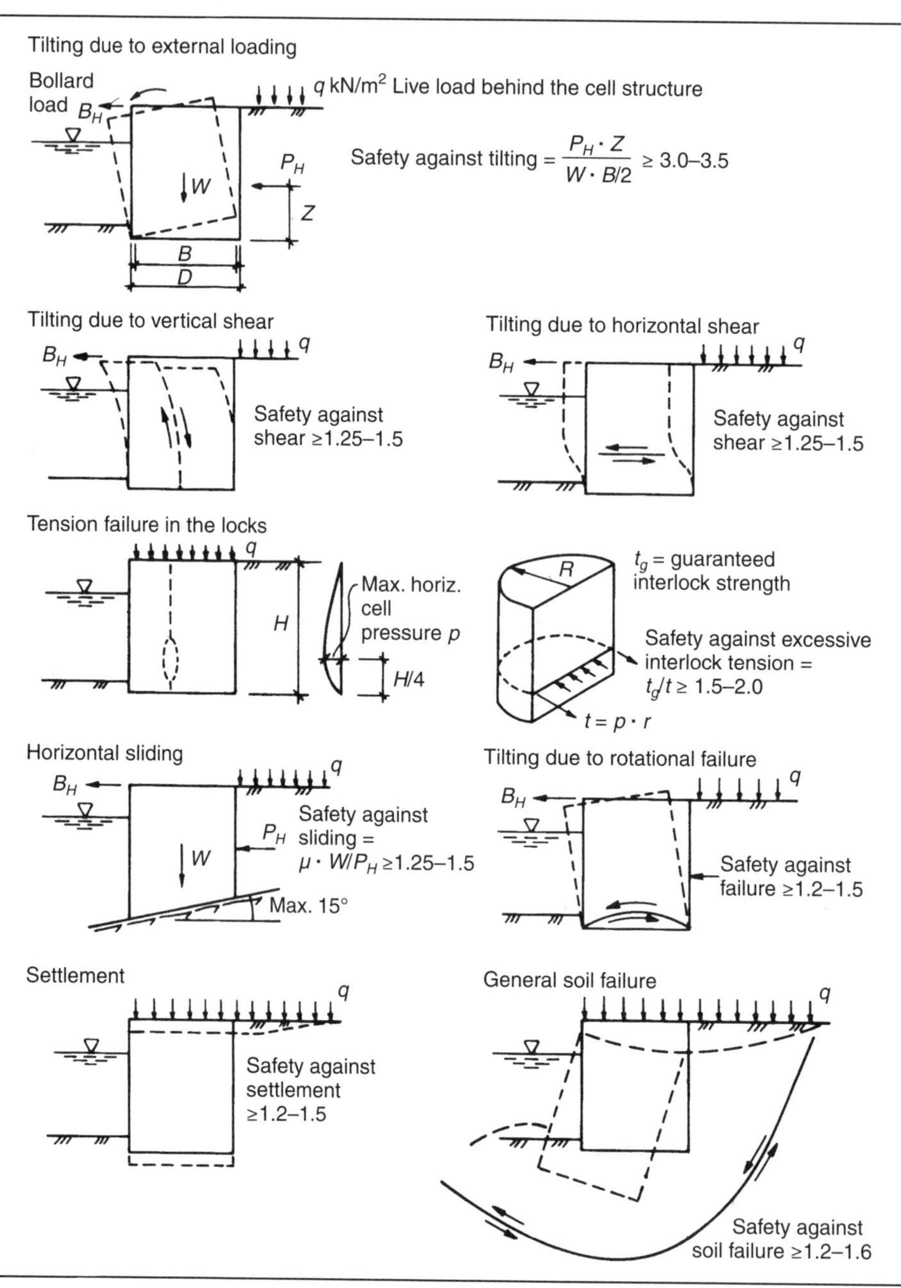

(a) The template for the first main cell is erected in position.
(b) The sheet piles for connecting the main cell to the arched cells are put in position.
(c) The sheet piles between the above connecting piles for the arched cells are put in position and the main cell is closed. In general, the whole cell must be set and closed before driving the piles, and it is presupposed that the initial slight penetration of the sheet piles into the soil is achieved by way of their own deadweight.

(d) The first main cell is driven into firm ground or rock. The driving should be carried out in several stages, proceeding around the cell circumference and always limiting the penetration to 1.0–2.0 m.
(e) The first main cell template is removed.
(f) The first main cell is filled immediately.
(g) Stages (a) to (f) are repeated for the second main cell while the first cell is being filled according to point (f).
(h) Stages (a) to (f) are repeated for the third main cell while the second cell is being filled.
(i) The template for the first arched cell (between the first and second main cells) is placed when the filling of the second main cell is finished.
(j) The sheet piles for the first arched cell are put in position and driven.
(k) The arched cell template is removed.
(l) The arched cell is filled.
(m) Points (i) to (l) are repeated for the second arched cell (between the second and third main cells) when the filling of the third main cell is finished.
(n) The cap (continuous concrete capping beam) is boarded, reinforced and concreted on top of the cells.
(o) Ordinary fill is placed behind the cap.
(p) The fill is levelled and the quay pavement, for instance asphalt, is laid.

It is extremely important in all types of cell construction that the template is adapted to the particular structure for which it will be used. The template diameter must be correct so that the placing of the last sheet pile will just close the cell. It is not critical whether an internal or an external template is used, but internal templates are the most common. Figure 9.18 shows the construction of an extension of a cell berth, and Figure 9.19 shows the detail of the internal template and a cell before filling.

The template must be sufficiently strong to resist any wind and wave pressure occurring at the site during construction. It must be able to absorb the loads and stresses transferred to it by way of the sheet piles when these are being placed and driven. Before these piles are firmly founded, the wind and wave forces must be absorbed in full by the template. It should be noted that driving cannot be started before the cell is closed. The contractor must, therefore, plan carefully how the forces will be absorbed without distortion or collapse of the template.

As far as possible, preparation should be made to shorten the time required to place of the sheet piles. The design engineer should also keep in mind that the smaller the cell diameter, the less construction time required and the less the forces from wind, waves and currents expected. Thus the contractor and the engineer should ensure the correct placing and smooth closing of the cell by careful planning of the work procedure. The diameter of the filled cell will be 1–2% greater than the theoretical diameter of the cell after the slack in the piling interlocks has been tightened due to filling of the cell.

Crane and driving equipment must be adapted to suit the lengths of the sheet piles. It is quite usual to have pile lengths of 20–25 m. The driving must be carried out continuously along the whole circumference of the cell, in several stages, so that the penetration of a pile is not more than 50–70 cm compared with the neighbouring pile. The energy needed per blow is normally 6–10 kNm, and is achieved by use of a vibration hammer or double-acting air hammer giving 100–200 blows/min. A geotechnical engineer should evaluate the driving criterion in each individual case.

Splicing of sheet piles should be minimised and should not be allowed above an elevation of 3.0 m. The sheet pile ends should be cut at right angles, and only two splices should be allowed in each sheet pile.

Figure 9.18 Construction of a cell berth

The vertical distance between splices in adjacent sheet piles should be a minimum of 1.0 m. After threading and soft driving has been completed, and before any filling takes place, the geometry of the cell should be controlled.

The geometry of the cell should comply with the following requirements:

(a) For all cells and half cells, the centre position should not deviate from the theoretical position by more than 200 mm.
(b) The ovality of any cell ($D_{max} - D_{min}$) should not exceed 500 mm.
(c) The inclination of any sheet pile in a radial or tangential direction should not exceed 40 : 1.
(d) If significant deviation from the required cell geometry is registered, provisions to correct the deviation should be made before the cell is filled at each stage of erection.

In soil containing sizeable stones it may be difficult to drive the sheet piles without damaging them. It is quite possible to place the cells directly on bare rock, but it is then necessary to secure a firm grip on the

Figure 9.19 Detail of the construction of a cell berth

rock surface for the sheet piles and to pack around the base to prevent washing out of the fill from inside the cell or sliding of the cell on the rock. As a rule of thumb, due to the sliding of the cell on the rock surface, the slope of the rock surface should be less than 15°. If necessary, a horizontal shelf is provided by blasting into the rock surface to secure a safe founding for the cell.

Sand, gravel or stone is used to fill the cells, as these undergo only a small specific settlement after filling and moderate soil pressure in the cells. In the case of filling with excavated rock, the block size should be restricted to a maximum of 300 mm. In countries like Norway, during winter the cell fill material should be free from ice and snow. If dredging is necessary in front of the berth, it should be investigated whether the dredged material is inorganic and suitable for filling the cells, for instance by suction dredging. The economy of cell berths, as compared with open columns or piled berths, depends very much on the cost of filling the cells with suitable material. In order to counteract the development of a possible hydrostatic head inside the cell, it is customary to make 6–10 cm weep holes around the cell circumference. Care should be taken to ensure that there is no washout of the fill material through the weep holes.

Cell berths have the following advantages and disadvantages:

(a) Unlike traditional sheet pile berths, which become unduly expensive for increasing water depth, cell berths become more competitive economically the greater the water depth. As mentioned above, the weight of the steel per linear metre of berth front is almost independent of the cell diameter. The minimum cell diameter is determined by the stability factor, and the allowable tensile load on the sheet pile locks determines the maximum diameter. Thus the latter determines the maximum water depth at the cell berth front.

(b) Diver work is only needed when the cell berth is founded directly on rock, and this requires the blasting of a horizontal shelf on the rock surface, and possibly also in situ casting of a concrete beam around the cell, or dowelling the sheet piles into the rock. In other cases, only inspection by a diver is needed.
(c) Cell berths can be used for most types of soil condition provided that the load-bearing strength and the stability of the ground are such that the load from a solid berth structure can be resisted.
(d) Cell structures are able to absorb considerable deformation, both vertical and horizontal, during construction. However, the sheet piles along the berth front must, in any case, be driven to a non-yielding stratum in order to secure a firm base for the capping beam. It is not unusual for a circular cell to be slightly transformed into an oval during the filling of the cell, but this is of no importance for the structure overall.
(e) The high load-bearing capacity of cell berths is one of their most important advantages. The cell structures are very well adapted to resist heavy horizontal loads from the backfill, as well as from berthing and mooring forces. Therefore they can also be used as the structural elements in pier structures, ensuring the transmission of horizontal loads down to the ground. Where there are favourable ground conditions, the load-bearing capacity with regard to vertical live loads can be as much as 250 kN/m^2, while 30–50 kN/m^2 is usual for column berths. The cell berths are also able to resist large point loads, such as wheel loads, which usually create difficulties in open berth structures.
(f) Compared with other types of berth structure, cell berths require shorter construction times.
(g) The risks of damage to the cells due to ship collision should always be considered. If the bow of a ship runs into the cell and penetrates it, the fill may run out of the cell and the stability of the structure may be greatly endangered. This applies in particular to pier head structures, which would depend solely on the stability of the cells. Repair of such holes in a cell is often time-consuming and costly.

REFERENCES AND FURTHER READING

BSI (British Standards Institution) (1988) BS 6349-2: 1988. Maritime structures. Design of quay walls, jetties and dolphins. BSI, London.

British Standards Institution (2000) BS 6349-1: 2000. Maritime structures. Code of practice for general criteria. BSI, London.

Bruun P (1990) *Port Engineering*, vols 1 and 2. Gulf Publishing, Houston, TX.

Centre for Civil Engineering Research and Codes (CUR) (2005) *Handbook of Quay Walls*. CUR, Gouda.

EAU (Empfehlungen des Arbeitsausschusses für Ufereinfassungen) (2004) *Recommendations of the Committee for Waterfront Structures, Harbours and Waterways*, 8th English edn. Ernst, Berlin.

Ministerio de Obras Públicas y Transportes (1990) ROM 02-90. Maritime works recommendations. Actions in the design of maritime and harbour works. Ministerio de Obras Públicas y Transportes, Madrid.

Norwegian Concrete Association (2003) *Guidelines for the Design and Construction of Concrete Structures in Marine Environments*. Norwegian Concrete Association, Oslo, Publication No. 5 (in Norwegian).

PIANC (International Navigation Association) (1978) *Standardisation of Roll On/Roll Off Ships and Berths*. PIANC, Brussels.

PIANC (1986) *Bulletin No. 54*. PIANC, Brussels.

Port and Harbour Research Institute (1999) *Technical Standards for Port and Harbour Facilities in Japan*. Port and Harbour Research Institute, Ministry of Transport, Tokyo.

Port Designer's Handbook
ISBN 978-0-7277-6004-3

http://dx.doi.org/10.1680/pdh.60043.215

Chapter 10
Sheet pile wall structures

10.1. General

Like gravity-wall berths, anchored sheet pile wall berths can be subdivided into the following three groups, depending on how the load-bearing elements are designed:

(a) simple sheet pile wall berths
(b) solid platform berths
(c) semi-solid platform berths.

The sheet pile wall itself is unable to resist the horizontal loads from backfill, moving forces, etc. A part of the horizontal force is absorbed by passive earth pressure in front of the sheet piles through their fixation in the seabed, and tied anchor rods to an anchor structure behind the sheet pile wall will transmit part of the force.

In general, sheet pile walls are not suitable where the bedrock is found at such a high level that the loose seabed material left above it is insufficient to secure the fixation of the sheet piles. However, where the seabed is bare rock, it is possible to fix the sheet piles to the rock by dowels or to provide an anchored concrete beam on the rock to prevent sliding of the piles from the horizontal forces. In some cases the sheet piles have been driven into a blasted trench in the rock surface along the berth front. It must always be checked that the ground under the piles is able to resist the load and that the stability of the whole structure is secured.

Sheet pile materials can be wood, reinforced concrete or steel. In earlier times, wood and reinforced concrete were often used in walls of modest height, but nowadays reinforced concrete is usually only used in small or secondary structures, while wood is hardly ever used. Steel sheet piles are today the most widely used sheet wall elements in berth structures. Therefore, only sheet piles made of steel are considered here.

The many manufacturers of deep-arch steel sheet piles produce a variety of profiles, but generally they can be divided into three main types:

(a) The **U profile** sheet pile, where the locks are located at the neutral axis of the profiles, as in, for instance, the Larssen profiles. Characteristics are good corrosion resistance due to the greatest steel thickness lying on the outer part of the geometry, and the ease of installing tie rods and swivelling attachments, even under water.
(b) The **Z profile** sheet pile, where the locks are located at the flange of the profiles, as in Hoesch, Arcelor and Peiner profiles, etc. The main characteristics of the Z profile sheet pile are the continuous form of the web and the specific symmetrical location of the interlock on both sides of the neutral axis.

(c) The **H pile**, box pile and tubular pile, which has either interlocking sheet pile elements or separate interlocks.

All the above profiles are produced in various sizes and with various moments of inertia, section modulus and weights/m^2. Figures 10.1 and 10.2 give the main specifications for some of the most common deep-arch steel sheet pile profiles from ArcelorMittal, Luxembourg. Also, special sheet piles for use at wall corners, at transitions from one type of profile to another, etc., are produced. The steel sheet piles have a very large moment of resistance due to their weight, compared with sheet piles of reinforced concrete.

The sheet pile bending moment capacity $M = W \times \sigma$ can be found from Figures 10.1 and 10.2.

For sheet piles having their locks at the neutral axis, like the U profiles, the useful moment of resistance must in some cases be reduced due to the risk of sliding of the locks. In order to achieve such a reduction, two and two sheet piles welded together can be ordered from the manufacturer, or the welding can be done at the site.

Steel pipe piles can be jointed together into a continuous retaining wall by connecting them with weld-on interlocking sections or into a combined wall comprised of conventional sheet piles as intermediate sections. Such pile walls and combined walls have an enhanced resistance to vertical and horizontal loads compared with conventional sheet pile sections. The pipe piles in a retaining wall are usually open ended, but can be equipped, if necessary, with pile shoes. An advantage of steel pipe pile walls is that they can be driven into most hard-to-penetrate soils. Figures 10.3 and 10.4 show different types of steel sheet pile used as sheet pile wall. The advantages of these systems are low weight and a high moment capacity. These systems have been used, for example, in the Netherlands since around 1980. The principle of these systems is that stiff primary steel elements are driven at a fixed distance from each other, and then the spacing between these elements is filled by standard sheet pile elements.

The steel grades employed for steel sheet piling should be in accordance with DIN EN 10248-1 (Table 10.1) or an equivalent.

10.2. Driving of steel sheet piles

The driving of steel sheet piles of corrugated steel sheet piling of Z and/or U profiles should always be driven in pairs. Driving of single sheet piles should, if possible, be avoided. Driving with triple or quadruple piles may have technical and economic advantages in specific cases.

For driving, slow-stroke drop hammers, diesel hammers, hydraulic hammers and rapid-stroke hammers may be used. The efficiency of the sheet pile driving generally increases when the ratio of the driving hammer weight to the weight of the driving element including the pile cap is increased. For free-drop hammers, hydraulic hammers and single-acting steam hammers, a ratio of the hammer weight to the weight of the driving element of 1 : 1 is preferred. For rapid-stroke hammers, a ratio of 1 : 4 is preferred. Slow-stroke heavy hammers are recommended in cohesive soils, while in non-cohesive soils, rapid-stroke hammers are recommended.

During driving, Z shaped sheet piles have a tendency to lean backwards due to the driving direction, while U shaped sheet piles tend to lean forward in the driving direction.

Figure 10.1 U steel sheet pile profiles. (Courtesy of ArcelorMittal, Luxembourg)

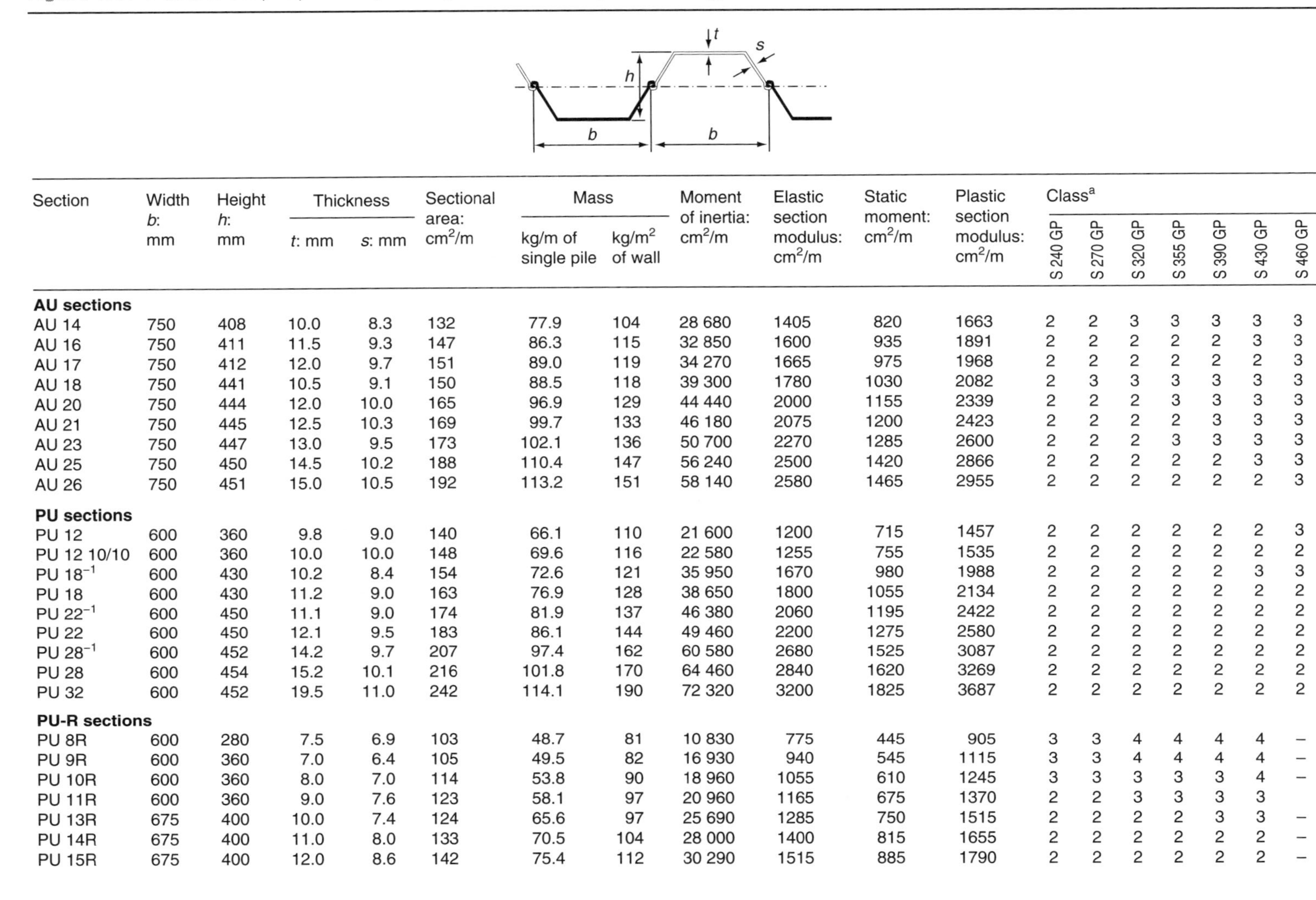

Section	Width b: mm	Height h: mm	Thickness t: mm	Thickness s: mm	Sectional area: cm²/m	Mass kg/m of single pile	Mass kg/m² of wall	Moment of inertia: cm²/m	Elastic section modulus: cm²/m	Static moment: cm²/m	Plastic section modulus: cm²/m	Class[a] S 240 GP	S 270 GP	S 320 GP	S 355 GP	S 390 GP	S 430 GP	S 460 GP
AU sections																		
AU 14	750	408	10.0	8.3	132	77.9	104	28 680	1405	820	1663	2	2	3	3	3	3	3
AU 16	750	411	11.5	9.3	147	86.3	115	32 850	1600	935	1891	2	2	2	2	2	3	3
AU 17	750	412	12.0	9.7	151	89.0	119	34 270	1665	975	1968	2	2	2	2	2	2	3
AU 18	750	441	10.5	9.1	150	88.5	118	39 300	1780	1030	2082	2	3	3	3	3	3	3
AU 20	750	444	12.0	10.0	165	96.9	129	44 440	2000	1155	2339	2	2	2	3	3	3	3
AU 21	750	445	12.5	10.3	169	99.7	133	46 180	2075	1200	2423	2	2	2	2	3	3	3
AU 23	750	447	13.0	9.5	173	102.1	136	50 700	2270	1285	2600	2	2	2	3	3	3	3
AU 25	750	450	14.5	10.2	188	110.4	147	56 240	2500	1420	2866	2	2	2	2	2	3	3
AU 26	750	451	15.0	10.5	192	113.2	151	58 140	2580	1465	2955	2	2	2	2	2	2	3
PU sections																		
PU 12	600	360	9.8	9.0	140	66.1	110	21 600	1200	715	1457	2	2	2	2	2	2	3
PU 12 10/10	600	360	10.0	10.0	148	69.6	116	22 580	1255	755	1535	2	2	2	2	2	2	2
PU 18^{-1}	600	430	10.2	8.4	154	72.6	121	35 950	1670	980	1988	2	2	2	2	2	3	3
PU 18	600	430	11.2	9.0	163	76.9	128	38 650	1800	1055	2134	2	2	2	2	2	2	2
PU 22^{-1}	600	450	11.1	9.0	174	81.9	137	46 380	2060	1195	2422	2	2	2	2	2	2	2
PU 22	600	450	12.1	9.5	183	86.1	144	49 460	2200	1275	2580	2	2	2	2	2	2	2
PU 28^{-1}	600	452	14.2	9.7	207	97.4	162	60 580	2680	1525	3087	2	2	2	2	2	2	2
PU 28	600	454	15.2	10.1	216	101.8	170	64 460	2840	1620	3269	2	2	2	2	2	2	2
PU 32	600	452	19.5	11.0	242	114.1	190	72 320	3200	1825	3687	2	2	2	2	2	2	2
PU-R sections																		
PU 8R	600	280	7.5	6.9	103	48.7	81	10 830	775	445	905	3	3	4	4	4	4	–
PU 9R	600	360	7.0	6.4	105	49.5	82	16 930	940	545	1115	3	3	4	4	4	4	–
PU 10R	600	360	8.0	7.0	114	53.8	90	18 960	1055	610	1245	3	3	3	3	3	4	–
PU 11R	600	360	9.0	7.6	123	58.1	97	20 960	1165	675	1370	2	2	3	3	3	3	
PU 13R	675	400	10.0	7.4	124	65.6	97	25 690	1285	750	1515	2	2	2	2	3	3	–
PU 14R	675	400	11.0	8.0	133	70.5	104	28 000	1400	815	1655	2	2	2	2	2	2	–
PU 15R	675	400	12.0	8.6	142	75.4	112	30 290	1515	885	1790	2	2	2	2	2	2	–

GU sections																
GU 6N	600	309	6.0	6.0	89	41.9	70	9 670	625	375	765	3	3	3	4	–
GU 7N	600	310	6.5	6.4	94	44.1	74	10 450	675	400	825	3	3	3	3	–
GU 7S	600	311	7.2	6.9	100	46.3	77	11 540	740	440	900	2	2	3	3	–
GU 8N	600	312	7.5	7.1	103	48.5	81	12 010	770	460	935	2	2	3	3	–
GU 12-500	500	340	9.0	8.5	144	56.6	113	19 640	1 155	680	1 390	2	2	2	2	–
GU 13-500	500	340	10.0	9.0	155	60.8	122	21 390	1 260	740	1 515	2	2	2	2	–
GU 15-500	500	340	12.0	10.0	177	69.3	139	24 810	1 460	855	1 755	2	2	2	2	–
GU 16-400	400	290	12.7	9.4	197	62.0	155	22 580	1 560	885	1 815	2	2	2	2	–
GU 18-400	400	292	15.0	9.7	221	69.3	173	26 090	1 785	1 015	2 080	2	2	2	2	–

[a] Classification according to EN 1993-5.
The moment of inertia and section moduli values given assume correct shear transfer across the interlock.
Class 1 is obtained by verification of the rotation capacity for a class 2 cross-section.
A set of tables with all the data required for design in accordance with Eurocode EN 1993-5 is available from the steel company ArcelorMittal. All PU sections can be rolled up or down by 0.5 and 1.0 mm. Other sections are available on request.

Figure 10.2 Z steel sheet pile profiles. (Courtesy of ArcelorMittal, Luxembourg)

Section	Width b: mm	Height h: mm	Thickness		Sectional area: cm^2/m	Mass		Moment of inertia: cm^4/m	Elastic section modulus: cm^3/m	Static moment: cm^3/m	Plastic section modulus: cm^3/m	Class[a]						
			t: mm	s: mm		kg/m of single pile	kg/m^2 of wall					S 240 GP	S 270 GP	S 320 GP	S 355 GP	S 390 GP	S 430 GP	S 460 GP
AZ 12	670	302	8.5	8.5	126	66.1	99	18 140	1200	705	1409	2	3	3	3	3	3	3
AZ 13	670	303	9.5	9.5	137	72.0	107	19 700	1300	765	1528	2	2	2	3	3	3	3
AZ 14	670	304	10.5	10.5	149	78.3	117	21 300	1400	825	1651	2	2	2	2	2	3	3
AZ 17	630	379	8.5	8.5	138	68.4	109	31 580	1665	970	1944	2	2	3	3	3	3	3
AZ 18	630	380	9.5	9.5	150	74.4	118	34 200	1800	1050	2104	2	2	2	3	3	3	3
AZ 19	630	381	10.5	10.5	164	81.0	129	36 980	1940	1140	2275	2	2	2	2	2	3	3
AZ 25	630	426	12.0	11.2	185	91.5	145	52 250	2455	1435	2873	2	2	2	2	2	2	2
AZ 26	630	427	13.0	12.2	198	97.8	155	55 510	2600	1530	3059	2	2	2	2	2	2	2
AZ 28	630	428	14.0	13.2	211	104.4	166	58 940	2755	1625	3252	2	2	2	2	2	2	2
AZ 46	580	481	18.0	14.0	291	132.6	229	110 450	4595	2650	5295	2	2	2	2	2	2	2
AZ 48	580	482	19.0	15.0	307	139.6	241	115 670	4800	2775	5553	2	2	2	2	2	2	2
AZ 50	580	483	20.0	16.0	322	146.7	253	121 060	5015	2910	5816	2	2	2	2	2	2	2
For minimum steel thickness of10 mm																		
AZ 1310/10	670	304	10.0	10.0	143	75.2	112	20 480	1350	795	1589	2	2	2	2	3	3	3
AZ 1810/10	630	381	10.0	10.0	157	77.8	123	35 540	1870	1095	2189	2	2	2	2	3	3	3
AZ-700 and AZ-770																		
AZ 12-770	770	344	8.5	8.5	120	72.6	94	21 430	1245	740	1480	2	2	3	3	3	3	3
AZ 13-770	770	344	9.0	9.0	126	76.1	99	22 360	1300	775	1546	2	2	3	3	3	3	3
AZ 14-770	770	345	9.5	9.5	132	79.5	103	23 300	1355	805	1611	2	2	2	2	3	3	3
AZ 14-770-10/10	770	345	10.0	10.0	137	82.9	108	24 240	1405	840	1677	2	2	2	2	2	3	3
AZ 17-700	700	420	8.5	8.5	133	73.1	104	36 230	1730	1015	2027	2	2	3	3	3	3	3
AZ 18-700	700	420	9.0	9.0	139	76.5	109	37 800	1800	1060	2116	2	2	3	3	3	3	3
AZ 19-700	700	421	9.5	9.5	146	80.0	114	39 380	1870	1105	2206	2	2	2	3	3	3	3
AZ 20-700	700	421	10.0	10.0	152	83.5	119	40 960	1945	1150	2296	2	2	2	2	2	3	3
AZ 24-700	700	459	11.2	11.2	174	95.7	137	55 820	2430	1435	2867	2	2	2	2	2	2	3
AZ 26-700	700	460	12.2	12.2	187	102.9	147	59 720	2600	1535	3070	2	2	2	2	2	2	2
AZ 28-700	700	461	13.2	13.2	200	110.0	157	63 620	2760	1635	3273	2	2	2	2	2	2	2
AZ 37-700	700	499	17.0	12.2	226	124.2	177	92 400	3705	2130	4260	2	2	2	2	2	2	2
AZ 39-700	700	500	18.0	13.2	240	131.9	188	97 500	3900	2250	4500	2	2	2	2	2	2	2

[a] 2 sides, inside of interlocks excluded

Figure 10.3 Steel tubes in a sheet pile wall

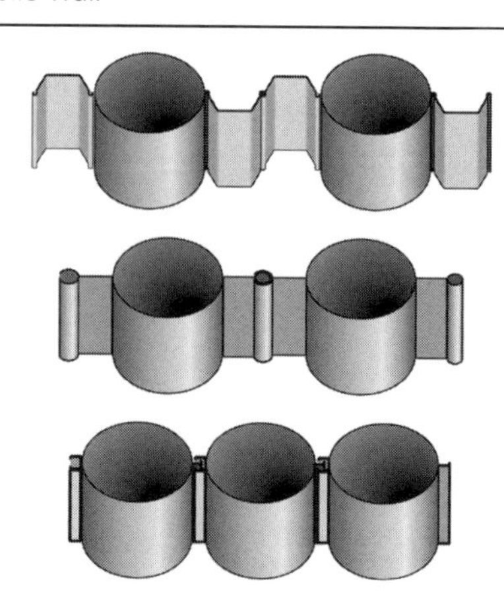

The EAU (2004) recommends that the following driving deviations and tolerances should be included in the calculations at the planning stage:

(a) 1.0% for normal soil conditions and driving on land
(b) 1.5% of the driving depth for driving on water
(c) 2.0% of the driving depth with difficult subsoil.

With increasing driving depth of the sheet piles, the deviation from the vertical will increase.

Figure 10.4 Different combined sheet pile wall systems

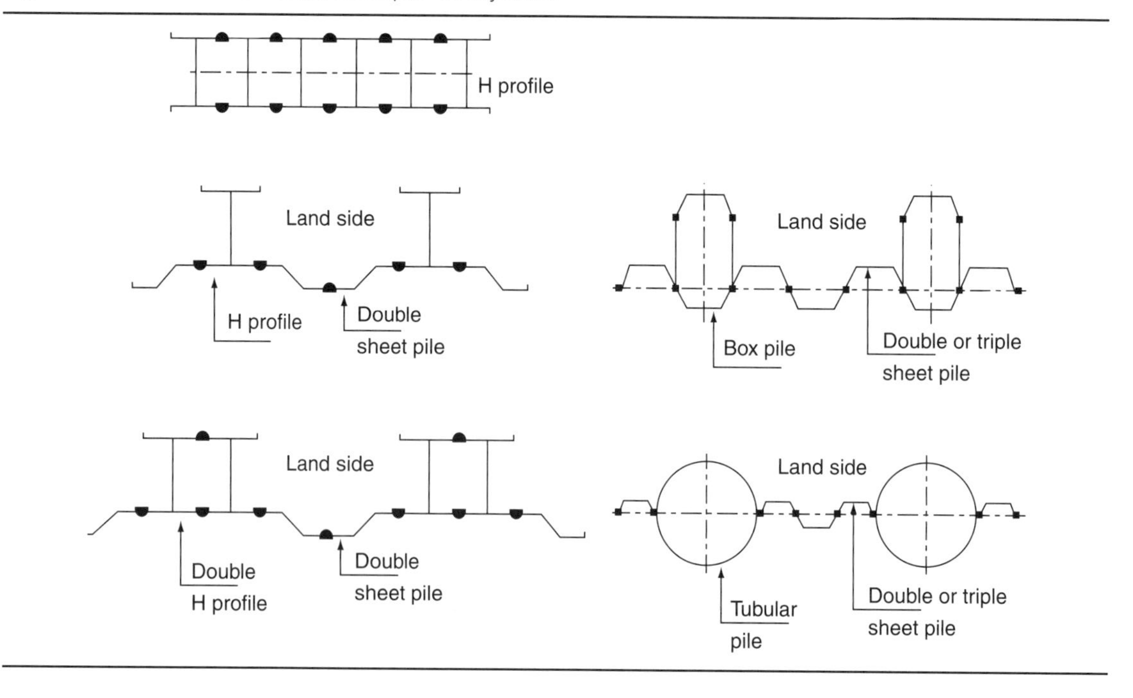

Table 10.1 Steel grades according to DIN EN 10248

Grade, EN 10248	Minimum yield point: N/mm^2	Minimum tensile strength: N/mm^2	Minimum elongation $L_0 = 5.65\sqrt{S_0}$: %	Chemical composition: % (maximum)					
				Carbon	Manganese	Silicon	Phosphorus	Sulphur	Nitrogen
S 240 GP	240	340	26	0.25	–	–	0.055	0.055	0.011
S 270 GP	270	410	24	0.27	–	–	0.055	0.055	0.011
S 320 GP	320	440	23	0.27	1.70	0.60	0.055	0.055	0.011
S 355 GP	355	480	22	0.27	1.70	0.60	0.055	0.055	0.011
S 390 GP	390	490	20	0.27	1.70	0.60	0.050	0.050	0.011
S 430 GP	430	510	19	0.27	1.70	0.60	0.050	0.050	0.011
Mill specification									
S 460 GP	460	550	17	0.27	1.70	0.60	0.050	0.050	0.011

For details, see EN 10248, grade S 460 AP (following mill specification) upon request

Figure 10.5 A simple sheet pile wall

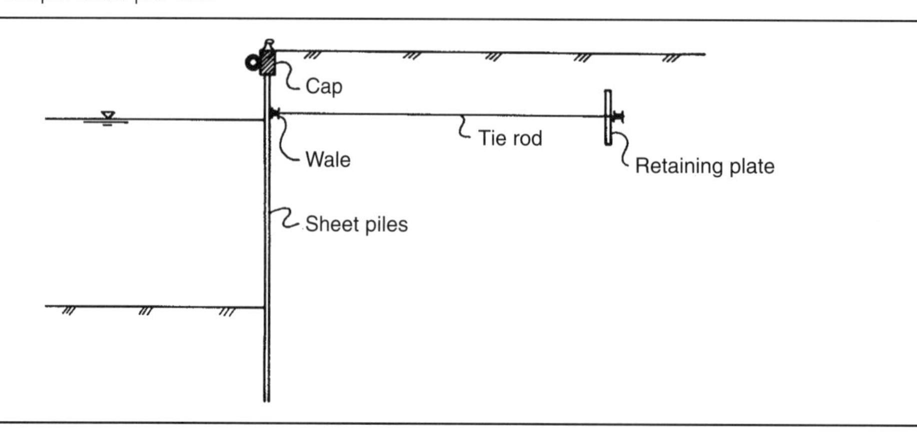

10.3. Simple anchored sheet pile wall berths

When the water depth at the berth front is moderate and the soil conditions are favourable, a simple sheet pile wall (Figure 10.5), can be an economically good solution. The sheet piles are driven down to sufficient depth, one after the other. The horizontal anchor force is transmitted from the wale (anchoring beam) by way of a tie rod to a retaining plate or other type of anchorage. It is advantageous if the wall can be driven and anchored first, and the dredging, to the prescribed depth, in front of the wall can be done afterwards.

The most economical free wall height of a simple sheet pile wall berth with a simple anchoring system varies from about 7 to 10 m, depending on the soil conditions, the price of steel, the driving equipment and the useful load on the quay surface.

The berthing loads normal to the berth structure are transmitted to the earth behind the structure by the fender system either to the capping beam or to the sheet wall itself. The horizontal loadings in the longitudinal direction from friction between the berthing ship and the fender system are taken by the sheet pile wall itself.

For detailed geotechnical design calculation of the steel sheet pile structures, see 'References and further reading' at the end of this chapter.

The maximum horizontal loading on the wall arises when an extremely low water level in front of the berth and an extremely high water level behind the wall occur together with maximum live load and mooring forces on the structure. A low water level in front of the berth, due to low tide, wind, air pressure, etc., cannot be avoided, but a high hydrostatic pressure behind the wall can normally be avoided by using coarse-graded fill combined with a special drainage system to prevent an excessive inside water pressure. Nevertheless, in the design of the sheet piles, provision should be made for a periodical difference in water levels between the two sides of the wall.

The active soil pressure on the wall determines the cross-sectional dimensions of the sheet piles. The passive soil pressure determines the support of the wall at its lower end below the dredge line. The horizontal tie rod tension has to be absorbed either by the passive soil pressure in front of the retaining plate, as illustrated in Figure 10.5, or by friction between a friction plate and the soil. The sheet pile

Figure 10.6 Effects of the flexibility of an anchored sheet pile wall

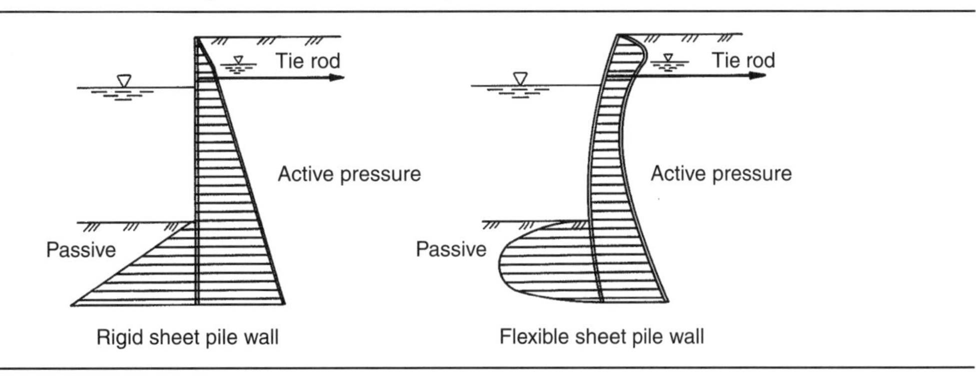

and the wale are designed primarily to resist bending stresses, while the tie rod cross-section is determined by the maximum tensile stress it has to resist, and also by its length.

Thus, the dimensions of the structural elements of the wall are determined by the soil conditions, the water level variations, the duration of the construction work, the method of filling behind the wall, the filling period, etc. Taken together, these factors imply a rather complicated design pattern, in which the soil mechanics aspects must be given particular attention. Figure 10.6 illustrates the distribution of the active and passive pressure on single-anchored rigid and flexible sheet pile retaining walls.

The wall is finished on top by the installation of a special steel beam or by the casting of a reinforced concrete capping beam (cap). The latter is usually made wide enough for the installation of bollards and a quay-front kerb on top, and high enough for the installation of fenders on the front. In major berths, the bollards are usually anchored through their own tie rods to the retaining structure. If the berth has cranes on rails, the capping beam is often also utilised as a support for the outer rail. Account must then be taken of the necessary free clearance between the crane and berthing ships and possible settlement of the wall.

Expansion joints in the capping concrete beam are placed at intervals of 15–30 m, depending on the structure and the amount of reinforcement in the capping beam to absorption of the increased tensile forces in the longitudinal direction of the capping beam.

The different methods and principles applied in the design of anchoring for sheet pile walls are shown in Figure 10.7, and some of the most used anchoring systems are described in the following. In principle, these systems can also be used for the anchoring of other types of berth structure. Within harbour construction, it generally applies that underwater work should, if possible, be avoided and that anchorage or other work should be achieved in the simplest possible way. This implies that the anchor points and the tie rods should be placed as close to the tidal zone as possible. The bending moment in the sheet piles will then be moderate, and the underwater work can be avoided or made easier. But it is necessary to realise that all of the systems will give a negative skin friction downwards to the sheet piles due to the earth pressure behind the wall.

Different ways of designing sheet pile wall anchoring, as shown in Figure 10.7, are as follows:

(a) Retaining or anchoring wall supported by the passive soil pressure in front of the wall. The retaining wall must in theory be pulled some small distance in the direction of the sheet pile

Figure 10.7 Different anchoring systems

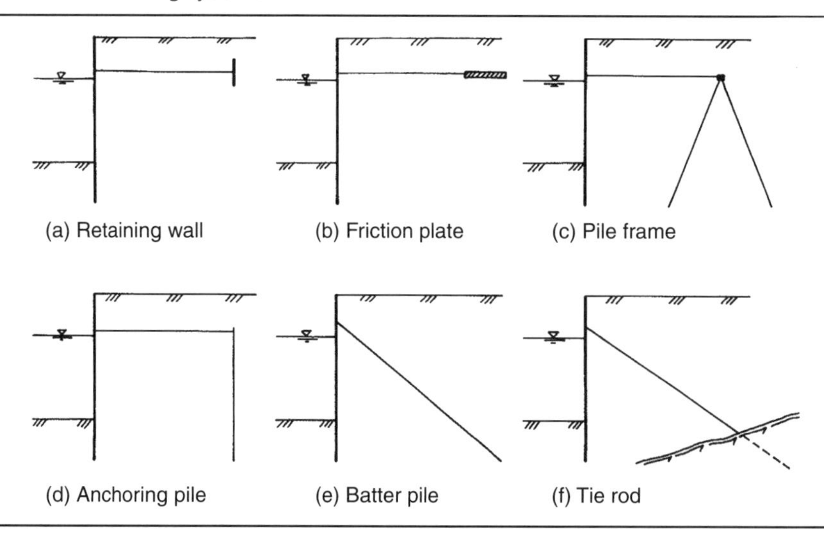

wall before the passive soil pressure is fully mobilised. The soil in front of the retaining wall should therefore be thoroughly compacted to minimise this movement before the tie rod is fixed.

(b) Retaining plate or friction plate transmitting the force only by way of friction to the soil under the friction plate. This takes place when the width of the plate is greater than the thickness of the soil layer above the plate. The force is then transmitted to the underlying soil without any movement of the plate.

(c) The force is transmitted to pile frames, each consisting of one tension pile and one compression pile. The piles can be rather long and thick in order to provide the necessary contact area to the soil. The piles should, if possible, be positioned behind the active soil wedge, to allow frictional resistance to be developed along the full length of the piles.

(d) The force is absorbed through the bending of anchoring piles. This is a convenient method when the distance from the berth front to the anchoring point is limited and the force is moderate.

(e) Direct anchoring by way of batter friction piles behind the wall that absorbs the force. The batter friction piles will be rather long and thick in order to provide the necessary contact friction area. This method sets up vertical downward-directed forces in the sheet piles, which can cause settlements of the wall. This system will also give an additional downward load on the sheet pile system.

(f) The wall is anchored by way of tie rods down to firm rock, or to anchors in the soil. This is a suitable method of temporary anchoring during repair of the wale, etc., and where none of the above methods can be utilised. This system will also give an additional downward load on the sheet pile system. For a permanent anchor, there must be no settlement in the soil above the anchor tie rods.

The tie rods should be designed to accommodate any settlement of the ground under the tie rods. If noticeable settlements of the fill behind the wall are expected, the connection between the sheet piles and the tie rod should be formed as a hinge, and wooden poles, to prevent damage due to settlement in the filling, should support the rods. Usually, the tie rods are placed just above the upper level of the

Figure 10.8 Details of one type of steel sheet wall anchoring

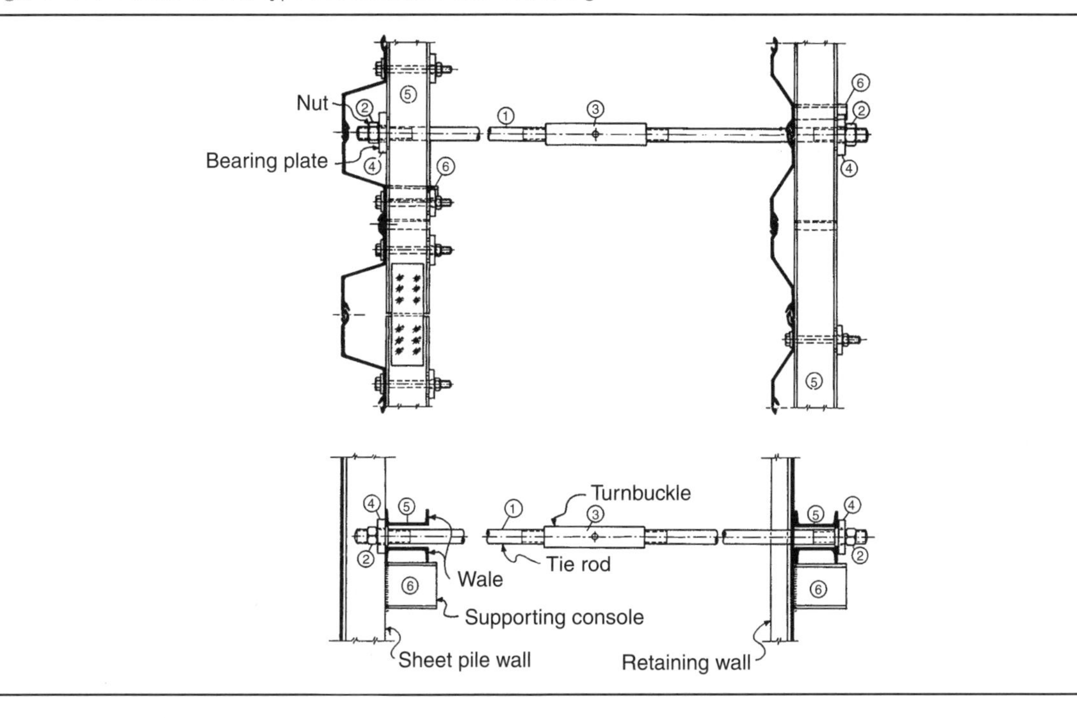

tidal zone. When there are also great tidal variations, a lower tie rod can be installed just above the lower level of the tidal zone. Thus, the bending moments in the wall can be considerably reduced. Figure 10.8 shows details of a steel sheet pile wall, a tie rod and a retaining wall.

The tie rods are usually made of round bars of diameter 5–10 cm and quality S 355. Frequently used types of end fastening and splicing of tie rods are shown in Figure 10.8. Very often, the tie rod ends are jolted so that their diameter at the bottom of the thread is the same as the diameter of the unthreaded rod. Due to difficulties in designing the tie rods exactly according to the tensile force, the rods are usually over-designed by 50–100%. To protect the tie anchor system from additional loading by the weight of the backfill, the tie rods should be placed with a negative sag of approximately 5–15 cm, depending on the length of the tie rod and the estimated settlement, to compensate for the settlement of the backfill. The tie rod with its connection should have a safety factor of at least 2 under all loading conditions. Joints, in which stress concentrations may occur, are also over-designed by 50–100%. This philosophy should also be adopted in the design of tie rods made of reinforced concrete.

The consequences of a possible failure of the anchorage should always be taken into account. Buckling or overloading of the sheet piles will usually not be as serious as an anchorage failure.

Steel tie rods are also normally protected against corrosion when they are located above the tidal zone. Most frequently, a bituminous material is used, and the tie rods are wrapped in canvas or similar and embedded in sand. If concrete is used as the protective material, it must be reinforced against shrinkage cracks.

To obtain a rough estimate of the sizes of the forces and dimensions in simple steel pile wall berths for different live loads, diagrams of the type shown in Figure 10.9 may be used. The diagrams are based on the assumption that the soil at both sides of the sheet pile wall consists of dense, sharp-grained sand or gravel and no corrosion. The angle of internal friction, ϕ, is therefore estimated to be 38°. The diagram should not be used for weaker soils (i.e. soils of $\phi < 36°$). The main assumptions for the calculations are shown in the figure, and the simplified diagram may be used for rough estimations.

The diagrams in Figure 10.9 are calculated according to Publication 16 from the Norwegian Geotechnical Institute (Janbu *et al.*, 1956). The diagrams are based on classical earth pressure theories. As an illustration, the length of the sheet pile, the design moment, the anchor force, the length of the tie rod and the length of the friction slab are shown for a sheet pile wall with a water depth of 8 m and a live load of 50 kN/m^2. In recent years, data programs have been developed that account for several other factors, such as the rigidity of wall and anchors.

10.4. Solid platform berths

When the height of the berth front exceeds about 8–10 m, the simple sheet pile wall structure will normally not be the most economical solution. Better solutions can be obtained by using one or more of the following adaptations:

(a) Tie rods are placed at two different levels. This can also be done where the tidal variations are great.
(b) Use of special sheet pile profiles designed to resist large bending moments.
(c) Use of lightweight fill material behind the sheet pile wall.
(d) A relieving plate, placed on the wall and on piles behind the wall, transmits useful loads and the weight of the fill on top as axial loads to the wall and the piles, and reduces the horizontal load acting on the wall. This type of structure is called the solid platform type of berth (Figure 10.10).

Solid platform berths are probably most interesting from an economic point of view in cases where the live load and/or the quay-front height are relatively high. Also, where the berth-front height is relatively small, this type of structure can be economically justified provided very great live loads are acting on the berth deck. This type of structure has also been used in cases where the utilisation of sheet piles with small moments of resistance has been pursued.

Horizontal forces are absorbed either by batter piles under the relief plate or by an anchorage further back. As mentioned in the discussion of the anchoring of simple sheet pile walls, a vertical retaining wall must be pulled a small amount in the direction of the sheet pile wall before the passive soil pressure is fully mobilised, while a friction plate transmits the force directly to the underlying soil. This implies that batter piles and retaining plates can be utilised simultaneously.

The absorption of the horizontal load by a combination of batter piles and retaining wall is not equally simple. In this case, special precautions should be taken to make the piles and the wall act together, such as stretching of the tie rods. The choice of anchoring system depends on the method of construction to be utilised. There are two principal methods of constructing these berths:

(a) **Offshore**: the berth is built independently in the water, and the shore connection is filled afterwards. Figure 10.11 shows the principle of this procedure. It is of decisive importance that the sheet pile wall is sufficiently anchored against the soil pressure during the construction.

Figure 10.9 Estimate diagrams for steel sheet piles

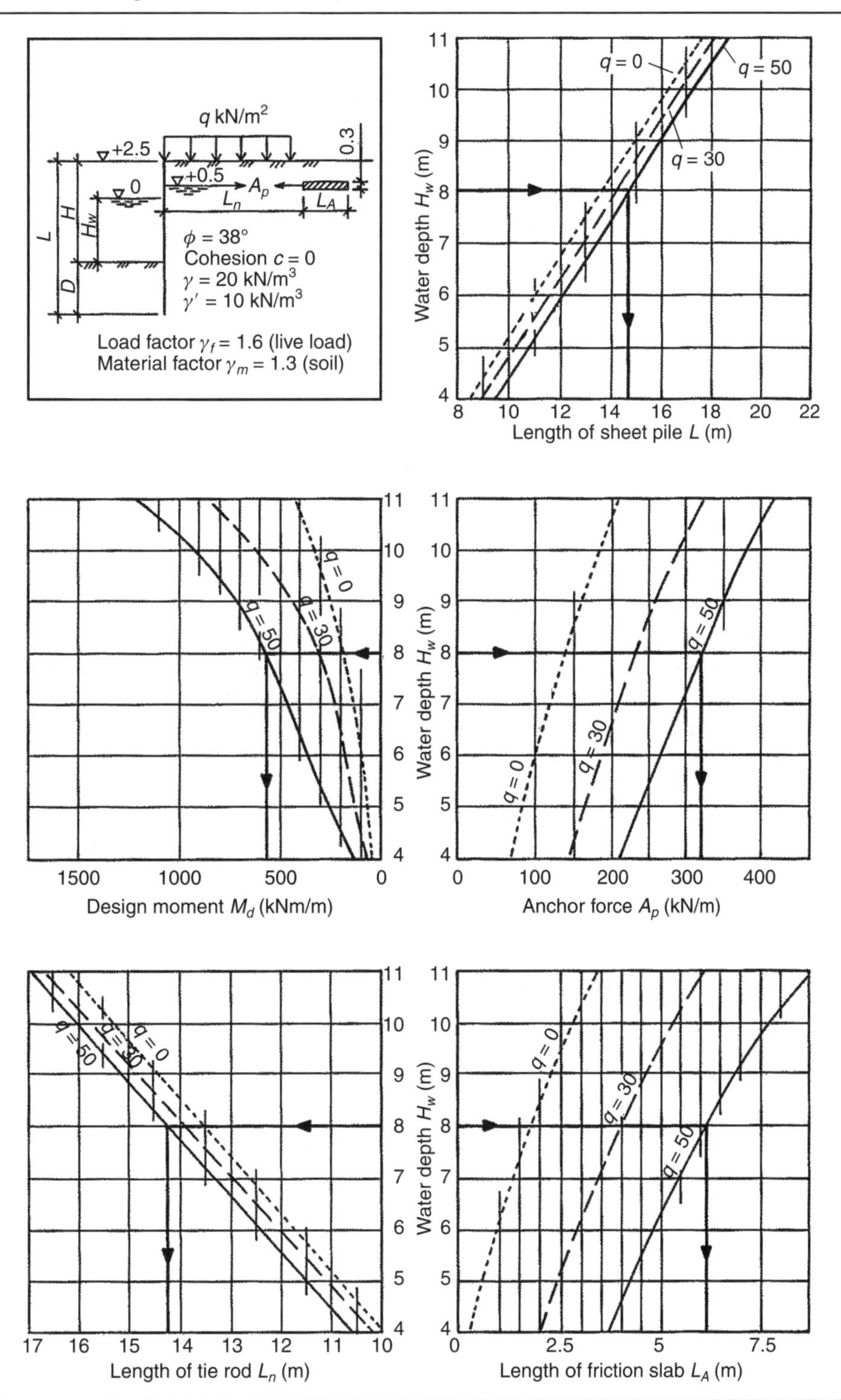

Figure 10.10 A solid platform berth

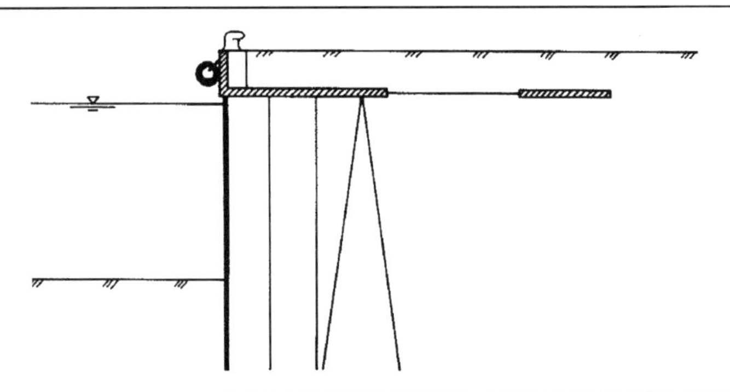

When using the working sequence indicated in the figure, the platform can be cast directly on the filled area. Usually, batter piles absorb the horizontal forces and/or retaining plate, as shown in the figure.

(b) **Onshore**: the berth is built onshore, and the basin in front of the berth is dredged after the installation of the anchoring and the filling on top, as shown in Figure 10.12 and in Figures 10.13–10.15. Batter piles are not needed in this case.

If it is desired to avoid the construction of a retaining or friction plate, the structure can be designed as shown in Figure 10.16. The platform is supported by the sheet pile wall and the pile frame, and acts as an anchorage between the wall and the frame. In order to get the largest possible stabilising load on the pile frame and to reduce the bending moment in the platform span, the platform should form a cantilever to the inner side of the pile frame.

Figure 10.11 Offshore construction

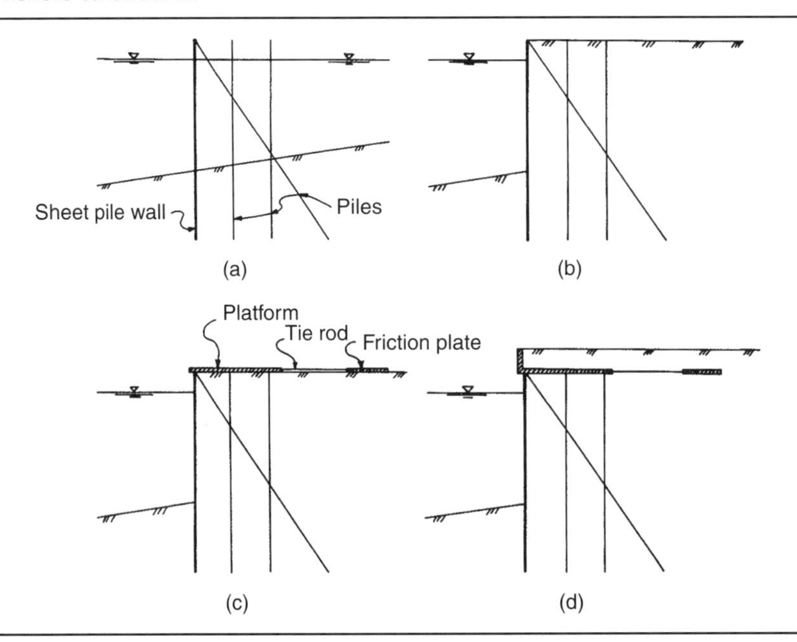

Figure 10.12 Onshore construction

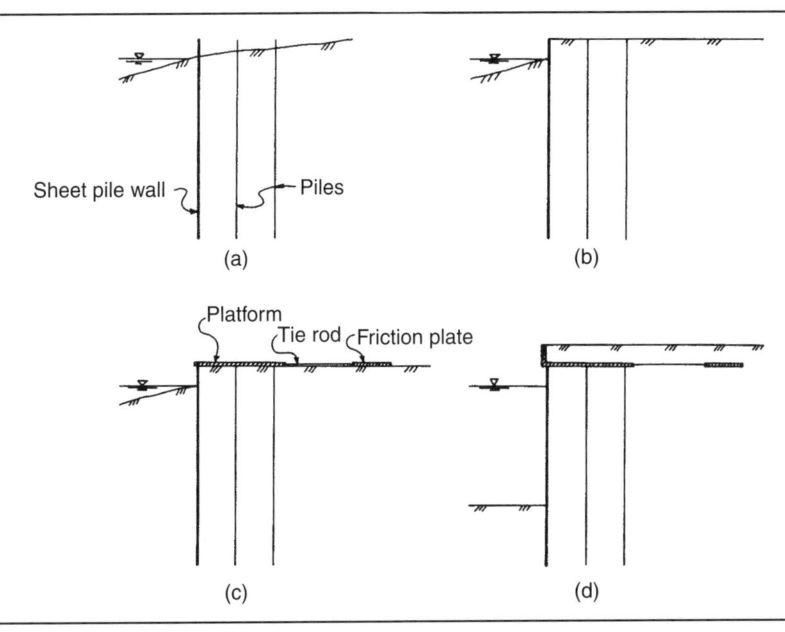

Neither in the offshore method nor in the onshore method can any friction occurring between the platform and the underlying soil be assumed because settlements under the platform will arise as time goes by. The platform and the retaining plate should be placed at such a high level that they can be constructed above water. Both wooden and concrete piles can be used under the platform. Wooden

Figure 10.13 Onshore construction phases (a) and (b)

Figure 10.14 Onshore construction phase (c). The construction of tie rods and the friction slab

Figure 10.15 Onshore construction phase (d). The backfilling of sand

Figure 10.16 A solid platform berth

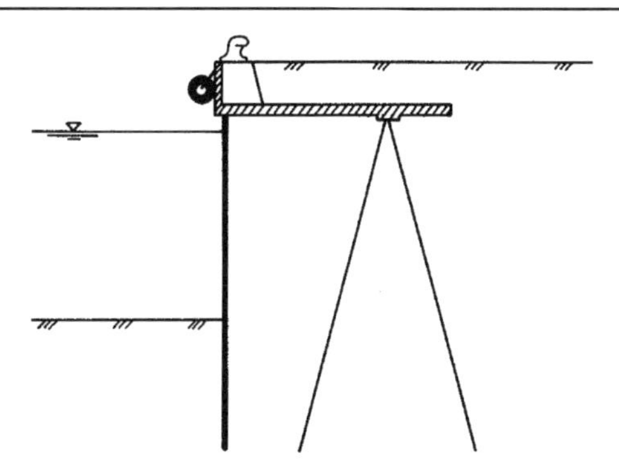

piles must be protected against rotting, and also against wood borers (sea worms) because settlements can take place underneath the platform.

The economically most favourable berth-front height for this type of berth is about 14–18 m, depending on the geotechnical conditions on site and the cost of sheet piles and driving.

10.5. Semi-solid platform berth

If a greater quay-front height than about 14–18 m is needed, the type of berth structure shown in Figure 10.17 may prove to be suitable. In principle, it is very much like the solid platform type, but the soil pressure against the sheet pile wall is further reduced by omitting some of the fill under the platform, as shown in the figure. This implies that a formwork has to be made for the platform, instead of casting it directly on the soil, as for the solid quays. To facilitate in-situ work, the platform can be made of precast, reinforced concrete elements. Many such elements, of more or less sophisticated designs, have been introduced.

The main advantage of this semi-solid type, as compared with the solid type, is that the berth-front height can be made substantially higher. The disadvantages are that untreated wooden piles cannot be used, that the piles must have greater cross-sectional dimensions due to the risk of buckling and that corrosion may occur on both sides of the wall.

Figure 10.17 Semi-solid platform berth

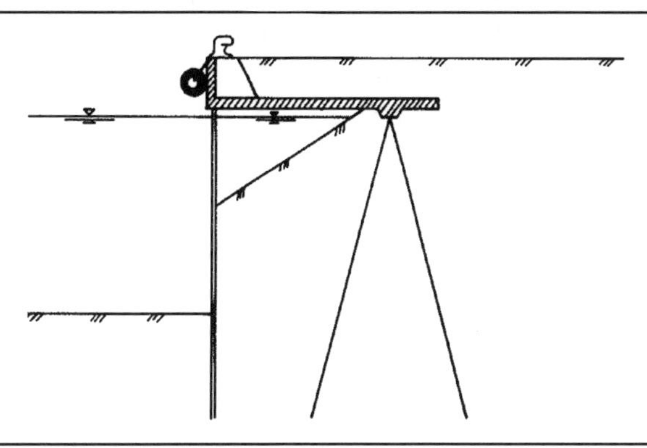

Figure 10.18 Earth pressure against the pile

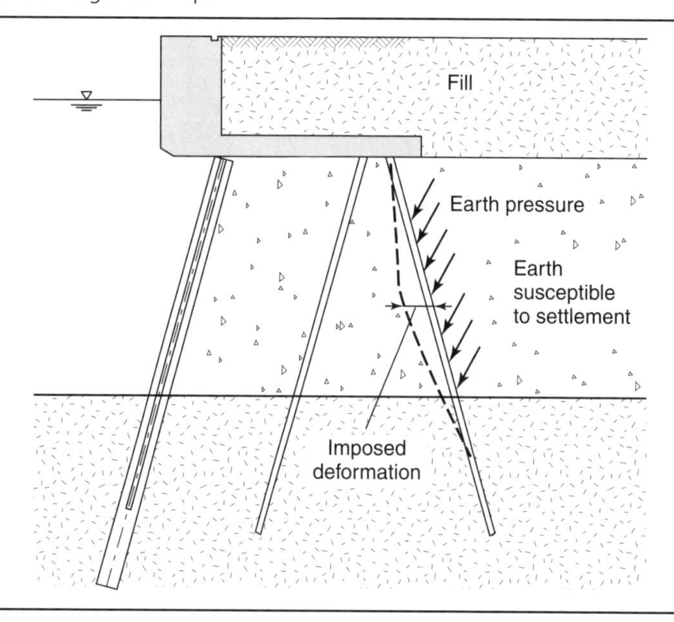

The semi-solid platform berth may be characterised as a transition from the solid berth structure to the open column and lamella berth structures.

It is necessary to be aware of the earth pressure against the sloped piles (Figure 10.18). This can impose high deformation, from both the earth pressure and the live loads on the apron.

10.6. Drainage of steel sheet piles

The best form of drainage behind a steel sheet pile wall, to avoid extra pressure behind the wall during low tide or after a heavy and long period of rain, is to have graded coarse friction materials approximately 1 m behind the wall, well below the groundwater level. Effective drainage is only possible in non-cohesive soils.

In addition, weep holes should be installed below the mean water level, if possible. The weep holes should be located above the low-water level to allow maintenance, and in such a way that they cannot be damaged by ships or floating debris. The weep holes should be sealed with a suitable filter, to prevent loss of fill materials. Weep holes should not be used if there is danger of heavy growth of barnacles. The design of weep holes should take into consideration the possibility of blockage due to marine growth.

REFERENCES AND FURTHER READING

Bruun P (1989) *Port Engineering*, vols 1 and 2. Gulf Publishing, Houston, TX.

BSI (British Standards Institution) (1988) BS 6349-2: 1988. Maritime structures. Design of quay walls, jetties and dolphins. BSI, London.

BSI (2000) BS 6349-1: 2000. Maritime structures. Code of practice for general criteria. BSI, London.

CUR (Centre for Civil Engineering Research and Codes) (2005) *Handbook of Quay Walls*. CUR, Gouda.

EAU (Empfehlungen des Arbeitsausschusses für Ufereinfassungen) (2004) *Recommendations of the Committee for Waterfront Structures, Harbours and Waterways*, 8th English edn. Ernst, Berlin.

Janbu N, Bjerrum L and Kjarrnsli B (1956) *Veiledning ved løsning av fundamenteringsoppgaver*. Norwegian Geotechnical Institute, Oslo. Publication 16.

Ministerio de Obras Públicas y Transportes (1990) ROM 0.2-90. Maritime works recommendations. Actions in the design of maritime and harbour works. Ministerio de Obras Públicas y Transportes, Madrid.

Port and Harbour Research Institute (1999) *Technical Standards for Port and Harbour Facilities in Japan*. Port and Harbour Research Institute, Ministry of Transport, Tokyo.

Port Designer's Handbook
ISBN 978-0-7277-6004-3

http://dx.doi.org/10.1680/pdh.60043.235

Chapter 11
Open berth structures

11.1. General

Open berth structures constitute, together with their berth platforms, a prolongation over the slope from the top of the filled area out to the berth front. In this chapter only berth structures made of reinforced concrete, or platforms made of reinforced concrete founded on concrete-filled tubular steel piles, will be described. Open berth structures made of wooden materials will not be described, but the construction principles, load-bearing capacity, etc. are largely the same as for structures made of reinforced concrete.

In the same way as for solid berth structures, open berth structures can be divided into the following main types, depending on the principles according to which the front wall and the platform are designed to resist the loading, so that the berth structures have the necessary stability.

(a) **Column or pile berths**: the berth platform and the berth front wall are founded on either columns or piles, or a combination of both, which do not have a satisfactory stability against external forces. Therefore, the berth structure must be anchored, for instance by a friction plate in the filling. The structure must then be built simultaneously with the filling, or preferably after the fill has been established.
(b) **Lamella berths**: the berth platform and the berth front are founded on vertical lamellas, which provide the loaded berth structure with a satisfactory stability. The berth structures are stable enough in themselves to resist loads from ships, live loads, possible pressure from fill at the rear of the structure, etc. without anchoring of the structure. In the same way as for gravity-wall structures, the lamella berth itself can be built first, and then the fill can be placed behind and close up to the structure.

Very often the two types are combined. For instance, in a pier or jetty type of berth structure, the shore base is anchored by a retaining wall, while at the head of the jetty the horizontal loading is resisted by lamellas. The berth platform between the shore base and the head of the structure is founded on columns and/or piles.

The open berth structure is more suitable than the massive berth structure in the following circumstances:

(a) The seabed is too weak to carry a massive berth structure.
(b) The ground condition below the seabed is suitable for bearing piles.
(c) The water depth is large.
(d) There is a need to minimise the hydraulic regimen.
(e) There are difficulties in obtaining suitable backfill materials for a retaining wall berth structure.

Figure 11.1 Characteristics of an open berth structure

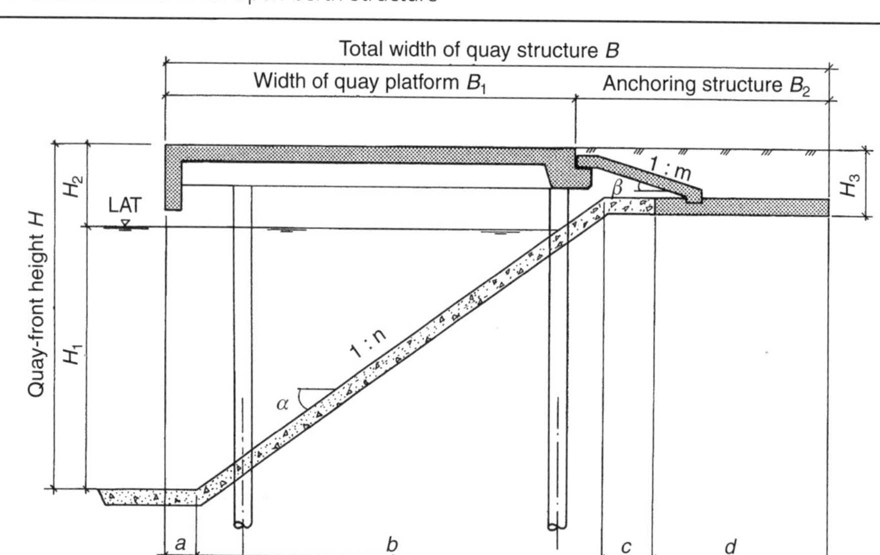

Figure 11.1 shows the cross-section of a pile berth anchored at the rear embankment, and indicates the main characteristics of open berth structures. This method of accommodating the horizontal forces by use of an anchor and friction slab is a typical Norwegian design. In general, the best solution economically for an open berth structure will be obtained when the total width *B* is as small as possible in relation to the height of the berth front. The different characteristic dimensions influencing the structural dimension and design are discussed below.

H is the height of the berth front, and is determined by the necessary water depth and the height of the berth surface above the lowest astronomical tide (LAT).

H_1 is the depth between the LAT and the bottom of the harbour basin, and is determined by the draft of incoming ships when fully loaded plus an over-depth to account for the trim of the ship, the wave height and a safety margin against sea bottom irregularities.

H_2 is the elevation of the top of the berth platform above the LAT, and is determined by the elevation of the area behind the berth and/or by the types of ship which will call at the berth structure. The top of the platform should preferably not be at a lower level than the highest observed water level plus 1.0 m.

H_3 is a distance determined by the location of the rear wall or the retaining slab above the LAT. The bottom of the slab should not lie lower than the mean sea level. The rear structure should be placed at such a level that concreting can be carried out on dry land. If it is necessary for the rear structure to lie lower, construction can be carried out using prefabricated elements placed on a levelled base underwater (Figure 11.2). By increasing the height *H* and the angle of the slope, the total width *B* will be reduced, but the economic benefit achieved by reducing *B* can easily be eliminated by higher installation costs due to the necessary underwater work.

Figure 11.2 A precast anchoring structure on a levelled underwater base

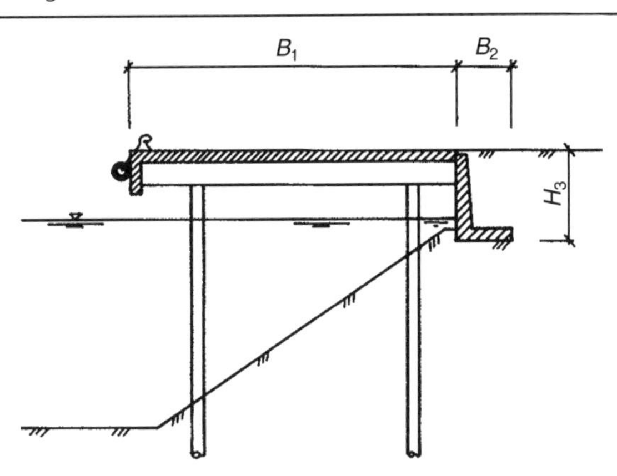

a is the slope, which should start about 1.0 m behind the berth front so that the toe of the slope is kept away from the turbulence caused by propellers. The possibility of rocks falling off the slope and outside the berth front line will then probably be prevented.

b is a distance that is determined by the steepness of the slope. The slope *a* will normally vary between 38.7° (1 : 1.25) and 29.7° (1 : 1.75), depending on which the materials the slope is made of, and whether it is covered with blasted rock, etc. Usually the angle is 33.7° (1 : 1.5). The stability of the slope itself, the coarseness of the materials and the danger of erosion from waves and propeller turbulence determine the angle of the slope.

c is a very exposed area because of the danger of slides in front of the rear wall or retaining friction slab. The width *c* of the shoulder must be at least 3.0 m, and the shoulder itself must be well covered. The width *c* must also be sufficient to ensure stability in front of the retaining slab.

d is the width of the anchor or friction slab, which needs to resist the horizontal load, and depends on the frictional angle in the ground, the safety factor required against sliding and the vertical load acting on the retaining slab.

e is the distance between the berth line and the centreline of the first row of supporting columns or piles, and is determined by the possibility that a ship with a long bulbous bow protruding well forward of the forepeak below the water line and with a markedly flared hull at the bow, like larger container ships, could hit the columns or piles if berthing with a large berthing angle. To avoid this possibility, the centreline of the columns or piles should lie at least 2 m behind the berth line.

B_1 is the width of the berth platform, and is determined by the other main characteristics of the berth structure.

B_2 is the dimension determined by the thickness of the rear wall or the slope of the relief plate and by the width of the anchor slab. The angle β should not be larger than 15° (1 : 3.75) because of the magnitude of the vertical forces that can be exerted when the horizontal load is transmitted by way of the sloping or settlement slab to the anchor slab. This settlement slab should have an upper horizontal part lying about 30 cm under the top of the filled apron.

Figure 11.3 A possible precast anchoring structure with minimal ground settlements

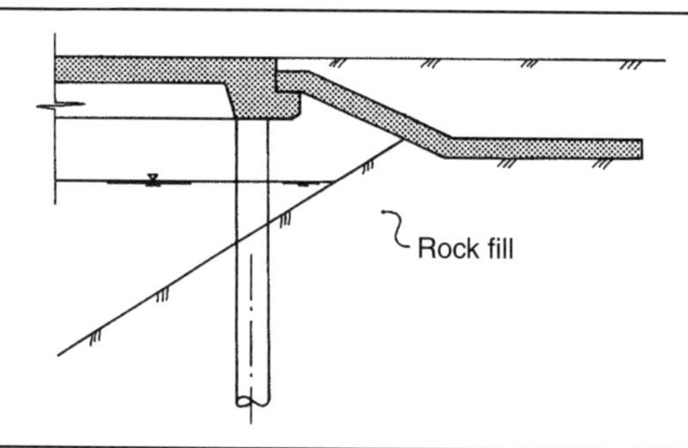

Even if it is generally desirable that the width B be reduced as much as possible, it should not be forgotten that the horizontal forces must be transmitted to the anchor slab in a satisfactory way. It is necessary to design the transitions between the berth platform and the settlement slab and between the settlement slab and the anchor slab in such a way that the settlements of the fill under the anchor slab are taken care of. As a rough guideline, the settlement of a filling consisting of stone–sand friction materials should be 1–3% of the height of the fill. Fill made of rock and stones will have a volume ratio between fill placed with good compaction to solid rock of about 1.3–1.5, and between fill placed without any compaction underwater to solid rock of about 1.5–1.7.

If the settlements are expected to be minimal, the anchoring structure can be shaped as shown in Figure 11.3. Where greater settlements are expected, the anchoring structure must be shaped as shown in Figure 11.4.

If the anchoring structure is shaped as shown in Figure 11.5, where a part of the rear wall is founded on piles, the anchor slab must be shaped in such a way that the forces can be transferred through friction between the slab and the fill.

If the anchoring structure is shaped as shown in Figure 11.6, the horizontal forces are taken up by batter piles, while the anchor slab only makes possible the transfer of adequate stabilising weight to the batter piles.

In cases where the width B_2 of the anchoring structure is made very short, it can be shaped as shown in Figure 11.7. The rear wall and the berth platform are cast in one, and batter piles under the rear wall

Figure 11.4 An anchoring structure where considerable ground settlements are expected

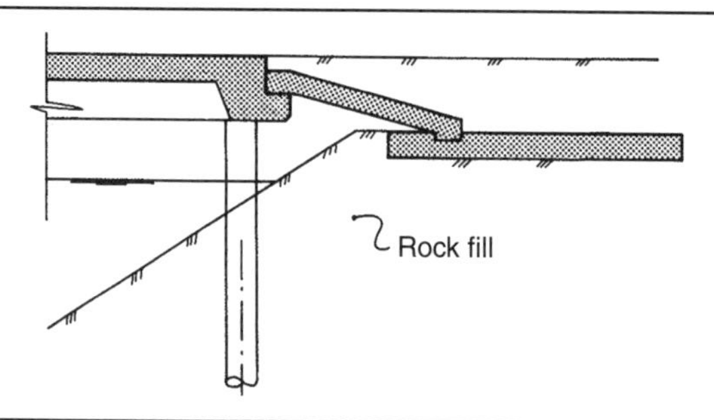

Figure 11.5 An anchoring structure placed partly on piles

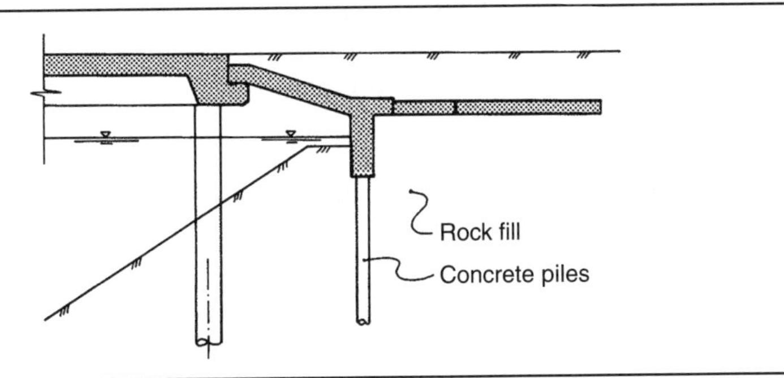

Figure 11.6 Anchoring by use of batter piles

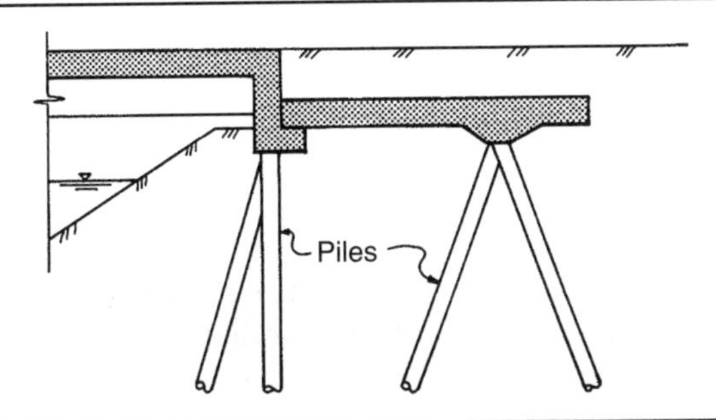

take up the horizontal forces. This design is labour intensive and costly. It should also be noted that a vertical wall would attract forces onto the structure from the fill behind.

B is the total width of the berth structure, and therefore B depends on several variable factors. The designer should try to choose a berth design that will give a clearly set out load bearing and the easiest possible construction work for the contractor. Sophisticated solutions should be avoided. In piers and jetties the total width B is equal to the berth platform width B_1, and the horizontal loads must be taken

Figure 11.7 Anchoring by means of very short structure anchoring

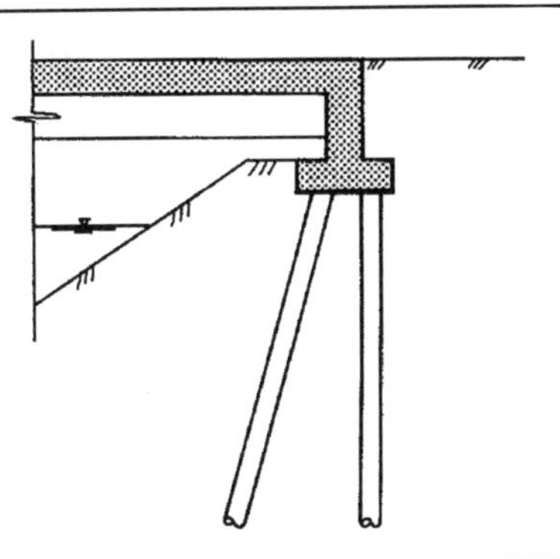

up by lamellas or by batter piles under the berth. The deciding factor for determining the width B will, therefore, be the port activities that will be carried out on the berth structure. Where the pier is so wide that fill can be used for the middle part, or where the quay lies parallel to the shore, the total width B will be determined by the dimensions shown in Figure 11.1.

The foundations for column, pile and lamella berths, respectively, are described in the following sections. The berth platforms are similar for the three types of structure, and are described in a separate section.

11.2. Column berths

A column berth has its berth platform founded on columns cast in situ.

The column berth is a type of berth that is specific to Norway, where contractors have specialised in the construction of long and slender columns cast in situ underwater. In other countries this type of construction is met with scepticism, and the majority of berths are built as solid berths or pile berths. This fact is confirmed by the non-Norwegian literature on open berth structures, where very little, if anything, is said about slender concrete columns cast underwater using the tremie pipe method. Figure 11.8 shows a cross-section through a large open berth anchored in the rock fill by an anchor slab.

In recent years, the ratio between the amount of column berths and pile berths constructed has changed because, nowadays, open berth structures on tubular steel piles are preferred if the ground conditions permit. This is not because of scepticism about the durability and performance of concrete column berths, but because structures on tubular steel piles are economically more advantageous, especially as expensive underwater work is avoided.

In column berths the berth platform is supported on in situ cast columns, which are either founded directly on moraine or rock, or penetrate through loose deposits in vertical wells down to rock. Alternatively, each column is supported on a group of friction or point-bearing piles. The various methods are described below.

Figure 11.8 A column berth

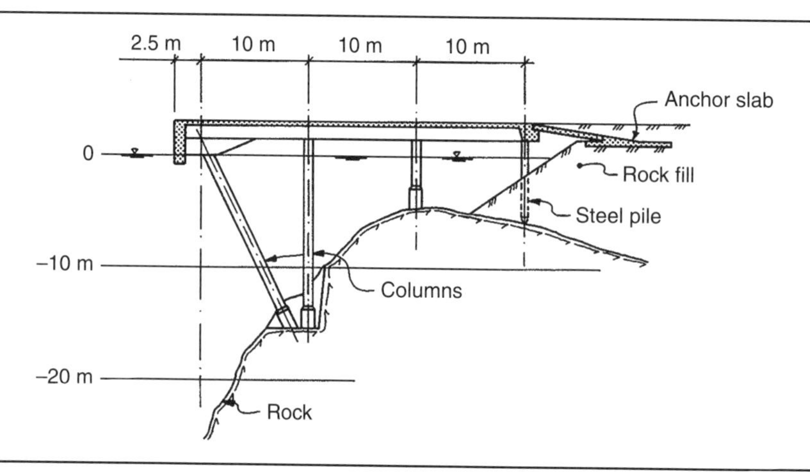

Figure 11.9 A foundation on hard moraine or rock

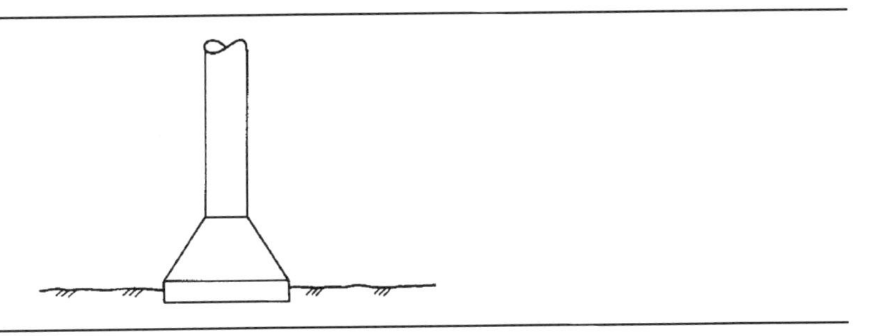

Moraine

As shown in Figure 11.9, the column with a widened foot is placed directly on the moraine. To make certain that the moraine has sufficient load-bearing capacity, it is recommended that, in addition to removing any loose deposits, the upper layer of the moraine should be removed. However, this must be done carefully to avoid disturbing the underlying material and causing settlement of the column. The size of the foundation will depend on the allowable compressive stress under the foundation and on considerations related to the casting technique.

As a rough guideline, the following allowable pressures can be assumed when making a rough estimate of the required size of a foundation:

Rock	5000–30 000 kN/m^2
Hard moraine above groundwater level	500–800 kN/m^2
Hard moraine below groundwater level	400–600 kN/m^2
Compacted rock/stone filling above groundwater level	300–500 kN/m^2
Compacted rock/stone filling below groundwater level	200–300 kN/m^2
Dense sand and gravel above groundwater level	300–500 kN/m^2
Dense sand and gravel below groundwater level	200–300 kN/m^2
Loose to medium dense sand and gravel above groundwater level	150–250 kN/m^2
Loose to medium dense sand and gravel below groundwater level	100–150 kN/m^2
Hard clay (dry crust)	150–250 kN/m^2
Medium-soft clay	80–150 kN/m^2
Soft clay	20–80 kN/m^2

It should be borne in mind that in some cases settlement may limit the allowable bearing pressure.

Rock

Where rock is uncovered, or where the seabed layer above rock is such that by dredging the rock easily becomes visible, the column is founded directly on the rock. Whether the rock surface is horizontal or sloping, one should always blast a shelf or a niche, as shown in Figure 11.10. This is especially important if the column is to be placed in a backfill slope.

The column foot should be shaped according to the code of practice for load-bearing underwater concrete constructions, as described in Chapter 17. Before casting, all loose deposits and mud should be removed. Anchoring bolts between the column and the rock are recommended. However, the foundation should be designed in such a way that the load pressure is borne by the concrete alone and not

Figure 11.10 A foundation on rock

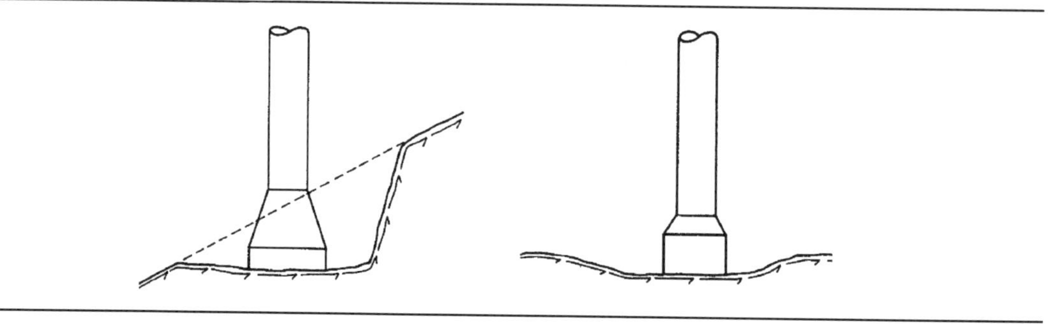

by the anchoring bolts. To achieve adequate stability of the columns during the construction phase (i.e. prior to the beams and the slab being cast), it is advisable to place an anchor bolt at each corner of the column to avoid tilting.

Wells

The decisive factors when selecting the method to be used (i.e. shaft a well or dredge) are the dredging characteristics of the soil and the contractor's equipment. General practice indicates that, if the thickness of the soil is about 1.5–4.0 m, it could be worthwhile using vertical wells, as shown in Figure 11.11. Dredging inside the well can be done either by grabbing or by means of a large suction pump. As soon as the dredging work is finished, a recess must be either blasted or chiselled into the rock surface, and the rock cleared. The widening of the column at the bottom is achieved by stopping the formwork about 0.5–0.7 m above the rock. The well is formed by steel cylinders on manhole elements of at least 1.5 m diameter, allowing the diver to operate inside them.

Piles

If the depth from the seabed to the rock or the load-resisting stratum is more than about 4.0–5.0 m, it will probably pay to use piles as the foundation for the columns. The simplest, and often most

Figure 11.11 A well foundation

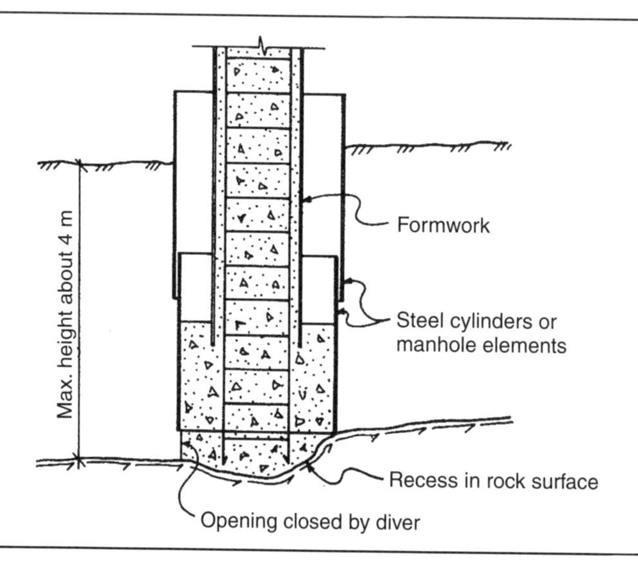

Figure 11.12 A pile foundation

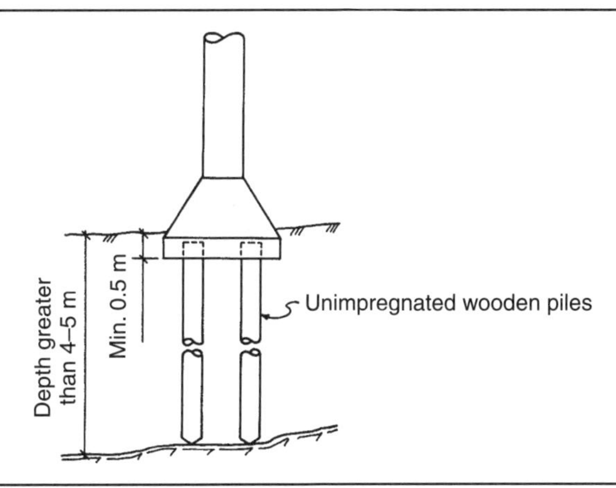

economical, method is to use untreated wooden piles, as shown in Figure 11.12. In order to prevent the wooden piles being attacked by wood borers, the foundation should be dredged at least 0.5 m down into the seabed. In order to avoid a construction joint between the column foot and the column, the column and the foot must be cast in one and the same operation.

With regard to the precision required in placing the columns, the centre of a column should be placed within a ± 5 cm from the theoretical column centre.

Formwork

Berth columns should be shaped in such a way that a multi-use formwork system can be applied, thus requiring little diving work for installation underwater. The cross-sections of the columns can be either rectangular or circular, but with the advanced formwork systems in use nowadays the circular cross-section is probably the simplest and most economical. A circular cross-section will, when using the same amount of concrete as a square cross-section, have about 13% less formwork surface, and therefore about 13% less surface exposed to frost and chemicals than will a square cross-section. A well-designed formwork system with a circular cross-section is simple to erect and dismantle, which is very important, particularly when divers have to be used. Furthermore, circular stirrups are easy to bend and install inside the form.

Usually the formwork for columns, with the reinforcement installed inside, is made ashore, as shown in Figure 11.13, and then towed to its place in the structure. Alternatively, the formwork can be lifted into place using a crane, as shown in Figure 11.14. It is preferable to standardise the column cross-sections and vary the amount of reinforcement within them in relation to the column length and load. When cast underwater, a column should not have a diameter of less than 70 cm. The cross-section, the reinforcement and the shaping of the columns should be as recommended in Chapter 17.

All formwork placed under the tidal zone should be removed, so that the cast concrete can be examined closely by divers. In the tidal zone the formwork should remain for the protection of the concrete. Experience has shown that it is only in the tidal zone that the concrete is severely exposed to frost destruction, wave action, etc., as shown in Figure 11.15.

Figure 11.13 Column formwork with reinforcement

Figure 11.14 Column formwork with reinforcement being lifted into place

Figure 11.15 A column with an hour-glass shape

Permanent formwork in the tidal zone should be made of material that is fully impregnated and held in place by copper ties and copper pegs. Instead of fully impregnated wooden material, plastic pipes and thin steel tubes have been used in the tidal zone, with good results. Such permanent formwork covers the area from the bottom of the berth platform beams down to 50 cm under the LAT (Figure 11.16).

Figure 11.16 General permanent formwork on a column structure

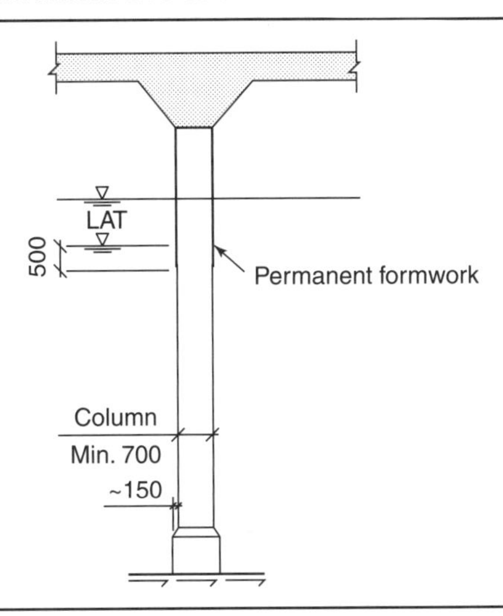

Figure 11.17 A pile berth

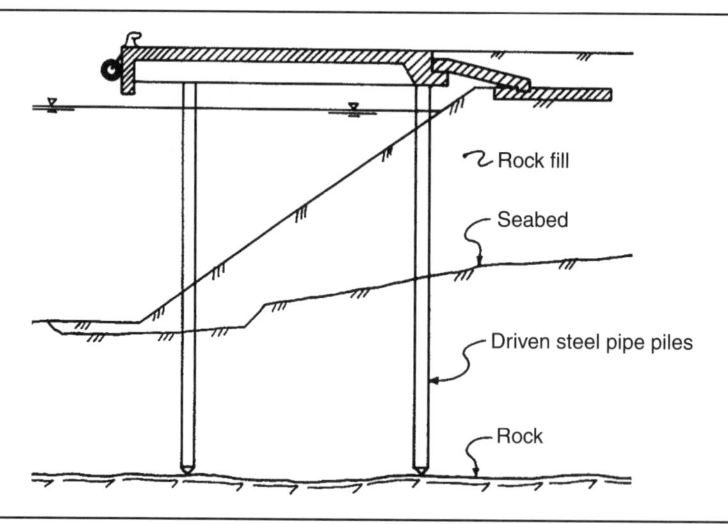

11.3. Pile berths

In smaller open berth structures where the water depth is small and the load-bearing capacity of the berth platform is limited, the platform is usually supported by concrete or timber piles. If the platform is required to withstand greater live loads, the piles must be placed so closely together that other structural solutions are more economical. The usual alternative is concrete columns of a minimum 70 cm diameter and founded as described above for column berths.

Where the thickness of the seabed material is more than about 3–4 m, and where foundation treatment will require extensive use of divers and underwater works, driven piles of much greater cross-sectional dimensions could be used. The piles that reach up to the berth platform (Figure 11.17) are either concrete piles or tubular steel piles filled with concrete or reinforced concrete. In general, concrete piles are the more expensive alternative per linear metre of berth, due to a higher transportation cost if the pile factory is not located near the site, and the fact that the concrete piles must be strengthened and/or protected in the tidal zone. The construction of a pile berth using tubular steel piles is shown in Figure 11.18.

Tubular **steel pipes** can be driven into harder and more difficult strata than can concrete piles. For instance, a 70 cm diameter tubular steel pile can penetrate a 20 m thick rubble fill consisting of stones with diameters as large as 50 cm. Piles of diameter 500–800 mm are the most commonly used.

The load on a steel pile filled with concrete or reinforced concrete is taken by three possible methods:

(a) The steel pile itself and the concrete without reinforcement are just used to stiffen the pile.
(b) The load is taken by the reinforced concrete only, and the steel tube acts merely as formwork because of the corrosion risk. In this case the entire axial load on the pile is taken by the reinforced concrete, and the steel pile should just be strong enough to be driven down to the bearing layer or the bedrock. As the pipe is emptied prior to reinforcing and concreting, the pile can be regarded as a reinforced concrete pile cast in the dry. Usually the pile is reinforced only 2–4 m below the sea bottom, depending on soil conditions.
(c) By a combination of the two cases above.

Figure 11.18 Construction of a pile berth

Table 11.1 shows spirally welded steel pipe piles with outside diameters in the range 60.3–1220.0 mm. Prefabricated pile units complete with fittings are available in lengths up to 26 m for delivery direct to the construction site.

The steel piles are made of general structural steel and high-strength gas pipeline steel in accordance with BS EN 10219, as shown in Table 11.2. The steel grade most commonly used for pipe pile is between S355J2H and S550J2H.

Steel pipe piles with diameters in the range 600–900 mm are usually used. The pipes should have a minimum wall thickness of 10 mm and should be delivered to the site in lengths of 12 m minimum. The material should be normalised steel and the quality of the steel should be S355J2H. A certificate from the manufacturer is required. Due to the fact that the pipes could be joint welded on site, the tolerances should be:

(a) diameter ± 2 mm
(b) ovality ($d_{max} - d_{min}$) 5 mm
(c) deviation from 90° end cuts 2.5 mm
(d) linearity 5 mm over a 5 m length.

The steel piles must withstand heavy pile driving. The steel thickness depends on the driving conditions and how much driving energy is required, but usually 50–80 cm diameter piles have a steel thickness of more than 12 mm. Pile driving requires heavy equipment. During driving, the tubular steel piles are filled with water.

The pile sections can be spliced in the rammer as shown in Figure 11.19. If the lower pile has been driven into hard soil, the upper 100 mm of the pipe should be cut off and the contact surface

Table 11.1 Steel pipe piles. (Courtesy of Ruukki, Finland)

D: mm	t: mm	M: kg/m	A: mm^2	A_u: m^2/m	A_b: mm^2	W_{el}: cm^3	I: cm^4	EI: kN m^2	Z: kN s/m	$A_{1.2}$: mm^2	$A_{2.0}$: cm^3	$I_{1.2}$: cm^4	$I_{2.0}$: cm^4	$EI_{1.2}$: kN/m^2	$EI_{2.0}$: kN/m^2
60.3	6.3	8.4	1 069	0.19	2 856	13.1	39.5	83	43.4	846	703	29.8	23.9	63	508
76.1	6.3	10.8	1 382	0.24	4 548	22.3	84.8	178	56.1	1 099	916	65.0	52.8	137	111
88.9	6.3	12.8	1 635	0.28	6 207	31.6	140.2	295	66.4	1 304	1 089	108.4	88.7	228	186
114.3	6.3	16.8	2 138	0.36	10 261	54.7	312.7	657	86.8	1 711	1 432	244.5	201.4	514	423
114.3	8.0	21.0	2 672	0.36	10 261	66.4	379.5	797	108.5	2 245	1 966	311.3	268.2	654	563
139.7	5.0	16.6	2 116	0.44	15 328	68.8	480.5	1 009	85.9	1 594	1 251	355.3	275.4	746	578
139.7	8.0	26.0	3 310	0.44	15 328	103.1	720.3	1 513	134.4	2 788	2 445	595.1	515.2	1 250	1 082
139.7	10.0	32.0	4 075	0.44	15 238	123.4	861.9	1 810	165.4	3 553	3 210	736.7	656.8	1 547	1 379
168.3	5.0	20.1	2 565	0.53	22 246	101.7	855.8	1 797	104.1	1 935	1 520	636.0	494.6	1 336	1 039
168.3	10.0	39.0	4 973	0.53	22 246	185.9	1 564.0	3 284	201.9	4 343	3 928	1 344.1	1 202.7	2 823	2 526
168.3	12.5	48.0	6 118	0.53	22 246	222.0	1 868.4	3 924	248.4	5 488	5 073	1 648.5	1 507.1	3 462	3 165
193.7	5.0	23.3	2 964	0.61	29 468	136.6	1 320.2	2 773	120.3	2 238	1 760	984.1	766.9	2 067	1 610
219.1	6.3	33.1	4 212	0.69	37 703	217.8	2 386.1	5 011	171.0	3 390	2 848	1 898.6	1 582.4	3 987	3 323
219.1	10.0	51.6	6 569	0.69	37 703	328.5	3 598.4	7 557	266.7	5 748	5 205	3 110.9	2 794.7	6 533	5 869
219.1	12.5	63.7	8 113	0.69	37 703	396.6	4 344.6	9 124	329.4	7 292	6 749	3 857.0	3 540.9	8 100	7 436
273.0	6.3	41.4	5 279	0.86	58 535	344.0	4 695.8	9 861	214.3	4 254	3 576	3 749.6	3 132.6	7 874	6 579
323.9	12.5	96.0	12 229	1.02	82 397	916.7	14 846.5	31 178	496.5	11 012	10 206	13 262.9	12 226.7	27 852	25 676
406.4	10.0	97.8	12 453	1.28	129 717	1 204.5	24 475.8	51 399	505.6	10 926	9 912	21 340.7	19 281.4	44 816	40 491
406.4	12.5	121.4	15 468	1.28	120 717	1 477.9	30 030.7	63 064	628.1	13 941	12 928	26 895.6	24 836.3	56 481	52 156
457.0	10.0	110.2	14 043	1.44	164 030	1 535.7	35 091.3	73 692	570.2	12 325	11 184	30 628.9	27 693.0	64 321	58 155
457.0	12.5	137.0	17 456	1.44	164 030	1 888.2	43 144.8	90 604	708.7	15 737	14 597	38 682.4	35746.9	81 233	75 068
508.0	10.0	122.8	15 645	1.60	202 683	1 910.2	48 520.2	101 893	635.2	13 735	12 466	42 386.1	38 344.9	89 011	80 524
508.0	12.5	152.7	19 458	1.60	202 683	2 352.6	59 755.4	125 486	790.7	17 548	16 279	53 621.3	49 580.1	112 605	104 118
508.0	14.2	172.9	22 029	1.60	202 683	2 645.6	67 198.6	141 117	894.4	20 118	18 849	61 064.5	57 023.3	128 236	119 749
559.0	10.0	135.4	17 247	1.76	245 422	2 325.6	65 001.1	136 502	700.3	15 145	13 748	56 822.5	51 428.6	119 327	108 000
559.0	12.5	168.5	21 461	1.76	245 422	2 868.0	80 161.8	168 340	871.4	19 358	17 961	71 983.2	66 589.3	151 165	139 838
559.0	14.2	190.8	24 304	1.76	245 247	2 781.9	84 846.6	178 178	765.3	16 554	15 029	74 213.3	67 194.1	155 848	141 108
610.0	12.5	184.2	23 464	1.92	292 247	3 434.6	104 754.7	219 985	952.7	21 169	19 644	94 121.5	87 102.3	197 655	182 915
610.0	14.2	208.6	26 579	1.92	292 247	3 869.0	118 003.9	247 808	1 079.2	24 284	22 759	107 370.6	100 351.4	225 478	210 738

610.0	16.0	234.4	29 858	1.92	292 247	4 320.7	131 781.4	276 741	1 212.3	27 563	26 038	121 148.2	114 129.0	254 411	239 671
660.0	10.0	160.3	20 420	2.07	342 119	3 268.8	107 870.5	226 528	829.1	17 937	16 286	94 396.3	85 495.1	198 232	179 540
660.0	12.5	199.6	25 427	2.07	342 119	4 039.6	133 306.4	279 944	1 032.4	22 944	21 293	119 832.2	110 931.0	251 648	232 955
660.0	14.2	226.2	28 810	2.07	342 119	4 553.4	150 263.1	315 553	1 169.7	26 326	24 675	136 788.9	127 887.6	287 257	268 564
660.0	16.0	254.1	32 371	2.07	342 119	5 088.5	167 921.2	352 635	1 314.3	29 347	17 568	118 449.4	107 309.5	248 744	225 350
711.0	12.5	215.3	27 430	2.23	397 035	4 707.3	167 343.2	351 421	1 113.7	24 754	22 975	150 491.3	139 351.4	316 032	292 638
711.0	14.2	244.0	31 085	2.23	397 035	5 309.0	188 735.2	396 344	1 262.1	28 409	26 630	171 883.3	160 743.4	360 955	337 561
711.0	16.0	274.2	34 935	2.23	397 035	5 936.4	211 039.8	443 184	1 418.4	32 259	30 480	194 187.9	183 047.9	407 795	384 401
762.0	10.0	185.5	23 625	2.39	456 037	4 383.9	167 028.4	350 760	959.2	20 757	18 850	146 276.7	132 551.0	307 181	278 357
762.0	12.5	231.1	29 433	2.39	456 037	5 426.0	206 731.0	434 135	1 195.0	26 565	24 658	185 979.3	172 253.7	390 557	361 733
762.0	14.2	261.9	33 360	2.39	456 037	6 122.6	233 271.2	489 870	1 354.5	30 492	28 585	212 519.5	198 793.9	446 291	417 467
762.0	16.0	294.4	37 498	2.39	456 037	6 849.7	260 973.3	548 044	1 522.5	34 630	32 723	240 221.6	226 496.0	504 121	339 086
813.0	12.5	246.8	31 436	2.55	519 124	6 195.8	251 860.3	528 907	1 276.3	28 375	26 340	226 649.4	209 966.0	475 964	440 929
813.0	14.2	279.7	35 635	2.55	519 124	6 994.2	284 314.9	597 061	1 446.8	32 575	30 539	259 103.9	242 420.6	544 118	509 083
813.0	16.0	314.5	40 062	2.55	519 124	7 828.3	318 221.7	668 266	1 626.6	37 001	34 966	293 010.8	276 327.4	615 323	580 288
813.0	18.0	352.9	44 956	2.55	519 124	8 741.7	355 350.0	746 235	1 825.3	41 896	39 861	330 139.1	313 455.7	693 292	658 257
914.0	10.0	222.9	28 400	2.87	656 119	6 349.0	290 147.2	609 309	1 153.1	24 959	22 670	254 307.1	230 570.5	534 045	484 198
914.0	12.5	277.9	35 402	2.87	656 119	7 871.1	359 708.4	755 388	1 437.4	31 961	29 672	323 868.3	300 131.7	680 124	630 277
914.0	14.2	315.1	40 138	2.87	656 119	9 959.3	455 141.8	955 798	1 832.7	41 697	39 408	419 301.7	395 565.1	880 534	830 687
914.0	18.0	397.7	50 668	2.87	656 119	11 130.5	508 664.8	1 068 196	2 057.2	47 226	44 937	472 824.7	449 088.1	992 932	943 085
1016.0	10.0	248.1	31 604	3.19	810 732	7 871.1	399 849.7	839 684	1 283.2	27 779	25 233	350 602.3	317 964.5	736 265	667 725
1016.0	12.5	309.3	39 407	3.19	810 732	9 766.2	496 123.1	1 041 858	1 600.0	35 582	33 036	446 875.7	414 237.9	938 439	869 900
1016.0	14.2	350.8	44 691	3.19	810 732	11 038.6	560 762.0	1 177 600	1 814.5	40 865	38 320	511 514.6	478 876.8	1 074 181	1 005 641
1016.0	16.0	394.6	50 266	3.19	810 732	12 371.6	628 479.4	1 319 807	2 040.9	46 440	43 894	579 232.0	546 594.2	1 216 387	1 147 848
1016.0	18.0	443.0	56 436	3.19	810 732	13 835.7	702 854.2	1 475 994	2 291.4	52 610	50 064	653 606.9	620 969.0	1 372 574	1 304 035
1220.0	10.0	298.4	38 013	3.83	1 168 987	11 405.5	695 737.9	1 461 050	1 543.4	33 419	30 360	610 420.2	553 821.4	1 281 883	1 163 025
1220.0	12.5	372.2	47 418	3.83	1 168 987	14 169.3	864 326.6	1 815 086	1 925.3	42 824	39 766	779 008.9	722 410.1	1 635 919	1 517 061
1220.0	14.2	422.3	53 792	3.83	1 168 987	16 028.9	977 764.6	2 053 306	2 184.0	49 197	46 139	892 446.9	835 848.1	1 874 139	1 755 281
1220.0	16.0	475.1	60 520	3.83	1 168 987	17 980.7	1 096 821.7	2 303 326	2 457.2	55 925	52 867	1 011 504.0	954 905.2	2 124 158	2 005 301
1220.0	18.0	533.6	67 972	3.83	1 168 987	20 128.6	1 227 843.9	2 578 472	2 759.8	63 377	60 319	1 142 526.3	1 085 927.4	2 399 305	2 280 448

A = cross-sectional area, A_u = external surface area, A_b = pile base area, Z = pile impedance, I = moment of inertia, W_{el} = section modulus. Cross-sectional values are reduced by corrosion allowances of 1.2 mm and 2.0 mm

Table 11.2 Steel pipe piles and steel grades of Ruukki steel pipe piles

Steel grade	Carbon equivalent, CEV max.: %	Chemical composition, max.				Mechanical properties				
		Carbon: %	Manganese: %	Phosphorus: %	Sulphur: %	f_y min: MPa	f_u: MPa	A_s min.: %	Impact strength	
									T: °C	kV min.: J
S355J2H	0.39	0.22	1.60	0.035	0.035	355	490–630	20	−20	27
S440J2H	0.39	0.18	1.60	0.020	0.018	440	490–630	17	−40[a]	27
S550J2H	0.39	0.12	1.80	0.025	0.015	550	600–760	14	−20[a]	27
X60	0.43	0.15	1.60	0.030	0.030	413	≥517	18	0	27
X70	0.43	0.15	1.70	0.030	0.030	482	≥565	18	0	27

[a]For material thicknesses exceeding 10 mm, the impact strength requirement must be agreed separately

Figure 11.19 Welding of steel piles in the rig

should be worked to even, right-angled planes before the upper pile is welded to the lower pile. Figure 11.20 shows the welding of steel piles being done on the ground.

The steel pile points are either of a conical type with dowel and ribs or of a flat type. The conical type with dowel and ribs is shown in Figures 11.21 and 11.22. Figure 11.23 shows the conical pile point and the construction details of the point with high-strength steel strengthened dowel with ribs. The flat type is shown among the different pile shoes in Figure 11.23.

The welding should be done using basic electrodes corresponding to, for instance, OK 48.30. The welding of higher strength steel requires the use of other electrodes. The electrodes must be stored in warm containers to prevent moisture accumulation. In order to secure the relative positions of the pipe ends, an inside pipe of 3 mm thickness and 60 mm length should be provided, as shown in Figure 11.22.

In order to facilitate control of the linearity, the splice should be carried out at least 1.5 m above the water level. The maximum allowable angular deviation after splicing should be 1 : 250, measured over a length of 3.0 m. This requirement is valid for the entire length of the pile. All splices should be prepared for V welds.

Figure 11.20 Welding of steel piles on the ground. (Courtesy of Altiba AS, Norway)

The piles are equipped with **pile shoes** according to the specified bearing capacity and the ground conditions. Open-ended piles will usually be equipped with either an external or an internal reinforcement ring. Bottom plates are often used where the piles are mainly end bearing in a boulder-free soil layer. When the piles are to be driven through rocky moraine or into bedrock, rock shoes fitted with a structural steel dowel are used to prevent damage to the pile end and to centre the pile load. The rock

Figure 11.21 A steel pile point or pile shoes. (Courtesy of Altiba AS, Norway)

Figure 11.22 Example of a detail of a pile point or pile shoes

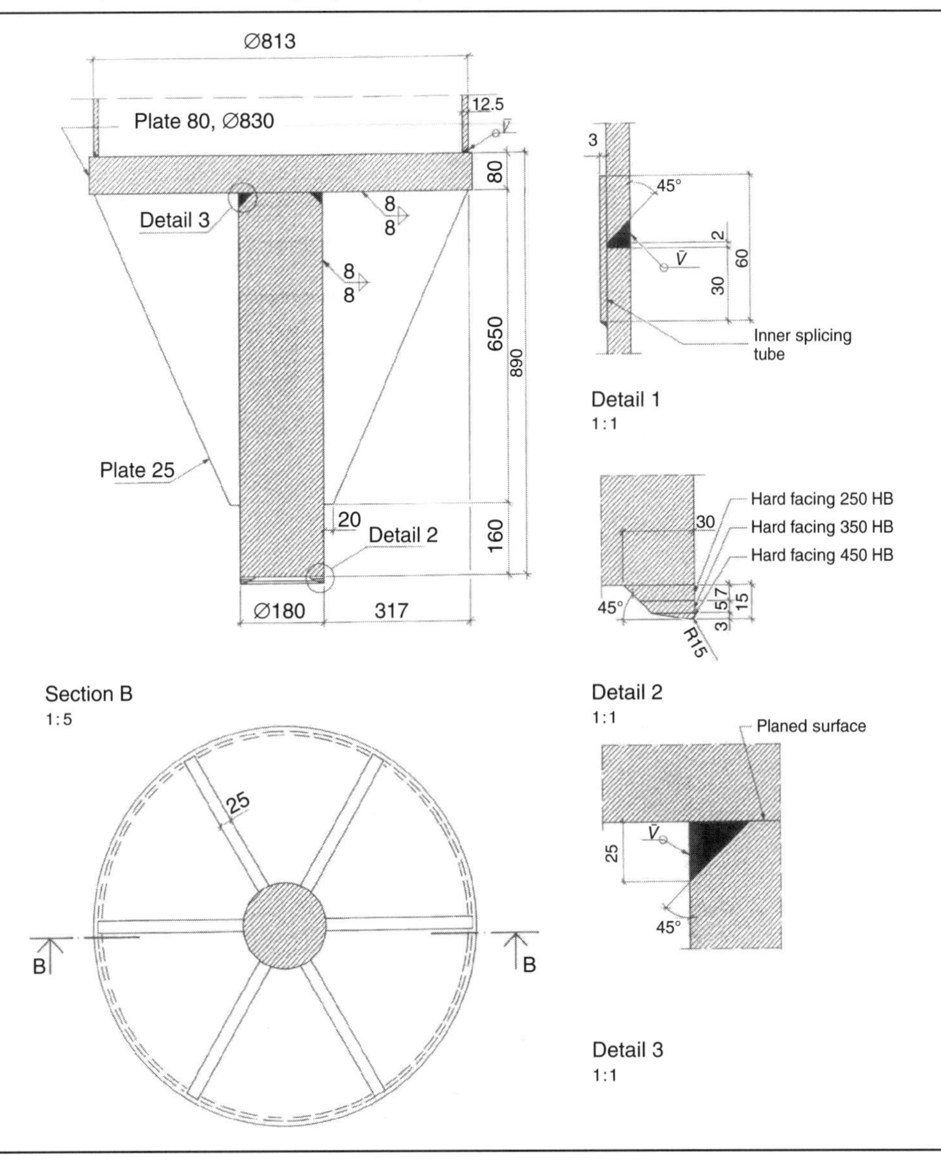

shoes with a hardened steel dowel are used especially to prevent the pile from skidding on the sloping rock surface and to get a better grip in the rock.

Before the welding, the steel parts should be preheated to 150°C, and a slow cooling after welding must be ensured. The hard facing of the pile shoe should be as shown in Figure 11.22, and should be made in the following steps:

(a) 2–3 welding layers on the base material with a hard weld electrode giving 200–250 HB
(b) 2 layers giving 300–350 HB
(c) 2 layers giving 400–450 HB.

Figure 11.23 Different pile shoes available from Ruukki, Finland

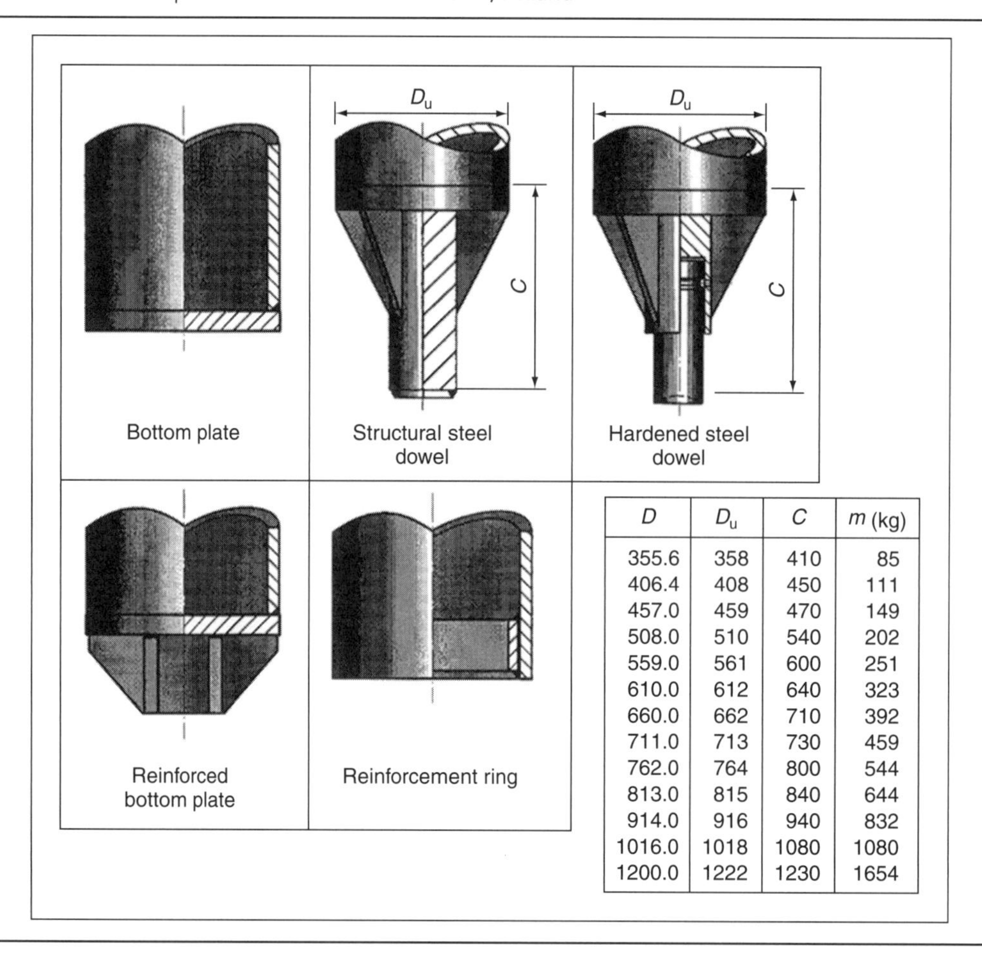

D	D_u	C	m (kg)
355.6	358	410	85
406.4	408	450	111
457.0	459	470	149
508.0	510	540	202
559.0	561	600	251
610.0	612	640	323
660.0	662	710	392
711.0	713	730	459
762.0	764	800	544
813.0	815	840	644
914.0	916	940	832
1016.0	1018	1080	1080
1200.0	1222	1230	1654

All welds should be checked using ultrasound equipment to detect hydrogen cracks. Tests using magnetic powder or penetration liquid are recommended for detecting surface cracks. The hard facing of the shoe should be controlled by the use of magnetic powder (with coil) or by the use of penetration liquid. The control of the hardness on test specimens can be accepted provided that the same welding procedure has been applied for both the hard facing and the specimen.

The **precision** with which the tubular **steel piles are placed** will depend on the characteristics of the fill material and the contractor's experience. However, the centre of the pile should be placed within ± 30 cm of the theoretical centre, and the deviation from the vertical shall not exceed 2° or approximately 30 : 1. The bottom side of the berth platform beams must have sufficient width to accommodate the tolerance of the piles. If the contractor can provide a flexible formwork system, there will be no difficulty in accommodating the position of the pile.

The **piles** should usually be **driven** using ordinary free drop gravity hammers, or single-acting pneumatic hammers with a weight of 80 kN or more. The equipment must be adjustable to ensure that the driving force acts along the axis of the pile.

A pile cap containing hard wood should, during the pile driving, protect the pile top. In the case of large deformations of the pipe top during hammering, the deformed section must be cut off and ground to an even plane before driving is continued. Because the submerged weight of an empty pile pipe is less than unity, the piles must be filled with water during driving.

If the piles are embedded in a thin soil layer above the bedrock, this layer may not give a sufficient lateral support of the piles, or sufficient anchoring against uplift when the piles are emptied before concreting. The contractor should, therefore, take the necessary measures to secure the positions of the piles in all directions.

A general preliminary pile driving procedure may be as follows. During driving through, for example, fill and moraine, the drop height should not exceed 600 mm. When the pile tip reaches the rock surface, the height should be reduced to 300 mm. The penetration per series of 10 blows should be measured, and when it has reached 5 mm/series and has become constant or started to decrease, the drop height should be increased to 500 mm. When the penetration again reaches 5 mm/series, the pile shoe–rock contact is regarded as satisfactory. In the case where the penetration is seen to increase, the drop height should be returned to a 300 mm and the whole procedure repeated. In order to ascertain that the last hammer blows have not caused any failure in the rock or pile, three series with a drop height of 300 mm should finish the driving. The penetration for each series should not exceed 3 mm.

The driving procedure and the criteria given may be changed depending on the conditions encountered on site and the contractor's equipment. The contractor should prepare a complete piling record for each pile driven. The contractor for the piling work should control both the quality of the piles and pile shoes, and the handling and the driving operation. The person responsible for this and the filling in of the piling records must be an experienced controller.

It should be possible to control the water tightness of the piles by emptying out or lowering the level of the water inside the pile pipe. If there is leakage, too great a curvature or other unfavourable conditions, the client shall decide if it is necessary to replace the pile with a new one. Replaced piles should be satisfactorily located and be included in the pile record.

When the driving of a pile is finished, the top of the pile should be levelled. Before concreting, the piles should be re-levelled in order to confirm that the pile tips are still in contact with the rock. If the upheaval is more that 3 mm, the pile should be driven further according to the requirements laid down in the preliminary driving procedures.

The tolerances on the cut-off levels of the top of the pile should be ± 5 mm. No final cut-off or concreting of the piles shall be performed until the pile has been inspected and found acceptable.

Some particular problems can arise with concrete-filled steel pipe piles during driving. The problems are due to the fact that the cross-sectional area of the steel pipe is too small to transmit a dynamic load of the same magnitude as the design load of the concrete-filled pipe. The maximum load that can be transmitted is equal to the yield stress of the steel multiplied by the cross-sectional area. The thickness of the pipe wall must, therefore, in many cases be increased in order to meet the requirements mentioned above.

If some of the seabed under the berth is bare rock, the piles can be put in place both with and without points or shoes onto a foundation, and fixed and filled with concrete as done in underwater concreting.

Figure 11.24 A pile in a concrete foundation

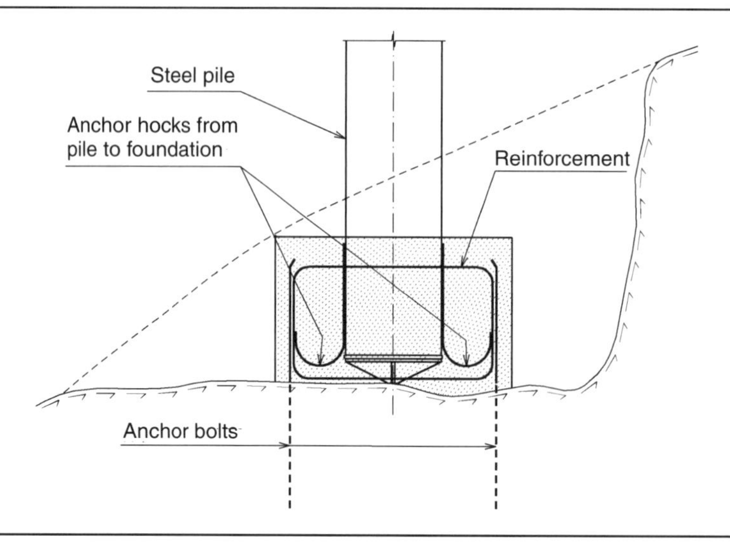

The piles are fixed by bolting them to the rock in a concrete foundation, sealed at the foot, emptied of water, and then filled with concrete, as for a dry formwork (Figure 11.24).

The load-bearing **capacity of piles** normally means the ability of the ground to bear the pile load rather than the structural capacity of the pile. The load-bearing capacity can be evaluated or calculated in different ways. There are many static formulas available for piles in soil (i.e. friction piles). Most of the formulas divide the capacity into shaft friction and point resistance. These are based on soil parameters derived from laboratory tests, and/or soundings, for instance standard penetration tests or cone penetration tests. In this way, the ultimate capacity and the necessary pile lengths can be calculated. The design load capacity is obtained by dividing the ultimate capacity by a material factor of 1.6–2.0 depending on the type and quality of soil data.

The typical formula for the ultimate bearing capacity is:

$$Q_u = Q_s + Q_p$$

where Q_s is the ultimate shaft friction and Q_p is the ultimate point resistance.

$$Q_s = \beta \times p' \times A_s$$

where

β = shaft friction coefficient. For rough estimates the value of this varies typically between 0.30 and 0.15, the lower value being for long piles in loose sand
p' = average vertical effective stress along the pile
A_s = pile shaft area.

$$Q_p = N_q \times p' \times A_p$$

where

N_q = traditional bearing capacity factor primarily depending on the friction angle of the soil
p' = effective overburden pressure at the pile tip elevation
A_p = pile cross-section area.

In addition to the static formula given above, there are dynamic formulas based on the driving resistance. The dynamic formula mostly recommended in Norway is the formula derived by Professor Janbu at the Norwegian University of Science and Technology, Norway. According to his formula, the bearing capacity is a function of the driving energy, the material and cross-sectional area of the pile, and the final penetration of the pile.

The dynamic formulas cannot be used for silt or clay where pile driving results in high pore-water pressures, and are not relevant for piles driven to bedrock. The evaluation of such piles should be based on the behaviour of the pile after reaching the rock surface. The drop height of the hammer is increased stepwise up to a height giving dynamic stresses in the pile of the same magnitude as the design load. Each step in the driving process should be continued until the set per blow reaches a predetermined minimum value, and then begins to decrease. The predetermined minimum value should be of the order of 2–5 mm per series of 10 blows. This driving procedure is widely recognised as preventing the tip from sliding on steep rock surfaces, and obtaining the necessary contact area of the pile tip.

11.3.1 Pile foundations in deep soils undergoing creep deformations

In deep soil sediments (e.g. more than 50–60 m) friction piles are normally preferred to point bearing piles. Both economical and technical evaluations will often come out in favour of friction piles. However, if the soil sediment is undergoing long-term creep deformations, friction piles may be an unsafe option. As the soil undergoes creep deformations, the friction pile will be subjected to an additional load due to negative skin friction between the pile surface and the sediment. If the thickness of soil susceptible to creep is considerable, say 10–20 m or more, the load due to negative skin friction may represent the dominant pile load. As a consequence, friction piles may undergo settlements that are unacceptable for the berth construction. The situation is illustrated in Figure 11.25.

A point bearing pile solution will, in these cases, often be preferred. The settlements will be under control and the pile material strength can be fully mobilised. However, with a depth to the point-bearing stratum of, for example, 70–80 m, a successful result is dependent on a good design and a detailed installation specification. Increased loading on a pile subjected to negative skin friction loads will, in the first instance, result in a reduced negative friction load. Short-term loads, smaller than the negative skin friction load, will not result in an increased point loading.

To minimise the negative skin friction, the pile cross-section should be kept as small as possible. A small pile cross-section will also keep the driving energy down; this is very important for long point bearing piles. The pile must, however, be able to withstand the required driving energy. In the case of a tubular steel pile a favourable design could be a combination of the following criteria:

(a) small-diameter piles (500–600 mm)
(b) large wall thickness (16–17 mm)
(c) high-strength steel (S440J2H).

Figure 11.25 Forces acting on friction piles and point bearing piles

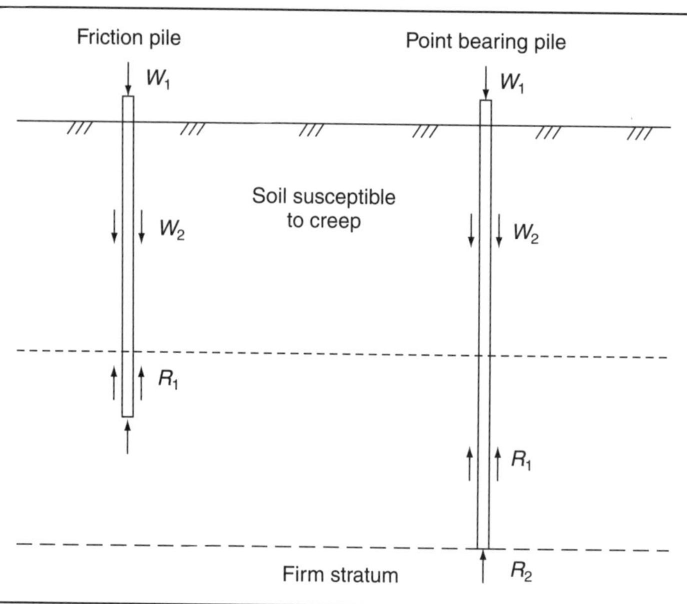

The installation of long piles will require a high driving energy. High driving energy is needed to drive through fill materials and stiff soils, such as gravel and moraine, but most importantly to document the design point bearing capacity.

If the soil deposit consists of clay soil, it is important that the preliminary driving should automatically be followed by the predetermined driving criteria. This is to avoid consolidation of the remoulded clay layer along the pile shaft until driving is terminated. When consolidation takes place, the cohesion between the pile and the soil will increase, and this effect can be considerable over a short period, say 2–3 days. The consequence is that the remaining driving will require increased energy. This is illustrated in Figure 11.26, where pile A was driven continuously through sand and clay layers and into the point-bearing moraine. The bearing capacity criteria were reached by using a driving energy of 120 kNm. Pile B was left 25 m into the clay layer for a month. As can be seen, the energy required to carry out the remaining preliminary driving is about 4 times as high as for pile A in the same depth interval. As a consequence, the driving criteria were not successful, with the hammer giving a maximum energy of 120 kNm. A larger hammer had to be used to reach the design point bearing capacity of the pile. The reason for this is the consolidation of the remoulded clay layer along the pile, giving rise to increased resistance to driving.

After a certain time, perhaps 3–4 weeks, the consolidated clay layer will reach a higher strength than the surrounding natural sediment. Pile driving will then result in a clay–clay failure, not a clay–steel failure as for a newly driven pile.

Documentation of the bearing capacity is prerequisite for all pile constructions. The bearing capacity can be calculated by using traditional bearing capacity formulas based on driving data and measurement of permanent and elastic deformations, or by dynamic testing. For a point bearing pile exposed to negative friction, the side friction will not contribute to the bearing of the pile. It is therefore essential that the point bearing can be documented separately from the side friction, and this can only be done

Figure 11.26 Resistance to driving increases with the amount of embedment due to consolidation of the remoulded clay layer

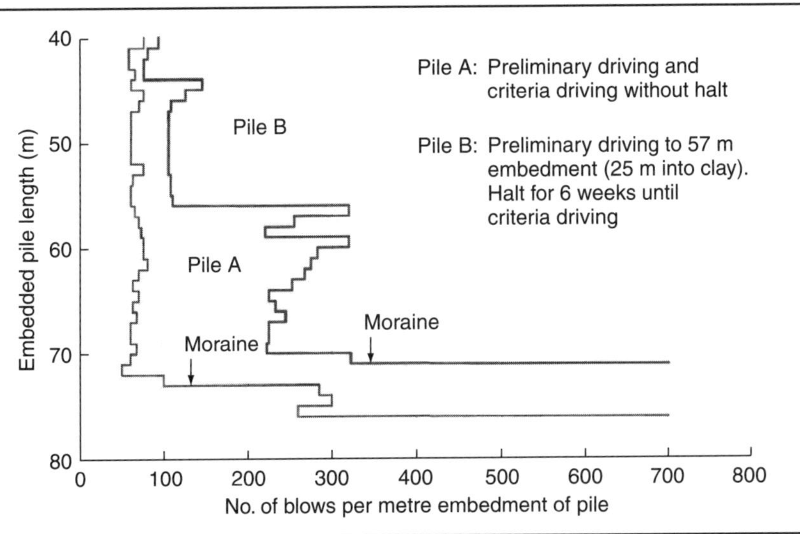

by using dynamic testing. Therefore, for these pile foundations, dynamic testing is required. The most used equipment for dynamic testing is the pile driving analyser (PDA).

The measurements are based on one-dimensional impact wave theory. Strain and acceleration are measured in a cross-section just below the pile top. Analysis of the test data, for example by using CAPWAP (CAse Pile Wave Analysis Program) software gives the following information:

(a) how the soil resistance is divided between side friction and point bearing
(b) how the side friction is divided along the pile
(c) the load–deformation curve for point resistance
(d) information about possible damage to the pile.

Dynamic testing is well documented in the literature.

Figure 11.27 shows the measured pile forces from PDA tests on two piles. For pile A, criteria driving took place directly after preliminary driving. For pile C, criteria driving was carried out 2 months after preliminary driving. For pile A, a point bearing of nearly 7000 kN is documented, while for pile C the documented point bearing is only 2800 kN. The PDA measurements confirm that piles left in the clay soil for a longer period before criteria driving will reach a much lower point bearing capacity than if criteria driving is carried out directly after preliminary driving.

11.4. Lamella berths

A lamella berth (Figure 11.28) can be a good alternative to a cell berth when the berth structure will have to accommodate ships before the filling of the cell structure can be completed.

The structure should be designed in such a way that its deadweight alone gives sufficient stability to avoid overturning due to the fill behind the berth. The deadweight plus the effect of anchoring bolts at the rear end of the lamellas provide the structure with its total stability against overturning moments

Figure 11.27 Measured pile forces from PDA tests on two piles The end bearing for a pile driven without halt is more than double that for a pile left for 2 months before criteria driving

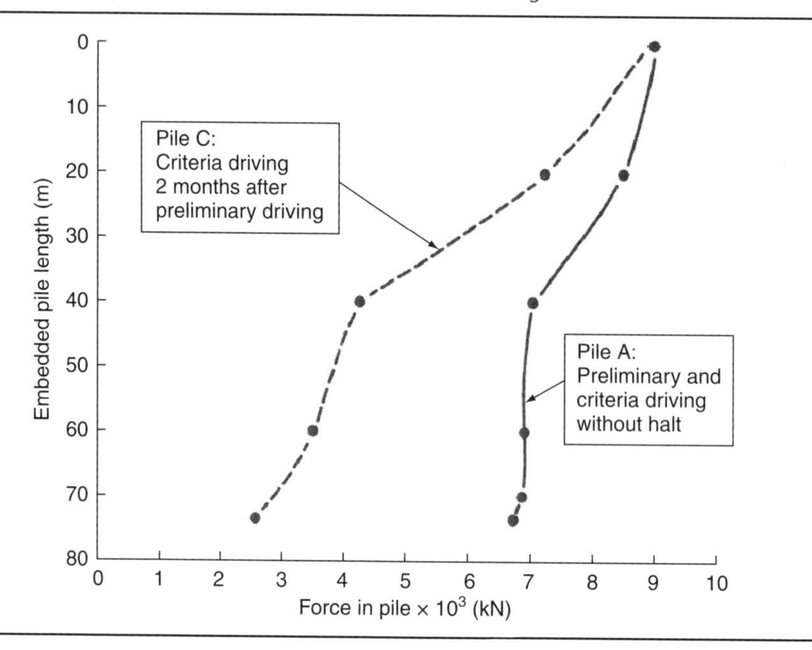

from the fill and the live loads. The dimensions of the bolts should also allow for possible corrosion. In order to increase the deadweight of the structure, the rear part of the platform can be shaped as shown in Figure 11.28, where a certain amount of fill adds to the stabilising weight.

Lamella berths are relatively expensive structures but require a lot of diver work and the use of heavy formwork and powerful cranes. This type of berth should, therefore, not be the first choice if other alternatives are acceptable. If a lamella berth is still the best alternative, much thought should be given

Figure 11.28 A lamella berth

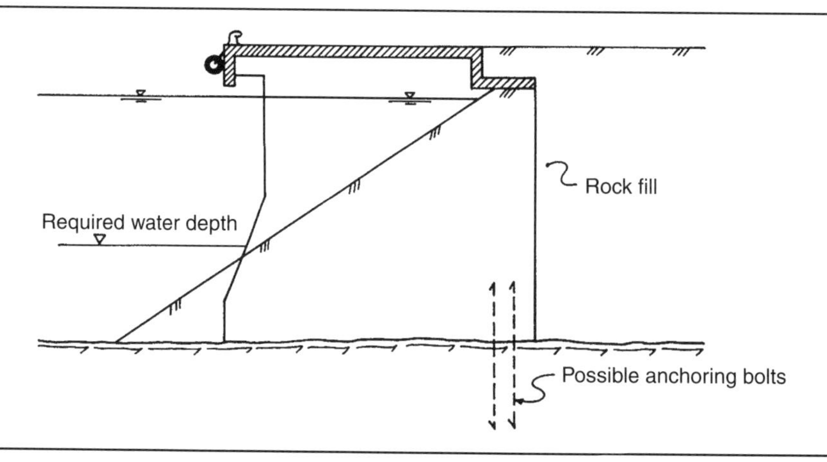

to finding a construction method that involves a minimum of underwater work. Such a method would imply onshore prefabrication of formwork and reinforcement for the lamellas in units of the maximum possible size for the available crane capacity. The bottom of the formwork must be shaped according to the rock profile. After placing the form, including the reinforcement, it must be anchored to resist waves, wind and current until the lamellas have been concreted and permanently bound together by the platform structure.

11.5. Open berth slabs

The berth slab must be designed in such a way that the vertical loads are transmitted safely by way of the beams to the columns, piles or the lamellas, and the horizontal loads from ship impacts, moorings, etc. are transmitted safely to those parts of the structure that are meant to absorb them. A typical cross-section of an Norwegian open berth is shown in Figure 11.29.

In the first reinforced concrete berth structures built in Norway, high and narrow rectangular beams were used to support the slabs. However, after 10–15 years of use these structures showed deterioration in the form of corrosion of the reinforcement at the bottom of the beams, and subsequent cracking and scaling-off of the concrete covering the reinforcement. The slabs had usually not deteriorated to the same extent. The reasons for this are many: the most important factors were that the beams came too close to the sea level, and were densely reinforced, had inadequate concrete cover and generally were more difficult to concrete satisfactorily.

To avoid these disadvantages, beamless slabs were built, which proved very durable. However, this type of structure is more costly to build, particularly because of the necessary formwork support system, and the slab–beam type structure is again used nowadays. However, modern quay structures are different from the old slab-and-beam structures in that they have low beams and broad, trapezoidal cross-sections (Figure 11.30). Thus most of the disadvantages experienced with the old beams are avoided. In Norway the trapezoid is now the normal shape of the beam cross-section in open berth structures.

In general, the formwork should have an over-height at midspan, corresponding to the deflection due to the deadweight of formwork and concrete. The top of the slab should, with a view to the practical use of the berth, lie about 50 cm above the highest high water observed, and it must also be put at a level high enough to permit the beams under it to be concreted in a dry form.

The load-bearing capacity of columns of 80 cm diameter or more is seldom fully utilised, and it is instead the dimensions of the beams and the slabs themselves that determine the lengths of beam-and-slab spans. To avoid very high formwork costs, the slab spans are usually about 6–7 m and the beam spans about 8–10 m. Preferably only one span length and beam cross-section for each beam should be maintained throughout the berth structure. This makes it possible to use the same formwork over and over again for many spans. Possible differences in loadings and/or moments should be reflected in variations in the amounts of reinforcement used, rather than in different span lengths or beam cross-sections. The formwork cost usually amounts to only 10–15% of the total construction cost, but nevertheless the planning and building of the formwork are very important for a successful construction of the works.

Three different types of rational berth formwork system for construction are described in the following sections.

Figure 11.29 Cross-section of an open berth on a steep rock slope

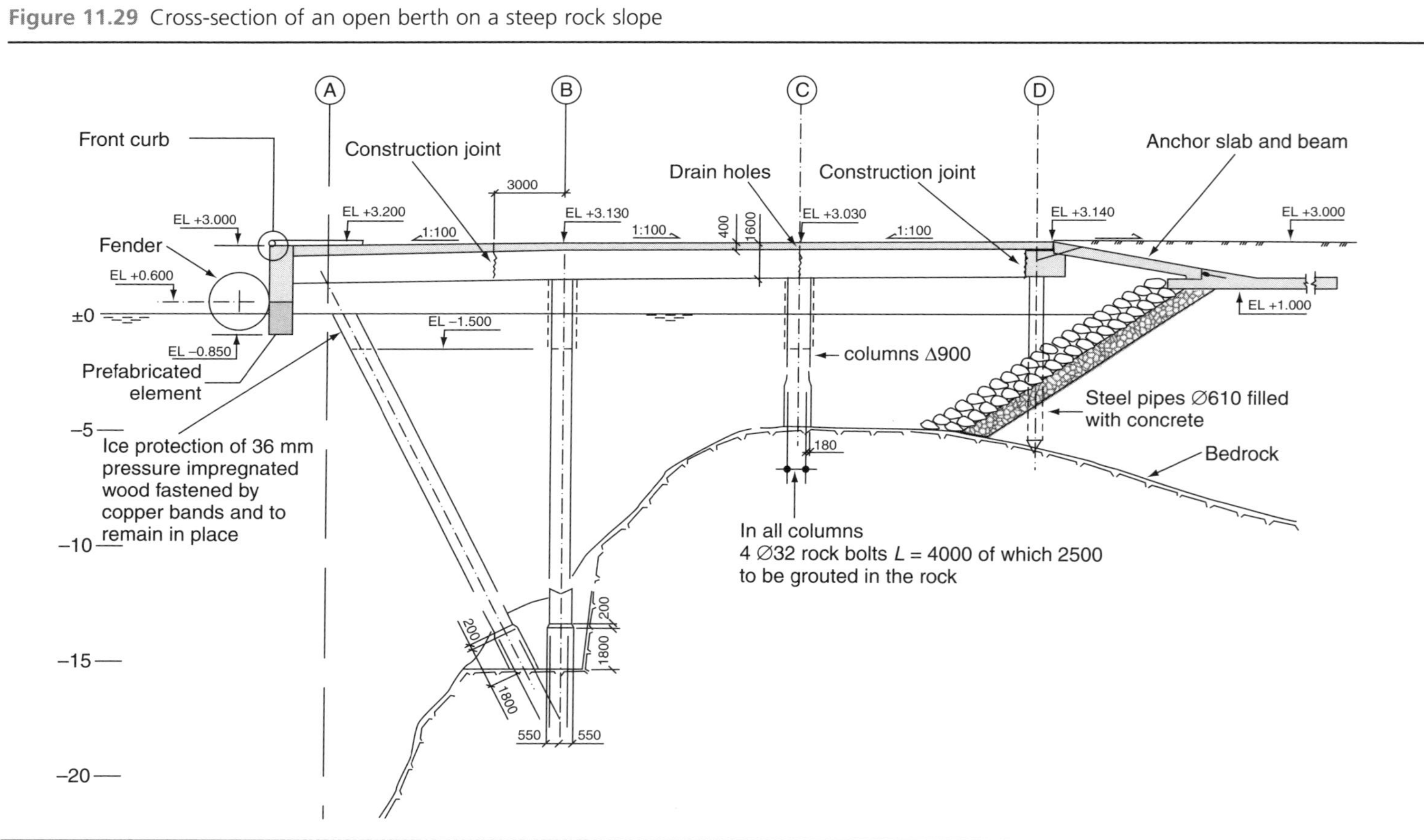

Figure 11.30 Cross-section of a general berth slab and beams

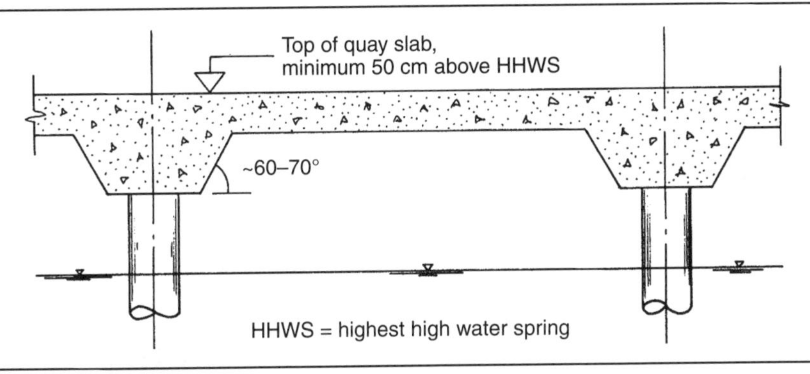

11.5.1 Jacket form system

In jacket form systems the concreting of the beams and the slab is done in one operation. The forms are supported by steel beams resting on column brackets, as shown in Figures 11.31 and 11.32. The support for the beam formwork can be either concrete brackets on the columns (Figure 11.32) or steel brackets on the steel piles (Figures 11.33 and 11.34).

Figure 11.31 Jacket forms

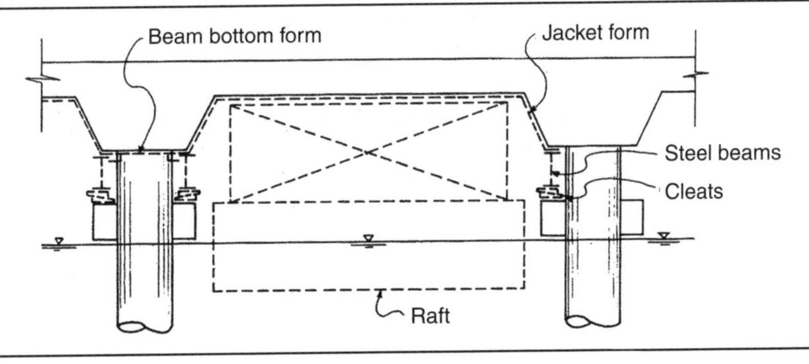

Figure 11.32 Detail of jacket forms

Figure 11.33 Support by steel brackets

The system requires great precision in the placing of the columns and supporting brackets on the columns or piles. If a column is somewhat out of place, problems are likely to arise. Rafts are used for the installation, moving and dismantling of the formwork, and the system is therefore best suited to situations where the under-platform clearance is large or there are high tidal variations.

Great attention should be paid to the dismantling of the formwork at the planning stage, because access from above is not possible after concreting. All details must be designed so as to allow the lowering of the formwork onto the raft without difficulties. The raft must be designed for lifting and lowering by pumping water in and out of it, or by using the tidal variations.

Figure 11.34 Detail of steel brackets used as support

11.5.2 Girder systems

When using a girder system for the slab span concreting, the beams are first formed and concreted up to the bottom of the deck slab. The formwork for the beams is either supported as shown in Figure 11.33, or the beam support can be hung from the column itself (Figure 11.35). Figure 11.36 shows a beam during the installation of the reinforcement, and Figure 11.37 shows the finished concrete beams before the use of either girder system or precast deck slab elements. Once the beams have been concreted, then girders spanning between them are installed to serve as supports for the slab span formwork (Figures 11.38 and 11.39).

The concrete beams are formed in the same way as when jacket forms are used, but the supporting steel beams for the concrete beam itself will have smaller dimensions. The girder system is a very flexible system, and very useful if, for instance, adjustments due to out-of-place columns are needed. This flexibility is very welcome particularly in pile berth construction, because the precision required in placing or driving the piles is less than that required in the placing of columns.

Figure 11.35 Support by hanging from the top of the columns

Figure 11.36 A beam during installation of the reinforcement, with the beam formwork hanging from the column

11.5.3 Precast concrete element system

Figure 11.40 shows the five types of precast or prefabricated elements generally used in a berth structure: anchor slab element, settlement slab element, deck slab element, front beam element and front wall element. A berth properly built with precast elements will be of the same quality and have the same design life as an ordinary monolithically built berth.

Figure 11.37 Beams before installation of the girder system or the deck slab elements

Figure 11.38 Girders

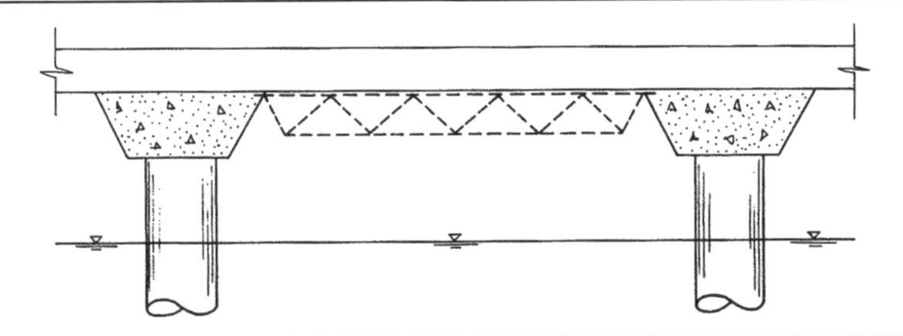

The elements can be either non-prestressed reinforced concrete elements or prestressed concrete elements, and are installed using a mobile or floating crane. A berth built with prestressed elements (pre-tensioned or post-tensioned) will not have the same monolithic strength against impact from loads as will a berth built with non-prestressed elements (Figure 11.41). It should be borne in mind that prestressed elements cannot be easily repaired after damage to or deterioration of the concrete, or corrosion of the reinforcement.

Precast or prefabricated non-prestressed concrete elements for the construction of berth structures are commonly used around the world. The prefabrication of elements is an effective means of reducing both the time and the costs of construction. The advantages of precasting or prefabrication are as follows:

Figure 11.39 The girders spanning between the beams

Figure 11.40 A berth with non-prestressed elements

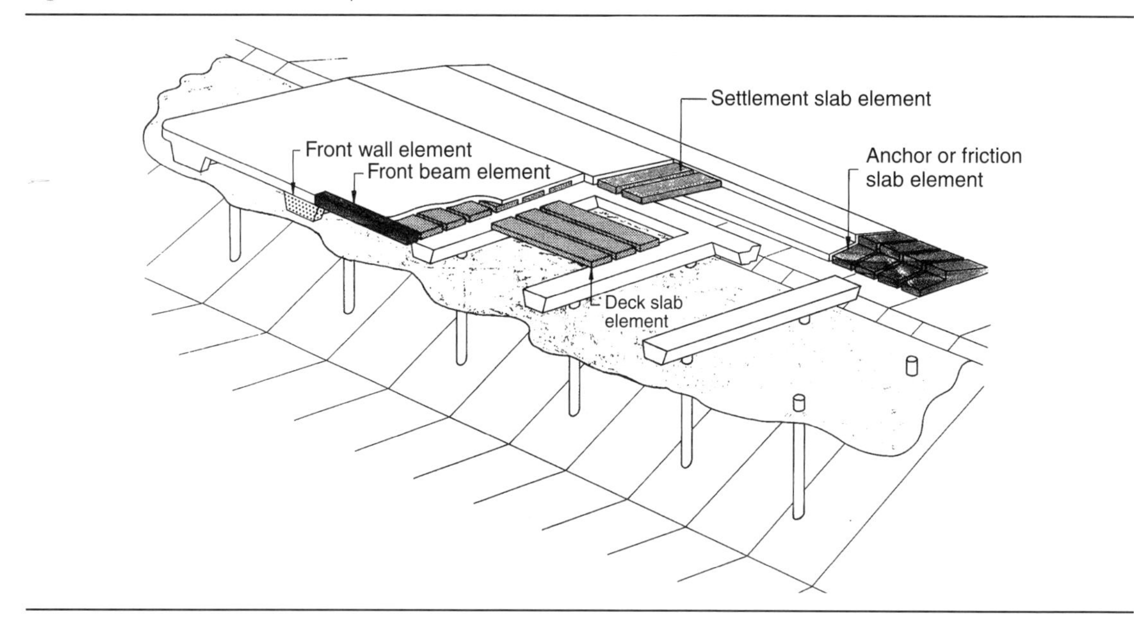

(a) reduction of construction time
(b) minimisation of costly formwork and cast-in-place concrete
(c) generally less dependent on the weather conditions
(d) good quality of the concrete produced.

Figure 11.41 Non-prestressed elements

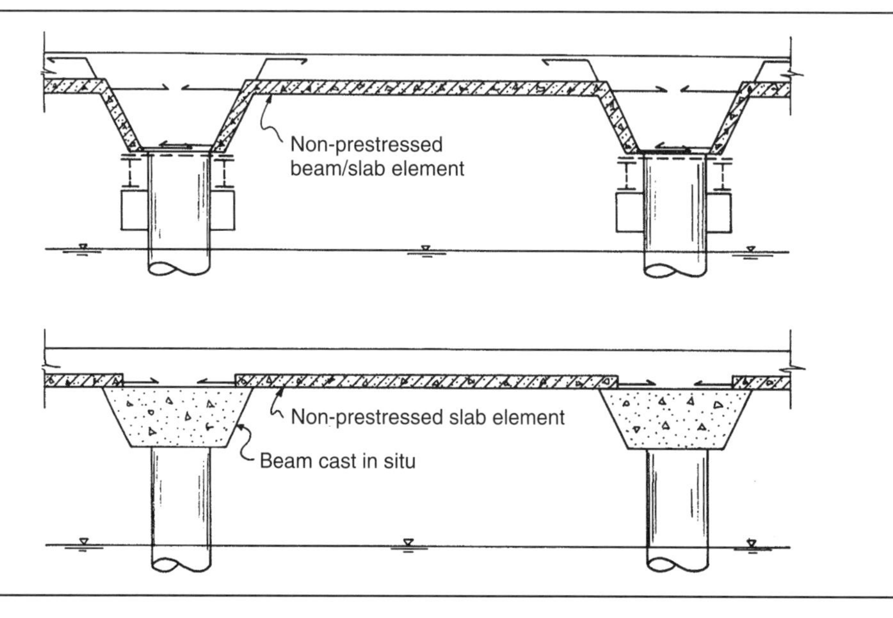

The disadvantages of precasting or prefabrication are:

(a) sensitive to the weather conditions during the installation process
(b) usually necessary to have large floating-crane capacity
(c) small tolerances for installation.

The use of precast or prefabricated elements in berth construction can be advantageous due to the shorter construction time, as the elements can be made while other works (e.g. installation of piles, columns) are being done, better uniform concrete quality can be achieved, etc.

Figure 11.42 shows non-prestessed beam-and-slab elements installed at the formwork, and Figures 11.43, 11.44 and 11.45 show the cross-section of the beam and deck elements with the reinforcement and the support on the beams. The amount of reinforcement in the elements shown in the figures will depend, among other factors, on the total loading on the berth slab.

In many cases it could be economical to prefabricate the berth beams, as shown in Figure 11.46. The slabs between prefabricated beams are usually deck slab elements (see Figures 11.44 and 11.45).

Figure 11.47 shows the lifting of the deck slab element into position by a mobile crane with a 40 kN lifting capacity, and the slab being placed on the berth beams before the mounting of the top reinforcement in the deck slab. Figure 11.48 shows an aerial view before all the deck elements have been installed, and in Figure 11.49 the deck element has been installed.

Figure 11.42 A beam side and a deck slab element

Figure 11.43 Cross-section of the beam and the deck elements with reinforcement

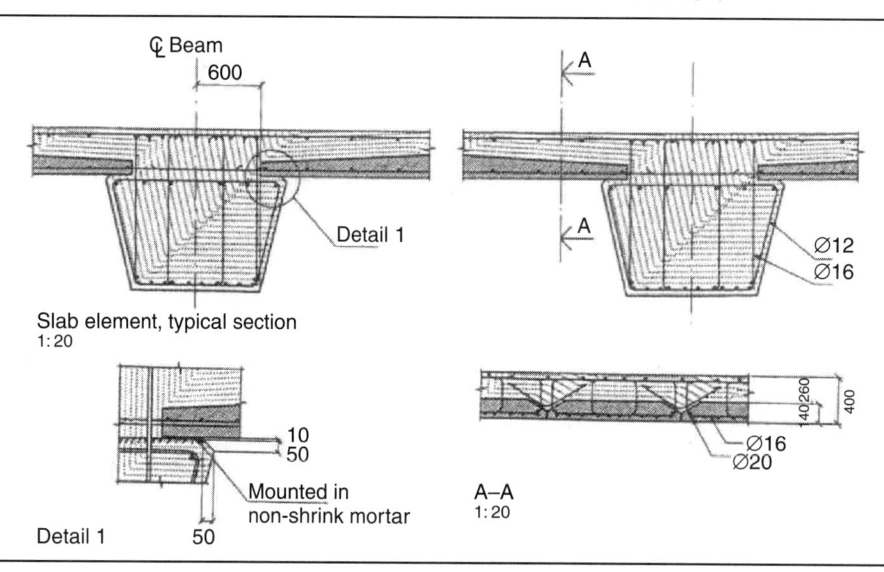

Figure 11.44 Layout and principle of reinforcement in a deck slab element

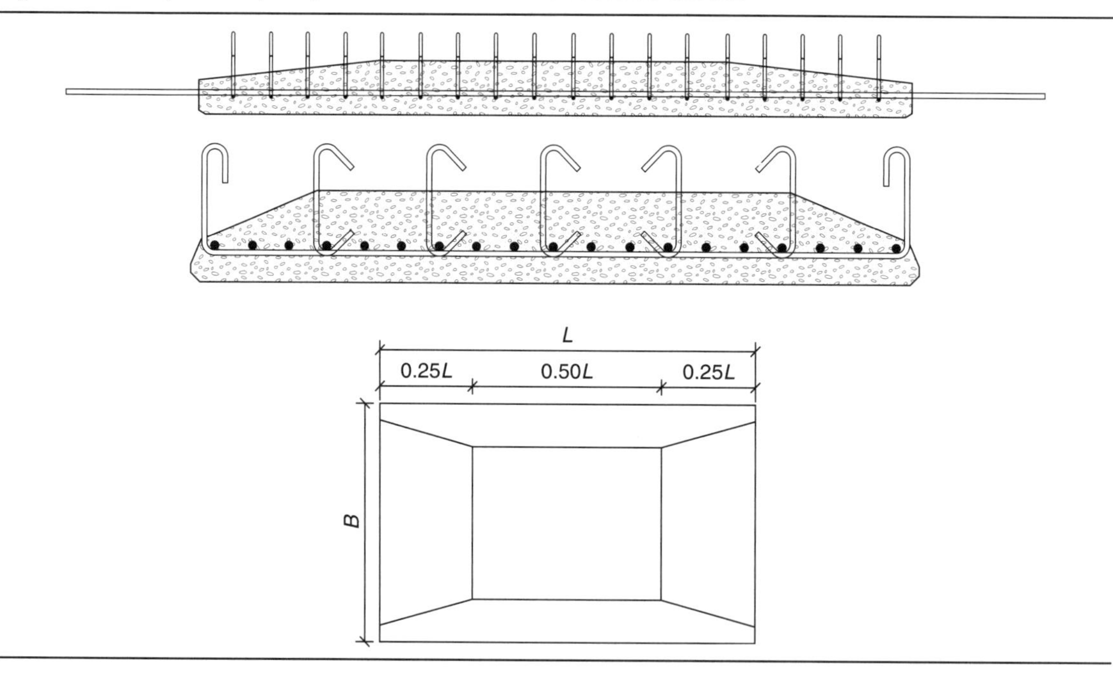

Figure 11.45 A deck element with reinforcement

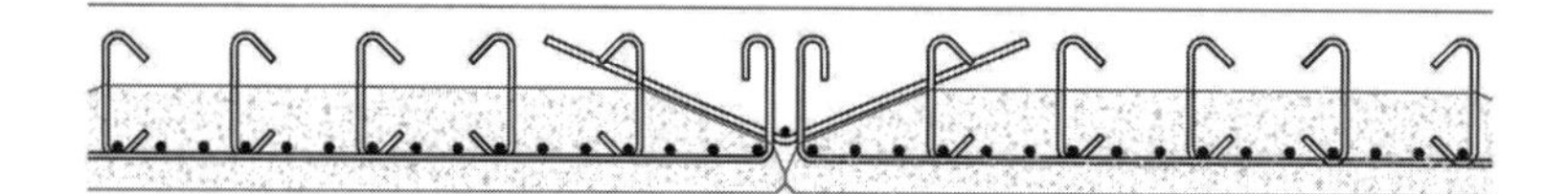

Figure 11.46 Prefabricated beam elements

Figure 11.47 A deck slab element being lifted into position by a crane of 40 kN capacity

Figure 11.48 An aerial view before all the deck elements have been installed

Figure 11.49 The deck slab element

Figure 11.50 A crude-oil jetty

For the crude-oil jetty shown in Figure 11.50 the loading platform slab was constructed using small deck slab elements (see Figure 11.49). The crude-oil jetty was designed for the berthing of 300 000 DWT oil tankers.

Figure 11.51 shows a combined beam-and-slab element with a weight of approximately 4000 kN being lifted into its final position over the steel piles. With this system the beams and the deck slab are installed in one operation.

Figure 11.51 A large combined beam-and-slab element of approximately 4000 kN being lifted into its final position

Figure 11.52 The quay construction, reinforcement and equipment are all covered with thick layers of ice. (Photo courtesy of Birken & Co., Norway)

During the construction of a large beam-and-slab element quay founded on driven steel piles during autumn and winter 2011 in the northernmost city in the world (Hammmerfest, Norway), the weather conditions gave rise to severe challenges. Figure 11.52 shows the result of a combination of a strong onshore wind and an air temperature of approximately −15°C – the quay construction, reinforcement and equipment are all covered in thick layers of ice.

The construction of the beam-and-slab element quay was supposed to be implemented in such a way that it involved as little winter work and weather risk as possible. It was planned that the following berth site work would be undertaken before winter arrived: the deepening of the harbour basin in front of the quay to a water depth of −22 m below sea level, the driving of 37 steel 812 mm diameter piles with a thickness of 14 mm filled with reinforced concrete, the casting of the friction slab and the construction of 300 ton bollard foundations on land (Figure 11.53).

This work was successfully concluded before the start of the winter, and the production of the large concrete elements was then started at a location approximately 1850 km (1000 nautical miles) further south. The casting of the quay beam and slab elements is shown in Figure 11.54. Each element was

Figure 11.53 The steel piles and friction slab. (Photo courtesy of Birken & Co., Norway)

Figure 11.54 The casting of the elements. (Photo courtesy of Birken & Co., Norway)

Figure 11.55 Installation of the last element. (Photo courtesy of Birken & Co., Norway)

constructed to fit with the underlying piles and the adjacent elements. The weight of each element was approximately 360 tons. The installation of the last element is shown in Figure 11.55.

Whether small or large elements are used in a construction will depend mainly on the following factors:

(a) the size of crane available
(b) the weather conditions on site
(c) construction site availability for producing small or large elements.

After installation and casting of the beams on top of the slab, gravel masses were compacted on the slab and concrete stones were placed on the surface. These will result in a quay top surface that can withstand large point loads, as well as a horizontal concrete surface facing the sea. This offers a minimum surface area, which is beneficial with regard to seawater and chloride penetration. Cables and pipes for water, electricity and bunker lines to the front of the quay were placed within the crushed stone masses.

Figure 11.56 shows an open berth structure constructed of large concrete slab-and-beam element founded on steel tube piles filled with reinforced concrete, and the settlement and anchor slab. Between the deck slab and the settlement slab, and between the settlement slab and the anchor slab element, there must be a hinge so that the anchor or friction slab can absorb any possible vertical settlement in the soil beneath the anchor slab (Figures 11.57 and 11.58). The hinge is usually designed for a settlement of the anchor slab of at least 50–60 cm. Figures 11.58 and 11.59 show the details of the hinge between the deck and the settlement slab, and the settlement slab and the anchor slab. Figure 11.60 shows the hinge reinforcement between the deck and settlement slab.

Figure 11.56 The principle of large elements. (Courtesy of Skanska, Norway)

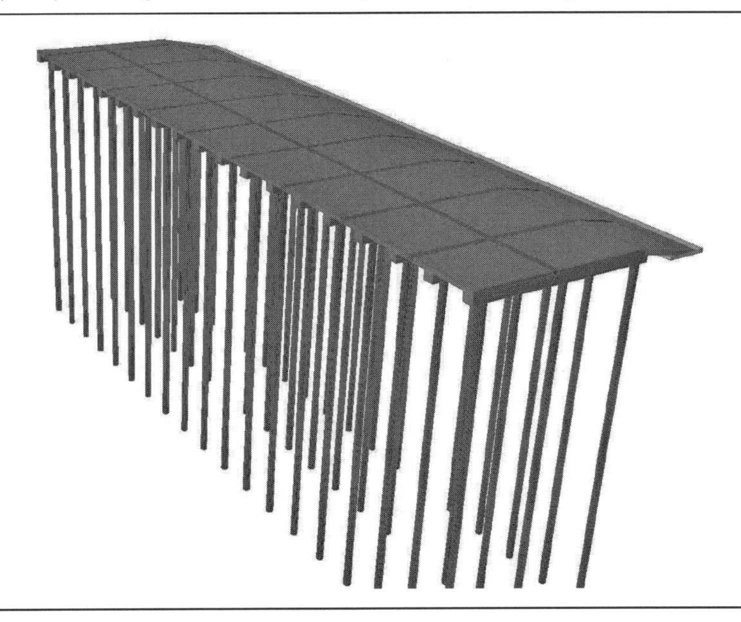

Figure 11.61 shows the settlement slab between the finished berth deck and the anchor slab being lifted into position. The settlement slab can also be constructed as a large finished concrete element. Figure 11.62 shows a finished concrete element with a total weight of approximately 3000 kN being lifted into position between the berth slab and anchor slab.

Figure 11.57 A deck slab element and a settlement slab element

Figure 11.58 The settlement slab

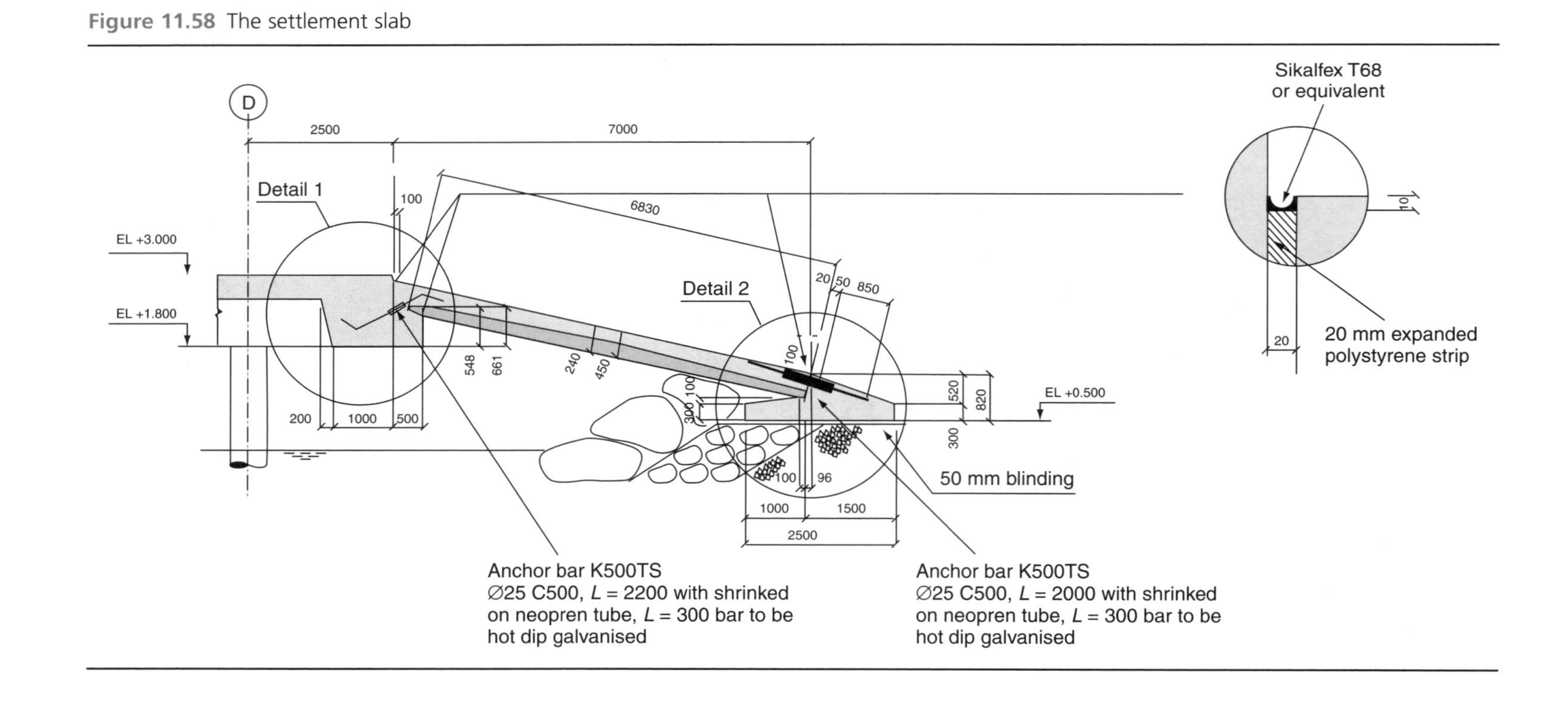

Figure 11.59 Detail of the hinge between deck and settlement slab and anchor slab

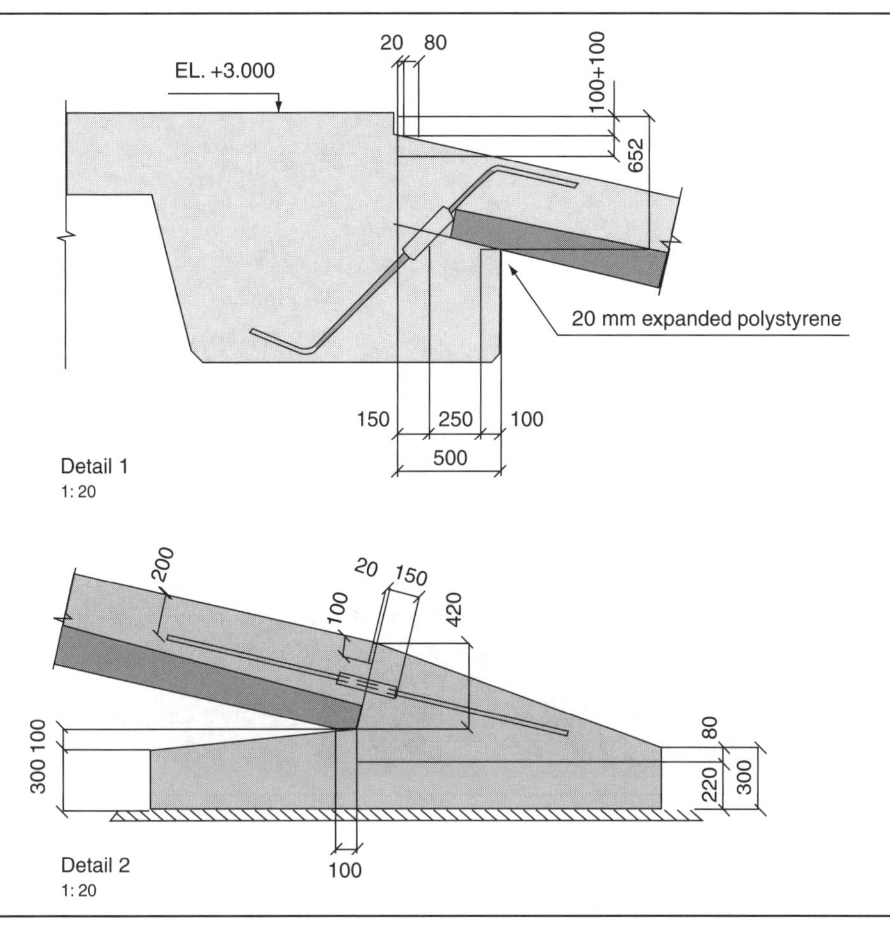

Figure 11.60 Detail of the rear beam and the hinge reinforcement to the settlement slab

Figure 11.61 The settlement slab element being lifted into position

Figure 11.62 A large settlement element of approximately 3000 kN being lifted into its final position

Figure 11.63 A special foundation. (Courtesy of AF Gruppen ASA, Norway)

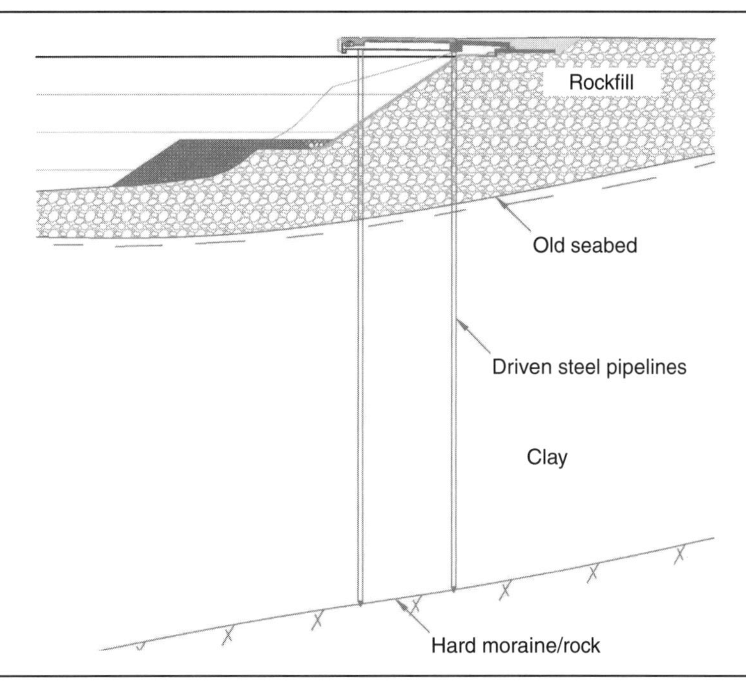

Figure 11.63 shows a section through an open berth structure founded on approximately 75 m long steel tube piles filled with reinforced concrete, and the settlement and anchor slab. As the beam-and-slab structure is founded onto very hard moraine or rock, it will have no settlement, but the settlement slab and the anchor slab, due to the soft clay and the rockfill beneath, could undergo settlement of up to 1.5–2 m.

Figures 11.64–11.66 show a system consisting of a replaceable settlement slab and anchor slab. Figure 11.64 shows the original system, and Figures 11.65 and 11.66 show the settlement slab being lifted into position.

Figure 11.64 The original system. (Courtesy of AF Gruppen ASA, Norway)

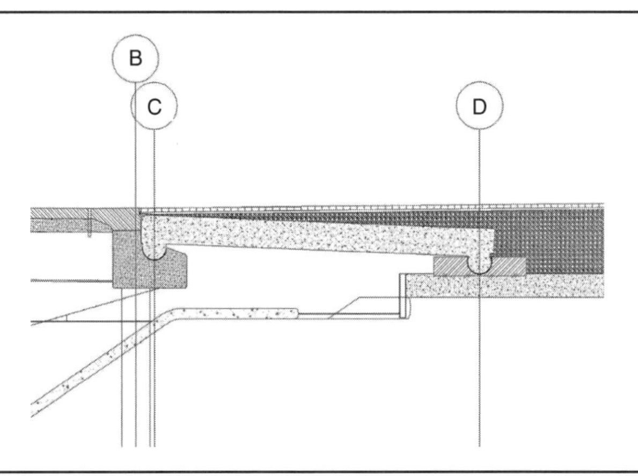

Figure 11.65 Lifting of the settlement slab

Figure 11.66 The settlement slab in position

Figure 11.67 Settlement of the achor system. (Courtesy of AF Gruppen ASA, Norway)

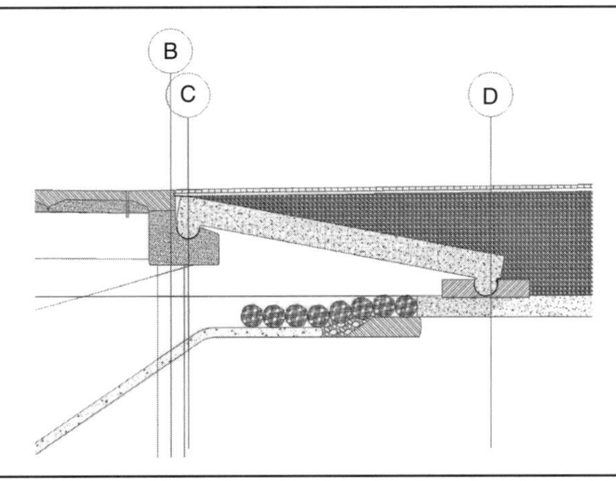

Figure 11.67 shows a settlement of approximately 1.5 m. Figure 11.68 shows that the settlement slab and anchor slab have been lifted up, and the ground under the anchor slab has been adjusted approximately 1.5 m upwards; the anchor and settlement slabs have been placed back again with the filling above the system, and with new erosion protection. Figure 11.69 shows new continuous settlement before new adjustment.

As can be seen from Figure 11.40, most structural elements in a berth or quay structure can be prefabricated. The choice of precast elements to be used in the superstructure will mainly be based on economic considerations. With the increasing availability of equipment for transportation and heavy lifting, prefabrication is now a common procedure. The advantages of prefabrication will be a reduction in construction time, more efficient quality control, and standardised design and construction, and the

Figure 11.68 Adjustment of the system. (Courtesy of AF Gruppen ASA, Norway)

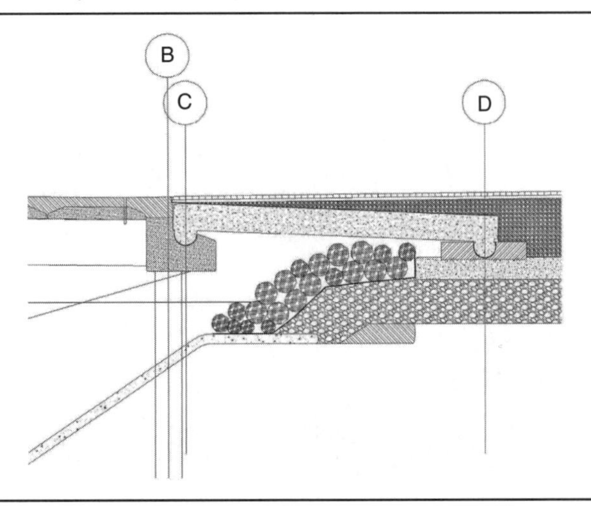

Figure 11.69 New settlement of the system. (Courtesy of AF Gruppen ASA, Norway)

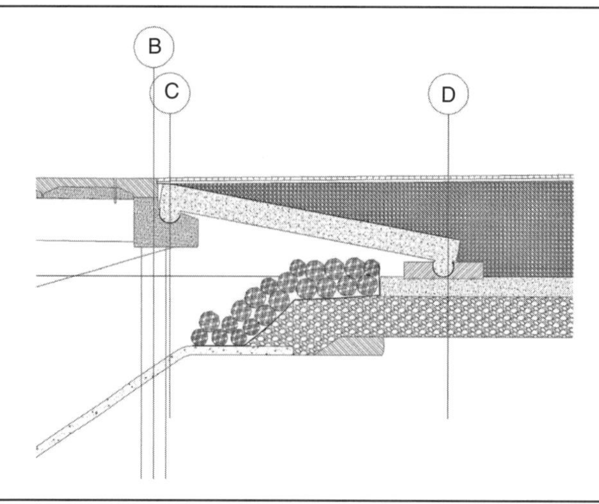

Figure 11.70 Construction of the berth front

Figure 11.71 The finished berth front

method will be favourable in places where the land space available for construction is very small. The disadvantages of prefabrication will be, among other things, lack of availability of suitable lifting equipment, small tolerances and stability during the construction period.

Which of the above systems is chosen is a question that has to be solved in each individual case. Important factors are then the expected lifetime of the berth structure, durability requirements and the contractor's experience and equipment.

The construction of the berth front, with suspension of the formworks from the deck and the reinforcement, is shown in Figure 11.70, and the finished berth front, equipped with fenders, front curb and rescue ladders, is shown in Figure 11.71.

A common element berth is shown in cross-section in Figure 11.72 and the details are shown in Figure 11.73. Reinforced concrete cap units are installed on top of the concrete or steel piles to support the prefabricated concrete beams and deck-slab elements. The final deck slab is concreted after the top slab reinforcement has been installed.

To resist traffic wear on the slab, a protective pavement should be placed on top. If made of concrete, this top layer can either be placed together with the concreting of the slab itself, thus constituting a 3–5 cm additional part of the cast in situ monolithic slab, or it can be made separately, after the curing of the slab, as an 8–10 cm reinforced top slab. In general, the first method is rccommended, but if very difficult weather conditions can be expected during the concreting the top layer should be placed at a later stage under more favourable weather conditions.

The maximum use of prefabricated berth elements may be adopted to achieve an earlier completion date, but the size and weight of the different concrete elements have to be within the handling capacity of the available crane.

Figure 11.72 Cross-section of an open element berth

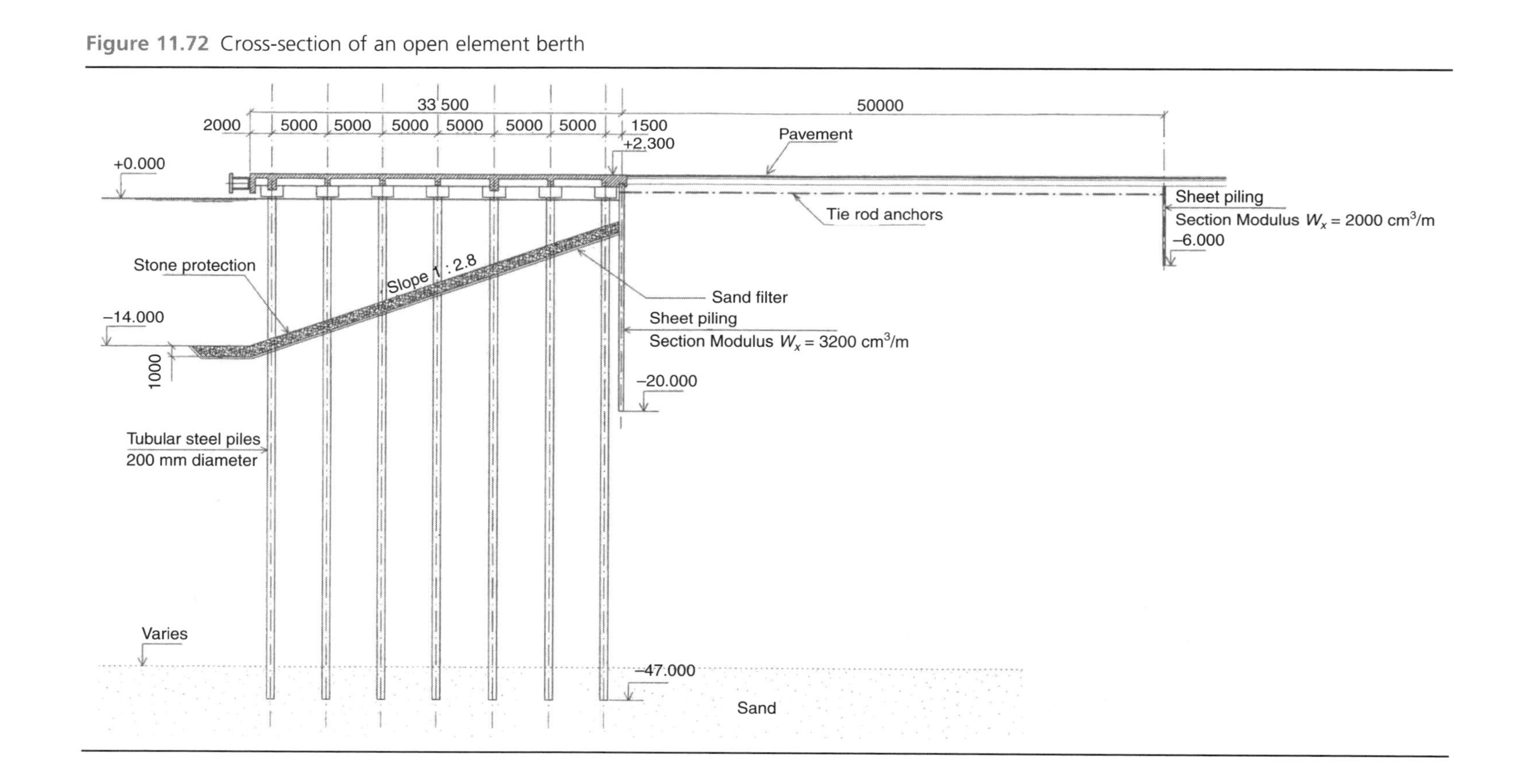

Figure 11.73 Element details

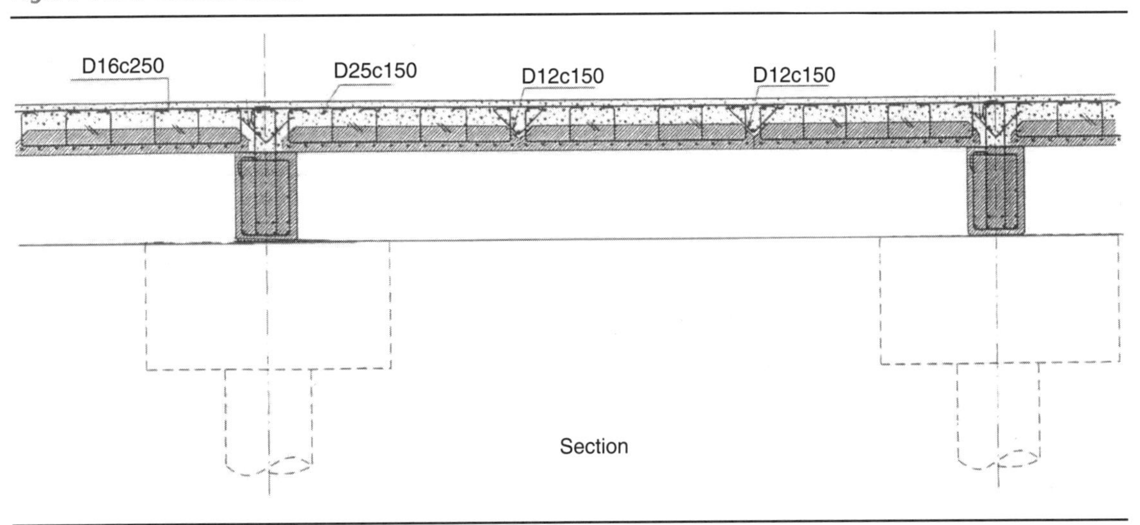

REFERENCES AND FURTHER READING

Bratteland E (1981) *Lecture Notes on Port Planning and Engineering*, Norwegian Institute of Technology. Kluwer Academic, London.

BSI (British Standards Institution) (1988) BS 6349-2: 1988. Maritime structures. Design of quay walls, jetties and dolphins. BSI, London.

BSI (2000) BS 6349-1: 2000. Maritime structures. Code of practice for general criteria. BSI, London.

Bruun P (1989) *Port Engineering*, vols 1 and 2. Gulf Publishing, Houston, TX.

CUR (Centre for Civil Engineering Research and Codes) (2005) *Handbook of Quay Walls*. CUR, Gouda.

EAU (Empfehlungen des Arbeitsausschusses für Ufereinfassungen) (2004) *Recommendations of the Committee for Waterfront Structures, Harbours and Waterways*, 8th English edn. Ernst, Berlin.

Norwegian Concrete Association (2003) *Guidelines for the Design and Construction of Concrete Structures in Marine Environments*. Norwegian Concrete Association, Oslo, Publication No. 5 (in Norwegian).

Tsinker GP (1997) *Handbook of Port and Harbor Engineering*. Chapman and Hall, London.

Port Designer's Handbook
ISBN 978-0-7277-6004-3

http://dx.doi.org/10.1680/pdh.60043.289

Chapter 12
Berth details

12.1. General

When working out the berth and slab details, consideration must be taken regarding the installation of berth equipment such as fenders, bollards, sockets for power and telephone, water outlets, etc.

The planning of supply facilities at the berth structure and at the terminal area should include at least the following supply facilities:

(a) mooring systems
(b) lighting
(c) electric power
(d) potable and raw water.

It should also include the following discharge facilities:

(a) water drainage
(b) sewage disposal
(c) oil and fuel interceptors.

12.2. Traditional mooring system

The term 'mooring' refers to the system for safely securing the ship to the berth structure. The mooring of the ship must resist the forces due to the most severe combination of wind, current, waves or swells, seiches, tides and surges from passing vessels.

The berth structure should be provided with mooring facilities such as to permit the largest ship using the berth to remain safely moored alongside the berth structure. In addition, the berth should be equipped with sufficient mooring points to provide a satisfactory spread of mooring for the different ranges of ship sizes that could use the berth.

To assist a port authority to plan the mooring arrangements of a ship arriving at a port, the arriving ship should be requested to send information about its mooring equipment prior to arrival. Information concerning the type, number and breaking strength of the mooring lines and the winch brake capacity is generally requested. For special berths, such as gas and oil berths, the mooring diagrams will normally be made available to the ship's captain by the pilot when boarding.

The mooring forces most difficult to predict are those caused by waves acting along the berth lines. Such forces are probably also the most common reason for broken moorings. The forces on the structure when the ship is moored include the tension in the bollards due to wind and/or current trying to

Figure 12.1 Mooring by hawsers and bollards

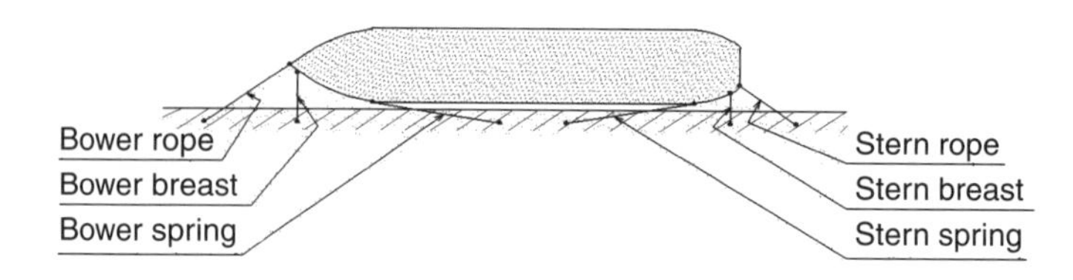

move the ship out from or along the berth structure. Other forces are the horizontal pressure caused by wind and current against the berth structure, and vertical forces caused by the ship chafing on the front (fenders) under vertical movements.

The ship mooring system and configuration must effectively restrain the mooring forces expected to be encountered over the design life for the berth structure while preserving both the operational capabilities and the maximum extreme forces expected on the moored ship and the berth structure.

When wind, waves and current hit a ship between the bow or stern and the beam from any quartering direction, they will exert both a longitudinal and transverse force. The resultant wind or current forces will not have the same angular direction as the wind or current itself. For the evaluation of the mooring layout and arrangement, the wind, wave and current forces must be added.

A ship coming alongside the berth is usually stopped partly by reversing the engine and partly by retarding using the spring hawser, so that the total design force transmitted to the berth structure through the bollard will be at least equal to the breaking load of the spring hawser. Materials for hawsers are steel wire, manila rope, nylon rope, etc. – in other words, different materials, implying great variation in the breaking loads and ductility of the various hawsers. Figure 12.1 shows a general mooring arrangement for a ship to a berth by way of bollards.

A fundamental principle is that all forces from the ship normal to the berth front (e.g. wind from land) should be taken by the breast lines, while the forces along the berth front (wind, current, etc.) shall be taken by the spring lines. The forces from wind, waves and current against a ship at the berth structure will have an influence in the following ways:

(a) **Breast mooring lines**: in the evaluation of the forces, which have to be absorbed by the breasting lines, it must be taken into account that the topography and the jetty itself can reduce the wind area of the ship. The breast lines are used to reduce the sway and yaw motions, and should be perpendicular to the ship. They should be connected to the bow and stern.

(b) **Spring mooring lines**: in the evaluation of the capacities of the spring lines, the full effect from the wind and the current will normally be obtained along the tanker. The spring lines are used to reduce the surge motion of the ship along the berth front. The spring lines should be as parallel as possible to the berth front. The angle between the berth front and the shipside should be equal to or less than 10°.

(c) **Head and stern lines**: these can be used in addition to the spring and breast lines, to reduce the ship's motions.

(d) **Fender system**: in addition to the berthing of the tanker, the fender system must also withstand the forces from the wind, waves, etc., against the tanker.

Table 12.1 Bollard load P and approximate spacing

Ships with a displacement up to: tonnes	Bollard load P: kN	Approximate spacing between bollards: m	Bollard load normal from the berth: kN/m berth	Bollard load along the berth: kN/m
2 000	100	10	15	10
5 000	200	15	15	10
10 000	300	20	20	15
20 000	500	20	25	20
30 000	600	25	30	20
50 000	800	25	35	20
100 000	1000	30	40	25
200 000	1500	30	50	30
>200 000	2000	35	65	40

It is necessary to specify minimum loadings that the bollards should be able to resist for ships of various tonnages (this has been done in most port engineering standards and recommendations). Thus, the bollards, their dimensions and anchoring, and the berth structure itself, should be designed for a certain minimum loading. The idea is that if a ship has too strong a hawser compared with the design load of the bollard, only the latter will break at its footing, without the berth structure itself being much affected.

Bollards should be provided at intervals of approximately 5–30 m, depending on the size of the ship along the berthing face. The bollard load P and the approximate spacing between bollards should be as shown in Table 12.1.

For larger ships, specific calculations must be carried out to determine the maximum bollard load, taking into account the type of ship and the environmental loading.

Bollard loads are assumed to act in any direction within 180° around the bollard at the sea side, and from horizontally to 60° upwards (Figure 12.2). Figure 12.3 shows different types of bollards.

If the berth structure is exposed a lot to wind of more than approximately 17 m/s and a current of more than approximately 1 m/s, the above bollard loads should be increased by 25%. When the ship is

Figure 12.2 Bollard load directions

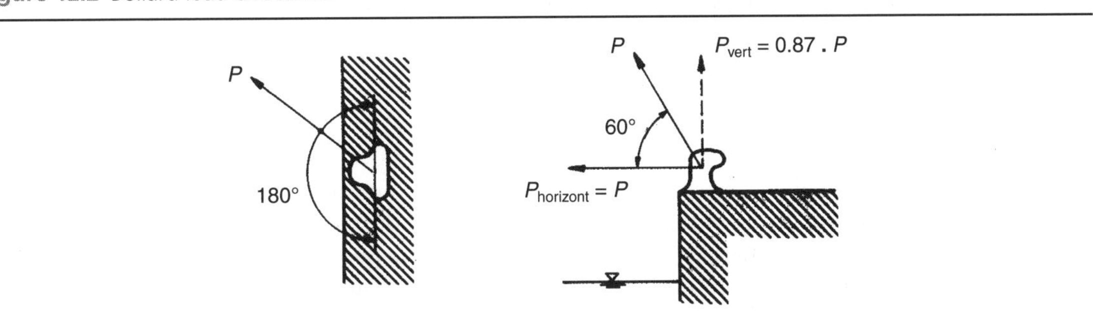

Figure 12.3 Different types of bollards. (Courtesy of Trellebog)

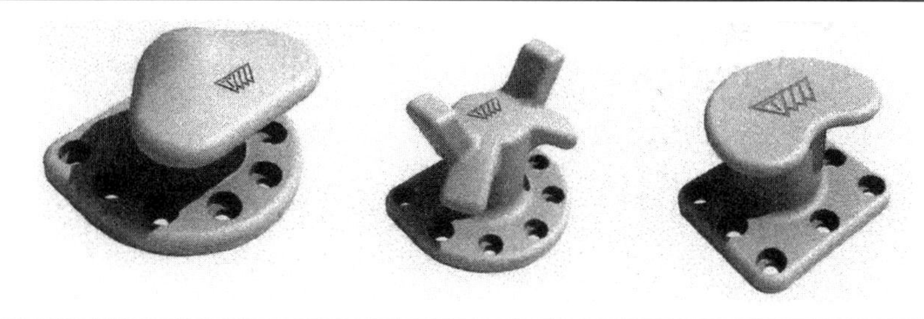

moored, the berth structure should be designed for a minimum vertical force of $0.87 \times P$. The bollard foundation itself should be designed for a force 20% greater than the capacity of the bollard.

Mooring dolphins should be designed for the same loads as the bollards. In addition to the usual berth bollards, storm bollards are often installed behind the apron, designed for twice the above bollard loadings.

If the same bollard accommodates more than one hawser, some standards recommend that the bollard should still be designed for the tabulated load only. This is because it is most unlikely that all the hawsers are fully loaded and pulling in the same direction at the same time. But it is recommended that, if it is possible for two ships to use the same bollard point, either two bollards or one double bollard should be installed. This is because if two ships use the same single bollard and the first ship leaves, the second ship may be reluctant, in windy conditions, to temporarily slacken its mooring rope to enable the first ship to unberth.

Generally, the mooring lines should be symmetrical about the centreline of the ship, to obtain an equal load distribution over them. It is important that the normal or transverse forces from the stern and aft mooring lines are symmetrical around the centreline of the ship if it is exposed to ship motion. All the mooring lines and fenders should ideally have the same stiffness. To prevent impulse shock on the mooring lines and to create sufficient friction between the ship and the fenders, the mooring lines should always be kept taut.

To all **mooring points**, mooring lines of the same size and materials should be used. For an oil or gas tanker the mooring lines should be arranged as symmetrically as possible about the centreline of the piping manifold or transverse centreline of the ship. The spring lines should be oriented as parallel as possible to the longitudinal centreline of the ship. The breast lines should be oriented perpendicular to the longitudinal centreline of the ship and as far aft and forward as possible (Figure 12.4). All the mooring lines for large tankers should be between 35 and 50 m.

The breast line horizontal angles should, if possible, be less than 15° between the ship and the shore mooring point. The head and stern line should be about 15°. The spring horizontal angle should be less than 10°.

The maximum vertical angle of the spring and breast mooring lines should be as small as practicable, and preferably not exceed 25° from the horizontal throughout the entire range of the ship loading or unloading conditions. These criteria will therefore determine the position of the mooring points. This means that, for example, the breast mooring structures will be located approximately 35–50 m behind the berthing face (Figure 12.5).

Figure 12.4 Layout of dolphins for berthing of tankers

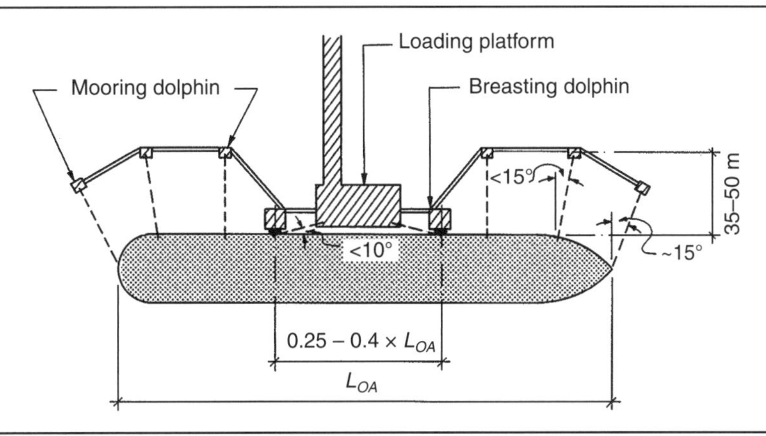

The distance between the breasting dolphins should be approximately 0.3 of the overall length L_{OA} of the ship. If the breasting dolphins need to accommodate a range of different ship sizes, the distance should be within 0.25–0.4 of the ships' overall length. It is very important that the distance between the breasting dolphins is within the flat bodyline of a ship's sides (Figure 12.6).

The mooring arrangement described above will be suitable for oil tankers and for LNG and LPG tankers with flush deck structures (membrane type, etc.). For cylindrical tank LPG carriers and

Figure 12.5 Two tankers moored at the same jetty. (Photograph courtesy of Øyvind Hagen, Statoil Mongstad, Norway)

Figure 12.6 The distance between the breasting dolphins should be within the flat bodyline of the ship side

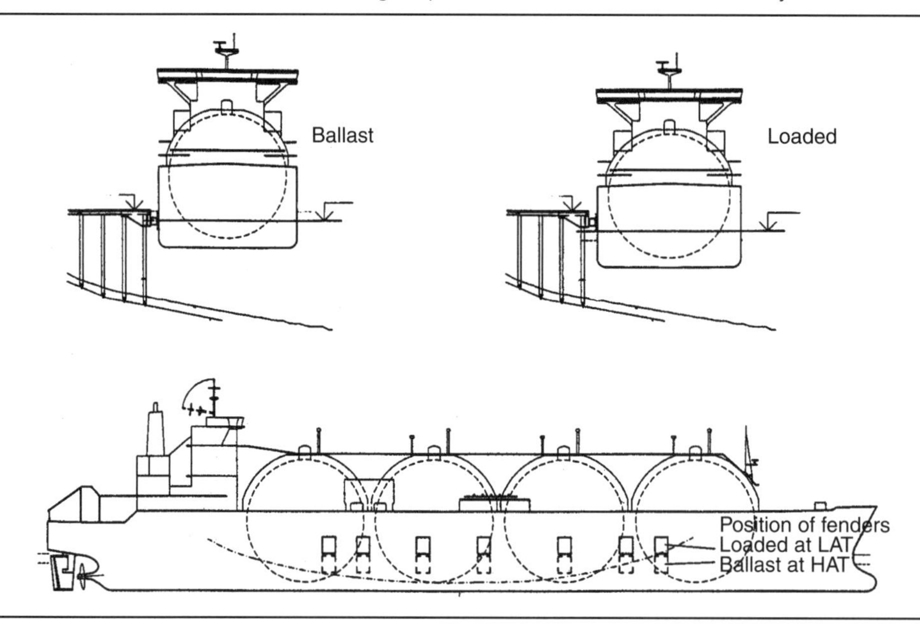

spherical tank LNG carriers, it is usually not practicable to have the main deck winches for the spring lines on the main deck. The spring lines must, therefore, be accommodated from the forward main deck and from the aft, as suggested by the Oil Companies International Marine Forum (OCIMF) in Figure 12.7. For more details, see the OCIMF recommendations.

Figure 12.7 Mooring arrangement for cylindrical and spherical tankers. (Note: circled numbers represent mooring lines)

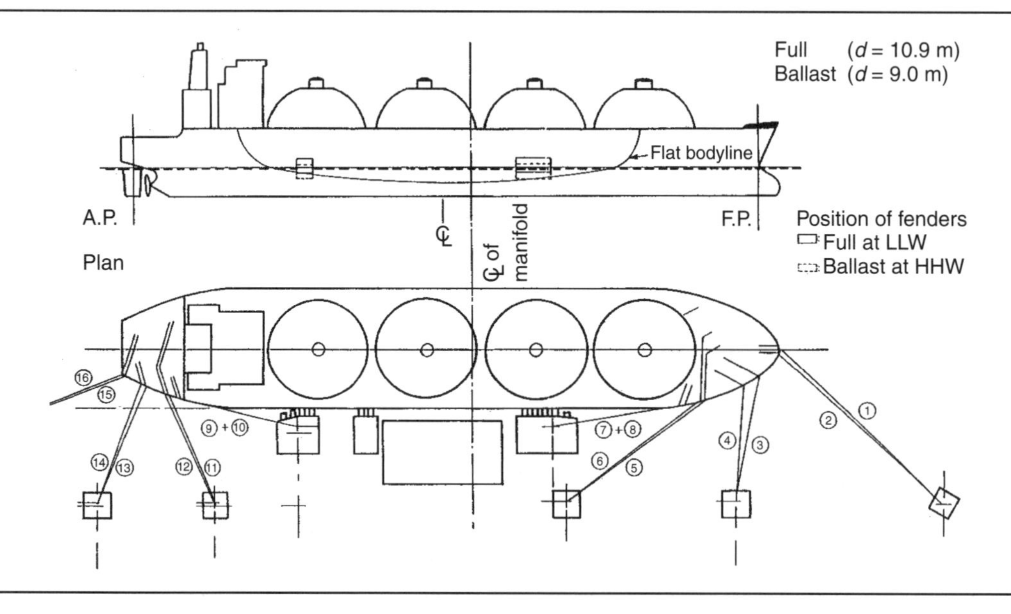

At a berth structure for oil and gas tankers, the mooring line should be fixed to **mooring hooks**, instead of a bollard, with capacities of up to 3000 kN, manual or oil hydraulic release devices, and remote control systems for the quick release of the mooring lines if a ship leaves in an emergency. Fixed shore bollards that require a line to be physically lifted from the mooring point when deberthing are not recommended for VLCC berths, but may be used for smaller tankers.

To simplify the running of the mooring lines from tankers to the mooring hooks and pulleys, the hooks and pulleys should have motor-driven capstans behind the mooring gear to haul in the heavy mooring lines for a double quick-release hook (QRH) from Mampaey, or an equivalent (Figure 12.8). Figure 12.9 shows a sextuple hook assembly.

The capstans should have about a 30–40 cm diameter barrel and a standard line-pull capacity range of 1–3 t, with a motor geared to give a pull rate of about 25–30 m/min.

For gas or oil berths, each of the mooring points should be equipped with QRHs with capstans. The QRHs would usually have a safe working load (SWL) of 600–2000 kN.

The design load for each hook support structure should be defined as the total number of hooks times the safe working load per hook.

The proof load (PL) is 1.5 × SWL. In addition, a material factor of 1.3 should be used. The QRHs should be provided with an SWL of not less than the MBL of the largest rope anticipated to enable the handling of the mooring of the largest ship.

If required, the QRH locking mechanism can be designed to fail and/or release the mooring line at any predetermined line load. The QRHs can also be equipped with a radio-controlled remote system. The hook should have a built-in safety-locking device to prevent accidental or unauthorised release.

All mooring dolphins or mooring structures should be designed on the principle that the breaking limit of each mooring point or dolphin must be at least 20% greater than the total breaking limit of all mooring lines that can be put out to the mooring point. In other words, the design load should be 1.2 times the sum of the breaking loads of all the mooring lines.

Winter conditions, with low temperatures, snow and ice, can be challenging, as shown in Figure 12.10 for the operation of a jetty from both a health and safety and an effectiveness point of view. Snow will cover surfaces and equipment, and needs to be removed before the jetty can be used. It is difficult to clear all the snow, and the result is often a slippery/icy surface. Strong winds combined with a low temperature can result in sea spray icing on jetties and equipment located near the shore. Splash from waves in front of exposed sheet pile jetties can increase the icing problems on the jetty deck and equipment. A build-up of more than 1 m ice thickness at near-shore structures has been experienced on the coast of northern Norway.

Snow and ice will create a slippery surface with an increased risk of falling into cold seawater, and combined with strong winds the risk will further increase. Also, the QRHs, as well as other jetty equipment, may be covered by snow and ice, and malfunctions may occur. Icy hawsers may be difficult to haul due to slippery steel capstans.

Figure 12.8 An example layout of a double QRH with a capstan

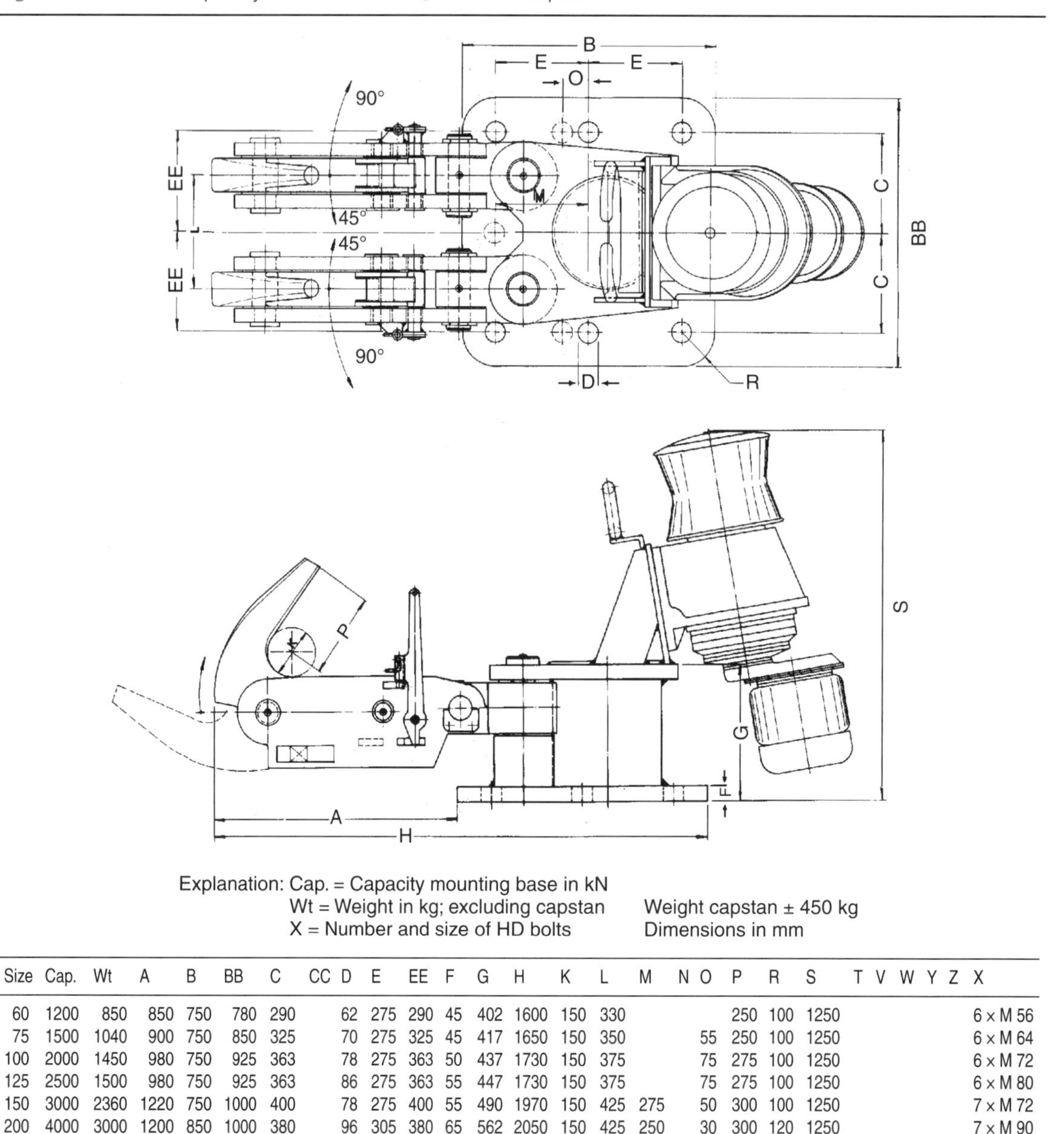

Explanation: Cap. = Capacity mounting base in kN
Wt = Weight in kg; excluding capstan
X = Number and size of HD bolts

Weight capstan ± 450 kg
Dimensions in mm

Size	Cap.	Wt	A	B	BB	C	CC	D	E	EE	F	G	H	K	L	M	N	O	P	R	S	T	V	W	Y	Z	X
60	1200	850	850	750	780	290		62	275	290	45	402	1600	150	330				250	100	1250						6 x M 56
75	1500	1040	900	750	850	325		70	275	325	45	417	1650	150	350			55	250	100	1250						6 x M 64
100	2000	1450	980	750	925	363		78	275	363	50	437	1730	150	375			75	275	100	1250						6 x M 72
125	2500	1500	980	750	925	363		86	275	363	55	447	1730	150	375			75	275	100	1250						6 x M 80
150	3000	2360	1220	750	1000	400		78	275	400	55	490	1970	150	425	275		50	300	100	1250						7 x M 72
200	4000	3000	1200	850	1000	380		96	305	380	65	562	2050	150	425	250		30	300	120	1250						7 x M 90

Several measures can be used to mitigate these problems. The use of salt on surfaces is a possibility; however, this is increases the risk of corrosion of jetty structures and equipment. Sanding of surfaces can be undertaken, but this is not effective in snowfall, and it can easily be blown away from icy surfaces. Further, sanding is problematic for moving parts on equipment on the jetty. The use of steam on snowy/icy surfaces is work-intensive, but may be an option if the materials of the equipment/surfaces can handle it.

Figure 12.9 Sextuple mooring hook assembly. (Courtesy of Mampaey, the Netherlands)

Heat tracing of the jetty surfaces, structures and equipment may be the best technical solution. Such heat tracing can be distributed either by electrical cables or by heated water/glycol in buried pipes, and both systems are used with great success on jetties in northern parts of Norway. Capping structures/sheds to protect critical equipment and to give shelter could further improve the working environment on exposed jetties.

Each mooring hook member is usually capable of taking up to three separate mooring lines of approximately 50 mm diameter each. For gas tankers, the SIGTTO recommends that multiple hook

Figure 12.10 Winter conditions with ice on QRH mooring equipment

assemblies should be provided at those mooring points where multiple mooring lines are deployed so that not more than one mooring line is attached to each single hook.

The following are types of mooring lines or hawsers that are generally used for the various types of ships:

(a) Freighters and coasters, which use domestic and/or short sea routes, are typically less than 10 000 DWT: generally moored with polypropylene lines.
(b) General cargo ships, ranging from 5000–10 000 DWT: generally moored with polypropylene lines. Larger ships are generally equipped with nylon lines and/or steel wires.
(c) Large tankers: generally moored with steel wires and steel wires with nylon tails.
(d) Bulk carriers: mainly moored with synthetic lines, and have steel spring lines. All mooring lines are attached to bollards along the berth front rails for the loading and unloading of the ship.
(e) Container ships: generally moored with steel spring lines to reduce surge motions and with polypropylene mooring lines.

OCIMF recommends that if there is no knowledge of specific berth geometry, the ship's general mooring line requirement should be based on the maximum components of the environmental forces and assuming an efficiency of 90% of the spring lines and 70% of the breast lines. The necessary number of mooring points should be designed on the basis that the maximum allowable loads in any one of the mooring lines should not exceed 55% of its MBL or 100% of the ship's rated winch-brake capacity.

The wind and current forces on a tanker should at least be designed per the OCIMF recommendation, but be aware that other different standards and recommendations, as indicated in Chapter 2, can give other design forces.

The elasticity of the mooring lines will depend on the following:

(a) the diameter of the mooring line
(b) the length of the line from the ship to the mooring point
(c) the material and construction of the line.

The following materials are used in the mooring lines:

(a) Natural-fibre lines manufactured from manila, sisal, etc., and which are the traditional mooring lines. These lines have low load/diameter factors and cannot easily absorb peak loads. Their lifetime is relatively short.
(b) Synthetic fibre lines manufactured from polypropylene, nylon, etc. Compared with natural fibres, they have a high load/diameter ratio, are relatively light and are easier to maintain. Their lifetime is relatively long, and they are relatively cheap to buy.
(c) Steel wires, which are stiff, have a high load/diameter ratios and low elongation. Their lifetime is long, and they are relatively cheap to buy.
(d) Combi-lines, which are a combination of steel wires with synthetic tails. The tails should generally not be longer than 10 m. Because of the synthetic tails, these lines have excellent shock absorption characteristics. They are used for the mooring of large tankers in particular.

Figure 12.11 Typical mooring line force characteristics

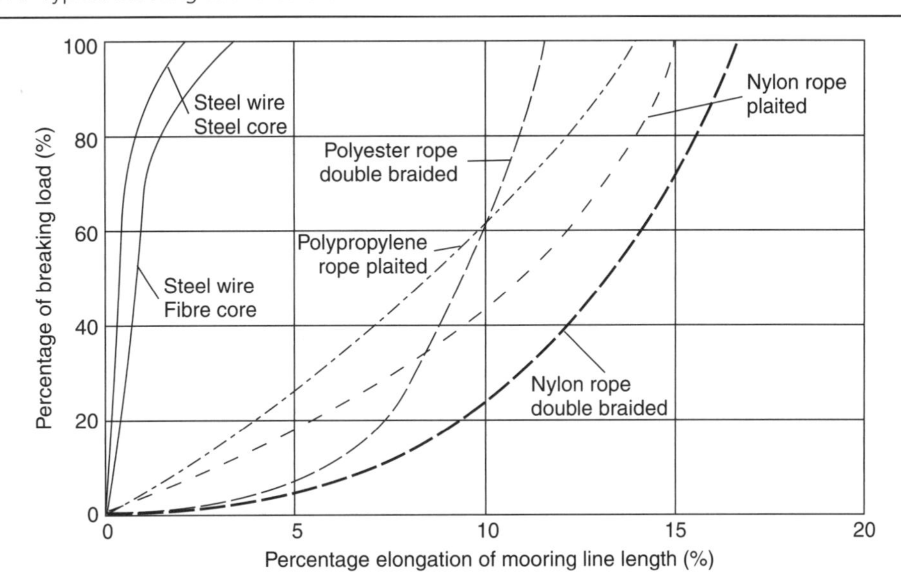

From the mooring point of view, it is important that all breast mooring lines and spring lines have the same length and are of the same materials. A mooring line with high elasticity is desirable for ship-to-ship mooring and for mooring at a berth subjected to swell but, on the other hand, this type of line can cause problems with gantry cranes or loading arms. OCIMF recommends that the safety factor for steel lines should be 1.82, and for nylon tails 2.5. Figure 12.11 shows typical mooring-line force characteristics. Synthetic tails are often used on the end of wire lines to permit easier handling and to increase the line elasticity.

The release of the mooring lines can be done locally at the hook or pulley, or by remote control. Continuous control of the tension of the mooring lines should be maintained by remote-reading tension meters. The mooring lines should be adjusted by picking up the slack and readjusted by the mooring winches during loading or unloading of the ship. The mooring line tension must be continuously adjusted, but automatic tension is not allowed.

Some terminals are fitted with the ability to release the QRH remotely from a shore control room in the event of an emergency, and a few terminals have the ability to release all of the QRHs simultaneously. This practice is not recommended by either SIGTTO or OCIMF, as it may result in a vessel drifting off the berth, unable to use its engines to manoeuvre due to mooring ropes in the water in the vicinity of the propellers.

For safety reasons, the terminal mooring supervisor should oversee the mooring from the terminal control centre and by regular inspections at the berth, for as long as the tanker is alongside. The tanker should be notified if the moorings are not properly maintained and tightened. The mooring lines will require to be tightened due to changes in the tide, freeboard or weather, to prevent them from being overloaded or going slack. All moorings on self-tension winches should be secured with winch breaks in the locked position.

The movement of moored ships that have synthetic mooring lines should not exceed the design envelope of the loading arms, hose or gangway structure. Where synthetic tails are used on the end of the wire-mooring lines to reduce dynamic peak loading, these should be examined to ensure that the design envelope is not exceeded.

In deteriorating weather conditions, the ship's captain may have to decide whether to use additional mooring lines, request standby tugboats to hold the ship alongside the berth, or leave the berth for open sea.

After loading or unloading operations are completed, operators should check the berth area for any local restrictions or hazards. The unberthing is monitored by the terminal mooring supervisor from the terminal control centre, but is directed by the ship's captain assisted by the pilot and tugs. It is always the ship's captain who decides when, and in which order, mooring lines will be released.

For design safety, weather conditions and commercial criteria, the operation of oil and gas tankers will generally require that a tanker should be loaded or unloaded in approximately 12 h, so that the tanker can turn around in 24 h.

If a fire occurs – ashore, at the berth or on the ship – that cannot be extinguished with the fire-fighting facilities immediately available, a decision may be taken as to whether the ship should remain at berth or should be removed by tugboats to a safe distance from the berth.

The **mooring forces** against the tanker from wind, waves and currents are difficult to estimate accurately. In the evaluation of the forces against the berth structure and the forces to the mooring system, it is recommended by OCIMF that the design wind velocity should be 30 m/s against the tanker, and the current velocity should be 1.0–1.5 m/s parallel to the tanker. The reason for this is that if the tanker cannot deberth before a storm, the berth structure and the mooring system must be able to absorb all the force from the storm. In countries with very rough and exposed coastlines, it can be justifiable, for environmental reasons, to use a designed wind velocity of 40 m/s without a gust factor.

As an example, the forces during a full storm of 40 m/s without any gust factor acting against a ballasted 200 000 DWT oil tanker, will, in line with the OCIMF recommendations for the ship wind area for 95% confidence limits, be as indicated in Table 12.2.

This clearly shows that, with a full storm blowing normal to the tanker at the berth, it will be nearly impossible for the tanker to leave the berth safely, even with extended tugboat assistance.

Table 12.2 Forces in kN against a tanker in the loaded condition

Size of the tanker: DWT	Wind 40 m/s against the tanker				Current 1.0 m/s along the tanker	Wind and current parallel to the tanker
	Normal to the tanker	Wind 45° to the tanker		Parallel to the tanker		
		Normal to the tanker	Parallel to the tanker			
200 000	6403	4246	707	1481	50	1531

12.3. Automatic mooring system

Instead of the traditional mooring rope system, a system has been developed by Cavotec comprising large mooring units with specially designed vacuum pads to hold the ship moored against the berth structure (Figure 12.12). One model of the system is the MoorMaster 400, which has two vacuum pads and can hold loads up to 400 kN.

The mooring system can consist of several units, each equipped with either one or two vacuum pads to create a fast and secure attachment between the berth structure and the ship. Each pad represents 200 kN holding force and 100 kN shear force that can be used for dampening ship movements caused by the swell, tide or passing ships, as well as for actively changing the ship's position alongside/shifting. The vacuum pads are positioned via a robotic mechanical structure, and monitored and controlled by remote controls. The pads can seal over welding unevenness up to 25 mm at the ship's hull without significant loss of efficiency.

The system can accommodate tidal and draft changes either by moving the pads vertically on the rail set or, if the change exceeds the length of the rail set, by automatically detaching and moving the pads up or down the hull before reattaching in a patented process referred to as 'stepping'. The system uses hydraulic cylinders for the extension operation towards the hull. Hydraulic cylinders are also used to control the vertical and horizontal positions of the pads.

When the ship arrives at the port, the captain manoeuvres the ship to its berth position. Once the ship is in the correct berth location, an operator pushes a button, and the mooring units will reach out until the vacuum pads attach to the hull.

A single operator, by using a wireless remote-control system, can moor the ship within approximately 30 s (Figure 12.13), and release the ship in 10 s.

Figure 12.12 The automatic vacuum mooring system. (Photographs courtesy of Cavotec MoorMaster, New Zealand)

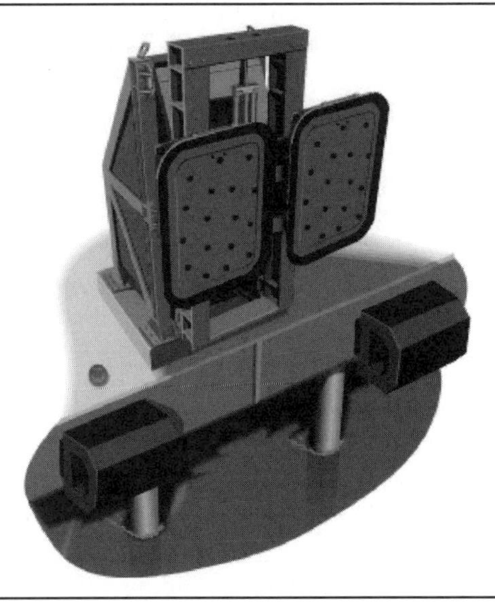

Figure 12.13 Mooring of a container ship. (Photograph courtesy of Cavotec MoorMaster, New Zealand)

For most applications, pairs or groups of mooring units will be used to secure the ship. In a typical instance, a ship of 150 m length may use four units, rated at 400 kN for each mooring unit, which combined will give an overall holding capacity of 160 t.

MoorMaster systems are suitable for the very fast mooring of any kind of ship, as long as it has enough flat parallel hull body for the vacuum pads to attach to. This applies to roll-on/roll-off ships, ferries, bulk carriers, general cargo ships, container ships, and oil and gas tankers.

12.4. Lighting

The berth structure and access roads and the terminal area should be equipped with sufficient suitable lighting during all berth operations and at night as a defence against crime. The following are therefore recommended:

(a) the lighting during terminal operation and during the loading and unloading of the ship should be 100 lux
(b) the lighting for security of the port area should be 30 lux.

12.5. Electric power supply

12.5.1 General

Only underground cables for low- and high-voltage supply systems to the port installations, crane installations, lighting, etc., should be used. The earth cover of the supply system should be approximately 0.8–1.0 m. The power connection points along the berth front should be at intervals of approximately 50–200 m, depending on the type of berth activities.

12.5.2 High-voltage shore connection

Delivery of electric power to vessels in port

The running of cables from an onshore electricity grid to small ships at berth is not a new phenomenon. Onshore power has been used for lighting, heating and for charging batteries on ferries and tugboats that are berthed overnight.

The power supplied makes it possible to shut down the engines while still having electricity on board. This is in most cases up to 50–100 kW, which is about the equivalent for a large residential house. This power has the same voltage and frequency as the regular grid, either 230 or 400 V at 50 Hz.

Delivery of power to ships in port has thus been common practice for smaller ships for other reasons than environmental ones.

To reduce the pollution from, especially, very large cruise ships, a new ISO standard was adopted in 2012: **ISO/IEC/IEEE 80005-1: 2012**, on high-voltage shore connection (HVSC) systems.

The high-voltage connection to larger ships can supply sufficient electricity through only one cable. The principle is shown in Figure 12.14. In almost all European countries the electricity is delivered at a frequency of 50 Hz, but approximately 99% of all cruise ships are based on 60 Hz. The ISO standard also sets a standard for the plug, as shown in Figure 12.15.

Figure 12.14 Onshore power supply for large ships

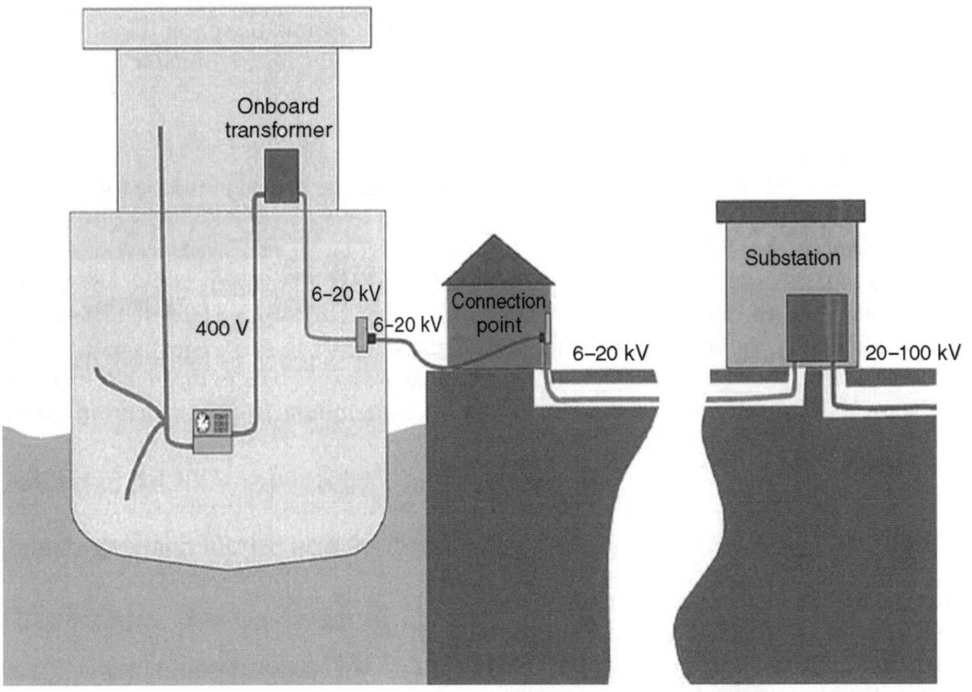

Figure 12.15 A plug for the power supply to a large ship

12.6. Potable and raw water supply

To safeguard the delivery of potable and raw water to the port or terminal area, at least two delivery lines are required, independent of each other, to each port or terminal section. Hydrants should be installed at approximately 100–200 m intervals. In cold areas, the water pipeline system should be placed with sufficient earth cover to be protected against frost.

12.7. Water drainage system

The drainage system for berth structures and terminal areas can be divided into the following systems.

12.7.1 The open system

The top surface of the berth structure and the terminal area should be designed to allow spray from waves and rainwater to be drained away directly to the harbour. For areas where differential settlements can be anticipated, the cross-falls should be as high as 1 : 40. For surface areas with no settlement risk, the cross-fall should be between 1 : 60 and 1 : 100. The drainage system, through the berth deck, is shown in Figure 12.16. The figure shows the drainage pipe at minimum of 50 mm below the deck, but it is advisable for it to be 100 mm below.

Figure 12.17 shows a concrete BASAL drainage channel system, which effectively collects the surface water through longitudinal slots. The benefits of this system are:

(a) simple runoff of areas with permanent cover
(b) runoff from one or both sides of the horizontal drain
(c) a maximum of 50 m between the outlets from the drain to the stormwater pipeline
(d) robust, with easy maintenance.

Figure 12.16 Drainage detail

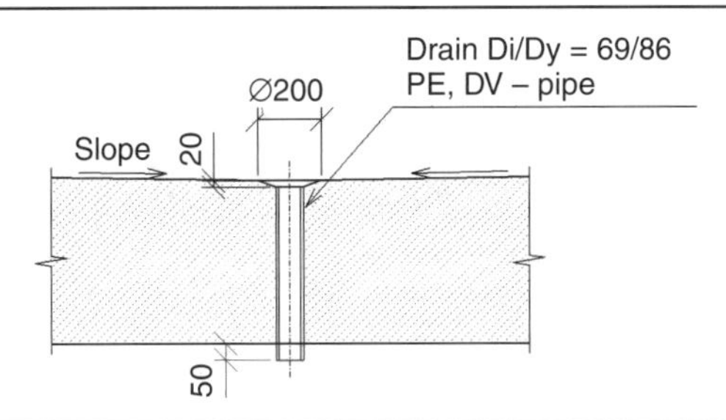

The function of this drainage channel system with slits made of concrete according to the standard EN 1433 (Drainage channels for vehicular and pedestrian areas) is to direct surface water away from large areas with permanent cover such as terminal areas, airports, etc., with container traffic and industrial areas. The drainage channel should be built into the bearing layer aligned with the cover; it collects water through the longitudinal slots, and drains it away in a simple and safe way.

The system consists of a drainage channel, an inspection element, an outlet element, and slots. The outlet element is equipped for connection to a manhole for stormwater.

A type 200/275 BASAL drainage channel is used primarily for large areas and where the requirement is to strength class F900 according to the standard EN 124. This provides a flow of 35 litres per second at a maximum of 50 m between the outlets.

Figure 12.17 The BASAL drainage system. (Courtesy of Basal AS, Norway)

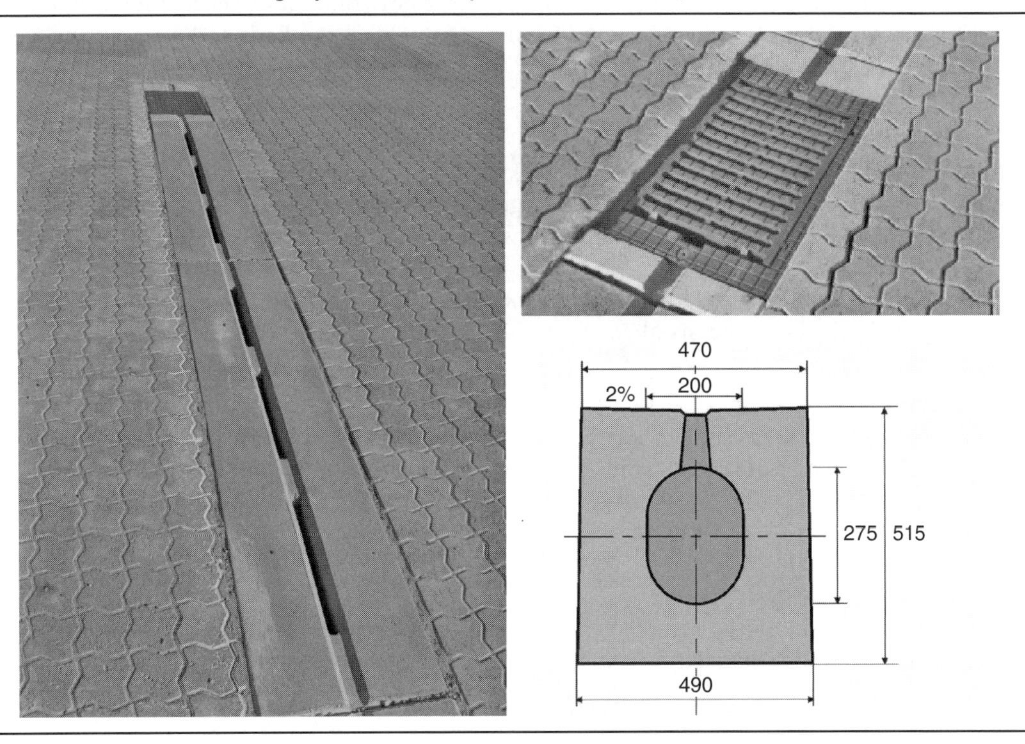

BASAL drainage channels are produced in high-strength concrete (dry cast). This gives a very dense material with very low water absorption and particularly high resistance against aggressive water. Resistance to salting and freezing will also be very good. The inspection and outlet elements are equipped with a lockable and hinged grate of cast iron. This makes it easy to remove unwanted objects from the flow system and to perform high-pressure jetting. These types of drainage channels have a significant advantage compared with other types of channels with a continuous grate, where high trafficking creates considerable maintenance needs.

12.7.2 The closed system

With this system the water can be polluted, for example by possible oil spillage during the loading of oil products at an oil berth. The surface water must then be collected by a separate drainage system for treatment.

12.8. Sewage disposal

Any sewage disposal in the port area should be fed through a special pipe sewerage system to the municipal system or to a dedicated treatment facility.

12.9. Oil and fuel interceptors

All oil and fuel waste should be collected in special interceptors.

12.10. Access ladders

Access ladders should be placed at 50 m intervals along the front of the berth structure. In order to be accessible from the water, the ladder must extend down to 1 m below the LAT. In order to give the ladders sufficient strength, they should be designed for a horizontal and vertical load of 1.0 kN/m. Figure 12.18 shows a flexible ladder hanging from the front steel rail kerb and a stabilising weight – an old rubber tyre filled with concrete.

12.11. Handrails and guardrails

Handrails should be provided on both sides of walkways and on part of the berth structure itself if they do not interrupt cargo handling or mooring arrangements for ships. The top of the handrail should be at least 1.0 m above the berth deck, and the walkway elevation is shown in Figure 12.19.

Along, for example, an access bridge out to an oil berth structure or along the terminal area against the waterfront, guardrails should be installed (Figure 12.20).

12.12. Kerbs

Around the berth edges, kerbs should be provided to prevent, for example, trucks from sliding into the water. The kerbs should be at least 200 mm high and designed for a horizontal point load of 15–25 kN, depending on the type of traffic. The kerbs can be either of concrete (Figure 12.21) or constructed from used rails (Figure 12.22).

12.13. Lifesaving equipment

Lifesaving equipment should be installed on all berth structures, especially jetty heads. It is recommended that chains are suspended at the seaward side between the ladders. The chains should be extended to 1.0 m below the LAT. Lifebuoys with approximately 30 m of buoyant line should be installed along the berth structure at 50 m intervals.

Figure 12.18 A flexible ladder

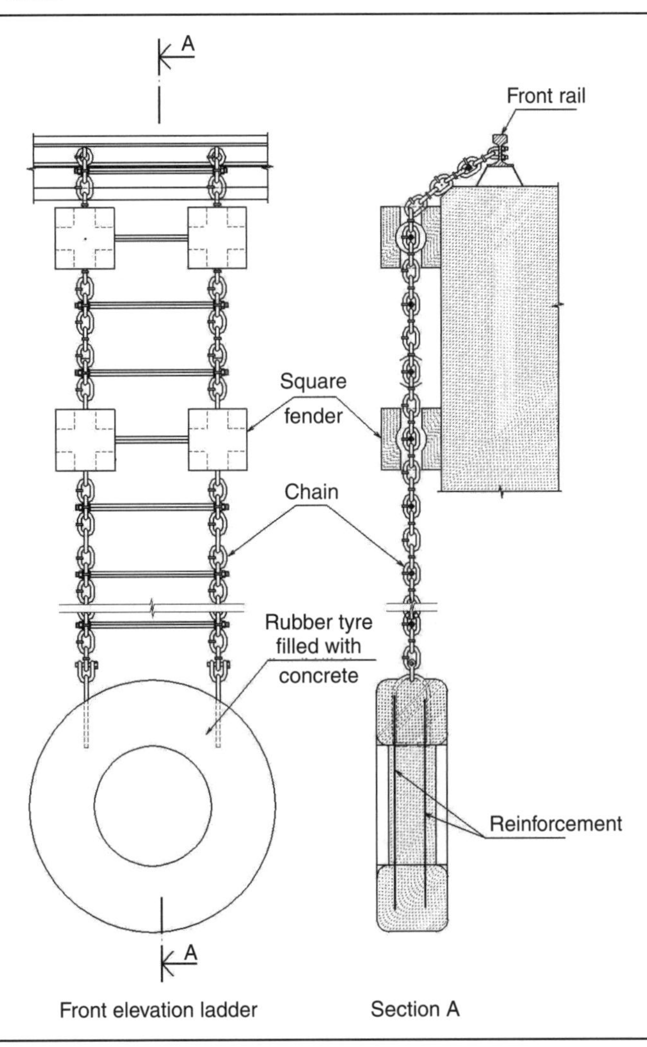

12.14. Pavements

12.14.1 General

A durable pavement area with high performance is vital to container and port terminal operations. Nowadays, there are different kinds of area pavements, and the most common types are asphalt, cast concrete and concrete block pavers. Figure 12.23 shows a concrete block paver area.

Area pavements of concrete block pavers have proved to be beneficial in areas where heavy equipment is used, such as large fork-lifts. Figure 12.24 shows concrete pavers in detail. The geometric shape and quality of the pavers are of vital importance in terms of the performance of the pavement. Due to the various methods of engineering design and traditional choices of materials around the world, the following can only be a general guideline for the design of different base-courses, depending on the subgrade. The general recommendations given in this chapter are applicable to heavy duty areas in harbours. Correctly constructed pavements have the following advantages:

Figure 12.19 A typical handrail

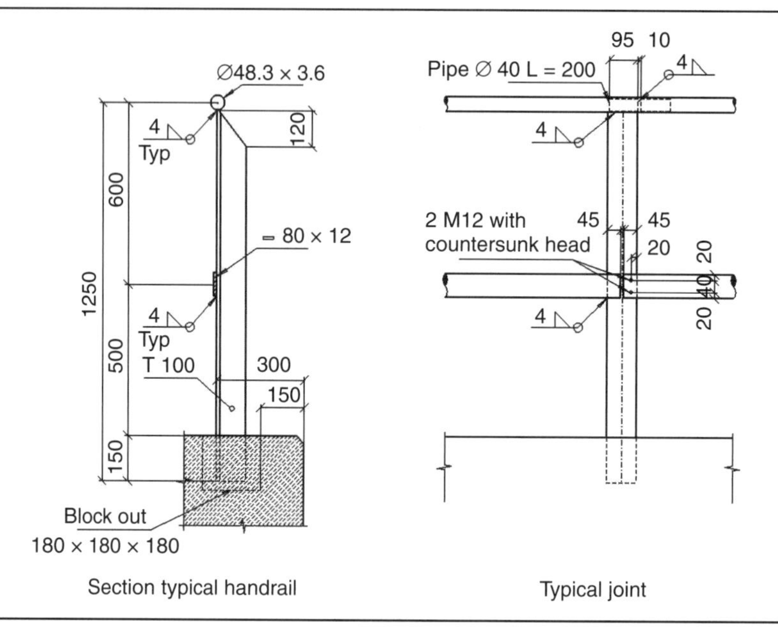

(a) good performance
(b) economical
(c) low maintenance
(d) high durability.

The first concrete block paver was invented around 1880, and nowadays there are more than 250 different types. There is an increasing number of harbours worldwide where the pavement is made

Figure 12.20 A typical guardrail

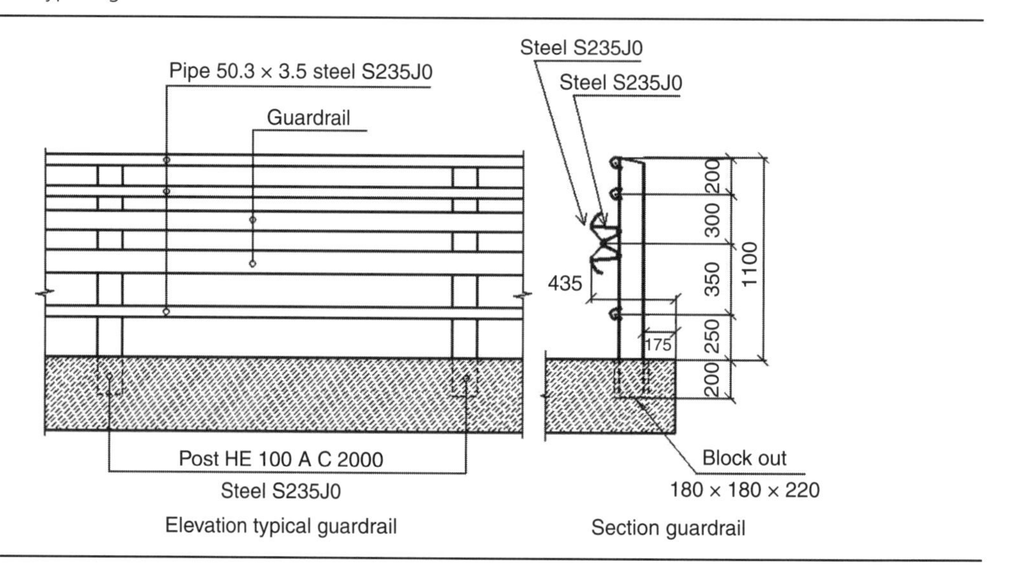

Figure 12.21 A concrete kerb

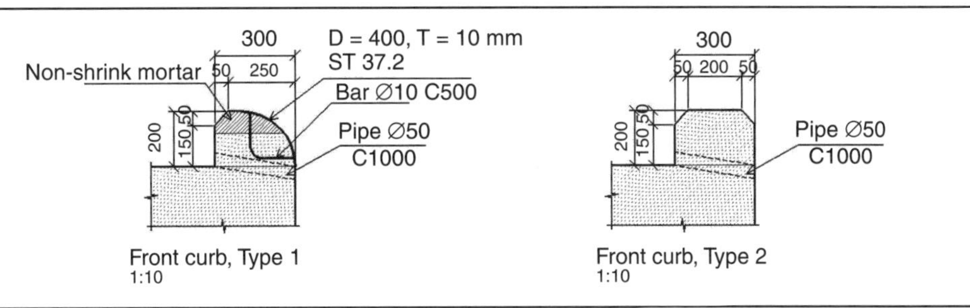

Figure 12.22 A rail kerb

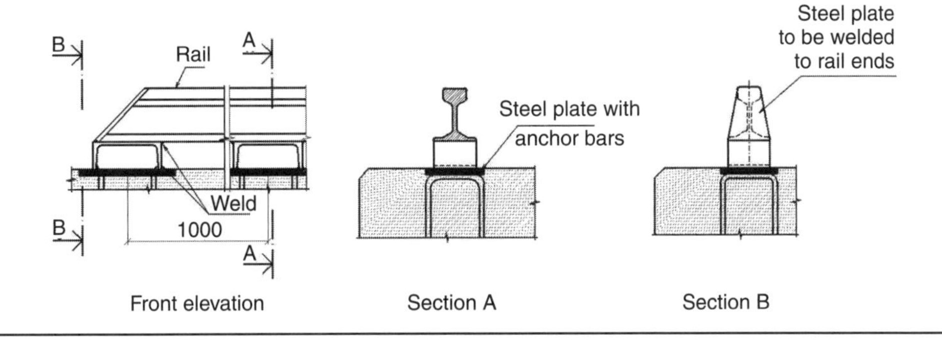

Figure 12.23 A concrete block paver

Figure 12.24 Detail of concrete pavers

of concrete blocks. One of the first projects was one of the world's largest container terminals in Rotterdam in 1965, which is still in service.

12.14.2 The construction components

Base-course (sub-base and base)

The complete base-course consists of different material layers, as shown in the following figures. The required dimensions and materials used in the different layers will depend on the subgrade condition and the estimated traffic and loads. Within economical limits, best-quality materials should be used. In the following three cases, the design loads are a live load of 100 kN/m^2 and an axle load of 1000 kN.

The design of the base-course for a concrete pavement should be conservative. The base-course for an asphalt pavement should be in accordance with a country's national standards. Only minor changes to the top of the base are required, namely tight elevation tolerances for the top of the base, and the use of materials of small grading to prevent the bedding sand escaping into the base.

The following guidelines and recommendations should generally be followed for the construction of the base-course:

(a) Base-courses should be designed and constructed as for asphalt pavements according to national standards or by the use of specific computer programs.
(b) Mechanically stabilised base/sub-base: the material in the base should be constructed with crushed rock grading approximately 0–30 mm (maximum 0–60 mm). The thickness of this layer should be 100–150 mm. When using a sub-base, the material should be approximately 0–60/200 mm.
(c) Other materials for the base: asphalt or cement stabilised.

Figure 12.25 Typical construction for subgrade for CBR ≈ 25% or more

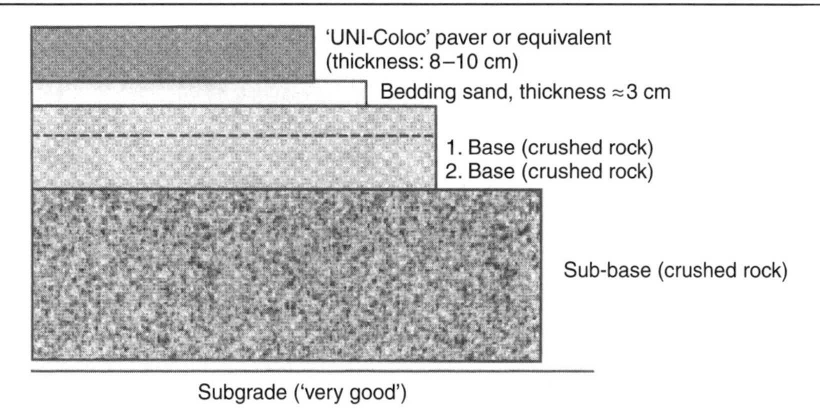

(d) Elevation tolerance: to comply with the thickness requirements for the thin bedding layer, the tolerance of the upper base should be approximately ±10 mm for maximum stability.
(e) Materials should meet the quality requirements in national standards.

Where the subgrade soil condition is very good and has a California bearing ratio (CBR) of approximately 25%, the construction and use of the materials should be as shown in Figure 12.25 and Table 12.3. The total base-course, as shown, will be approximately 45 cm.

Where the subgrade soil condition is moderate and has a CBR of approximately 10%, the construction and use of the materials should be as shown in Figure 12.26 and Table 12.4. The total base-course, as shown, will be approximately 80 cm.

Where the subgrade soil condition is very poor and has a CBR of approximately 5%, the construction and use of the materials should be as shown in Figure 12.27 and Table 12.5. The total base-course, as shown, will be approximately 100 cm.

Bedding layer

Between the concrete paver and the base-course, a layer of bedding sand is required. The bedding should be constructed in a thin layer to give maximum stability. The following guidelines and

Table 12.3 Construction layer and materials for CBR ≈ 25% or more

Construction layer and type of materials	Thickness: cm
Pavement: interlocking pavers	8–10
Pavement: rectangular pavers	10–12
Bedding layer (crushed rock: 0–8 mm)	3
1. Base (upper) (crushed rock: 0–30 mm)	5
2. Base (lower) (crushed rock: 0–60 mm)	10
Sub-base (crushed rock: 0–150 mm)	30
Other	–
Subgrade	Existing

Figure 12.26 Typical construction for subgrade for CBR ≈ 10%

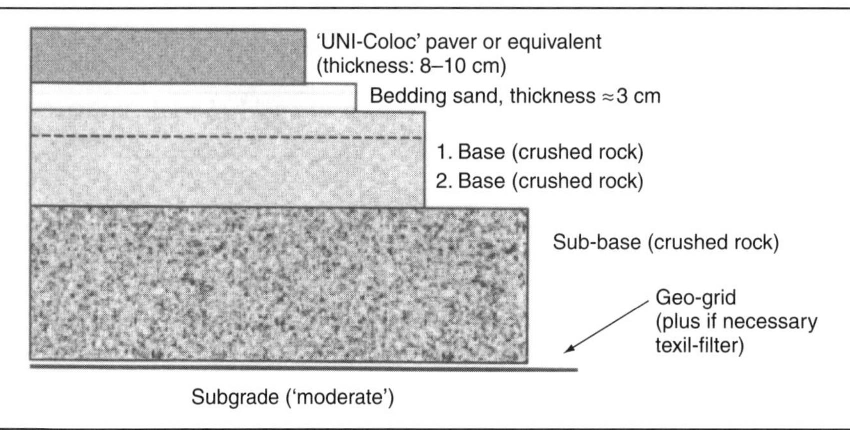

Table 12.4 Construction layer and materials for CBR ≈ 10%

Construction layer and type of materials	Thickness: cm
Pavement: interlocking pavers	8–10
Pavement: rectangular pavers	10–12
Bedding layer (crushed rock: 0–8 mm)	3
1. Base (upper) (crushed rock: 0–30 mm)	5
2. Base (lower) (crushed rock: 0–60 mm)	15
Sub-base (crushed rock: 0–150 mm)	60
Geo-grid	(e.g. 'Tensar SSLA30')
Subgrade	Existing

Figure 12.27 Typical construction for subgrade for CBR ≈ 5%

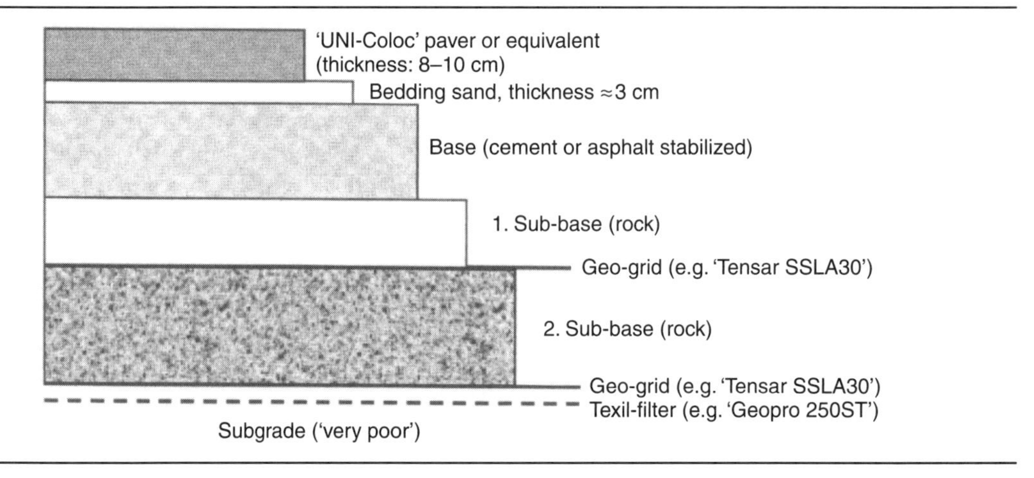

Table 12.5 Construction layer and materials for CBR ≈ 5%

Construction layer and type of materials	Thickness: cm
Pavement: interlocking pavers	8–10
Pavement: rectangular pavers	10–12
Bedding layer (crushed rock: 0–8 mm)	3
Base (cement or asphalt stabilised)	20
1. Sub-base (upper) (rock: 0–80 mm)	20
Geo-grid	(e.g. 'Tensar SSLA30')
2. Sub-base (lower) (rock: 0–150/200 mm)	60
Geo-grid	(e.g. 'Tensar SSLA30')
Texil-filter	(e.g. 'Geopro 250ST')
Subgrade	Existing

recommendations should generally be followed for the construction of the bedding layer:

(a) Material to be crushed rock grading 0–8 mm (maximum 0–11 mm).
(b) Compressed (mill) layer with an average thickness of approximately 30 mm. Local deviation for thickness: maximum ±10 mm (with reference to a theoretical thickness of 30 mm).
(c) Material to be moistened during finishing of the bedding layer.
(d) Geometric tolerance for the top bedding layer: same as for the top pavement.
(e) Suitable materials for good drainage.
(f) Materials should meet the quality requirements in national standards.

For all types of pavements, there could be a problem with settlement close to solid structures, such as foundations and concrete slabs, drains, etc., due to difficulties in compacting the base-course. The problem usually appears after a long period of use. To maintain a proper cross-fall, the pavement should be constructed with an increased elevation close to the structures. The increase in the thickness of the bedding layer could be approximately ±10 mm gradually over 1–2 m.

Types of block pavers

In heavily trafficked areas, exposed to heavy loads, it is important to choose a block paver of suitably robust design and geometry.

In the international literature, pavers are classified in the following three categories:

(a) category A comprises dentate pavers that key into other pavers on all vertical faces, such as UNI-Coloc (UNI-Anchorlock in the USA) or an equivalent
(b) category B comprises dentate pavers that key into other pavers on only two faces, such as SF-Paver or an equivalent
(c) category C comprises non-dentate pavers that do not key into other pavers, such as hexagonal and rectangular pavers or similar shapes.

Experience has shown that interlocking concrete pavers have a higher load capacity than asphalt, particularly on warm days, where supports for the container or the container corners can penetrate the asphalt (Figure 12.28). Based on test results, concrete paver layers with a thickness of 80 mm

Figure 12.28 Damage to an asphalt paver

have a relatively high modulus of elasticity of $E \approx 6000$–7000 MPa (category A), compared with an 80 mm thick asphalt layer of $E \approx 3000$–4000 MPa.

An alternative to the storage of containers on asphalt in some container terminals is the use of thick steel plates under the corners of the containers as support (Figure 12.29).

Figure 12.29 Support for containers

Generally, the thickness of the concrete block pavers used in harbours worldwide is between 8 and 12 cm. Some of the largest harbours, for example in the Netherlands, are constructed with ordinary rectangular block pavers up to 12 cm thick. Harbours in Norway are mainly constructed with an 8 cm thickness of interlocking concrete pavers. Therefore, it is possible to reduce the thickness of concrete pavers in category A, due to better performance and less movement and rotation than for pavers in categories B and C. The following guidelines and recommendations should generally be followed for the selection of pavers:

(a) Quality. The quality of concrete pavers should meet the minimum requirements in national standards.
(b) Type of paver. It is recommended, for example, that container terminals use category A. If rectangular pavers are chosen, it is recommended that the thickness of the paver is increased and pavers are laid in a herringbone pattern to increase the stability of the pavement.
(c) Thickness. The thickness of the paver should be 80 mm (100 mm for extreme loads) for category A and 100–120 mm for category B or C.

Laying pattern

There are many different laying patterns for various kinds of concrete block pavers for heavy duty areas such as container terminal areas. Some types of paver may require a specific laying pattern. Rectangular pavers in a stretcher pattern in category C will have an increased pavement performance if laid in a herringbone pattern instead, but will never achieve the same level of performance as the best designed pavers in category A.

The best result from an area pavement is achieved by block pavers of category A with an interlocking pattern. Comprehensive neutral tests and long experience prove that these kinds of block pavers have the best characteristics. They are also designed for effective machine laying.

The pavement's performance will be improved if the direction of the pattern system is twisted 45° to the main traffic direction. To achieve the most benefit from interlocking, the joints between block pavers need to be less than 5 mm. The following guidelines and recommendations should generally be followed for the laying of pavers:

(a) use pavers in an interlocking system as for category A
(b) if the pavers are not designed for a specific pattern, use a herringbone pattern if possible
(c) the pattern should be twisted 45° to the traffic direction
(d) the joint width should be an average of 2–3 mm and should be less than 5 mm.

Performance of a pavement

Block pavers for areas larger than 2000–3000 m^2 are normally laid by machine (Figure 12.30). One machine can carry out up to 1000 m^2 of pavement/10 h. The guidelines and recommendations listed below should generally be followed:

(a) use interlocking pavers in category A with high concrete quality
(b) use a well-recognised contractor with relevant references
(c) before the start of paving, check the level and evenness of the bedding layer
(d) the paving pattern should follow straight lines with an average jointing width of 2–3 mm
(e) use dry jointing sand and a vibrator to fill the joints completely
(f) survey and document the completed pavement level and elevations.

Figure 12.30 Concrete pavers laid by machine

12.15. Crane rails

Figure 12.31 shows the general principles of the installation of crane rails for ship-to-shore or gantry cranes at the berth surface.

In the past, for protection of the electrical cable to the crane, which must always be waterside of the crane rail, steel plates were often used to cover the cable tray (Figure 12.32). This was a heavy

Figure 12.31 General principle of crane rail installation

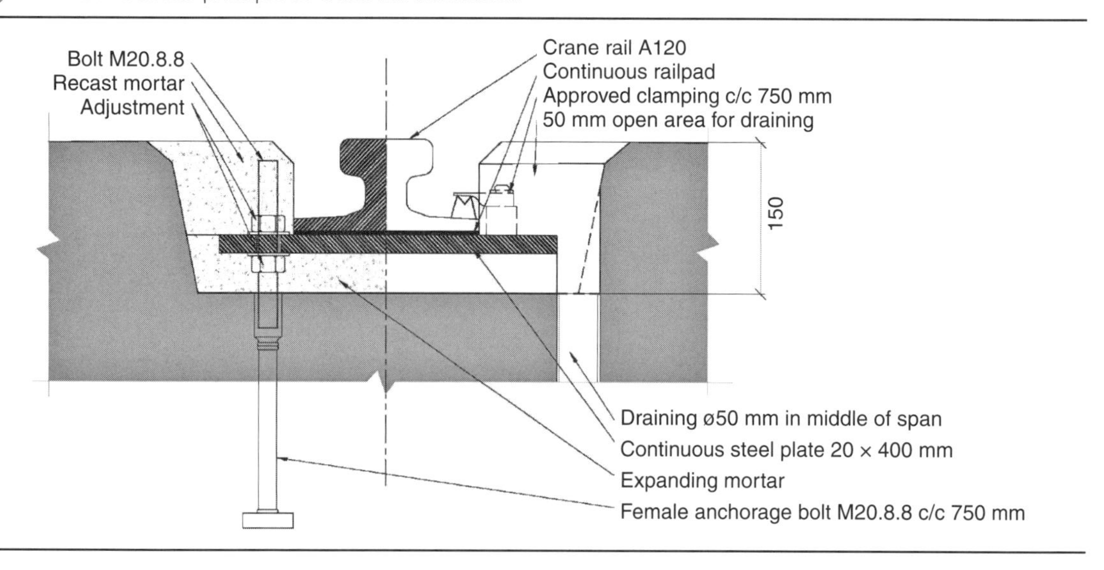

Figure 12.32 The old system for protection of the electrical cable with steel plates

system that needed much maintenance; for example, all hinges needed to be greased. The system itself required strong guides at the cranes to open and close the plates. Quite often, the plates remained open and were damaged by the crane itself when it returned, or by other vehicles driving on them and breaking them.

A newer system, the Panzerbelt, is a rubber belt made with different layers to keep it strong and to ensure safe operation in all weather conditions. The system is shown in Figure 12.33, and its principal and parts illustrated in Figure 12.34.

The channel profile is easy to install in the concrete during the construction of the berth. It is ready for the installation of the belt, and the installation time is reduced. The belt and channel are made in different sizes, depending on the size of cables to be placed in the channel.

The following benefits accrue by using this system:

(a) safe protection of cables
(b) easy installation of the belt itself and the belt-lifting device on the crane
(c) limited noise when the crane opens and closes the belt
(d) cranes can move fast
(e) no problem for trucks, cars and fork-lifts to pass over it
(f) flexible in cold weather
(g) no maintenance is required.

Figure 12.33 The Panzerbelt system. (Photograph courtesy of Cavotec MSL)

Figure 12.34 The principal and parts of the Panzerbelt system. (Courtesy of Cavotec MSL)

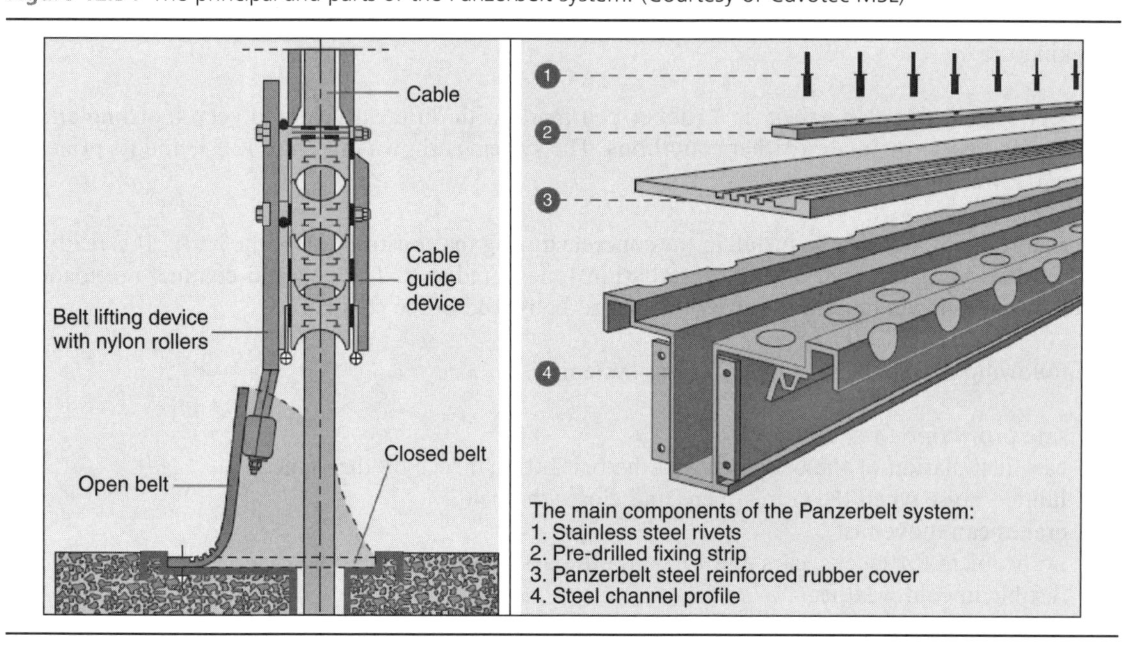

Figure 12.35 A crane stopper

The main limitation in use is that if a fork-lift with small wheels places the wheels on the belt and turns through 90° with heavy weights, this can damage the belt.

A stopper for a gantry crane is shown in Figure 12.35.

REFERENCES AND FURTHER READING

Bruun P (1989) *Port Engineering*, vols 1 and 2. Gulf Publishing, Houston, TX.

BSI (British Standards Institution) (1988) BS 6349-2: 1988. Maritime structures. Design of quay walls, jetties and dolphins. BSI, London.

BSI (2000) BS 6349-1: 2000. Maritime structures. Code of practice for general criteria. BSI, London.

BSI (2003) BS EN 1338: 2003. Concrete paving blocks. Requirements and test methods. BSI, London.

CUR (Centre for Civil Engineering Research and Codes) (2005) *Handbook of Quay Walls*. CUR, Gouda.

EAU (2004) *Recommendations of the Committee for Waterfront Structures, Harbours and Waterways*, 8th English edition. Ernst, Berlin.

Ministerio de Obras Públicas y Transportes (1994) ROM 4.1-94. Guidelines for the design and construction of port pavements. Puertos del Estado, Madrid.

Shackel B (1990) *Design and Construction of Interlocking Concrete Block Pavements*. Elsevier, London.

Yager T (1992) *Friction Evaluation of Concrete Paver Blocks for Airport Pavement Applications*. SAE, Warrendale, PA. Technical Paper 922013.

Port Designer's Handbook
ISBN 978-0-7277-6004-3

http://dx.doi.org/10.1680/pdh.60043.321

Chapter 13
Container terminals

13.1. Site location

The planning and evaluation of a container terminal can be a very complex task. The designer must make the most of available local resources to meet the required level of productivity, while trying to reach a balance between the needs of the port authorities, port operators, stevedoring companies and container shipping lines. The capacity of a port is commonly expressed as the amount of cargo throughput, and its efficiency is usually expressed as its ability to handle cargo or containers at a minimum cost.

A survey must be carried out to identify existing and potential sites that will be suitable to meet the activity requirements of the port. A port plan must also indicate areas earmarked for future port expansion, and establish guidelines for such development. The goal of all port developments, and especially for a container port, should be the possibility of working 24 hours a day, 7 days a week and 365 days a year.

Therefore, the following requirements should be satisfied:

(a) a sufficient approach channel to the port area
(b) a sufficient harbour area, turning basin and water depth
(c) a sufficient berth construction and a large terminal area
(d) the possibility for expansion, including new berths and larger terminal areas.

Port improvements frequently enable the shipping using the port to be turned around more quickly, either through reduced waiting time or through more efficient cargo-handling operations, which result in a reduced berth service time. The quicker handling of cargo, whether transfers from ship to berth, from berth to storage, or to and from land transport systems, usually results from improved mechanisation of the berth facilities.

From the moment a container arrives at the port, either at the port gate or at the berth side, it should be logged into a computerised system that can track the container through each stage of its transit through the container terminal. In this way customers can know the status of their container at any time.

Sometimes quite significant improvements can be made by reorganisation and improved management systems, for instance by establishing one terminal operator. Therefore, improvements to the port facilities and organisation, together with improvements in the port layout, will, in most cases, result in more efficient handling and storage of cargo.

13.2. Existing areas

The capacity of existing berth facilities and port areas has to be assessed. New loading and unloading methods usually have the result that the bottleneck in port efficiency is no longer a lack of berth capacity but a lack of areas and installations ashore. In many older port terminals the area close behind the berth front contains too many sheds and buildings, so that there are few open areas for the handling and storage of containers and large units of cargo. Therefore it should be evaluated whether a relocation of facilities within the existing port area could increase port output to that of a modern container port. Figure 13.1 shows a possible solution, from an area point of view, for an old general cargo jetty, which has been converted into a modern container terminal.

When assessing the effective output of existing port areas, the following points must be considered:

(a) technical level
(b) operational level
(c) storage capacities
(d) ownerships
(e) possibility of relocating existing facilities
(f) environmental considerations.

Improving the available area of land behind the berth and the addition of new terminal equipment can increase the port capacity and efficiency. This means that the capacity of a modern port is more dependent on efficient management, the amount of available space on land behind the berth (than on the length of the berth front itself) and on new terminal equipment (Figure 13.2). Figure 13.3 shows the Altenwerter container terminal in Hamburg, which is fully automatic.

13.3. Potential areas

For many ports it can be difficult to find suitable areas for expansion adjacent to the existing port area. Very often the near surroundings of existing ports are so crowded and restricted, owing to town development, that direct expansion of the existing port facilities is more or less impossible. Therefore,

Figure 13.1 The conversion of an existing port area into a container port

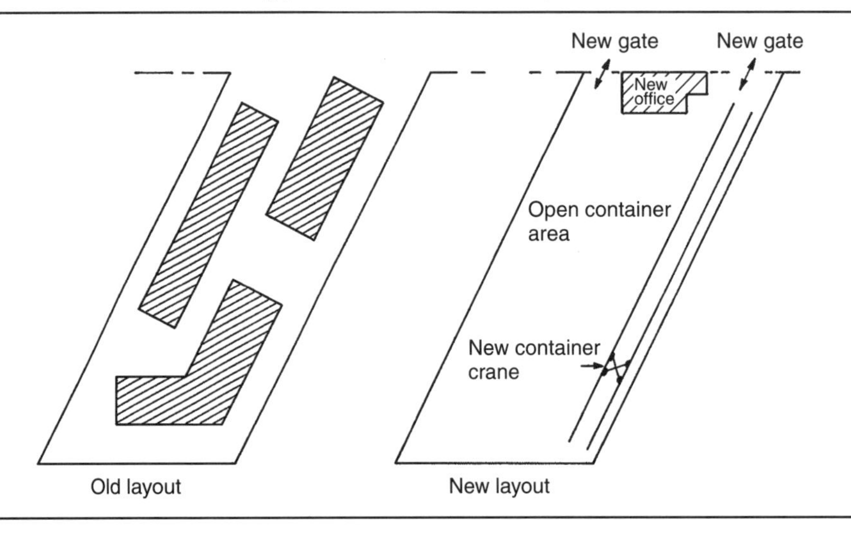

Figure 13.2 A modern container port

Figure 13.3 The Altenwerter container terminal, Hamburg. (Photo courtesy of CTA)

one must survey the stretches of the coast where port development may be possible and include them in the overall port plan. This survey must take into account the following factors:

(a) availability of sufficient area
(b) possibility of future extension
(c) availability of hinterland connections
(d) accessibility and distance from sea
(e) nature of the subsoil and the risk of settlements or geotechnical problems
(f) shelter from waves/wind/current
(g) earthquake danger
(h) environmental assessment.

In particular, the transhipment operations of containers and roll-on/roll-off cargo is very sensitive to ship movements due to, for example, wind and waves, and this can lead to considerable downtime. The location of a port or terminal should, therefore, be chosen with the utmost care.

13.4. Container ships

Since Sea-Land Service introduced the metal shipping container in 1956, not just the size but also the design of a containerships have totally changed. Prior to that time, the Panamax ship size (the limiting size for ships travelling through the Panama Canal) was widespread. Today, the maximum size of ship that can pass through the locks of the Panama Canal is: beam 32.30 m, length 294.10 m and draught 12.00 m. Therefore, most of the harbours on the east coast of the USA have a maximum draught of approximately 12.50 m.

In 2015, the new locks on the Panama Canal will be opened, and these will be able to handle ships having the dimensions: beam 50 m, length 385 m and draught 15.2 m. This will have a great influence on the transport corridors between east and west, and the routes between Asia and America will have alternative routes.

Figure 13.4 shows the development of the containership from the first one, Ideal-X in 1956, which was a converted World War II T2 tanker, up to now. The following approximate grouping of container ships is used.

Panamax-size container ships with a width of up to approximately 32 m:

(a) first generation, with a capacity up to about 1000 twenty-foot equivalents (TEU) (see Figure 13.4)
(b) second generation, with a capacity up to about 1600 TEU
(c) third generation, with a capacity up to about 3000 TEU.

Post-Panamax-size container ships:

(a) fourth generation, with a capacity up to about 4250 TEU
(b) fifth generation, with a capacity up to about 5000 TEU
(c) sixth generation, with a capacity up to about 6000 TEU (Figure 13.5).

Post-Panamax-Plus-size (PPP) container ships:

(a) seventh generation, with a capacity over 7000 TEU.

Figure 13.4 The development of container ships

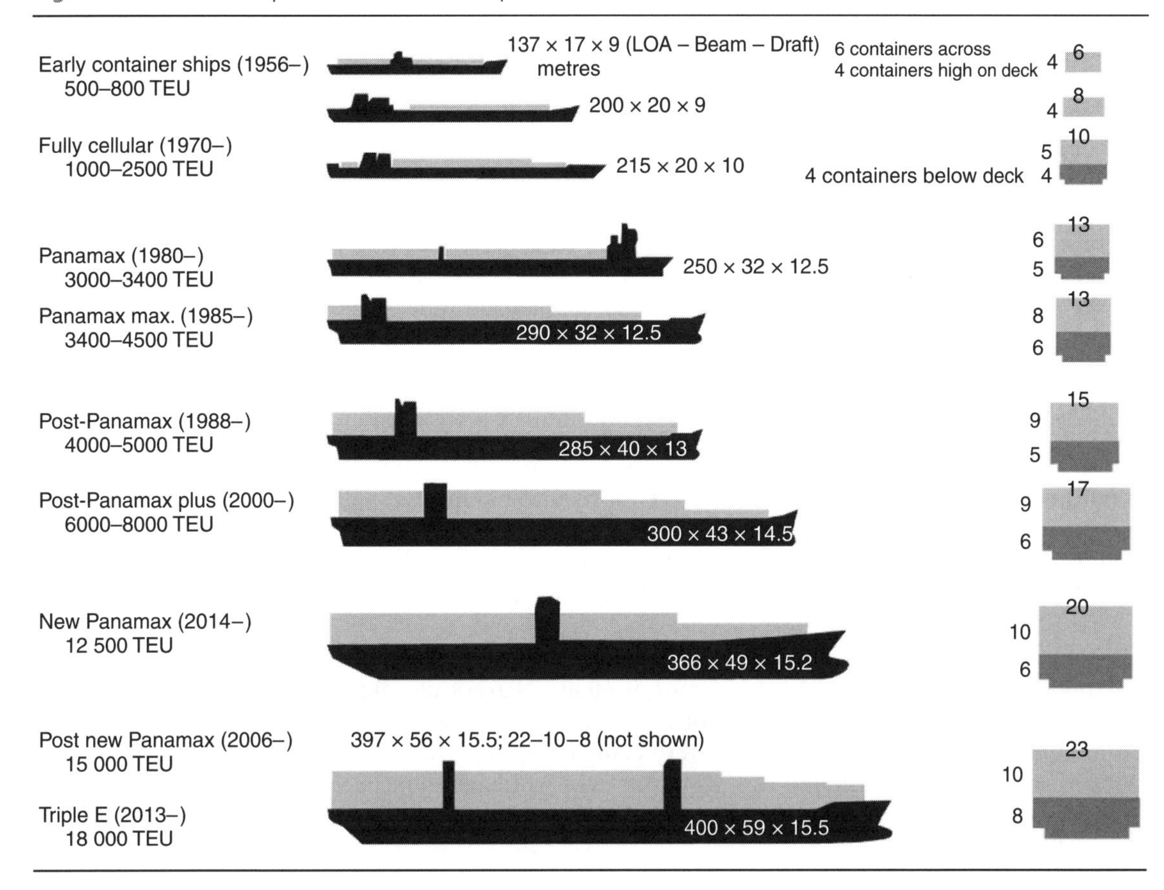

Figure 13.5 A post-Panamax container ship

Ultra-large container ships (ULCSs):

(a) These container ships have an overall capacity of 12 500 TEU or more, a length of 360 m or more, a ship beam of 50 m or more and a maximum design draft of 14.5 m or more. The design speed is in the range 23–25 knots, but, due to high bunker costs, most of these ships operate at a so-called Eco speed of about 14 knots. For this reason the newest ULCS have a lower design speed of between 18 and 20 knots.

There are now container ships in service that are capable of carrying approximately 18 000 TEU, and there are ULCSs in the design stage that will have a carrying capacity of approximately 22 000 TEU. These ships have a cell width of 24 containers on deck and 22 containers below, with a length of approximately 410 m, a beam of approximately 60 m and a maximum design draft of 16.5 m.

The aim of a terminal is to accommodate the largest container ship (the design ship for that specified terminal) without waiting to berth. It is very important to determine the design ship, but it is also important to the have an overview of the fleet of containerships that will dock at the terminal today and possibly in the future. It is more common nowadays to have different berths for different sized container ships, and this is reflected in different sizes of cranes, depths along the berth, berth construction etc. A cost–benefit analysis should be carried out the find the optimum solution.

The largest container ship in service today needs berth facilities that have a quay length of almost 450 m and a draft of 16.5 m. With a beam of 60 m (24 rows of containers), large cranes are also required. However, there are very few of these ships in service, although more will be built in the future and they dock at few ports. It is very demanding to facilitate such large ships, but container ships in general are increasing in size, so all facilities will have to meet these increased requirements in the future.

Feeder container ships collect containers from smaller ports and carry them to the main container port, and vice versa. Feeder ships have increased in size, especially the last decade, and there are now few such ships under 500 TEU, the largest having a capacity of more than 3000 TEU. Even container barges have a capacity up to 300–400 TEU. Some feeder container ships have cranes for docking at small ports and terminals. The feeder ships will become increasingly important when the large container ships became even larger. The largest container ships will dock at very few ports, and feeder ships will have to bring more containers to and from the hubs. Therefore the size of the feeder ships will also increase.

13.5. Terminal areas

In the evaluation of new potential port areas, it is useful to divide the new potential land area behind a new berth line into an apron, a primary and a secondary yard, and a storage area. The length of the land area or berth will depend on the type of ship and/or cargo expected. For medium container ships (second-generation container ships) and multi-purpose ships, a length of about 200 m will be sufficient for one berth.

The total terminal area is usually divided into the following:

(a) the apron, or the area just behind the berth front
(b) the primary yard area or container storage area

(c) the secondary yard area, which includes the entrance facility, parking, office buildings, customs facilities, container freight station with an area for stuffing and stripping, empty container storage, container maintenance and repair area, etc.

13.5.1 Apron

The width of the apron will vary between about 15 and 50 m, depending on the loading and unloading equipment, trucks, cranes, etc. The dimensions of the various sections of the apron for a berth with a crane will be as follows:

(a) The distance from the berth line to the waterside crane rail should not be less than 2.5 m and should contain the crane power trench, bollards, gangway and other ship utilities.
(b) The distance between the crane rails varies from about 10 m (general cargo crane) to a maximum of approximately 35 m (container crane).
(c) The traffic area or road behind the landside crane rail and the boundary between the apron and the primary yard can vary in width from 5 to 15 m or more, depending on both the operating and the container handling system.

13.5.2 Yard area

The yard, or the area behind the apron, may be divided into a primary yard and a secondary yard, with an entrance, parking, office building, custom facilities, etc. The primary yard, or the storage area, is the area immediately adjacent to the apron, and is used primarily for storing inbound and outbound cargo. The secondary yard is the area for storing empty containers, equipment, etc.

The yard area for a modern multi-purpose and container terminal, like the port shown in Figure 13.6, should have a depth of at least 300 m behind the apron. Preferably, the depth should be up to about 400 m for a multi-purpose terminal and up to about 700 m for a modern container terminal.

Therefore, the land requirements are related to the storage density and the time for which the cargo stays in the port. Where a substantial proportion of the cargo is handled by roll-on/roll-off methods, the back-up areas can be much larger than for cargo handled by load-on/load-off methods. As a rule of thumb, the area required for a multi-purpose terminal will vary between about 5 and 15 ha/berth, and that for a container terminal between about 10 and 100 ha/berth, depending on the generation of container ships that the port will serve. These figures include areas for offices, sheds, workshops, roads, etc.

Generally, the total yard area can be divided into:

$$A_T = A_{PY} + A_{CFS} + A_{EC} + A_{ROP}$$

where

A_{PY} = the primary yard area or container stacking area. The area is approximately 50–75% of the total area
A_{CFS} = the container freight station (CFS) with an area for stuffing and stripping, etc. The area is approximately 15–30% of the total area
A_{EC} = the area for empty containers, container maintenance and repair area, etc. The area is approximately 10–20% of the total area. Generally, in modern container terminals, empty containers stored and containers repaired outside the terminal area, if possible

Figure 13.6 The layout of a modern container terminal

A_{ROP} = the area for the entrance facility, office buildings, customs facilities, parking, etc. The area is approximately 5–15% of the total area.

When evaluating the total yard area, the area for the entrance, custom facilities, etc. will be affected by the proposal by the IMO to improve the security at ports.

When there is little knowledge about the cargo to be expected in the future, the storage area should have an additional area of 25–40% as reserve capacity. To provide a reserve capacity of less than 25% would be unwise under any circumstances.

In a roll-on/roll-off container terminal, roads may occupy up to about 50–60% of the total area, and in a load-on/load-off container terminal they may occupy up to about 40–50% of the total area. Records from major European container ports show that a minimum of 30% of the total port area is used for roads, port services, parking, rail rods, etc.

Figure 13.7 shows a general layout of a total yard area of a terminal area that contains most of the container terminal activities. The key to the numbers in Figure 13.7 is as follows:

1 = berth
2 = apron
3 = stacking and storage area
4, 7 = internal road and terminal transfer system

Figure 13.7 A general prototype of a container terminal yard area

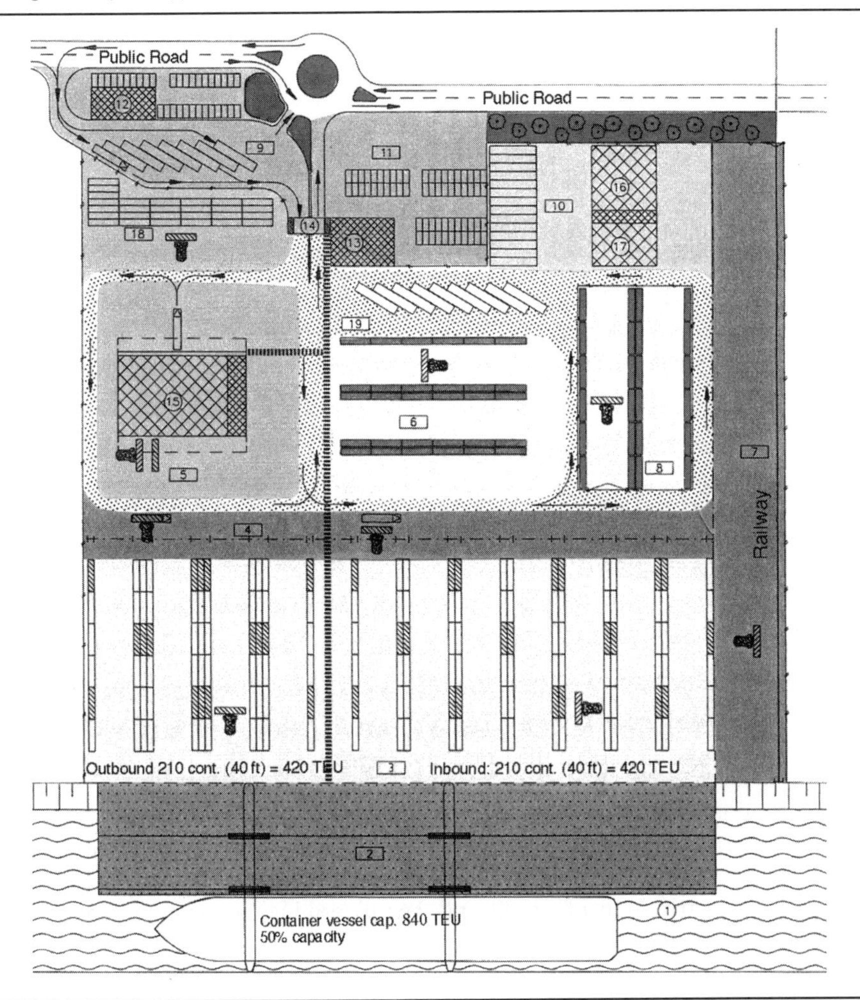

5, 15 = stuffing and stripping area and shed
6 = reefers area
8 = area for dangerous cargo, etc.
9 = terminal entrance
10, 16, 17 = service, repair, workshop equipment and workshop container area
11 = landside terminal
12 = load identification area
13 = office, canteen and convenience
14 = in and out checking
18 = depot for empty containers
19 = parking area for vehicles.

In the idealised scheme shown in Figures 13.7 and 13.8, the quay wall is shown as a single straight berth line, so that the berths can accommodate varying lengths of ships while maximising quay usage. The

Figure 13.8 The components of a container terminal

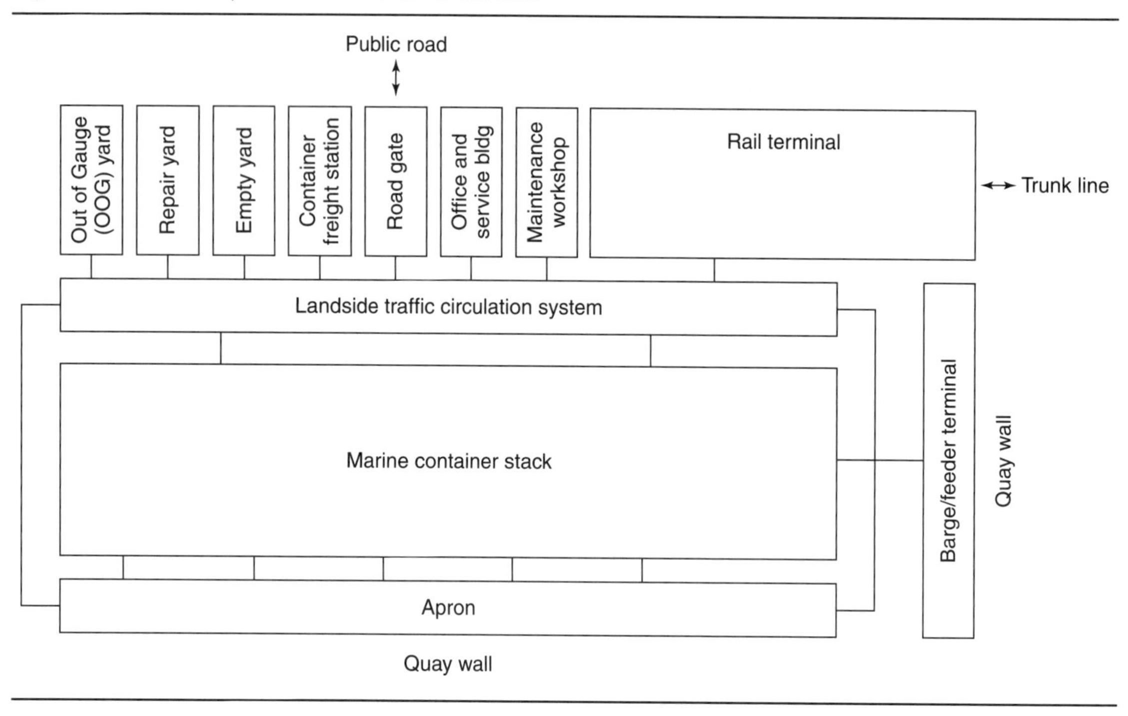

containers are stacked in a rectangular area directly behind the apron, allowing the stacking density, and hence utilisation of the stacking equipment, to be optimised. Other facilities are located at the back of the terminal, away from the quay and stack operations, where they do not impact on the productivity of the terminal. Similarly, landside traffic and the rail terminal are separate from the core terminal operations at the quay and the stack.

In practice, these idealised concepts are subject to compromise, as the available space for a new terminal is rarely of the ideal size or shape.

13.6. Ship-to-shore crane

The first specially designed container crane was completed in 1959, and development of the ship-to-shore (STS) crane has been a fantastic evolution in the design of container cranes.

A third-generation container ship can stow container boxes up to 13 rows across its width, while, for example, a sixth-generation post-Panamax ship can stow 16, a PPP ship can stow 18 and the largest ULCSs can stow 24. This increase in ship size demands increased crane capacity because, as ships become larger, their time becomes more precious, and time spent at berth is unproductive. Nowadays the larger ships use up to five or six container cranes simultaneously.

Traditional STS crane systems can be inadequate to properly serve existing, and especially future generations of PPP container ships and ULCS, efficiently. To increase the ability to serve the ships more quickly, one can either:

(a) generally increase crane efficiency
(b) increase the crane rate by lifting up to 4 TEU loaded containers simultaneously by using a tandem lift, or 2 TEU containers by using a twin lift
(c) introduce a dock system whereby it is possible to load and unload the ship from both sides.

The large PPP container ships and ULCSs are very expensive to run, so any delay in loading or unloading can be very costly and reduce the economic benefits that otherwise result from running a larger container ship. For example, if a 6500 TEU ship were to load and unload 80% of its capacity with three cranes of 30 lifts/h for 21 h/day, the total ship call would last more than 3 days.

Owing to the size of the largest container ships and their enormous costs per hour, they demand an optimum turnaround time of approximately 24 h. This means that the time for unloading and loading will be approximately 22 h and the time for sailing and manoeuvring in the port will be 2 h. Therefore, to retain or improve the productivity per berth per hour, for the larger container ships the terminal operator must use up to four or five STS gantry cranes.

In a traditional container berth system, the average berth production rate required for an 8000 TEU ship would, with a load factor of 0.85 and a TEU factor of 1.5, be approximately 250 moves/h, or five STS gantry cranes each with an effective capacity of approximately 50 moves/h.

The companies JWD and Liftech, in Holland, have evaluated an very interesting dock system for loading and unloading very large container ships from both sides (Figures 13.9 and 13.10). With this system the number of cranes working against the container ship can be doubled and the ship's stay in the port can therefore be reduced by approximately half. This dock system can use the existing berth crane technology and operating systems, but the dock should be sufficiently wide and deep to accommodate the larger container ships expected in the future. The size of a dock with a two-sided handling system will be approximately 70 m wide and a 380–400 m long.

Figure 13.9 Cross-section of the dock system

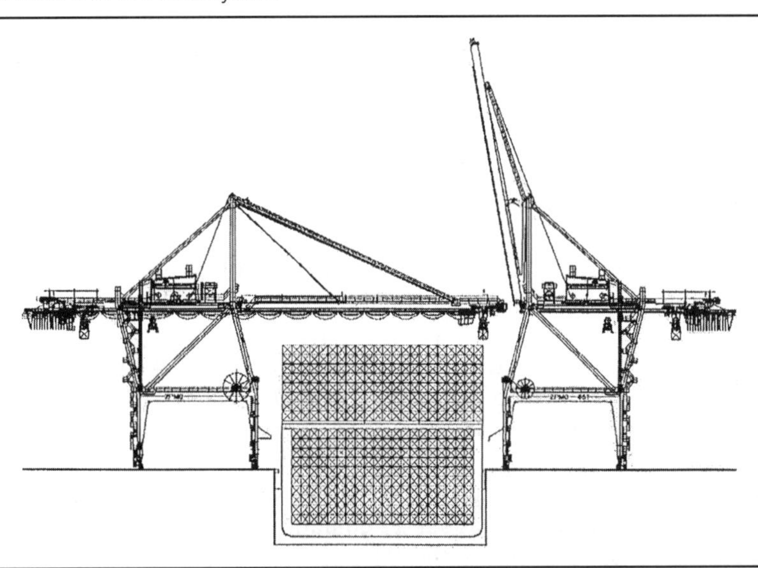

Figure 13.10 The layout of the dock container system

The design and layout of the container terminal surrounding the dock could be more difficult, as the terminal yard would need to wrap around the dock. Economically, the dock system would require the construction of a berth on both sides of the ship and at the end of the dock. The disadvantages dock with a two-sided loading and unloading system can be summarised as follows:

(a) The berthing and unberthing operations to and from the dock could take longer.
(b) Twice the berth length and twice the width of dock is needed per container ship.
(c) The orientation of the dock, due to the stacking area, could require more space or terminal area than one-sided berthing.
(d) The berth may not be suitable for all ship sizes.
(e) The container cranes cannot be transferred from one berth to another.

The conclusion is that the traditional berth with one-sided handling could be preferred, but due to the increase in container ship size one has to evaluate the advantages and disadvantages very carefully.

Some of the world's major container ports achieve an average of 3.2–3.4 TEUs per crane lift by using a tandem lift, meaning that 60–70% of containers are 40 ft containers, but the use of such lifts needs to be evaluated individually for each port.

The **crane capacity per hour** for handling containers can vary between 10 and, at the extreme, 70 containers, with an average capacity of about 25 containers/h per crane. The following figures on crane capacity can be used for guidance:

(a) pail-mounted harbour cranes, about 15 containers/h from ship to shore
(b) mobile container cranes, about 15–25/h
(c) STS gantry cranes about 30–40 containers/h; STS gantry cranes with a secondary trolley system 40–70 containers/h.

When undertaking a feasibility study a loss of output of about 10% of the basic output rate for starting and finishing the total crane operation should be take into consideration. The working crane time is about 80% of the time the ship is at the berth.

Modern STS gantry cranes achieve, with single container duty cycles taking an average of 90–120 s, about 30–40 container lifts/h, assuming a properly fed delivery service to and from the container stacking yard. The productivity of truck-based terminals is generally limited to about 28–35 lifts/h, while straddle-carrier-based terminals can support the highest STS crane productivity, with good management systems, of up to 40 lifts/h.

In the largest ports, up to six cranes can work simultaneously for a short period on the same side of the largest PPP container ships, but in practical terms it should be assumed that only four cranes can work simultaneously. In the dock system where cranes work simultaneously from both sides, the number of cranes working at one time can be high because the standard crane booms are narrower than their supporting frames, and the cranes on the opposite sides of the ship could theoretically be nested with opposite booms on alternate hatches.

An interesting survey into the average handling rates actually achieved around the world showed that crane performances are below the manufacturers' stated capabilities for moves per hour for modern cranes. The average handling speeds of container cranes based on a survey of 671 cranes worldwide were:

(a) up to 20 moves/h, 12% of sample
(b) 21–25 moves/h, 39% of sample
(c) 26–30 moves/h, 33% of sample
(d) 31–35 moves/h, 14% of sample
(e) over 35 moves/h, 1% of sample.

The most commonly used specialised STS container gantry cranes (Figures 13.11–13.13) feature the following:

(a) The minimum height between the lower part of the spreader and the berth level should be 30 m. Generally, the gantry crane must be able to stack 5 high on the deck of a large ship and must be able to access every individual container on the ship.
(b) For larger container vessels, the distance from the fender front to the front crane rail should be around 7.5 m, due to the shape of the bow and the berthing angle of the larger container ships during the berthing operation.
(c) The minimum outreach, measured from the fender front on the face of the berth, should be 35 m, with a maximum outreach of about 65 m and a minimum back reach of at least 15 m.
(d) The outreach must reach 13 containers wide for old Panamax ships, 20 containers wide for a new Panamax ships and up to 24 containers wide for ULCSs.
(e) A rail gauge of between 16 and 35 m. A rail gauge of about 35 m will allow six truck lanes between the crane legs. The distance between the rail gauges is usually not a crane stability

Figure 13.11 A general cross-section of a container berth

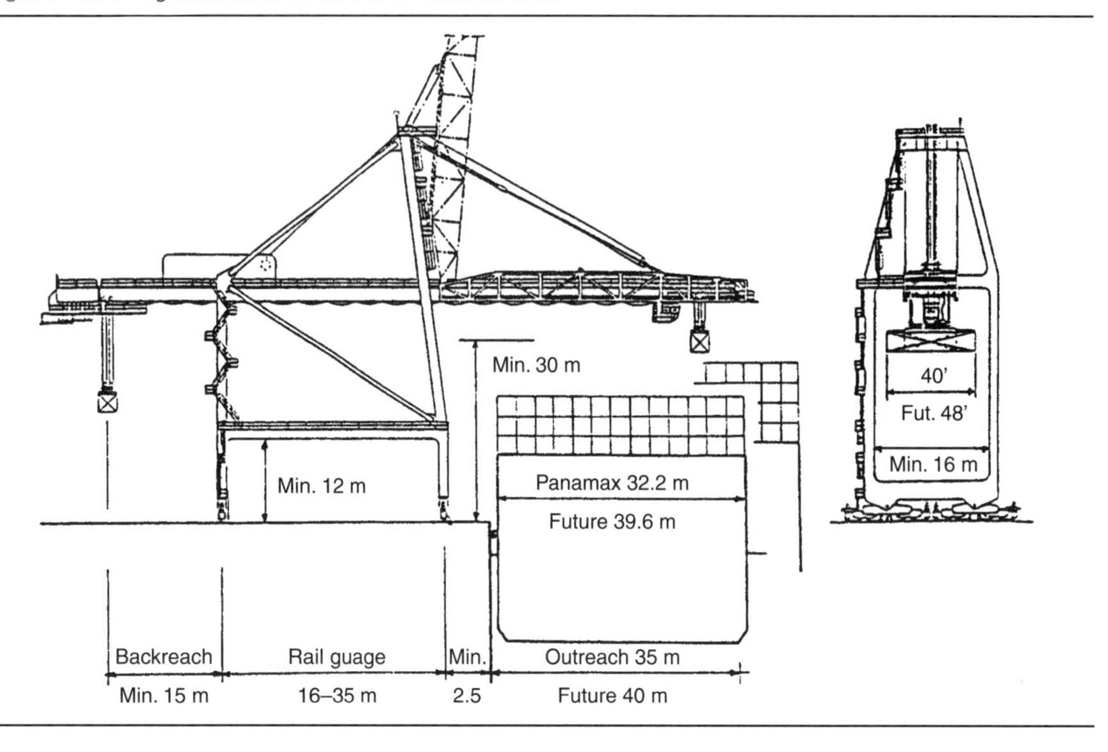

problem, but is determined more by the operating system between the apron area and the stacking area.

(f) A clearance of at least 16 m between the legs in order to leave room for containers and cargo hatches.

(g) The maximum width of the gantry crane (buffer to buffer) should not be more than 27.5 m.

(h) For a PPP gantry crane, the weight of the crane, the maximum container load, wind and dynamic effects and with four sets of a standard support of eight wheels, the weight on each wheel would be approximately 65 t, or 70 t/lin m crane rail.

Figure 13.12 A typical apron arrangement

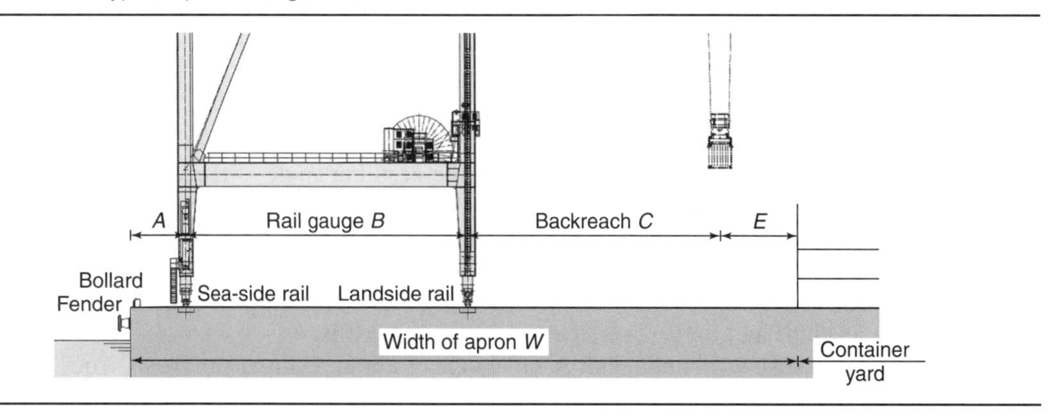

Figure 13.13 A sixth-generation container ship at berth

(i) At least 400 kN lifting capacity under the spreader, and for twin-lift or tandem-lift spreaders up to 660 kN lifting capacity. A tandem-lift spreader is shown in Figure 13.14 and Figure 13.15 shows a detailed diagram of the spreader and the twist lock.
(j) A lifting velocity of about 3 m/s with an empty spreader, adjustable to 1.50 m/s with a rated load.
(k) Programmable operation control and failure detection system.
(l) An alarm to indicate excessive wind speed and emergency shut-off.

Figure 13.14 A tandem lift in operation. (Courtesy of Professor Brinkmann, Germany)

Figure 13.15 Detail of a spreader and the twist lock. (Courtesy of Professor Brinkmann, Germany)

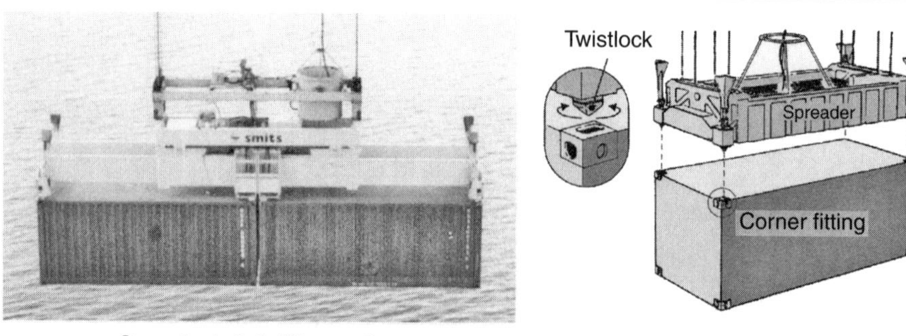

Spreader in twin lift operation

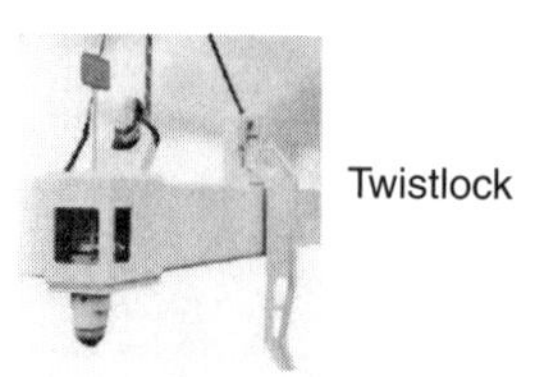

If the average crane capacity is increased to about 40–50 containers/h or more, the total terminal capacity (i.e. the area needed, storage and delivery capacity, etc.) must be increased tremendously. For this reason, the main benefits of a port improvement or development could be savings in ship waiting time and service time. Large costly ships require efficient ports that minimise the ship turnaround time, because improvements that reduce the waiting time to call at the port and the time spent at berth, etc., can save the ship owner large sums in operating expenses. Such savings will be reflected in the freight rates.

13.7. Container handling systems

The most commonly used container handling systems for stacking containers in the container stacking areas are:

(a) a fork-lift truck and reach-stacker system
(b) a straddle-carrier system
(c) a rubber-tyre gantry (RTG) and/or rail-mounted gantry (RMG) system
(d) a mixture of the above systems.

In general, all terminals should have a buffer storage area in front of the storage or stacking area. It should also be taken into account that all terminal equipment will have a reduction in capacity due to service and repair of approximately 5–30%.

13.7.1 Stack height

The stack height will affect both the total storage capacity of the stack and the accessibility of the individual boxes within the stack. With the need for storage and the limited space available, the tendency will be to adopt a high stack in order to maximise the storage capacity.

However, increasing the height of the stack also reduces the accessibility of individual containers within the stack as individual containers become buried deeper. This will increase the amount of

Figure 13.16 Stacking by reach stackers and by using terminal tractors between the STS gantry crane and the stacking area

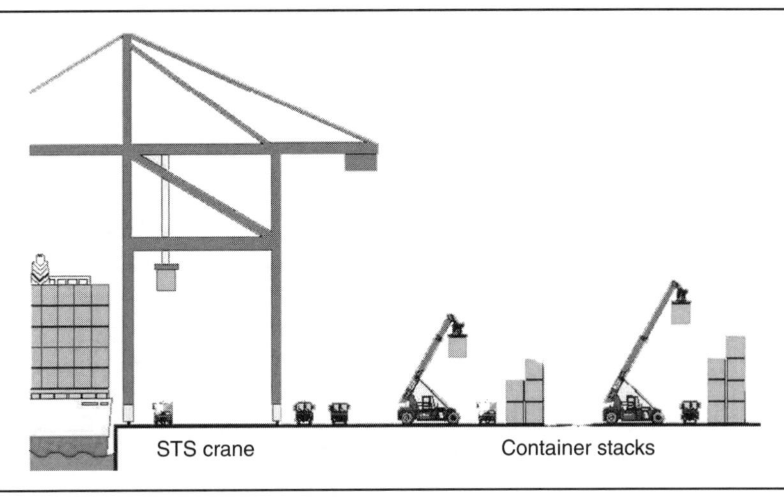

'digging' to retrieve containers and hence will lead to an overall reduction in terminal efficiency or an increase in equipment requirements. Therefore, stack height is limited by the need to allow enough space to place other containers during digging to locate an individual container. A compromise needs to be evaluated.

Digging for export containers can be reduced by careful planning of container placement within the stack, in line with shipping schedules. However, it is difficult to reduce the digging for the transfer of import containers, as the collection lorries are likely to arrive at random, making planning impossible. An optimum balance between these factors must therefore be found for the stack height.

13.7.2 Fork-lift truck and reach-stacker system

The STS gantry crane places the containers on a terminal tractor system and moves the container to the stacking area, where the container is stacked by a fork-lift or a reach-stacker system (Figures 13.16–13.19).

Heavy fork-lift trucks with top loaders have traditionally been used for container handling. Nowadays, more operators use a reach-stacker system, because of the higher productivity and higher stacking density that can be achieved. A system comprising fork-lift trucks and reach stackers can be the most economical, and is recommended for small terminals handling up to approximately 60 000–80 000 TEU/year and where the size of the terminal area is not restricted, while a reach-stacker system can be used economically for container handling at terminals with capacities up to approximately 200 000–300 000 TEU/year. A reach-stacker system can stack containers 4 deep and up to 6 containers high, but normally the stacking is 2 deep and 3–4 high to avoid too much reshuffling of the stack.

In general the following apply:

(a) Three to five terminal tractors and two reach stackers per STS gantry crane. The number of terminal tractors is dependent on the distance between the berth and the stacking area.

Figure 13.17 A terminal tractor

(b) Low-storage capacity with about 500 TEU/h, stacking the containers approximately 4 high.
(c) Medium STS crane productivity and no buffer zone under the STS crane.
(d) High labour, but low capital and operating costs.
(e) Low control; trucks allowed into stacking area.

Figure 13.18 A reach stacker

Figure 13.19 An empty-container fork-lift truck for block stacking empty containers

13.7.3 Straddle-carrier system

The STS gantry crane places the containers on the apron, where the straddle carrier moves the container to the stacking area and stacks it (Figures 13.20 and 13.21). The straddle-carrier system is an independent system, and does all the different handling operations between the STS crane and the stacking of the containers.

Figure 13.20 Stacking using straddle carriers and straddle carriers between the STS gantry crane and the stacking area

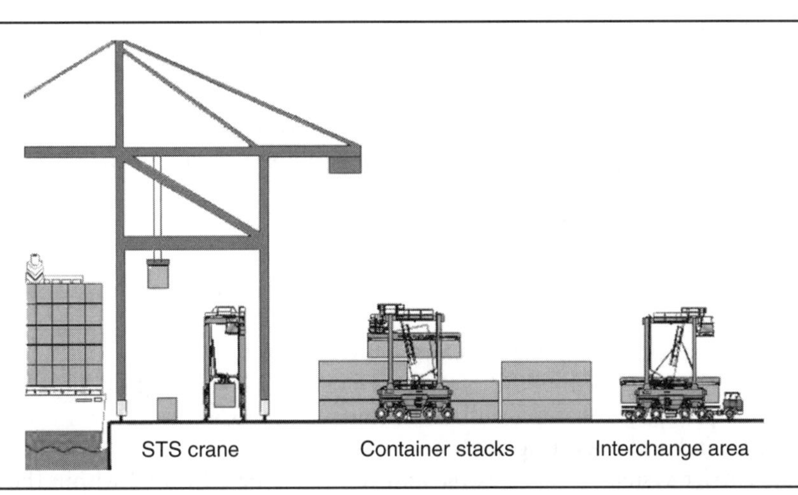

Figure 13.21 A type 1 over 2 straddle carrier

The straddle-carrier system is well suited for ports that have small terminal areas, and at a later date it is easy to alter the layout of the terminal. There is normally no need to reinforce the terminal pavement because the wheel loads of a straddle carrier are much lower than those of a reach-stacker.

The main benefits of a straddle-carrier compared with a reach-stacker system are: savings in labour costs, more ground slots in the same area, and easier and more direct access to the containers, resulting in improved selectivity and less unproductive moves.

Usually the straddle-carrier system stacks the containers two or three high. The system is usually the fastest system for terminals handling between 100 000 and up to about 3 000 000 TEU/year.

In general, the following apply:

(a) Three to five straddle carriers per STS crane depending on the distance between the berth and the stacking area.
(b) Generally, approximately 10 moves/h per straddle carrier.
(c) Medium stacking density, with about 500–750 TEU/ha with stacking of containers 3 high.
(d) High STS crane productivity and a buffer zone under the STS crane.
(e) High labour, capital and operating costs.
(f) The straddle-carrier system is very flexible and can easily be relocated within the terminal.

Figure 13.22 Stacking using an RTG and terminal tractors between the STS gantry crane and the stacking area

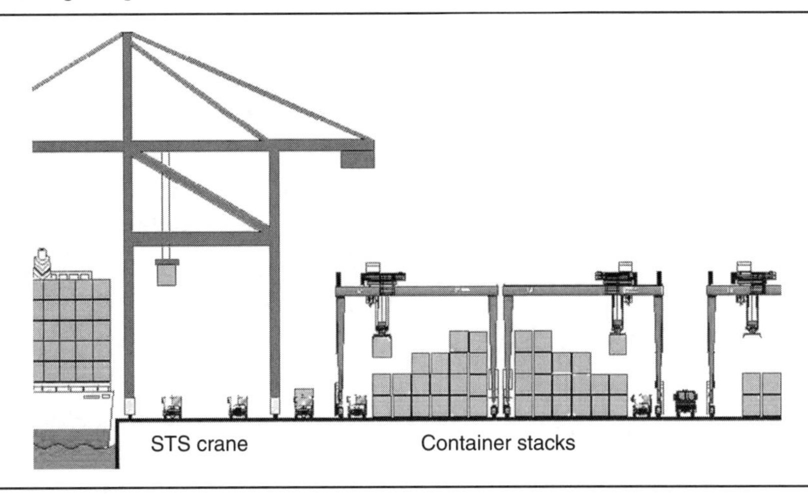

(g) The system lends itself to easy alteration of the terminal layout at a later date.
(h) High control; trucks are not allowed to load or unload in the stacking area.

Automated straddle carriers

Using similar technology to that used for automated guided vehicles (AGVs), a comparatively new innovation is the automated straddle carrier (ASC), a system which was first trialled at the Patrick Terminal in Australia. ASCs normally carry one container and, unlike the conventional tractor trailer or AGV systems, containers are placed on the pavement by either the STS crane or yard cranes. The containers are transported to/from the stack/ship to the shore crane back-reach area by the ASCs.

13.7.4 Rubber-tyre gantry (RTG) and/or rail-mounted gantry (RMG) systems

The STS gantry crane places the container on a terminal tractor (Figure 13.22) or a shuttle-carrier system (Figure 13.23) which moves it to the stacking area, where an RTG or an RMG system

Figure 13.23 Stacking using an RTG and shuttle carriers between the STS gantry crane and the stacking area

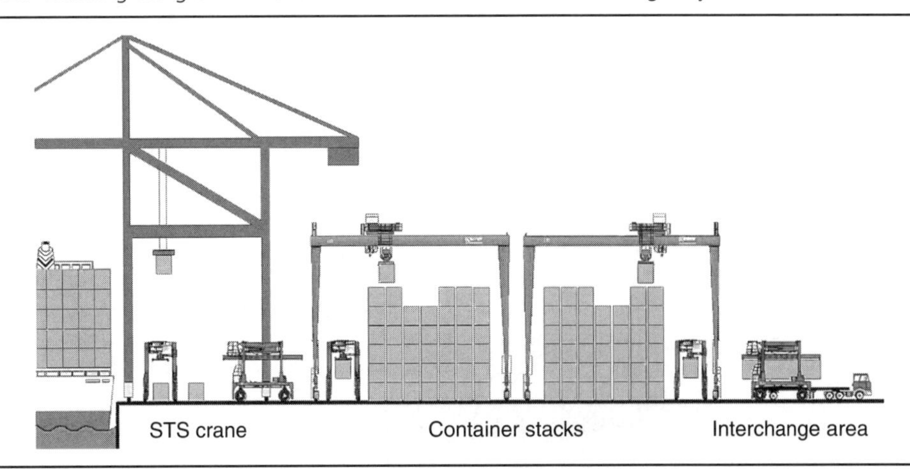

Figure 13.24 Detail of RTG stacking, 7 + 1 wide and 4 + 1 high

stacks it. The system usually stacks containers in blocks 5–9 wide and 4–6 high. The average handling capacity for one RTG crane can vary from 15 to 25 containers/h.

Figures 13.24, 13.25 and 13.26 show details of an RTG stacking system. The system is generally economical for terminals handling more than approximately 200 000 TEU/year. If the land for the terminal area is restricted in size or is very expensive, stacking using an RTG or RMG system can be the only practical solution to handling a large amount of containers.

Figure 13.25 A typical stacking arrangement when using RTGs

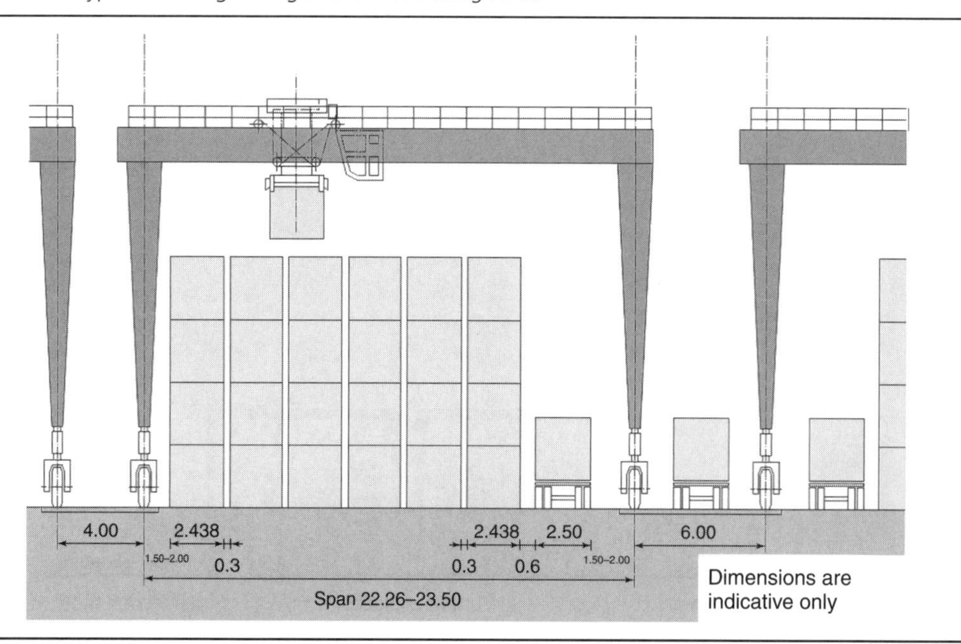

Figure 13.26 Detail of RTG stacking

Generally, with RTG and shuttle carriers the following apply:

(a) Two RTG cranes and two to three shuttle carriers per STS crane, depending on the distance between the berth and the stacking area.
(b) Good stacking density of about 800 TEU/ha when stacking containers four high.
(c) High STS crane productivity and a buffer zone under the STS crane.
(d) Low labour, but high capital and medium operating costs.
(e) Low control; trucks are allowed into the stacking area, and efficient traffic flow difficult to arrange. If trucks are not allowed into the stacking area, it is necessary to increase the number of shuttle carriers by at least one. This solution gives better control over the area, but the labour and operating costs will increase.

Generally, with RTG and terminal tractors the following apply:

(a) Two RTG cranes and three to five terminal tractors/STS cranes, depending on the distance between the berth and the stacking area.
(b) High stacking density of about 800 TEUs/ha when stacking of containers four high.
(c) Medium STS crane productivity and no buffer zone under the STS crane.
(d) High labour, but medium capital and operating costs.
(e) Low control; trucks are allowed into the stacking area.

Figure 13.27 shows a port where both straddle carrier and RMG systems are used.

Figure 13.27 A modern container terminal

Rubber-tyre gantry cranes

Rubber-tyre gantry (RTG) cranes are key tools and have been used all over the world for many years to stack containers at container terminals. Several manufacturers have been providing RTGs with a view to minimising both equipment maintenance and capital infrastructure costs. The main difference lies in the bogie design and the number of wheels.

16 wheels versus 8 wheels

The only difference between the traditional 8-wheel and the newer generation 16-wheel RTGs is in the bogie systems. The higher cost of the gear reducers and motors for a 16-wheel machine is compensated for by the elimination of separate wheel pivoting mechanisms. The cost of 16 standard wheels and rims is the same as the cost of 8 larger tyres and rims.

The wheels on a 16-wheel RTG are connected directly to the gear-reducer output shafts, so there are no wearing parts and no need for lubrication.

When moving between stacking areas, 16-wheel RTGs are up to 3 times faster than 8-wheel RTGs when it comes to wheel turning, as wheels mounted side-by-side roll when pivoted, which saves time and tyre wear. In addition, 8-wheel RTGs have lower tyre pressures and thus a higher coefficient of friction, and therefore additional power is required to provide optimal operating performance.

Perhaps the most significant savings can be achieved in the civil works infrastructure due to the low wheel loads imposed by 16-wheel RTGs. In many cases the need for expensive reinforced concrete runway beams can be eliminated, and the 16-wheel machines can run on the heavy-duty pavement using only painted runways, which can also provide flexibility in the yard layout.

The capital cost of the two types of RTG is comparable, but the 16-wheel design has high reliability, lower maintenance costs and can result in substantial civil works capital costs compared with the 8-wheel machine.

The Port of Felixstowe, the UK's largest container terminal handling 3 million TEU/year, has been successfully using 16-wheel RTGs for over 20 years.

13.7.5 Automated stacking cranes

As worldwide container traffic continues to grow, operators face challenges in boosting productivity at their terminals, either at existing facilities or at new sites. Environmental issues are also increasingly important, and not only should container handling equipment achieve high levels of performance but in doing so should be environmentally friendly.

The increasing demand for innovative handling solutions has led to the use of automated stacking cranes (ASCs) at an increasing number of terminals. Although these have a higher capital cost than RTGs and require dedicated crane rail systems, ASCs provide high stacking densities, high working speeds, high handling rates and an optimum use of space. As they are driverless, further benefits include improved productivity, lower labour and operating costs, a greater degree of safety than other container handling systems and dependable operation in virtually all weathers.

13.7.6 Wide-span gantry cranes

With spans of up to 80 m and cantilevers up to 40 m, wide-span gantry cranes (WSGCs) are used in inland waterway terminals for the loading and unloading of barges and small ocean-going vessels. WSGCs are also used in inland and port container rail terminals for the loading of containers, swap bodies and semi-trailers. In addition, WSGCs are used for intermodal handling between rail and road, and trimodal handling in barge terminals between vessels, freight trains and HGVs. In some instances they are also used in container yards in manual or semi-automated form.

13.7.7 Automated guided vehicles

Over the past 20 years or so the use of automated guided vehicles (AGVs) rather than conventional tractor/trailer systems as a means of moving containers to and from vessels and container yard stacking areas has grown. As they are software-controlled systems, the key benefits of ASCs include increased productivity, reduced labour and operating costs, increased safety, maximum use of space and predictable continuous operation. AGVs are normally used in combination with ASCs to provide optimum productivity and reduced operating costs.

13.8. The terminal area requirements

The required terminal area and annual container capacity will depend on, and are determined by, the choice of terminal handling equipment, operation system, available terminal area or land, and the forecast of throughput for inbound and outbound containers through the terminal. The aim should be for a total terminal working time of 24 hours a day, 7 days a week and 365 days a year.

For pre-engineering studies the following formulas will give sufficient accuracy to determine the necessary terminal areas and capacities, but for detailed design and container logistic evaluations an advanced simulation program should be used.

13.8.1 The terminal container capacity

The annual **terminal capacity** is usually expressed in terms of 20-foot-container equivalent units (TEUs). The **annual container TEU movement** is expressed as:

$$C_{\mathrm{TEU}} = \frac{A_{\mathrm{T}} \times 365 \times H \times N \times L \times S}{A_{\mathrm{TEU}} \times D \times (1 + B_{\mathrm{f}})} = \frac{A_{\mathrm{T}} \times 365 \times H}{A_{\mathrm{TEU}} \times D \times (1 + B_{\mathrm{f}})} \times N \times L \times S$$

The necessary **total yard area** A_{T} is given by:

$$A_{\mathrm{T}} = \frac{C_{\mathrm{TEU}} \times D \times A_{\mathrm{TEU}} \times (1 + B_{\mathrm{f}})}{365 \times H \times N \times L \times S} = \frac{C_{\mathrm{TEU}} \times D \times A_{\mathrm{TEU}} \times (1 + B_{\mathrm{f}})}{365 \times H} \times \frac{1}{N \times L \times S} = \frac{A_{\mathrm{N}}}{N \times L \times S}$$

where the following are the important parameters for determining the terminal capacity:

C_{TEU} = container movement/year
A_{T} = total yard area needed
A_{N} = net stacking area
H = ratio of average stacking height to maximum stacking height of the containers, varying usually between 0.5 and 0.8. This factor will depend on the need for shifting and digging of the containers in the storage area, and the need for containers to be segregated by destination
A_{TEU} = area requirement/TEU, which depends on the container handling system, as shown in Table 13.1
D = dwell time or the average number of days the container stays in the stacking area in transit. If no information is available, values of 7 days can be used for import containers and 5 days for export containers. For empty containers an average of 20 days' stay in a terminal can be used
B_{f} = buffer storage factor in front of the storage or stacking area, between 0.05 and 0.1
N = primary yard area or container stacking area compared to the total yard area, usually varying between 0.6 and 0.75 of the total yard area
L = layout factor due to the shape of the terminal area, varying usually between 0.7 for triangular area shape to 1.0 for rectangular area shape
S = segregation factor due to different container destinations, container maintnance system (CMS), procedures, etc. varying usually from 0.8 to 1.0

The area requirement A_{TEU} in m^2/TEU is dependent on the container handling system and the stacking density, the internal layout arrangement and type of equipment used to stack the containers, the internal access road system, and the maximum stacking height. Recommended approximate design estimates for area requirements A_{TEU}, including the stacking area, internal road system, etc., are shown in Table 13.1.

The area requirement A_{TEU} in m^2 will also depend on the size of the TEU ground slot. The ground slot will usually vary between approximately 15 and 20 m^2/TEU, depending on the container handling and stacking equipment.

The container stacking density is dependent on the container stacking layout (width and length), the stack height and the stack position. Therefore, the arrangement of the container stacks will directly affect the accessibility and storage of the containers, and will be of central importance to the throughput and efficiency of the container terminal.

Table 13.1 Approximate area requirement A_{TEU} (m^2/TEU), including internal roads

Handling equipment and method	Stacking height: no. of containers	Breadth or line of containers				
		1	2	5	7	9
Chassis	1	65				
FLT (front-lift truck) or RS (reach stacker)	1	72	72			
	2		36			
	3		24			
	4		18			
SC (straddle carrier)	1 over 1	30				
	1 over 2	16				
	1 over 3	12				
RTG or RMG	1 over 2			21	18	15
	1 over 3			14	12	10
	1 over 4			11	9	8
	1 over 5			8	7	6

The total number of **container slots** S_L at the stacking area will be:

$$S_L = \frac{A_T \times N}{A_{TEU}}$$

where

S_L = total number of container slots at the stacking area
A_T = total yard area
N = primary yard area or container stacking area compared to the total yard area
A_{TEU} = area requirement/TEU depending on the container handling system

Figure 13.28 shows the container movement per year C_{TEU}, container storage and stacking area, without the effect of the factors N, S and L.

***Example*:** From Figure 13.28, with a C_{TEU} of 65 000 TEU movements/year and an average dwell time D of 7 days in transit, the holding capacity required will be 1247 TEU. With an average area requirement A_{TEU} of 20 m^2/TEU, the net transit storage area will be 24 932 m^2. With a maximum average stacking height H of 0.8 and a reserve area capacity R of 25%, the required net stacking area A_N needs to be included for future expansion and would be approximately 38 955 m^2.

If the area for container stacking compared to the total container terminal area N is, for example, 0.65, the layout factor L is 0.9, the segregation factor S is 0.9 and the buffer storage factor B_f is 0.05; the total area of the primary container yard A_T would be approximately 77 700 m^2.

13.8.2 Berth container capacity

Due to the possible stochastic arrival of container ships each week, it is advisable to adjust the assumed containers handled per week using a peak factor. The total berth capacity in boxes/week

Figure 13.28 Container storage and stacking area design diagram

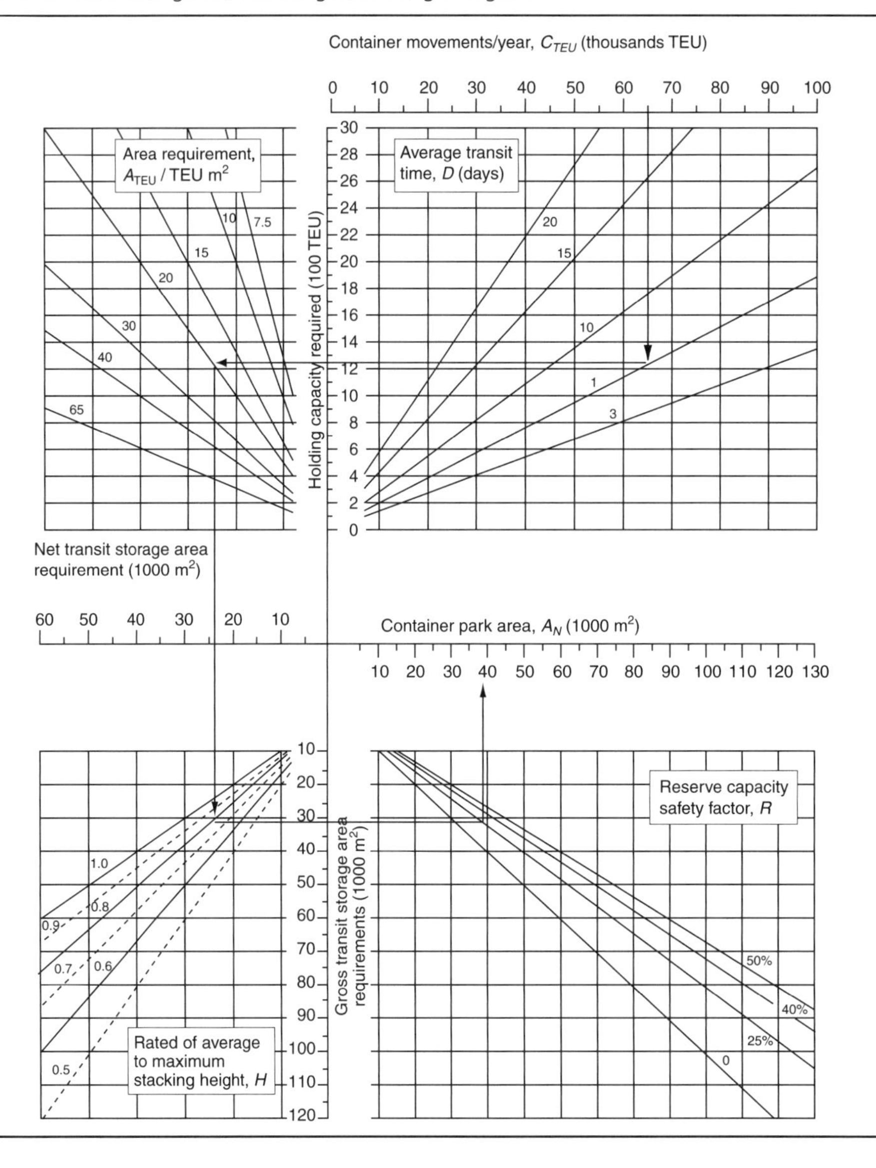

will be:

$$C_{BOX} = \frac{C_{TEU} \times P}{W_W \times R_{BT}}$$

where

C_{BOX} = container boxes handled/week
C_{TEU} = container movement/year
P = peak factor per week, normally varying between 1.1 and 1.3
W_W = number of working weeks/year; advisable to use 50 weeks/year

R_{BT} = ratio between the number of boxes (total number of 20 foot and 40 foot containers) and number of TEU containers; this normally varies between 1.4 and 1.7.

The total **number of container ships** needed to berth per week, including the peak factor, is expressed as:

$$S_{\mathrm{CS}} = \frac{C_{\mathrm{BOX}}}{S_{\mathrm{BCS}}}$$

where

S_{CS} = number of container ships berthing/week
C_{BOX} = number of container boxes handled/week
S_{BCS} = number of container boxes handled by one container ship

The crane **working time per container ship** for loading and unloading in hours is expressed as:

$$T_{\mathrm{WTC}} = \frac{S_{\mathrm{BCS}}}{C_{\mathrm{N}} \times G_{\mathrm{BH}} \times L_{\mathrm{SC}} \times W_{\mathrm{CT}}}$$

where the following are the parameters for determining the crane working hours/week:

T_{WTC} = total working time/container ship from berthing to unberthing in hours
S_{BCS} = number of container boxes handled by one container ship
C_{N} = total number of STS cranes working on each container ship
G_{BH} = number of container boxes handled per container crane per hour
L_{SC} = working time for starting and closing operations related to basic output time, normally varying between 0.8 and 0.9
W_{CT} = working crane time related to total ship berthing time, normally varying between 0.7 and 0.9

The total S_{TS} container crane working hours/week, including the peak factor, is expressed as:

$$G_{\mathrm{STS}} = S_{\mathrm{CS}} \times T_{\mathrm{WTC}}$$

where the following are the parameters for determining the gantry crane working hours per week:

G_{STS} = total STS container crane working hours/week, including the peak factor
S_{CS} = number of container ships berthing/week
T_{WTC} = total working time/container ship from berthing to unberthing in hours

13.8.3 Berth occupancy

The arrival of shipping at port is usually a stochastic process. The number of berths required will depend on the berth occupancy. Therefore, in order to calculate the number of berths required it is essential to know if the ships arrive randomly or if there are significant peaks, such as seasonal variations, in the arrival pattern. The berth occupancy ratio as a percentage of working time, including the peak factor/week, is calculated as follows:

$$B_{\mathrm{OR}} = \frac{T_{\mathrm{WTC}} \times 100}{B_{\mathrm{N}} \times \dfrac{W_{\mathrm{D}} \times W_{\mathrm{H}}}{S_{\mathrm{CS}}}}$$

Table 13.2 Berth occupancy

Number of berths	Control of arrival of ship to berth		
	None	Average	High
1	25%	35%	45%
2	40%	45%	50%
3	45%	50%	55%
4	55%	60%	65%
5	60%	65%	70%
6 or more	65%	70%	75%

or

$$B_{OR} = \frac{G_{STS} \times 100}{B_N \times W_D \times W_H}$$

where the following are the parameters for determining the berth occupancy ratio/week:

B_{OR} = berth occupancy ratio in percentage
T_{WTC} = total working time/container ship from berthing to unberthing in hours
B_N = number of berths
W_D = working days/week
W_H = working hours/day
G_{STS} = total STS container gantry crane working hours/week, including the peak factor
S_{CS} = number of container ships berthing/week

As a rough guide, the berth occupancies for container and conventional general cargo berth operations (multi-purpose berth) should be below the figures given in Table 13.2. The actual occupancy figures will depend on the port administration's control of the arrival of the ship at the berth.

For oil and gas berths a satisfactory occupancy factor, for instance for two berths, is 60%.

High berth occupancy factors can seem attractive because they yields the highest berth utilisation, but it is usual to assume a ratio of the average waiting time or congestion time to the average berth service time of not higher than 5–20%. The berth occupancy time will also depend on the type of berth, the type and size of ship, transfer equipment, environmental conditions, etc.

13.8.4 Terminal capacity

As indicated below, the annual berthside crane capacity, berth productivity/m and the stacking area capacity in TEU/m^2 stacking area/year will vary considerably between different container ports:

(a) The annual berth container crane capacity varies approximately from 50 000 to 350 000 TEU/year, with an average of 110 000 TEU/year.
(b) The annual berth productivity/m berth front varies approximately from 500 to 2500 TEU/berth m per year, with an average of 1000 TEU/berth m per year.

(c) The annual container stacking area capacity in TEU/m^2 per year varies approximately from 0.5 to 7.0 TEU/m^2 per year, with an average of 2.0 TEU/m^2 per year.

These figures for TEU per container crane per berth metre and per stacking area must be used with great care. It is almost impossible to compare different ports correctly, as the figures will depend on the type of container crane, the type of container stacking system used, the stacking height, the dwell time, etc. Therefore, when evaluating the capacity of a terminal the equipment, dwell time, etc. for that specific terminal should always be used, not any average figures for other terminals.

13.8.5 Hinterland

A serious restriction to improving the STS and terminal handling can be the ability of the landside back-up system or the hinterland road and/or train system to cope with the improved efficiency of the total terminal capacity.

The number of container box passes between the terminal and the hinterland road and/or train system per working hours per day will be:

$$C_{\mathrm{BTH}} = \frac{C_{\mathrm{TEU}}}{R_{\mathrm{BT}} \times W_{\mathrm{W}} \times W_{\mathrm{D}} \times W_{\mathrm{H}}}$$

where

C_{BTH} = number of boxes between the terminal and the hinterland/working hours per day
C_{TEU} = container movement/year
R_{BT} = ratio between number of boxes and number of TEU
W_{W} = number of working weeks/year; it is advisable to use 50 weeks/year
W_{D} = working days/week
W_{H} = the working hours/day

Whatever operating system is used, it will require close integration with the handling system between the container terminal and the associated intermodal yard and the hinterland. Improved handling at the STS interface and the terminal itself must, therefore, require the same handling capacity between the terminal and the hinterland.

Although the goal for all container terminals is to work 24 hours a day, 7 days a week and 365 days a year, experience from many large modern terminals has shown that in order to reduce the pressure on, for example, the road system, the activity between the terminal and the hinterland road system between midnight and 4.00 am is virtually negligible. In practice, the total traffic between the terminal and the hinterland occurs over approximately 10–12 h/day, including the peak factor, with maximum traffic in the morning and in the afternoon. Therefore, to reduce congestion the road system to the terminal needs to transfer the container distribution to other intermodal transport systems (e.g. trains).

One must, therefore, view the whole transportation system from the ship through the terminal and the gatehouses to the hinterland, or vice versa, as one operation. Space must be provided to accommodate the number of lorries passing through the gate per hour, for parking of lorries entering or leaving the terminal in front of the gatehouses, or coming from the hinterland and into the terminal.

13.9. The world's largest container ports

The 20 largest container ports in the world in 2012 are listed in Table 13.3.

Table 13.3 The largest container ports in the world in 2012

	Port	Country	Annual throughput: million TEU
1	Shanghai	China	32.53
2	Singapore	Singapore	31.65
3	Hong Kong	China	23.10
4	Shenzhen	China	22.94
5	Busan	South Korea	17.04
6	Ningbo-Zhoushan	China	16.83
7	Guangzhou	China	14.74
8	Qingdao	China	14.50
9	Jebel Ali Dubai	United Arab Emirates	13.30
10	Tianjin	China	12.30
11	Rotterdam	Netherlands	11.87
12	Port Kelang	Malaysia	10.00
13	Kaohsiung	Taiwan, China	9.78
14	Hamburg	Germany	8.86
15	Antwerp	Belgium	8.64
16	Los Angeles	USA	8.08
17	Dalian	China	8.06
18	Keihin	Japan	7.85
19	Tanjung Pelepas	Malaysia	7.70
20	Xiamen	China	7.20

REFERENCES AND FURTHER READING

BSI (British Standards Institution) (2000) BS 6349-1: 2000. Maritime structures. Code of practice for general criteria. BSI, London.

Cork S and Holm-Karlsen T (2002) Vessel growth and the impact on terminal planning and development. In *30th PIANC-AIPCN Congress 2002* (Cox RJ, ed.). Institution of Engineers, Sydney, pp. 1942–1956.

CUR (Centre for Civil Engineering Research and Codes) (2005) *Handbook of Quay Walls*. CUR, Gouda.

International Maritime Organisation (IMO) (2002) *Consideration of the Draft International Ship and Port Facility Security (ISPS) Code*. IMO, London.

Ligteringen H, van de Ruit GJ, van Schuylenburg M and Seignette RWP (1998) *Container Handling in Seaports, Recent Research in the Netherlands*. PIANC, Brussels.

PIANC (International Navigation Association) (1987) *Development of Modern Marine Terminals. Report of Working Group 09*. PIANC, Brussels.

PIANC (2010) *Design Principles for Container Terminals in Small and Medium Ports. Report of Working Group 135*. PIANC, Brussels.

PIANC (2014) *Design Principles for Container Terminals in Small and Medium Ports. Report of Working Group 135*. PIANC, Brussels.

UNCTAD (United Nations Conference on Trade and Development) (2003) *Modern Port Management, Port Training Programme*. UNCTAD, Geneva.
UNCTAD (2003) *Operation and Maintenance Features of Container Handling Systems*. UNCTAD, Geneva.
Ward T (1998) New ideas for increasing wharf productivity. *Ports 98*, vol. 1.

Port Designer's Handbook
ISBN 978-0-7277-6004-3

http://dx.doi.org/10.1680/pdh.60043.355

Chapter 14
Fenders

14.1. General

The PIANC Fender 2002 Committee made the following statement: 'There is a simple reason to use fenders: it is just too expensive not to do so.'

Marine fenders should provide the necessary interface between the berthing ship and the berth structure. The old wood fender system shown in Figure 14.1 does not have the principal function of the fender, which is to transform the impact load from the berthing ship into reactions that both the ship and berth structure can safely sustain. A properly designed fender system must therefore be able to gently stop a moving or berthing ship and absorb the berthing energy without damaging the ship, the berth structure or the fender. When the ship has berthed and been safely moored, the fender systems should be able and strong enough to protect the ship and the berth structure from the forces and motions caused by the wind, waves, currents, tidal changes and loading or unloading of cargo. The design of fenders should also take into account the importance of the consequences suffered by the ship and the berthing structure in the case of an eventual accident due to insufficient energy absorption fender capacity.

During the design of berth and fender constructions in the past and, even today, there has been a tendency to plan and design the berth structure itself first, and only later the type of fender that will hopefully satisfy the requirements as regards berth and ships. This approach to design has resulted in damage occurring quite frequently to berth and fender structures and, to a lesser degree, to ships.

The correct procedure should be to plan and design the fender and berth structures jointly. The choice of fender should be dependent on the size of the berthing ships and the maximum impact energy. After having identified the criteria for the fender, the design of the berth superstructure can be finalised. The following factors should therefore be considered in selecting the fender system:

(a) the fender system must have sufficient energy absorption capacity
(b) the reaction force from the fender system does not exceed the loading capacity of the berthing system
(c) the pressure exerted from the fender system does not exceed the ship's hull pressure capacity
(d) the capital construction costs and maintenance costs are considered during the design of both the berth structure and the fender system.

This procedure will lead to:

(a) correct structural solutions
(b) lower construction costs
(c) lower annual maintenance costs.

Figure 14.1 An old wood fender system

14.2. Fender requirements

A single or easy solution to all fender problems does not exist. Each type of berth structure has different demands. Factors having an impact on the choice of fender are the size of the ships, navigation methods, the location, tidal differences, water depths, etc. A ship berthing along an exposed berth structure will obviously have other demands on the fender system than if it were to berth along a sheltered berth structure.

We talk of the 'sensitivity' of a berth structure to impact from ships. Generally, a solid berth structure is more resistant to horizontal impact, whereas an open berth structure is less resistant or more sensitive. This means that the sensitivity of a berth structure to berthing impact increases with its 'structural slenderness', and with increasing slenderness the fender assumes greater importance. For instance, a berth structure of concrete blocks will be less vulnerable than, for example, an open-type berth supported by piles.

When selecting a fender system, it should be borne in mind the purposes of the berth structure. Structures with special functions are usually provided with fenders to accommodate certain types of ships, such as berths for oil tankers. But, on the other hand, if the berth should accommodate a large variation in ship sizes and types, such as a multi-purpose berth structure, the selection of a fender system is far more difficult and will require detailed consideration and possibly special design

Figure 14.2 A container ship during berthing

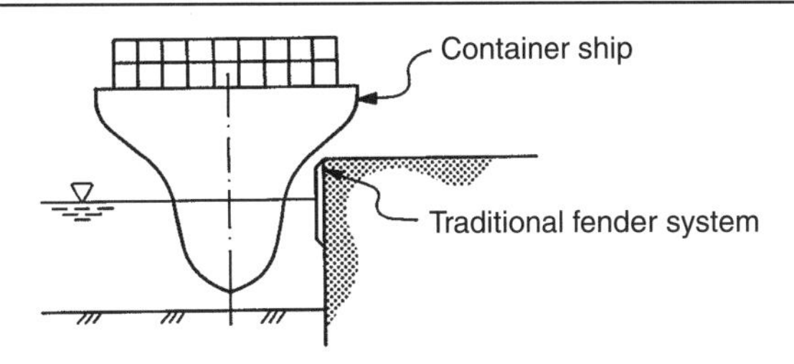

treatment. The problem of selecting the correct fender will be further complicated if the berth has an exposed location with difficult manoeuvring conditions and/or is subjected to extreme tidal variations.

The types of fendering provided at berth structures for general cargo ships are often unsuitable for use with specialised container ships due to their different hull and flare shapes. The large deck overhang of a container ship when berthing at an angle is illustrated in Figure 14.2. This overhang can impose high concentrated loads on a traditional fender system because of the very small contact area between the ship's hull and the fenders. Therefore, to solve these problems, the fender layout for a container berth should, in principle, be as illustrated in Figure 14.3.

The horizontal distance between the berth line and the fender line should generally be kept to a practicable minimum in order to reduce, for example, the required container crane outreach as much as possible. As shown in Figure 14.2, there must also be sufficient clearance to reduce the chance of the ship flare hitting the crane leg or the edge of the berth structure.

14.3. Surface-protecting and energy-absorbing fenders

The principal function of a fender placed between an approaching ship and the berth structure is to absorb the berthing energy or impact and transmit an acceptable load to the structure. Bearing these factors in mind, many designs of fender systems have been invented and tried out with varying degrees of success, from ordinary protecting fenders to sophisticated shock-resistant and energy-absorbing systems.

The great differences in the types of berth structures result in different requirements for the fender. Generally, a solid berth will be able to resist a high horizontal force, whereas an open-pier berth must have fenders that absorb energy and reduce the thrust on the structure. When a ship strikes a berth structure during berthing, it has a kinetic energy that must be absorbed and that results in a horizontal

Figure 14.3 Fenders for a container ship

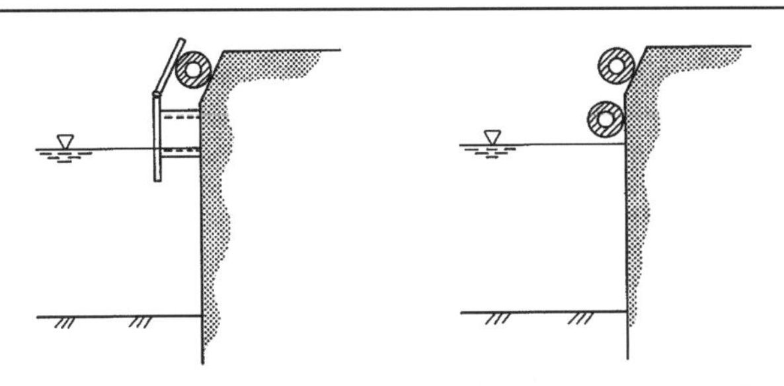

force, which the berth structure must resist. In other words, when choosing a fender system, the impact energy that the fender must absorb, E_f, and the force, P, the berth structure must resist have to be borne in mind.

This means choosing a fender with a **fender factor** that meets the berth requirements. The fender factor is the ratio between the force to be resisted and the energy absorption. This means that if the factor is 10 kN/kNm, a 10 kN horizontal force will be transferred to the berth structure for each 1 kNm of energy that the fender absorbs. If this fender were to absorb 100 kNm of energy, the resulting horizontal force to be resisted by the berth would be 1000 kN. The ideal fender is one that will absorb large amounts of kinetic energy and will transmit low reactive loads into the berth structure and the ship's hull.

Fenders can generally be divided into two groups:

(a) **Surface-protecting fenders** that transmit a high impact or reaction force to the berth structure for each 1 kNm of energy absorbed (i.e. the fender factor P/E_f is high).
(b) **Energy-absorbing fenders** that transmit a low impact or reactor force to the berth structure for each 1 kNm energy of absorbed (i.e. the fender factor P/E_f is low). From the performance point of view, the low reaction and the high-energy absorption with constant reaction over part of the deflection range are a distinct advantage, but for a large range of ships the fenders can be too hard for the smaller ship using the berth.

When checking manufacturers' catalogues to see if, for example, a cylindrical fender is of the type comprising surface-protecting or energy-absorbing fenders, it should be borne in mind that the capacity of such a fender is given when it is about 50% compressed.

As regards the force P, there are two factors that decide its magnitude and, consequently, the type of fender to be chosen:

(a) the horizontal force that the berth structure can resist
(b) the maximum pressure that the ship's side can withstand.

On the whole, the horizontal force on the berth structure is the decisive factor where smaller ships are concerned, and for larger ships it is the pressure on the ship's side. The latter depends, of course, on the contact area available for the pressure distribution between the berth and the ship's side during berthing.

From a berth designer's point of view, the purpose of a fender system is to reduce the reaction force and transmit a designed thrust to the berth structure that it can bear without difficulty. On the other hand, the impact energy increases with the ship displacement to a greater extent than does the strength of the hull. Therefore, the ship's hull also needs energy-absorbing fenders. Similar requirements for impact-reducing fender systems also arise where the berth structure is exposed to difficult berthing and weather conditions. In other words, the ideal aim for a fender system is to be able to absorb high-impact energy and transmit a low reaction force to the ship's hull and the berth structure.

Each type of fender system has its own characteristic force–deflection curve (Figure 14.4). The area under the curve represents the total energy absorbed in deflecting the fender. The shape of the curve gives an idea of the energy efficiency of the fender and the impact intensity. Generally, due to fender

Figure 14.4 General working diagram for fenders

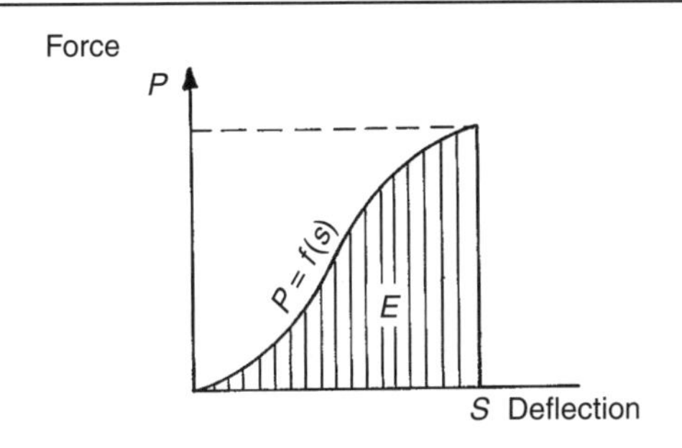

characteristics, three different curve patterns can be distinguished for the main types of fender on the market nowadays (Figure 14.5). Fender 1 is characterised as a hard fender, fender 2 as a medium fender and fender 3 as a soft fender. This is illustrated in Figure 14.6, where the areas under each of the curves are equal (area A = area B = area C). The different fenders (fenders 1, 2 and 3) have the same design reaction force and the same energy absorption capacity, but different deflection.

It is evident from Figure 14.6 that type 1 fenders, or the buckling-type fenders (e.g. cell fenders), require considerably less deflection to absorb the design energy than a side-loaded cylindrical rubber fender. The characteristics of a type 1 fender cause the maximum reaction force to occur during almost every berthing, even with ships smaller than the maximum design ship. Therefore, due to rather high contact pressures against the ship's hull, a panel or fender wall between the ship's hull and the fender itself is often needed to reduce this contact pressure. Type 1 fenders have, as illustrated in Figure 14.6, a higher performance from the energy point of view, but may not be recommended when the tonnage range of the ships likely to berth entails a very wide range of energies to be absorbed. The buckling of type 1 fenders is also susceptible to a significant reduction in the energy absorption capacity when subjected to impacts not perpendicular to the fender face.

The flexible fender piles, or type 2 fenders, are often an alternative where the soil conditions are suitable because they can combine the functions of a fender and breasting structure. Type 3 fenders,or the soft

Figure 14.5 Working diagram for three different types of fender

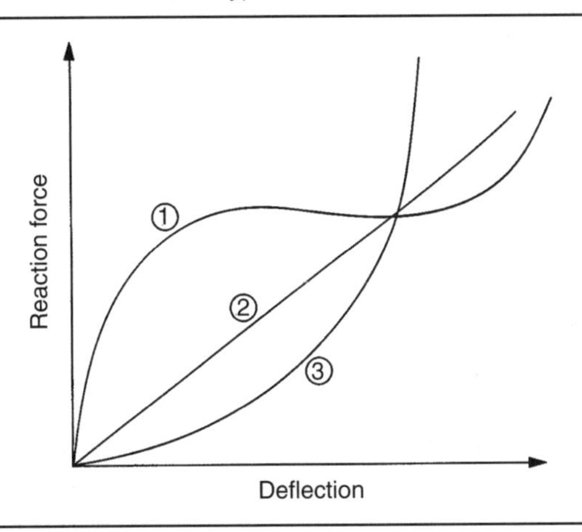

Figure 14.6 Reaction–deflection characteristics of various fender types

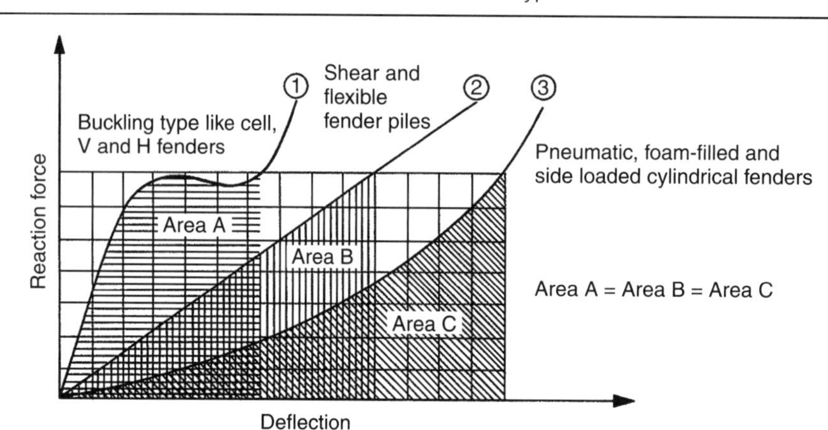

fenders, are very popular where the energy absorption requirements are not too high, but as can be seen from Figure 14.6 they must be larger than the corresponding type 1 fender, and thus require greater reach of the cargo-handling equipment.

After having estimated the energy to be absorbed by the fenders, the reaction forces against the berth structure can be read from the fender manufacturer's curves. If the fender manufacturer has not stated a tolerance on the figures quoted for reactions and energies, a tolerance of +10% should be taken into account in the design of the fender system. These curves are usually based on the uniform deflection of the fenders. As shown in Figure 14.7, a non-uniform deflection of the fender system can occur due to the following:

(a) the angle of approach between the ship and the fender line
(b) the curve of the ship's hull in plan where the ship makes contact with the fender
(c) the flare angle of the hull in section.

A study of a ship's hull curves will, as illustrated in Figure 14.8, show that the contact angles and the contact point will vary with the angle of approach and the height of the fender relative to the ship (Figure 14.9). The designer should, therefore, establish a maximum safe value of the berthing angle or angle of approach that can be economically achieved due to the ship's hull, flare angle and bulbous bow having considered both the fender and the berth structure layout.

Previously, the flare angle of the shipside was not considered, since most ships had a nearly vertical shipside at the contact point between the ship and the fender. But for modern ships, such as the third-generation or larger container ship, the flare angle has to be considered when selecting a fender system. The little available data on the flare angle at the point of contact with the fenders show that there is a wide range of flare angles for a given approach angle, both within a particular ship category and especially between different types of ships.

For general cargo ships, the flare angle at an approach angle of 5° is about 8–15°, and for an approach angle of 10° it is about 15–25°. For container ships, the flare angle at an approach angle of 5° is about 10–16°, and for an approach angle of 10° it is about 20–40°. It appears that general cargo ships, bulk

Figure 14.7 Angular compression of a fender

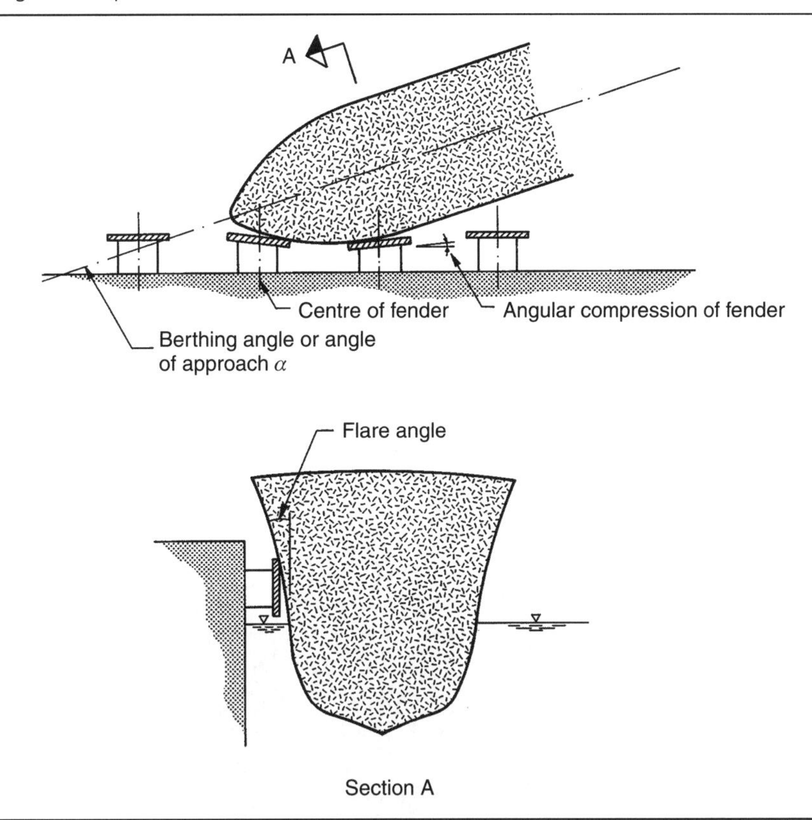

carriers and tankers have less variation in the flare angle than container ships. The ships that have less flare angle seem to have the largest block coefficients.

From these flare angle figures it can be shown that container ships exhibit greater flare angle than general cargo ships. In the future this will probably govern the design and layout of fenders in commercial ports. Consideration should also be given to ships with a high block coefficient, such as bulk carriers, which, because of their higher block coefficient, will have a hull with a small radius in plan

Figure 14.8 A ship's hull curves

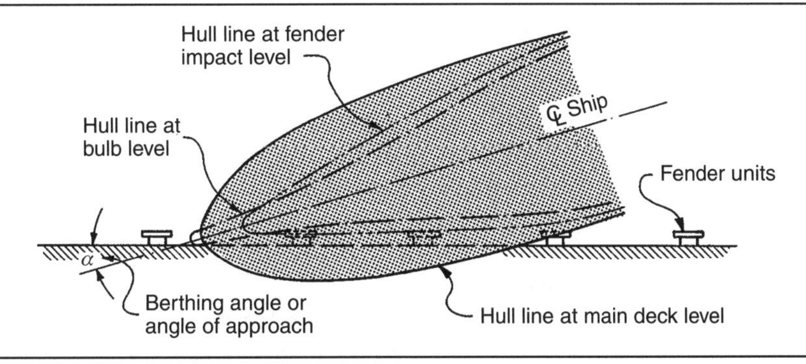

Figure 14.9 Contact between the ship and the berth

at the point of contact between the ship and the fender, and will therefore require a closer spacing between the fenders.

The choice of fender type will, in many cases, determine the design of the berth superstructure. Generally, the protective fender demands a solid berth structure because of the large force it exerts against it; in other words, a cheap fender system and an expensive berth construction. On the other hand, an energy-absorbing fender imparts a lesser force to the structure, thus demanding a less solid berth construction, which generally means an expensive fender and a cheaper berth construction.

These parameters will apply not only to new berth constructions but also very much to old berth structures being upgraded. Nowadays, there are many old berth structures still in use with sufficient water depth in front of the structure, but because of their structural design and their protective fenders of, for example, wood materials, they are not able to accommodate modern ships with increasing dimensions that exert greater horizontal loads than the structures were designed for. However, if energy-absorbing fenders replace the protective fenders, the structures will, in most cases, be able to also accommodate these larger ships.

14.4. Different types of fender

In the following, the various types of rubber fender will be discussed, of which currently the most used are prefabricated fenders. Since the first rubber fenders were made in the 1930s, they have proved resistant to aggressive and polluted water as well as wear and tear from ships, as long as they have been correctly installed. Their purchase price and maintenance costs are also below those of most other types of fenders. Rubber fenders are produced in many shapes and sizes, depending on their function. It is necessary to be aware of the fact that, for different manufacturers producing apparently identical fenders, the fender factors may differ significantly.

There are basically two types of fender:

(a) Fenders that in principle are **fixed or mounted to the berth structure**. The fixed fenders are again subdivided into buckling fenders (cell fenders, V type fenders, etc.) and non-buckling fenders (cylindrical fenders).
(b) **Floating fenders** between the ship and the berth structure. The floating fenders are again subdivided into pneumatic fenders and foam-filled fenders.

Figure 14.10 lists different types of rubber fenders, and Figure 14.11 indicates whether they are mainly surface protecting or mainly energy absorbing. As can be seen, for example, the different sizes of cylindrical fenders have, under radial loading, fender factors P/E_f varying from about 25 kN/kNm to about 1.3 kN/kNm.

There is not necessarily any connection between the fender factor P/E_f and the flexibility or the rigidity of a fender. There are fenders with low fender factors (energy-absorbing fenders), which are very rigid, and fenders with high fender factors (surface-protecting fenders), which are flexible. For instance, old car tyres used as fenders are very flexible but act as surface-protecting fenders. Even under small loads they are pressed flat and function only as solid fenders.

Most of the characteristics for the various fender types shown in Figure 14.10 are based on data published by different fender manufacturers. The actual fender performance may vary by as much as $\pm 10\%$, and the characteristics are based on normal or perpendicular impacts against the fender. The fender performance may vary considerably when subjected to angular impacts, which is the most common case.

Therefore, based on the manufacturers' performance curves, due to manufacturing tolerances, it is usual to recommend a tolerance of -10% for the energy absorption and $+10\%$ for the maximum reaction forces on the catalogue performance figures. This reaction force is a characteristic load, which should be used for the design of the berth structure.

14.5. Installation

The installation and mounting of the fender systems should be of robust and simple design, and only one type of metal should be used in order to avoid electrochemical corrosion, and no mountings should be allowed to touch the steel reinforcement in the concrete. This is especially important if the fenders are mounted on structures with cathodic protection against corrosion.

The fender system that is most commonly used worldwide is the installation of old used rubber tyres in front of the berth structures (Figures 14.12 and 14.13).

Cylindrical rubber fenders are the second most used fender system around the world. They are manufactured with outside diameters ranging from about 15 cm to approximately 2.6 m. The fender factor for a 40 cm-diameter radial load will be 20 kN/kNm, and for a 1 m-diameter radial load 5 kN/kNm.

Figure 14.10 Different types of rubber fenders

Type	Fender shape	Sizes in mm	Reaction kN	Energy kNm	Performance curve
Circular shape of the buckling fender with panel contact	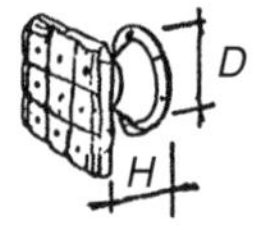 D, H	d/D/L 295/500/300 → 1765/2880/1800	60 → 3775	9 → 3530	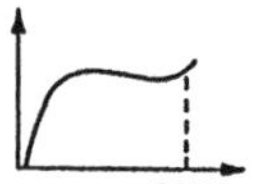 65%
	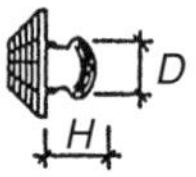 D, H	D/H 400/550 → 3000/3250	52 → 5800	8 → 6700	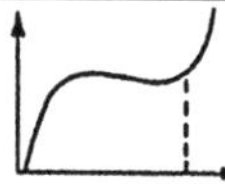 47.5 and 52.5%
Longitudinal shape of the buckling fender with panel contact	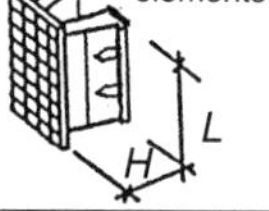 elements, L, H	H/L 300/600 → 1800/2000	66 → 1708	9 → 1260	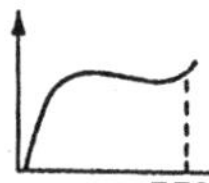 55%
	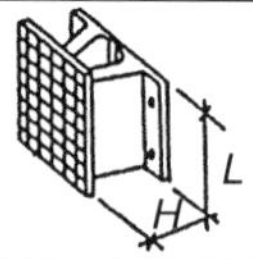 L, H	H/L 400/500 → 2500/4000	140 → 6900	22 → 7000	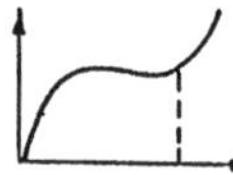 50, 52.5 and 60%
V type	 L, H	H/L 250/1000 → 1000/2000	150 → 2290	15 → 940	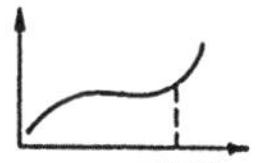 50%
	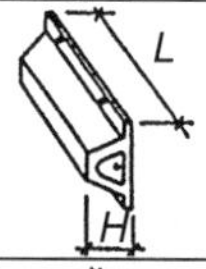 L, H	H/L 200/1000 → 1300/3500	150 → 3400	10 → 1500	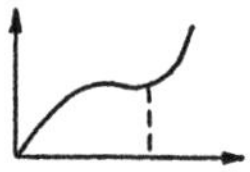 45%
	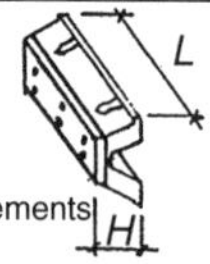 L, elements, H	H/L 300/600 → 1800/2000	66 → 1708	9 → 1260	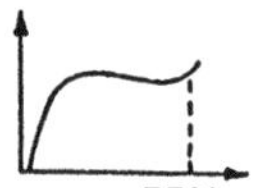 55%
Airblock	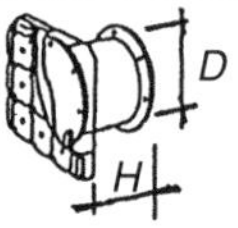 D, H	D/H 600/450 → 3200/3200	138 → 6210	15 → 4990	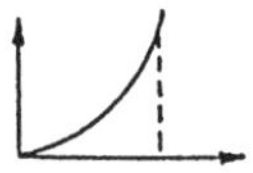 60 and 65%
Pneumatic	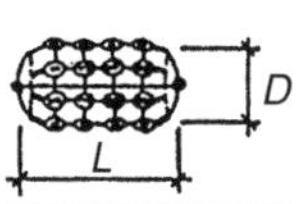 D, L	D/L 500/1000 → 4500/12 000	50 → 10 570	4 → 9080	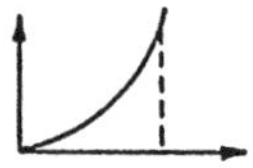 60%
Foam filled	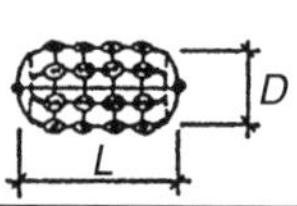 D, L	D/L 1000/1500 → 3500/8000	200 → 4050	41 → 3000	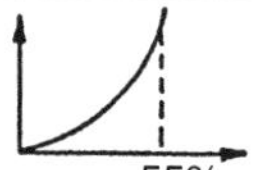 55%
Cylindrical	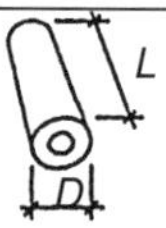 L, D	D/L 150/1000 → 2800/5800	80 → 6600	3 → 5000	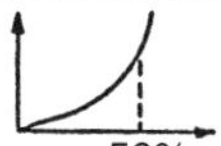 Reaction; 50%; Rated compression

Figure 14.11 Fender factor for different types of rubber fenders

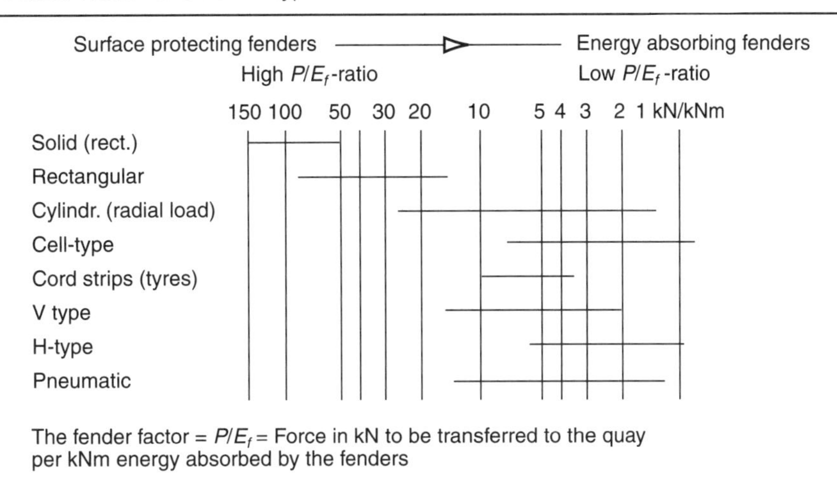

Figure 14.14 shows different ways of installing cylindrical fenders, which are the most common type of fender. The ladder and bracket installation systems are used for larger ships. Figure 14.15 shows a large cylindrical fender on a breasting dolphin.

Pneumatic floating fenders are available in sizes ranging from 50 cm outside diameter (OD) and 1.0 m length to 4.5 m OD and 12 m length. They are well suited as buffers between two tankers or between a tanker and a berth structure.

14.6. Effects of fender compression

After having calculated the probable impact energy a ship will have when berthing, the compression of the various fenders and the thrust the latter will transmit to the structure can be deduced from the

Figure 14.12 Old rubber tyres used as fenders

Figure 14.13 Installation of an old rubber-tyre fender

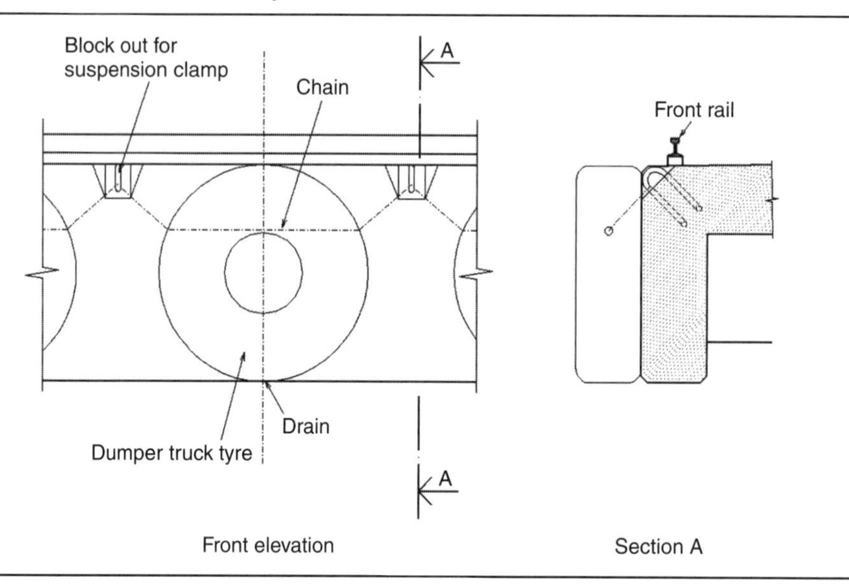

manufacturers' catalogues. Manufacturers always provide two diagrams for fenders, one showing the relationship between energy and compression and the other the impact force and compression relationship.

Two such diagrams, for a buckling fender and a side-loaded cylindrical fender, have been combined in Figure 14.16 to illustrate what happens to the two different fender types when a ship is berthing. As a more detailed illustration, the cylindrical rubber fender with 1500 mm OD and 800 mm inside diameter (ID) and a 1500 mm length will, with 50% compression, absorb impact energy of 330 kNm. The resulting force to be resisted by the berth structure will be 900 kN with a fender factor P/E_f $900/330 = 2.7$ kN/kNm. What is interesting about these large fenders that are designed for bigger ships is that they have a high fender factor with low compression: at 10% compression $P/E_f = 14.0$ kN/kNm. Where smaller ships are concerned, they will have little energy-absorbing effect, and function more as surface-protecting fenders. The curve shows that the fender factor decreases with increasing compression, by as much as 50% when it is 2.7 kN/kNm. Beyond this the factor increases with increasing compression.

It should be realised that for both fender types in Figure 14.16, the fenders can absorb more energy, even beyond approximately 50% compression, but the force to be resisted by the berth structure will then increase excessively. This is due to the fact that the fenders have been compressed to such an extent that they now function more like a surface-protecting fender. As a fender unit can only absorb a fixed amount of energy before failure, the fender structure can be provided with a device or an overload collapsible unit to prevent overload of the fender. The collapsible unit can be constructed in either concrete or steel, and installed between the fender and the berth structure. To prevent failure or damage to the fender, the collapsible unit can be designed to collapse for a reaction force equal to the fender reaction at about 55–60% compression of the fender.

The relationship between the OD and the ID for a cylindrical rubber fender under a radial load has a major influence on the fender factor. The usual OD/ID ratio is 2, but some manufacturers can produce

Figure 14.14 Different ways of installing cylindrical fenders

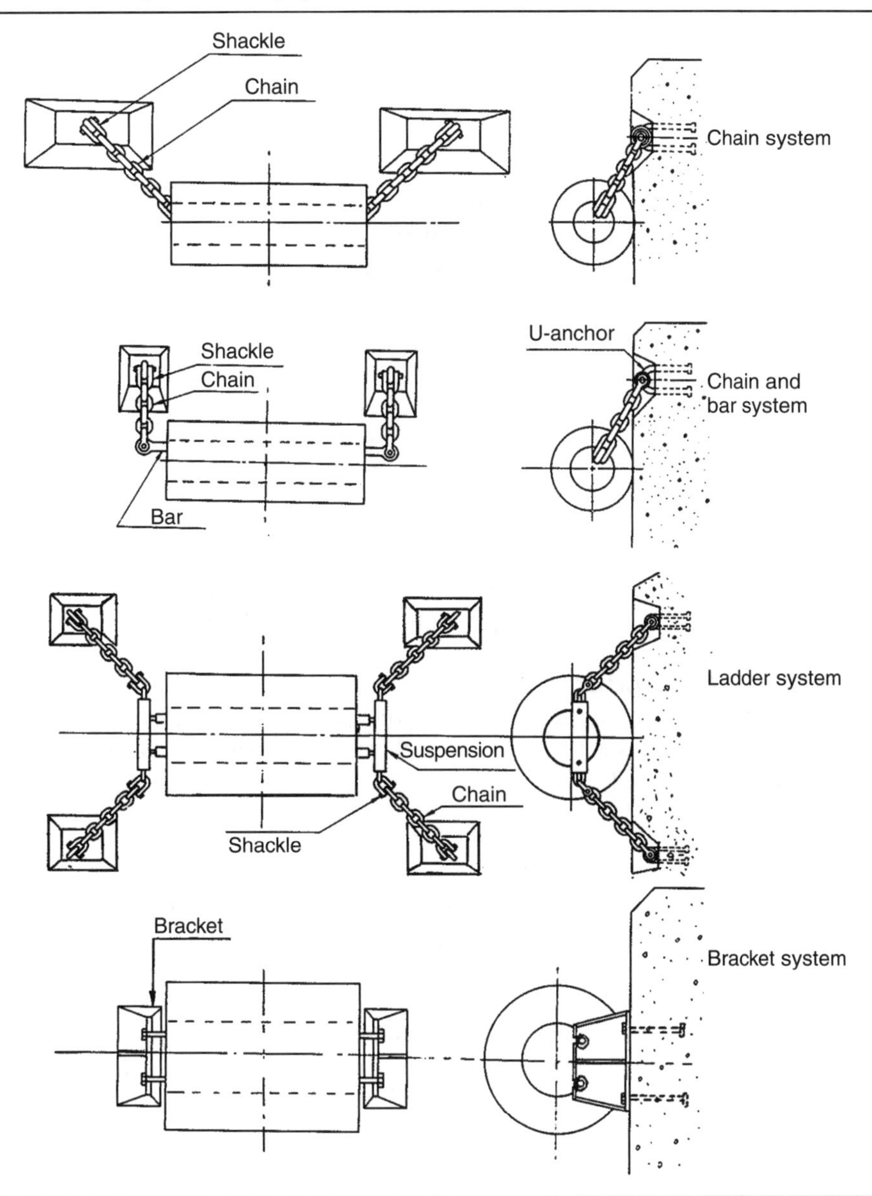

fenders down to a ratio of about 1.75. If there is a choice between several fenders with the same OD, the fender with the smallest diameter ratio will usually have the lowest fender factor and also be more economic.

The performance of a fender during angular loading is illustrated in Figure 14.17. During the actual berthing conditions, a fender will generally be loaded or compressed at an angle more or less equal to the approach angle of the berthing ship. This angular compression of the fender will change the characteristics of the fender reaction force and absorbed energy, compared with a fender compressed

Figure 14.15 Large cylindrical fender on a breasting dolphin

normal to the berth front. The correction factor for a single-cell fender unit compressed under different angles is shown in Figure 14.17. Therefore, the choice of fender will depend on the angular compression of the fender due to the curve of the ship's hull in plan and the flare angle.

14.7. Properties of a fender

In order to select the most proper and suitable fender for a particular berth, it is important to know the performance characteristics and the properties of each rubber fender type. Below, some of the more important factors will be discussed.

14.7.1 Design life

Although most rubber fender manufacturers produce fenders with a design life of about 20–30 years, the actual life of the fender will depend on the type of ships, the frequency of berthing and the influence of the natural environment, such as temperature, ozone density, sunlight hours and intensity, pollution and salt water as well as oils and fats. But, according to the manufacturers, most damage to fenders is the result of the fender being either under-dimensioned, bad manoeuvring or too high berthing velocities. For this reason, a higher safety factor should be used in designing the fender system. The different rubber manufacturers recommend a design life of about 5–15 years for fender systems installed on general cargo berths with a large range of different types of ships, and a design life of about 10–20 years for more particular berths, such as an oil berth.

14.7.2 Fender testing

The load deflection and energy deflection characteristics of a fender are only valid if the fender has been preconditioned by compression to the rated values at least three times before use. If not, the first maximum compression produced by the ship may well give higher than expected reactions. If the fenders are made of laminated rubber, the strength of the lamination should be equal to that of the material itself.

In line with the PIANC 2002 fender guidelines, the break-in deflection of the actual fender element should be at least the manufacturer-rated deflection, and at least one cycle should be performed. The break-in deflection should be mandatory for all fender types with a catalogue reaction rating of 100 t or

Figure 14.16 Reaction–deflection characteristics of various fender types

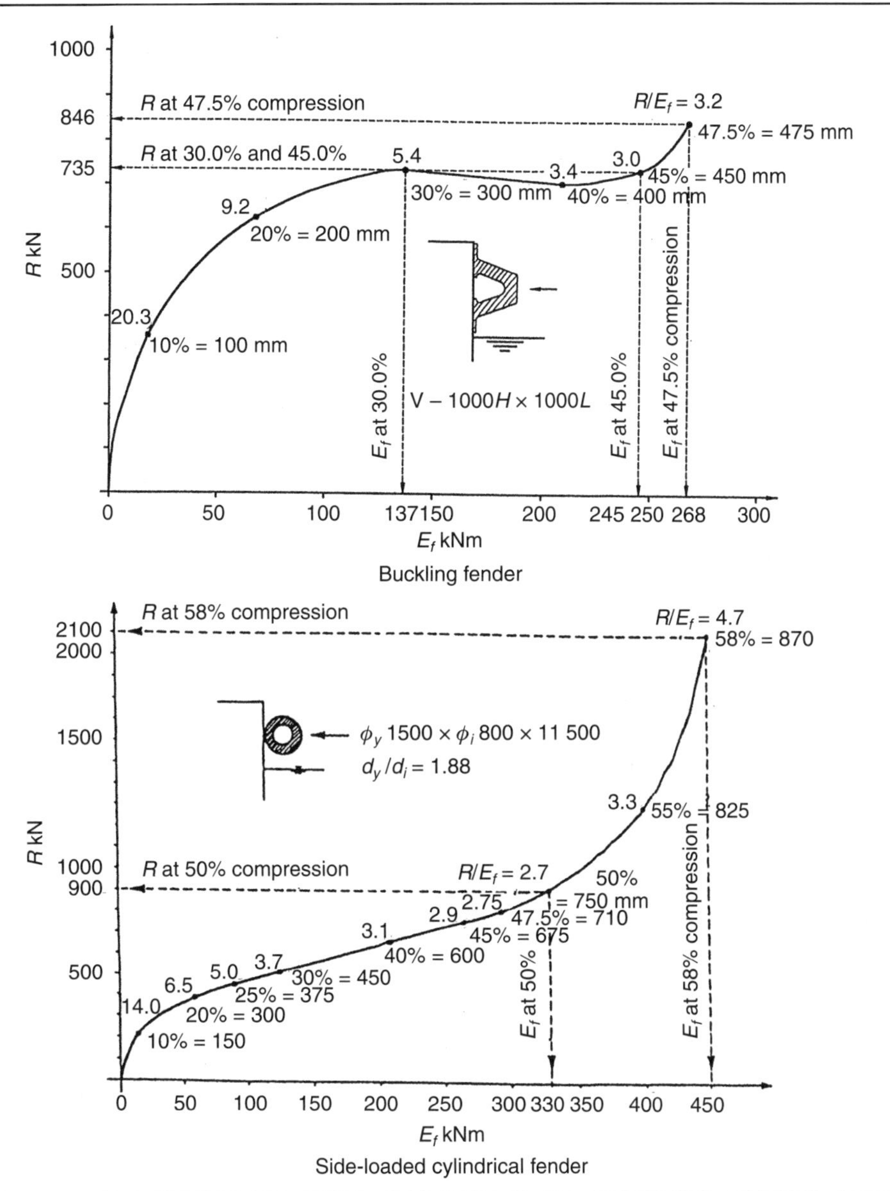

more, and they should be installed on a pile-supported berth structure. For other fender installations the customer should stipulate the break-in deflection.

The manufacturers' published performance curves and/or tables of the rated performance data (RPD) should be based on one of the following PIANC testing requirements:

(a) The traditional and widely used constant velocity (CV) method with constant slow velocity deflection: this method is the preferred method for the majority of manufacturers.

Figure 14.17 Correction factor under angular compression

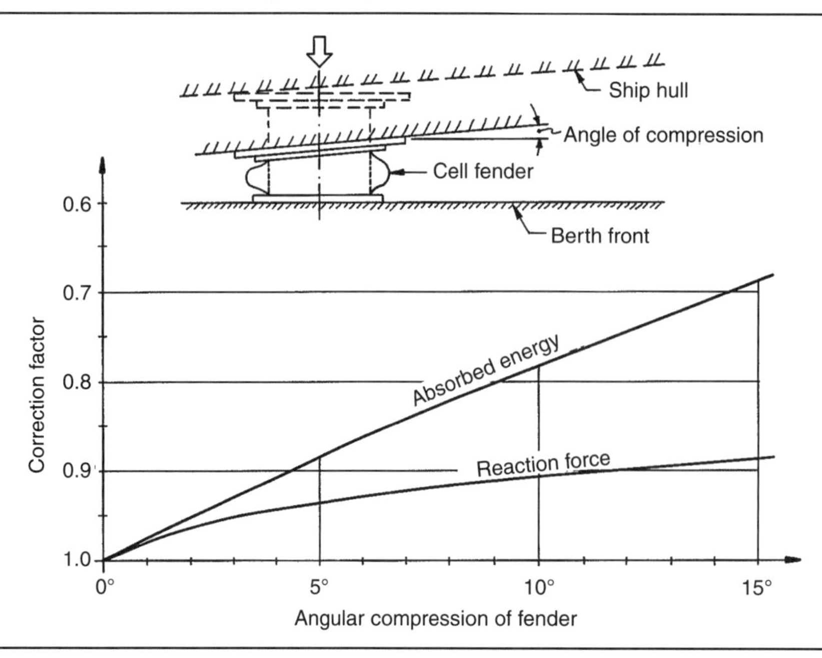

(b) The decreasing velocity (DV) method: an initial berthing deflection velocity of 0.15 m/s, decreasing to no more than 0.005 m/s at the end of the test.

The RPD should also be based on:

(a) testing of fully broken-in fenders
(b) testing of fenders stabilised at 23 ± 5°C
(c) testing of fenders at zero degrees angle of approach
(d) deflection (berthing) frequency of not less than 1 h.

All large fender manufacturers, such as Trelleborg, make their rubber fenders by using the highest-quality natural rubber (NR)- or styrene butadiene rubber (SBR)-based compounds that meet or exceed the performance requirements of international fender recommendations, such as those of PIANC and the EAU. All large fenders should be tested for their fender capacity. Trelleborg can also make fenders from other NR/SBR compounds or from materials such as neoprene, butyl rubber, EPDM and polyurethane. Figure 14.18 shows the testing of a large super cone fender.

Different manufacturing processes such as moulding, wrapping and extrusion require certain characteristics from the rubber. Table 14.1 gives usual physical properties for fenders made by these processes, which are confirmed during quality assurance testing. All test results are from laboratory-made and cured test pieces. The results from samples taken from actual fenders will differ due to the sample preparation process, so it essential to ask for details from the testing centre.

14.7.3 Hysteresis

The berthing energy or receivable energy that must be absorbed by a rubber fender during compression, as illustrated by curve 1 in Figure 14.19, is partially restored to the mass that acts on it during compression, as illustrated by curves 2 and 3, and partially dissipated in the form of heat within the

Figure 14.18 Testing a SCK 3000 fender

rubber material itself. This latter effect is called the hysteresis effect, and is represented by the area between curves 1 and 2 or 1 and 3. The ratio between the dissipated energy and the received energy will vary according to the type of rubber fender, and will be of the order of about 0.1–0.4.

The hysteresis effect can also be illustrated by dropping a rubber ball onto the floor. If the ball acts as a 'bouncing' ball, the hysteresis effect is very small, and a fender of this material will be a recoiling fender. Therefore, if a ship hits a fender of this type of rubber material, the ship will be thrown out from the fender after the fender has absorbed the berthing energy. On the other hand, if the ball acts as a 'dead' ball, the hysteresis effect is very large, and a fender of this material will be a non-recoiling fender, and a ship will not be thrown out from the fender. In this case, the ship will very slowly be pushed out from the fender. Therefore, a fender with a large hysteretic effect will have the best effect on the mooring of a ship and for the reduction of sway movements.

The principle of a permanent moored floating structure is illustrated in Figure 14.20 without the use of ordinary mooring lines. To reduce the ship movement as much as possible, this can be done as illustrated by Bridgestone, Japan, with cell fenders with a high hysteretic effect acting against each other.

14.7.4 Temperature

The influence of temperature on a cell fender is illustrated in Figure 14.21. As the temperature becomes lower, the reaction force will rise, and for a rise in temperature the reaction force will decrease. For

Table 14.1 Fender properties[a]

Property	Testing standard	Condition	Requirement
Moulded fenders			
Tensile strength	DIN 53504; ASTM D412 Die C; AS 1180.2; BS ISO 37; JIS K 6251	Original Aged for 96 h at 70°C	16.0 MPa (min.) 12.8 MPa (min.)
Elongation at break	DIN 53504; ASTM D412 Die C; AS 1180.2; BS ISO 37; JIS K 6251	Original Aged for 96 h at 70°C	350% 280%
Hardness	DIN 53505; ASTM D2240; AS 1683.15.2; JIS K 6253	Original Aged for 96 h at 70°C	788 shore A (max.) Original +8° shore A (max.)
Compression set	ASTM D395 Method B; AS 1683.13 Method B; BS 903 A6; ISO 815; JIS K 6262	22 h at 70°C	30% (max.)
Tear resistance	ASTM D624 Die B; AS 1683.12; BS ISO 34-1; JIS K 6252	Original	70 kN/m (min.)
Ozone resistance	DIN 53509; ASTM D1149; AS 1683-24; BS ISO 1431-1; JIS K 6259	50 pphm at 20% strain, 40°C, 100 h	No cracks
Seawater resistance	BS ISO 1817; ASTM D471	28 days at 95°C	Hardness: ±10° shore A (max.) Volume: +10/−5% (max.)
Abrasion	ASTM D5963-04; BS ISO 4649: 2002 BS 903 A9, Method B	Original 3000 revolutions	100 mm³ (max.) 1.5 cc (max.)
Bond strength	ASTM D429, Method B; BS 903.A21 Section 21.1	Rubber to steel	7 N/mm (min.)
Dynamic fatigue[b]	ASTM D430-95, Method B	15 000 cycles	Grade 0–1[c]
Extruded and wrapped fenders			
Tensile strength	DIN 53504; ASTM D412 Die C; AS 1180.2; BS ISO 37; JIS K 6251	Original Aged for 96 h at 70°C	13.0 MPa (min.) 10.4 MPa (min.)
Elongation at break	DIN 53504; ASTM D412 Die C; AS 1180.2; BS ISO 37; JIS K 6251	Original Aged for 96 h at 70°C	280% (min.) 224% (min.)
Hardness	DIN 53505; ASTM D2240; AS 1683.15.2; JIS K 6253	Original Aged for 96 h at 70°C	78° shore A (max.) Original +8° shore A (max.)
Compression set	ASTM D395 Method B; AS 1683.13 Method B; BS 903 A6; ISO 815; JIS K 6262	22 h at 70°C	30% (max.)
Tear resistance	ASTM D624 Die B; AS 1683.12; BS ISO 34-1; JIS K 6252	Original	60 kN/m (min.)

Table 14.1 Continued

Property	Testing standard	Condition	Requirement
Ozone resistance	DIN 53509; ASTM D1149; AS 1683-24; BS ISO 1431-1; JIS K 6259	50 pphm at 20% strain, 40°C, 100 h	No cracks
Seawater resistance	BS ISO 1817; ASTM D471	28 days at 95°C	Hardness: ±10° Shore A (max.) Volume: +10/−5% (max.)
Abrasion	ASTM D5963-04; BS ISO 4649: 2002	Original	180 mm (max.)

[a]Material property certificates are issued for each different rubber grade on all orders for SCN Super Cone, SCK Cell Fender, Unit Element, AN/ANP Arch, Cylindrical Fender, MV and MI Elements. Unless otherwise requested at the time of order, material certificates issued for other fender types are based on the results of standard bulk and/or batch tests that form part of routine factory ISO 9001 quality assurance procedures and are for a limited range of physical properties (tensile strength, elongation at break and hardness)

[b]Dynamic fatigue testing is optional at extra cost

[c]Grade 0 = no cracks (pass). Grade 1 = 10 or fewer pinpricks <0.5 mm long (pass). Grades 2–10 = increasing crack size (fail)

temperatures lower than about 35–40°C, the rubber compound should be specially designed, because the brittle point of the rubber is about 55°C. Some fender manufacturers have developed a special rubber type for arctic conditions (Figure 14.22).

14.7.5 Friction

The friction coefficient between the ship and the fender itself will depend upon the surface materials of the fender. To prevent damage to the fender due to the forward and/or backwards movements of the ship, the friction coefficient should be as low as possible. The following friction coefficients can be assumed:

(a) steel to special low friction materials, 0.1–0.2
(b) steel to steel, 0.2–0.3
(c) steel to timber, 0.4–0.6
(d) steel to rubber, 0.6–0.7.

Figure 14.19 Hysteresis effect

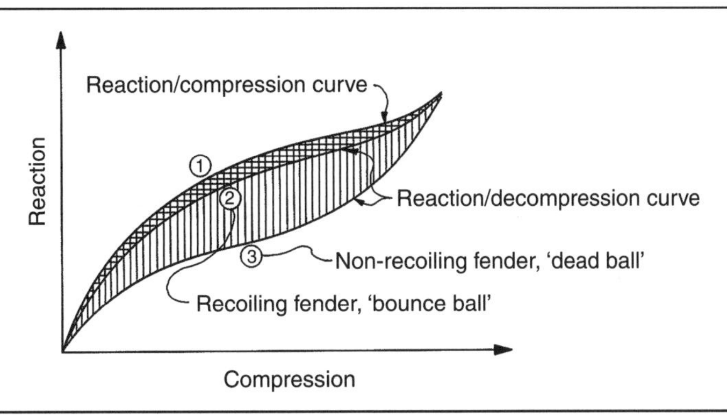

Figure 14.20 Permanent fender mooring

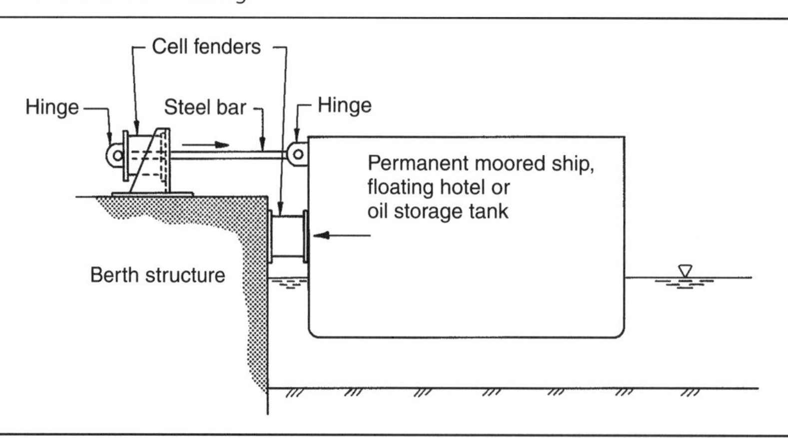

Generally, during tension or forced mooring, high lateral or horizontal resistance may be needed against surge movements due to long periodic waves such as seiches acting along the berth front, and a low vertical resistance against heave, roll and pitch. Fender walls covered with hardwood, such as azobe, have a horizontal friction factor of only about 0.3, and for vertical movement about 0.2. On the other hand, fender walls with small cylindrical rubber rollers with a horizontal axis can provide a horizontal friction coefficient of about 0.6–0.7, but only a vertical friction coefficient of about 0.1.

Since most fender systems are weak against forces acting parallel to the berthing face from berthing and moored ships, the fender wall must be designed for these parallel forces, which can act both vertically and longitudinally. The most common way to prevent failure of the fender wall and the fender is to anchor the fender wall by chains in such a way as to limit the vertical and longitudinal motion, and to use low-friction plastic pads on the fender walls.

Figure 14.21 The compressive performance of a cell fender at different temperatures

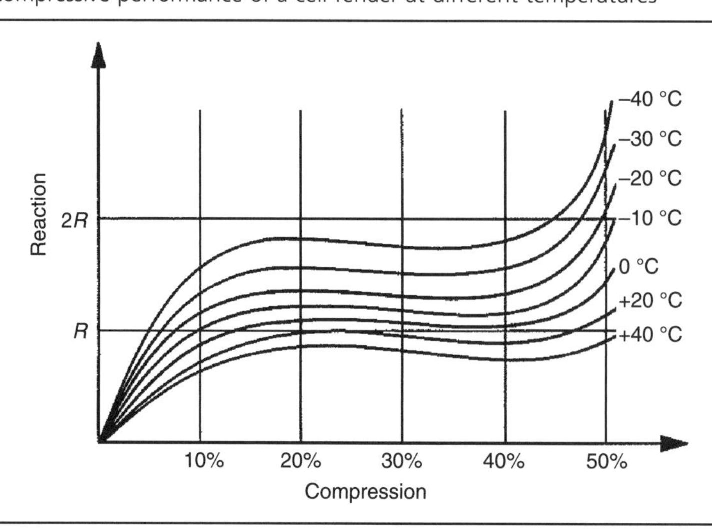

Figure 14.22 Fenders in artic conditions

14.8. Single- and double-fender systems

In the single-fender system, one layer of fenders is mounted on the berth front. Examples are shown in Figures 14.23(a) and 14.23(b). In the double-fender system, two layers of fenders are used, one outside the other, with a plate or wall in between, as shown in Figures 14.23(c) and 14.23(d).

The advantage offered by the double-fender system (Figure 14.23(c)) is that it can absorb twice as much impact energy for the same reaction forces. However, the impact force on the berth structure will remain the same with both systems. In other words, the fender factor will be halved when using this double-fender system. Under normal conditions, the impact loads during berthing are small, but upon maximum impact caused by a ship striking a berth structure, these double fenders are required to absorb abnormally high energy without causing damage to either the ship or the berth.

The double-fender system shown in Figures 14.23(c) and 14.23(d) is often called an ideal fender system because of its energy absorption and reaction force characteristics. When a cell fender and a cylindrical fender are combined in a double-fender system (Figure 14.24), the cylindrical fender will 'soften' the reaction/compression characteristics of the double-fender unit. This will make the double fender more useful, and it will act as an energy-absorbing fender also for smaller ships.

Figure 14.23 Single- and double-fender systems

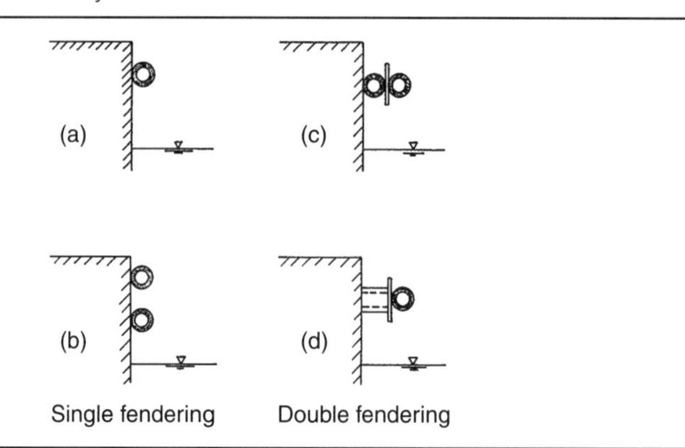

The cylindrical fender must be so large that the reaction force when the cylindrical fender is closed and the compression is equal to about 50% of the OD is equal to the reaction force needed to compress the cell fender. In order to prevent the cylindrical fender from being compressed by more than about 50%, a compression stopper or protector can be mounted (Figure 14.24).

The design of a double-fender system should be arrived at according to the trial-and-error method, and the procedure is as follows:

(a) calculate the ship's impact energy
(b) choose an impact force P equal to the horizontal force that the berth structure or ship's hull can resist divided by a safety factor
(c) check that the total fender energy absorbed is at least equivalent to the ship's impact energy
(d) check the fender factor.

It is very difficult to design a fender system that can be a soft fender system for smaller ships and at the same time is a good energy-absorbing fender system for larger ships. The Port of Reykjavik, Iceland,

Figure 14.24 Reaction/compression characteristics of double fender

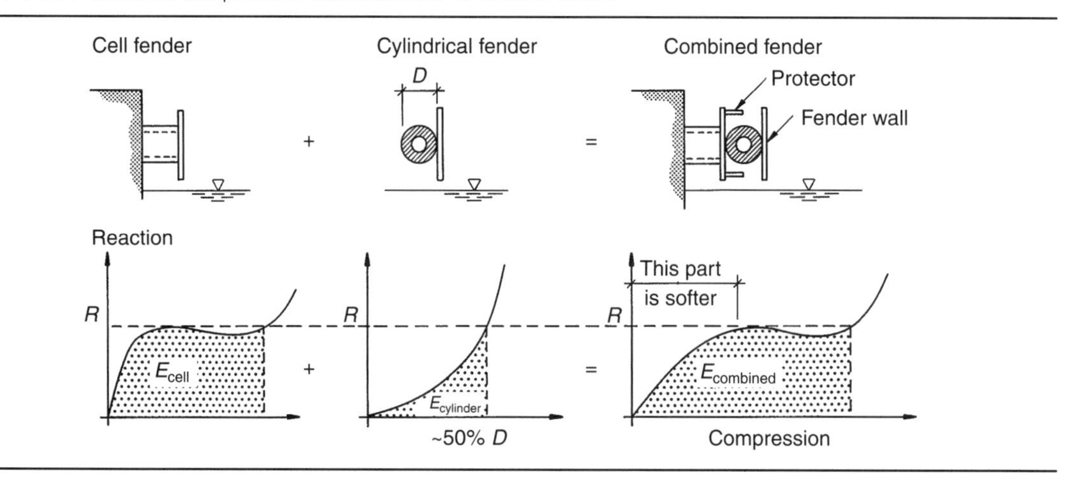

Figure 14.25 The RTT fender system

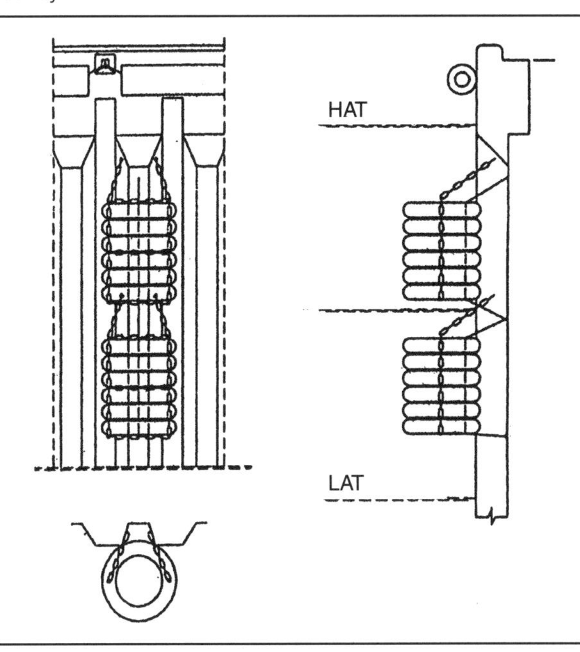

has developed a fender system called the Reykjavik Truck Tyre (RTT) fender system (Figure 14.25). It consists of six connected truck tyres with an outside diameter of about 1.1 m in a stack suspended on the outside berth structure constructed of steel sheet piles. On the top concrete cap beam, energy-absorbing fenders are mounted, as indicated in Figure 14.25, using cylindrical fenders.

The RTT system is a combination of buckling fenders at the top of the berth structure, and tyre fenders hanging along the lower front sheet pile wall with their fender line a little outside the fender line for the buckling fender. This fender system has good fender properties for the smaller ship but at the same time it is soft enough for the larger ship to compress the tyre fenders before it berths against the buckling fender. The system has proved suitable for ships in the range 4500–60 000 dwt.

With the concrete cap beam about 0.3 m outside the front of the steel sheet piles, while using a cylindrical or buckling fender protruding about 0.4–0.5 m outside the concrete berth line, the buckling fender will act together with the tyre fenders when the tyres have been compressed about 40–50%.

14.9. Fender wall

It is customary to berth the ship directly to the fender itself (Figure 14.26(a)), for both single- and double-fender systems. The fender here is suspended from the berth front, and the ships will have direct contact with the fender. Chains suspend the ordinary cylindrical rubber fenders, whereas the larger cylindrical fenders are suspended by a bracket or ladder system. While absorbing the berthing energy of a ship, the fender will give a reaction force to both the ship and the berth structure. Under normal berthing conditions, no plastic deformation of the ship's hull should take place.

To prevent excessive concentration of the ship's mooring forces, as well as berthing forces, both to the fendering system itself and to the ship's hull, a protection panel or wall should be provided as required to reduce the face pressure.

Figure 14.26 Berthing directly to a fender or fender plate

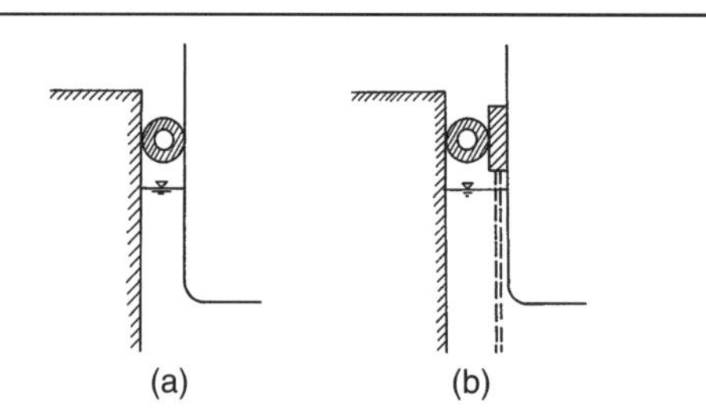

If there are large differences in the tidal range, or in the ship's waterline (loaded or in ballast), or if the ship's side requires a greater contact surface against the fender structure to restrict the hull-bearing pressure on the ship, a fender wall is usually placed between the fenders and the ship's hull (Figure 14.26). This fender wall method can also be applied when minimum or maximum friction between the ship and the berth structure is required. Fender walls, made of steel, azobe or greenheart, with a rubber fender behind, have proved to be economic as well as effective. When tankers berth, the fender system will have to absorb energy from about 1000 kNm (small tankers) up to 6000 kNm (larger tankers). To obtain the dimensions of the fender wall, divide the impact force by the permitted ship hull load/m^2. Figure 14.27 shows how two cell fenders are mounted behind a fender wall, and Figure 14.28 displays cell fenders with fender walls.

Below, different fender systems with a fender wall in front of the fender are shown. In principle, there are the following two wall systems:

(a) the fender wall can rotate in front of the fender (Figure 14.28)
(b) the fender wall moves parallel sidewise regardless of where the flare angle or the ship fender belting is hitting the fender wall (Figures 14.29–14.31).

Figure 14.27 Bridgestone cell fenders

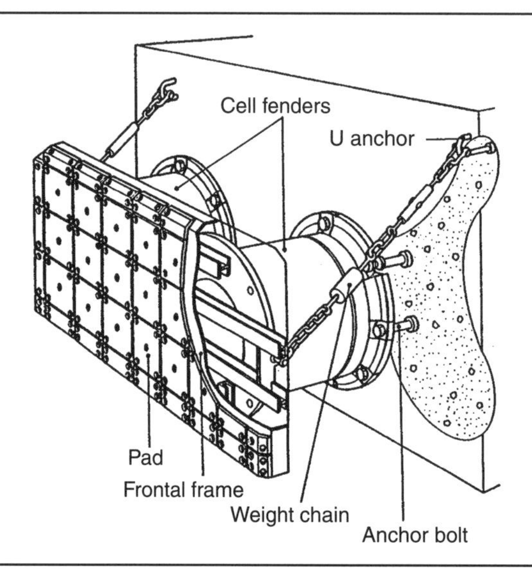

Figure 14.28 Bridgestone cell fenders with a fender wall

Figure 14.29 The Trelleborg parallel fender wall system

Figure 14.30 The Trelleborg parallel fender wall system with cone fenders

14.10. Hull pressure

Due to the variety in both the design and the type of ships, there are no firm or exact values for the allowable hull-bearing pressure associated with different types or sizes of ship, but as a very rough guideline the permissible hull pressures in Table 14.2 can be used. Ships with belting produce a line load on the fenders that can be considerably higher than the hull pressure indicated in this table.

The hull resistance to impact must be at least equal to the maximum hydrostatic pressure, which can act on the ship's hull.

Special attention should be paid to the horizontal chains on a fender panel. When chains are installed below the fender, the rotation of the fender panel, due to the ship's flare, can be restricted. Line loads may occur that exceed the permissible hull pressure.

Figure 14.31 The Trelleborg parallel fender wall system with unit element fenders

14.11. Spacing of fenders

The spacing of fenders varies from berth structure to berth structure, depending on the type of structure, the requirements to be met by the berth and the type of ships using the berth. For ships not using tugboat assistance during berthing, the fender spacing will, in most cases, be determined by the smallest ship using the berth and by the ship hull radius of curvature (Figure 14.32). The fender spacing will also depend on the fender height and the compression of the fenders.

Generally, to ensure that all ships can be supported at the berth, the fender spacing will be about 5–10% of the ship's length for ships up to about 20 000 dwt. For larger ships, the spacing can be about 25–50% of the ship's length, if the ship berths with tugboat assistance. For optimum effect, the fenders for larger ships should be located close to the ends of the straight-sided section of the ship.

Table 14.2 Hull pressures

Type of ship	Hull pressure: kN/m^2
Container ships	
1st and 2nd generation	<400
3rd generation	<300
4th generation	<250
5th and 6th generation	<200
Oil tankers	
<60 000 t displacement	250–350
>60 000 t displacement	<350
VLCCs	150–200
Bulk carriers	<200
Gas tankers (LNG/LPG)	<200

14.12. Cost of fenders

Figure 14.33 gives a general indication of the relative price/lin m of cylindrical rubber fenders for a single-fender system with respect to fender factor. As can be seen, the relative price remains approximately constant for surface-protecting fenders but increases sharply according to energy-absorbing demands. The cost of the fenders should take into account the frequency of the berthing operations. A high frequency of berthing will normally justify a greater capital expenditure for the fender system.

It should be noted that if energy-absorbing fenders with a fender factor of 3 or lower are required, it could be more advantageous to install a double-fender system instead of a single-fender system. This is in spite of the fact that a double-fender system involves higher maintenance costs. For instance, a single-fender system (see Figure 14.23(a)) with fender factor of 2 will have a relative price/lin m of about 25, whereas two single-fender systems, each with fender factor of 4, mounted as a double-fender system (as in Figure 14.23(c)), which will give the fender system a fender factor of 2, will have a relative price of about $2 \times 10 = 20$.

Figure 14.32 The spacing of fenders

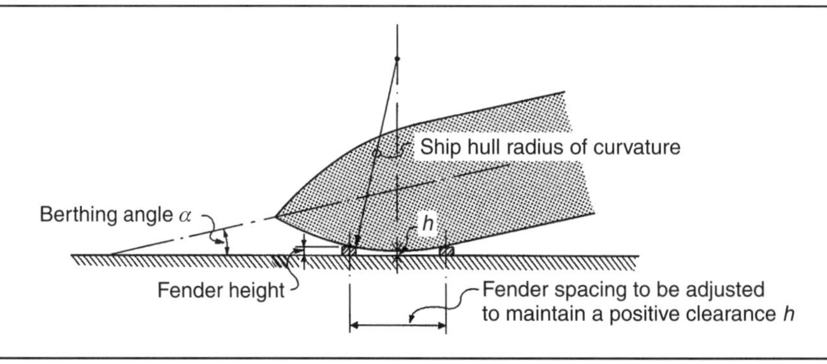

Figure 14.33 The relative price of cylindrical rubber fenders

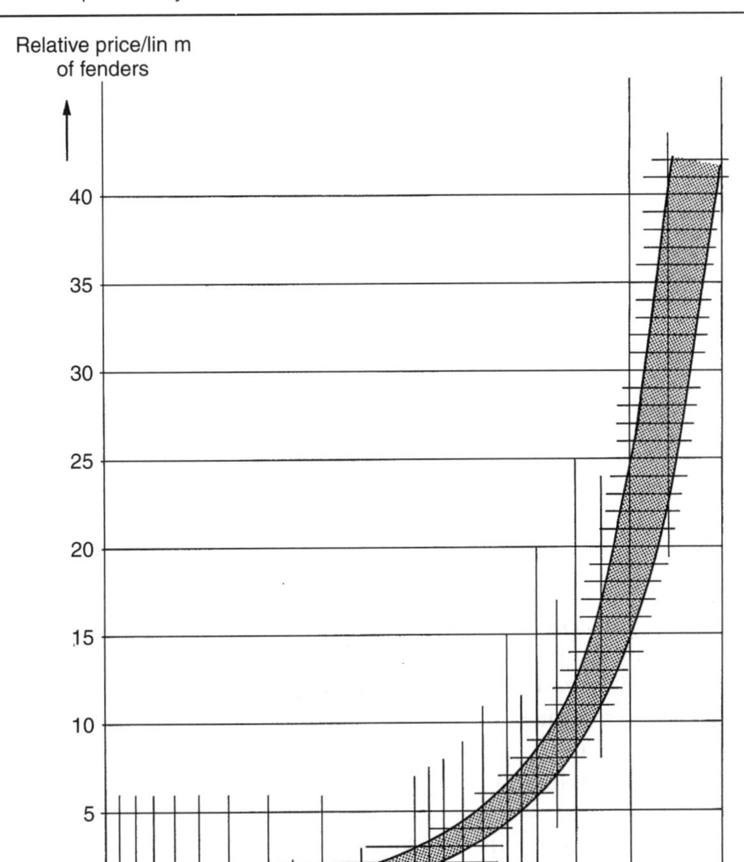

14.13. Damage to fender structures

In general, damage to fender structures can be divided into two groups:

(a) damage to be paid by the port owner, such as ordinary wear and tear by ships, or consequences suffered through incorrect type of fender, faulty mounting, etc.
(b) damage to be paid by the ship owner due to crashing into the berth structure during berthing, damage caused by ships' steel fenders, etc.

Apart from ordinary wear and tear, a port owner should be spared the damage mentioned under item (a). Of those mentioned under item (b), the damage from ships' steel fenders will cause annoyance and create extra work. Even if insurance companies compensate for damage inflicted, no payment is provided for the work involved in obtaining the compensation. More often, it pays to invest more money in the construction or upgrading of berth structures so that ships with steel fenders do not cause damage apart from ordinary wear and tear.

Figure 14.34 Ship's steel fenders

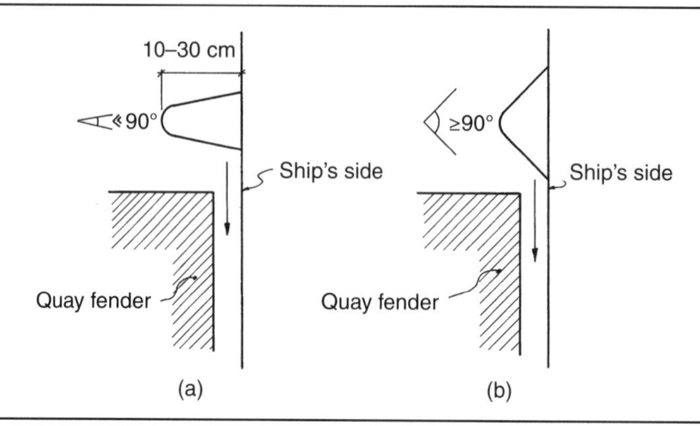

It has often been suggested that the use of steel fenders or belting should be prohibited. However, it would be better to urge ship owners to invest in steel fenders of such types that do less harm to the berth structures. Ship owners are undoubtedly right in stating that steel fenders protect the ship's sides when entering sluices, docks, etc.

Figure 14.34(a) shows the most common type of ship's steel fenders or belting. When the water level falls or while a ship takes on cargo, it is possible for a ship's fender to get stuck on top of the berth fender and that the ship will have difficulty in becoming detached. At worst, the berth fender may break. With the type shown in Figure 14.34(b), the ship will have no difficulty in sliding off the berth fender without damage to the latter. This type of ship's fender will also cause less chafing against wooden fender structures during loading and unloading.

Figure 14.35 illustrates three different types of fender as opposed to ship's steel fenders. Type (a) shows a pneumatic fender, (b) a hardwood fender wall and (c) a steel pile fender on the outside of the rubber fender. A wall made of hardwood has proved to be resistant to wear and tear, but in most cases a steel fender wall would be a better choice. The low maintenance cost of a fender wall often justifies its high initial cost.

When using a fender wall, the wall should always be constructed at the top or bottom, as shown in Figure 14.36.

Figure 14.35 Different ways of fendering the berth against a ship's steel fender

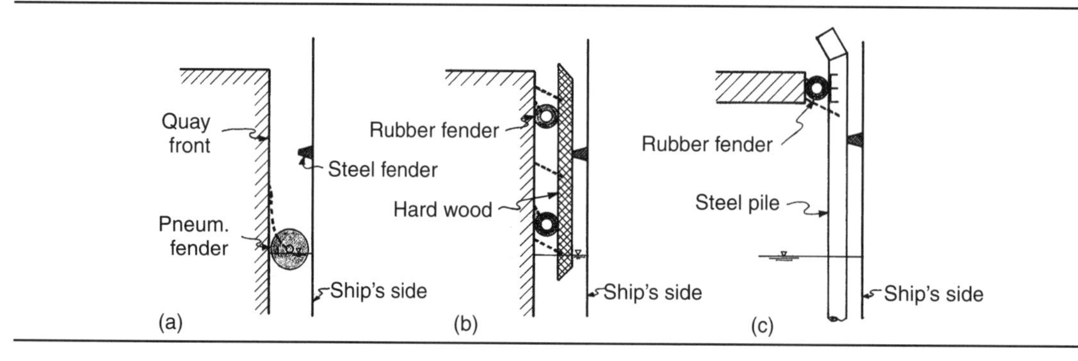

Figure 14.36 Detail of a fender wall

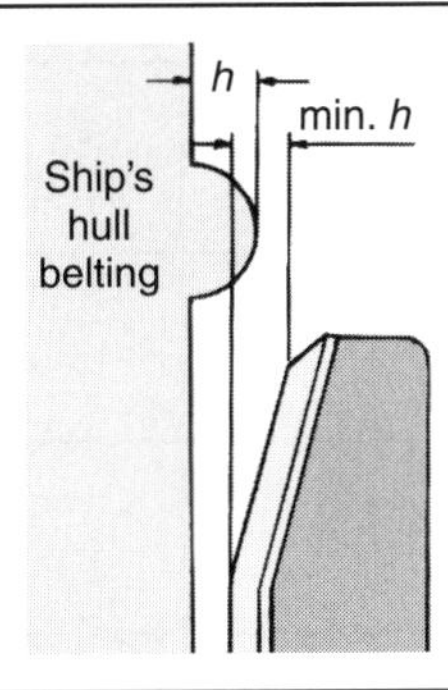

14.14. Calculation examples

The calculations are based on the theory described in Chapter 5.

Example 1

Figure 14.37 shows a cross-section of an open pier structure for ships of 6000 t displacement at the 95% confidence level. The ship's length, width and depth are approximately 105 m, 14.2 m and 6.4 m, respectively. It is assumed that $\phi = 60°$, and that the hydrodynamic mass coefficient is only about 70% because the ship does not move perpendicularly to its longitudinal axis. The ship's berthing velocity of approach is 0.25 m/s of the confidence level. Thus

$$C_H = \frac{6000 + (1/4 \times \pi \times 1.03 \times 6.4^2 \times 105) \times 70\%}{6000} = 1.4$$

It is assumed that r (the distance from the centre of gravity of the vessel to the point of contact) = 0.3 × 1 (i.e. $r/1 = 0.3$), which gives $C_E = 0.48$. Assume that $C_C = 1.0$ and $C_S = 0.95$ and the adjusting factor $C = C_H \times C_E \times C_C \times C_S = 0.64$.

The energy to be absorbed by the fender structure without an abnormal impact factor will be

$$E_f = 0.64 \times (0.5 \times 6000 \times 0.25^2) = 120 \text{ kNm}$$

Figure 14.37 Example cross-section of an open pier structure

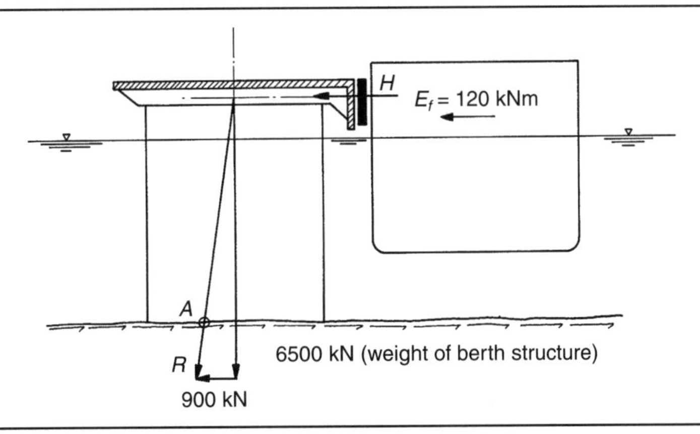

Figure 14.38 Curves representing different energy absorptions

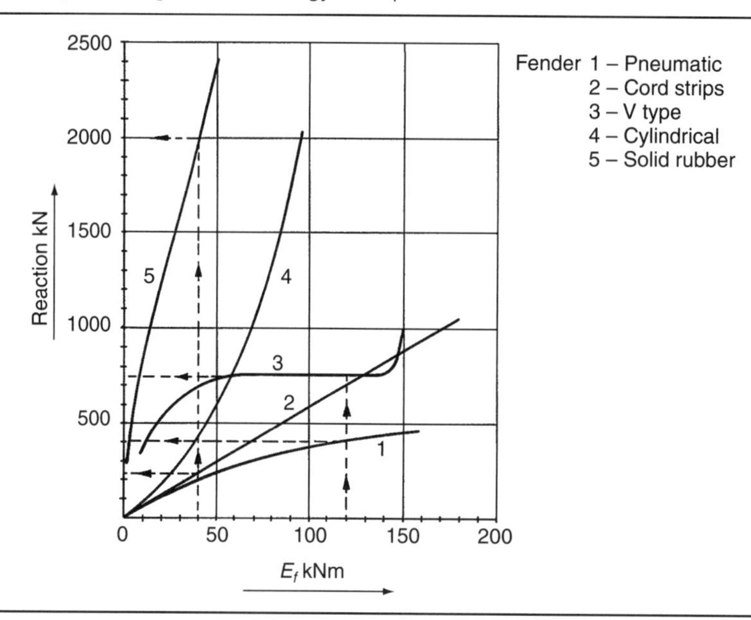

Out of the five different fenders represented in Figure 14.38, fenders 2, 4 and 5 absorb the impact energy over a fender length of 3 m, fenders 1 and 3 over a length of 3.5 m and 2 m, respectively. Curves 1 and 3 give the total reaction force, whereas curves 2, 4 and 5 give the reaction force, in linear metres, which has to be multiplied by 3 m to obtain the total reaction force. This is shown in Figure 14.39, where the type, the dimension, the cross-section, the load/lin m and the total reaction force are given.

The pneumatic fender, type 1, gives the lowest load factor. Because of its dimensions, it is best suited for oil harbours where the distance to berth front is of minor importance as ships are unloaded by means of pipes and loading arms.

Figure 14.39 Fender factors for different types of rubber fender

No.	Rubber fender (type and dimensions)	Cross-section	E_f and P noted	Loading (kN/lin m)	Total impact reaction (P kN)	P/E_f
1	Pneumatic $\phi = 2$ m $l = 3.5$ m	○	Total		400	3.33
2	Cord strips $\phi = 1$ m	○	Per lin m	230	690	5.75
3	V type $h = 0.5$ m 1 = 2 m	⏢	Total		750	6.25
4	Cylindrical $\phi = 0.61$ m	○	Per lin m	400	1200	10
5	Solid $h = 0.15$ m	▬	Per lin m	2000	6000	50

The other fenders, types 2, 3 and 4, have the same criteria where general cargo berth structures are concerned. Here, it is just a question of how much one is willing to, or has to, pay to absorb the impact energy on account of berth stability, etc.

Fender 5 is included to give a general idea of the various types of fenders. However, its purpose is not so much to absorb the energy as to protect both ships and berth structures against rubbing and damage.

If the resultant R in Figure 14.37 acts at A (i.e. within the mid one-third of the width of the structure), the horizontal force should not exceed 900 kN. With a safety factor of 1.3, the dimensional horizontal force is $900/1.3 = 700$ kN; that is, a fender with a lower fender factor than $P/E_f = 700/120 = 5.8$ kN/kNm must be found. According to Figure 14.39, fenders 1 and 2 with respective fender factors of 3.33 and 5.75 kN/kNm will meet these requirements. Fender 4, mounted as a double-fender system, will also meet these requirements.

Example 2

Figure 14.40 shows a calculation example for a double-fendering system where the berthing energy is 500 kNm and the horizontal force on the shipside or the berth structure should not exceed 1000 kN,

Figure 14.40 Calculation example for a double-fender system

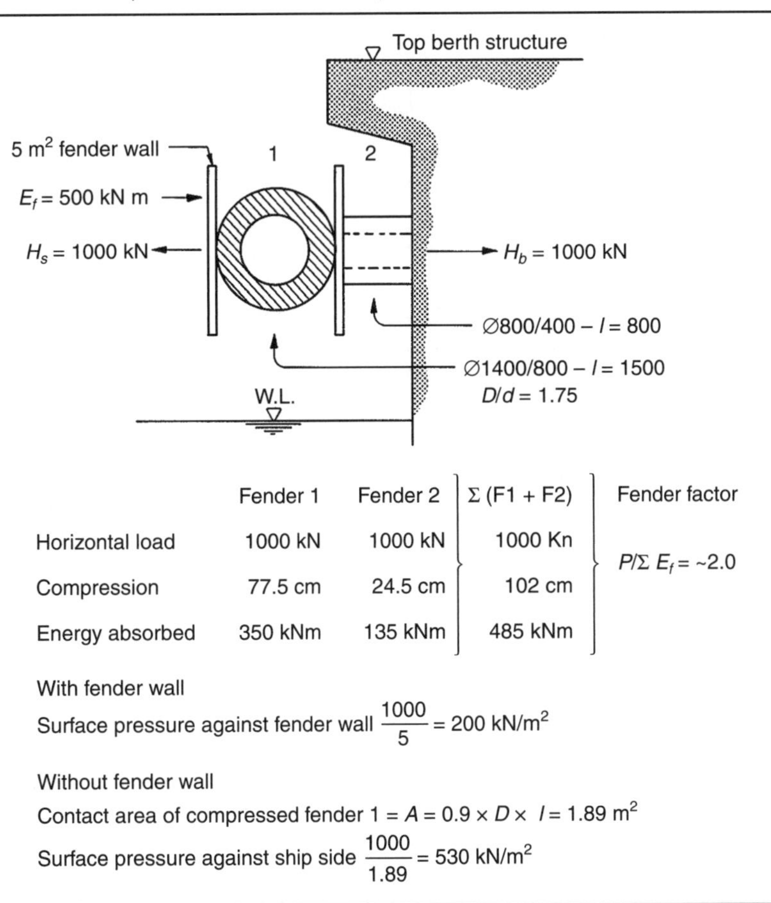

	Fender 1	Fender 2	Σ (F1 + F2)	Fender factor
Horizontal load	1000 kN	1000 kN	1000 Kn	$P/\Sigma E_f = \sim 2.0$
Compression	77.5 cm	24.5 cm	102 cm	
Energy absorbed	350 kNm	135 kNm	485 kNm	

With fender wall

Surface pressure against fender wall $\frac{1000}{5} = 200$ kN/m^2

Without fender wall

Contact area of compressed fender 1 = $A = 0.9 \times D \times l = 1.89$ m^2

Surface pressure against ship side $\frac{1000}{1.89} = 530$ kN/m^2

Table 14.3 Energy absorption and reaction forces on different sizes of 1000 mm-long cylindrical fenders

D: mm	*d*: mm	*R*: kN	*E*: kNm	*P*: kN/m^2	ε: *E*/*R*	Weight: kg/m
Extruded						
100	50	43	0.8	547	0.019	7.0
125	65	51	1.3	500	0.025	10.6
150	75	65	1.8	552	0.028	15.6
175	75	92	2.7	781	0.029	23.2
200	90	98	3.5	693	0.036	29.6
200	100	86	3.3	547	0.038	27.8
250	125	108	5.1	550	0.047	43.4
300	150	129	7.4	547	0.057	62.6
380	190	164	11.8	550	0.072	100.4
400	200	172	13.1	547	0.076	111.2
450	225	194	16.6	549	0.086	140.8
500	250	275	28	700	0.102	175
Wrapped						
600	300	330	40	700	0.121	253
700	400	325	52	517	0.160	309
750	400	380	61	605	0.161	377
800	400	440	72	700	0.164	449
875	500	406	81	517	0.200	482
925	500	461	93	587	0.202	567
1000	500	550	112	700	0.204	702
1050	600	487	117	517	0.240	695
1100	600	541	131	574	0.242	795
1200	600	660	162	700	0.245	1010
1200	700	542	151	493	0.279	889
1300	700	650	184	591	0.283	1122
1300	750	595	178	505	0.299	1055
1400	700	770	220	700	0.286	1375
1400	750	705	214	598	0.304	1307
1400	800	649	208	516	0.320	1235
1500	750	825	253	700	0.307	1579
1500	800	760	246	605	0.324	1506
1600	800	880	288	700	0.327	1796
1600	900	757	273	535	0.361	1637
1650	900	812	295	574	0.363	1789
1750	900	929	340	657	0.366	2107
1750	1000	811	325	516	0.401	1929
1800	900	990	364	700	0.368	2273
1850	1000	921	372	586	0.404	2266
2000	1000	1101	450	701	0.409	2806
2000	1200	871	415	462	0.476	2395
2100	1200	974	467	517	0.479	2778
2200	1200	1083	524	575	0.484	3180
2400	1200	1321	647	701	0.490	404

based on a 95% confidence level for the displacement, a 50% confidence level of the berthing velocity and a factor for abnormal impact of 1.5.

14.15. Information from fender manufacturers

The performance values supplied by fender manufacturers for different fender types are given in the following sections.

14.15.1 Different marine fendering systems

Cylindrical fenders

The energy absorption and reaction forces for cylindrical fenders are shown in Table 14.3. The fender performances in the table are for 1000 mm length and for a rated deflection equal to the ID. For a stable installation, the length *L* should be larger than the outside diameter *D*.

Super cone fenders

All energy absorption and reaction force values are at a rated deflection of 72%. The maximum deflection of the fender is 75%, and where the compressive loads may exceed the maximum fender reaction, an overload stopper may be used. The cone fenders are shown in Figure 14.41, and the energy and reaction forces are shown in Table 14.4.

Figure 14.41 Trelleborg super cone fenders

Table 14.4 Energy absorption and reaction forces on different sizes of cone fenders

Super cone

Energy index	Fender size	SCN 300	SCN 350	SCN 400	SCN 500	SCN 550	SCN 600	SCN 700	SCN 800	SCN 900	SCN 1000	SCN 1050	SCN 1100	SCN 1200	SCN 1300	SCN 1400	SCN 1600	SCN 1800	SCN 2000
E0.9	*E*: kNm	7.7	12.5	18.6	36.5	49	63	117	171	248	338	392	450	585	743	927	1382	1967	2700
	R: kN	59	80	104	164	198	225	320	419	527	653	720	788	941	1103	1278	1670	2115	2610
E1.0	*E*: kNm	8.6	13.9	20.7	40.5	54	70	130	190	275	375	435	500	650	825	1030	1535	2185	3000
	R: kN	65	89	116	182	220	250	355	465	585	725	800	875	1045	1225	1420	1855	2350	2900
E1.1	*E*: kNm	8.9	14.4	21.4	41.9	56	72	134	196	282	385	447	514	668	847	1058	1577	2244	3080
	R: kN	67	91	119	187	226	257	365	478	601	745	822	899	1073	1258	1459	1905	2413	2978
E1.2	*E*: kNm	9.2	14.8	22.1	43.2	58	74	137	201	289	395	458	527	685	869	1085	1618	2303	3160
	R: kN	68	93	122	191	231	263	374	490	617	764	843	923	1101	1291	1497	1955	2476	3056
E1.3	*E*: kNm	9.5	15.3	22.8	44.6	59	76	141	207	296	405	470	541	703	891	1113	1660	2362	3240
	R: kN	70	96	125	196	237	270	384	503	633	784	865	947	1129	1324	1536	2005	2539	3134
E1.4	*E*: kNm	9.8	15.7	23.5	45.9	61	78	144	212	303	415	481	554	720	913	1140	1701	2421	3320
	R: kN	72	98	128	200	242	276	393	515	649	803	886	971	1157	1357	1574	2055	2602	3212
E1.5	*E*: kNm	10.1	16.2	24.2	47.3	63	80	148	218	310	425	493	568	738	935	1168	1743	2480	3400
	R: kN	74	100	131	205	248	283	403	528	665	823	908	995	1185	1390	1613	2105	2665	3290
E1.6	*E*: kNm	10.4	16.7	24.8	48.6	65	82	151	223	317	435	504	581	755	957	1195	1784	2539	3480
	R: kN	75	102	133	209	253	289	412	540	681	842	929	1019	1213	1423	1651	2155	2728	3368
E1.7	*E*: kNm	10.6	17.1	25.5	50.0	67	84	155	229	324	445	516	595	773	979	1223	1826	2598	3560
	R: kN	77	104	136	214	259	296	422	553	697	862	951	1043	1241	1456	1690	2205	2791	3446
E1.8	*E*: kNm	10.9	17.6	26.2	51.3	68	86	158	234	331	455	527	608	790	1001	1250	1867	2657	3640
	R: kN	79	107	139	218	264	302	431	565	713	881	972	1067	1269	1489	1728	2255	2854	3524
E1.9	*E*: kNm	11.2	18.0	26.9	52.7	70	88	162	240	338	465	539	622	808	1023	1278	1909	2716	3720
	R: kN	80	109	142	223	270	309	441	578	729	901	994	1091	1297	1522	1767	2305	2917	3602

E2.0	*E*: kNm	11.5	18.5	27.6	54.0	72	90	165	245	345	475	550	635	825	1045	1305	1950	2775	3800
	R: kN	82	111	145	227	275	315	450	590	745	920	1015	1115	1325	1555	1805	2355	2980	3680
E2.1	*E*: kNm	11.8	19.0	28.3	55.4	74	93	169	252	355	488	565	652	847	1074	1341	2003	2851	3904
	R: kN	84	114	149	233	283	324	462	606	765	945	1042	1145	1361	1597	1853	2418	3060	3778
E2.2	*E*: kNm	12.1	19.4	29.0	56.7	76	96	173	258	364	501	580	669	869	1102	1376	2056	2926	4008
	R: kN	86	117	153	239	290	332	474	621	785	969	1069	1174	1396	1638	1901	2480	3139	3876
E2.3	*E*: kNm	12.4	19.9	29.7	58.1	77	99	177	265	374	514	595	686	891	1131	1412	2109	3002	4112
	R: kN	89	120	157	246	298	341	486	637	805	994	1096	1204	1432	1680	1949	2543	3219	3974
E2.4	*E*: kNm	12.7	20.3	30.4	59.4	79	102	181	271	383	527	610	703	913	1159	1447	2162	3077	4216
	R: kN	91	123	161	252	305	349	498	652	825	1018	1123	1233	1467	1721	1997	2605	3298	4072
E2.5	*E*: kNm	13.0	20.8	31.1	60.8	81	105	185	278	393	540	625	720	935	1188	1483	2215	3153	4320
	R: kN	93	126	165	258	313	358	510	668	845	1043	1150	1263	1503	1763	2045	2668	3378	4170
E2.6	*E*: kNm	13.3	21.3	31.8	62.2	83	108	189	284	402	553	640	737	957	1216	1518	2268	3228	4424
	R: kN	95	129	169	264	320	366	522	683	865	1067	1177	1292	1538	1804	2093	2730	3457	4268
E2.7	*E*: kNm	13.5	21.7	32.5	63.5	85	111	193	291	412	566	655	754	979	1245	1554	2321	3304	4528
	R: kN	97	132	173	270	328	375	534	699	885	1092	1204	1322	1574	1846	2141	2793	3537	4366
E2.8	*E*: kNm	13.8	22.2	33.2	64.9	86	114	197	297	421	579	670	771	1001	1273	1589	2374	3379	4632
	R: kN	100	135	177	277	335	383	546	714	905	1116	1231	1351	1609	1887	2189	2855	3616	4464
E2.9	*E*: kNm	14.1	22.6	33.9	66.2	88	117	201	304	431	592	685	788	1023	1302	1625	2427	3455	4736
	R: kN	102	138	181	283	343	392	558	730	925	1141	1258	1381	1645	1929	2237	2918	3696	4562
E3.0	*E*: kNm	14.4	23.1	34.6	67.6	90	120	205	310	440	605	700	805	1045	1330	1660	2480	3530	4840
	R: kN	104	141	185	289	350	400	570	745	945	1165	1285	1410	1680	1970	2285	2980	3775	4660
E3.1	*E*: kNm	15.9	25.4	38.1	74.4	99	132	226	341	484	666	770	886	1150	1463	1826	2728	3883	5324
	R: kN	114	155	204	318	385	440	627	820	1040	1282	1414	1551	1848	2167	2514	3278	4153	5126
Efficiency ratio (ε)		0.138	0.163	0.186	0.232	0.256	0.290	0.364	0.414	0.466	0.518	0.544	0.571	0.622	0.674	0.725	0.830	0.932	1.036

Figure 14.42 Installation of cell fenders

Cell fenders

All energy absorption and reaction force values are at a rated deflection of 52%. The fenders are, in principle, similar to the super cone fenders. The installation of the cell fenders is shown in Figure 14.42, their dimensions in Figure 14.43, and the energy and reaction forces are in Table 14.5.

Unit element fenders

Unit element fenders are shown in Figure 14.44, installed behind a fender wall.

All energy absorption and reaction force values are at a rated deflection of 57.5%. The maximum deflection of the fender is 62.5%. The performance values are for a single element, 1000 mm long. The unit element fenders are shown in Table 14.6.

Arch fenders

All energy absorption and reaction forces values are at a rated deflection of 51.5% for the AN fender and 54% for the ANP fender. The arch fenders are shown in Figure 14.45, and the energy and reaction forces are shown in Table 14.7.

MV and MI element fenders

The principle of the Trelleborg MV and MI element fender system is shown in Figures 14.46–14.48.

The rated performances for one element of the MV and MI element fenders are shown in Tables 14.8 and 14.9.

Figure 14.43 The dimensions of the cell fenders

Dimensions

	H: mm	ØW: mm	ØB: mm	D: mm	d: mm	Anchors/ head bolts	Weight: kg
SCK 400H	400	650	550	25	30	4 × M22	75
SCK 500H	500	650	550	25	32	4 × M24	95
SCK 630H	630	840	700	25	32	4 × M27	220
SCK 800	800	1050	900	30	40	6 × M30	400
SCK 1000H	1000	1300	1100	35	45	6 × M36	790
SCK 1150H	1150	1500	1300	40	50	6 × M42	1 200
SCK 1250H	1250	1650	1450	40	50	6 × M42	1 500
SCK 1450H	1450	1850	1650	42	61	6 × M48	2 300
SCK 1600H	1600	2000	1800	45	61	8 × M48	3 000
SCK 1700H	1700	2100	1900	50	66	8 × M56	3 700
SCK 2000H	2000	2200	2000	50	76	8 × M64	5 000
SCK 2250H	2250	2550	2300	57	76	10 × M64	7 400
SCK 2500H	2500	2950	2700	70	76	10 × M64	10 700
SCK 3000H	3000	3350	3150	75	92	12 × M76	18 500

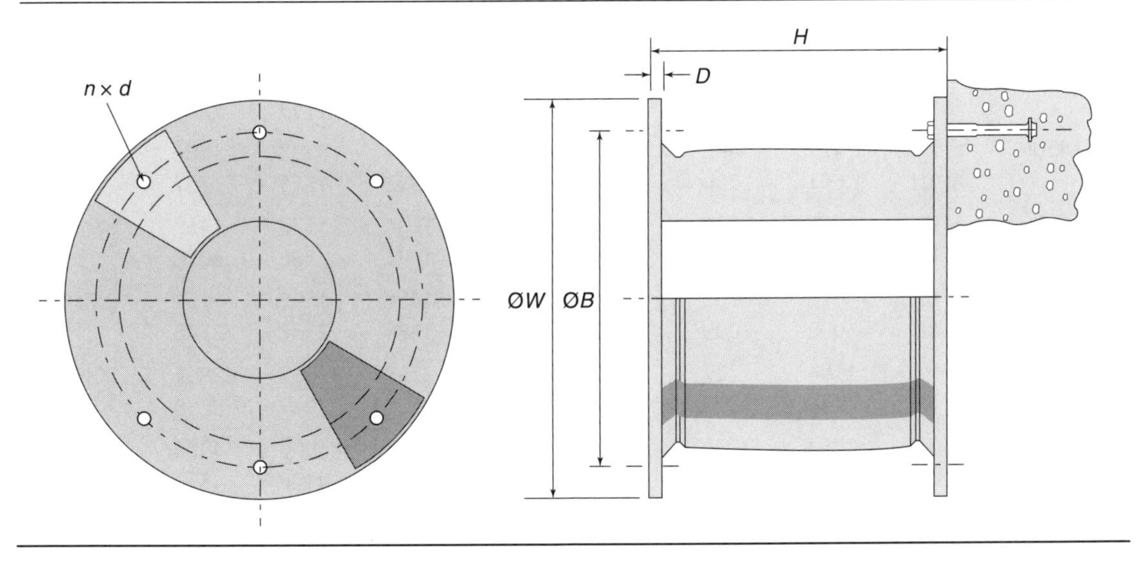

Table 14.5 The rated performance of the cell[a]

		0.9	1	1.5	2	2.5	3	3.1	E/R
SCK400H	*E*: kJ	9	10	13	16	18	21	23	0.174
	R: kN	50	56	74	91	104	118	129	
SCK 500H	*E*: kJ	17	19	25	30	35	39	43	0.213
	R: kN	79	87	115	142	163	184	203	
SCK 630H	*E*: kJ	34	38	50	62	71	80	88	0.277
	R: kN	124	137	180	224	257	290	319	
SCK 800H	*E*: kJ	67	75	100	124	144	163	179	0.351
	R: kN	190	211	283	355	409	464	510	
SCK 1000H	*E*: kJ	138	153	201	249	286	324	356	0.438
	R: kN	314	349	458	568	653	737	811	
SCK 1150H	*E*: kJ	210	233	306	379	436	492	541	0.505
	R: kN	416	462	606	750	863	976	1073	
SCK 1250H	*E*: kJ	269	299	393	486	559	633	696	0.548
	R: kN	491	545	716	887	1020	1153	1269	
SCK 1450H	*E*: kJ	421	468	614	760	874	988	1086	0.637
	R: kN	661	734	969	1193	1372	1551	1707	
SCK 1600H	*E*: kJ	566	629	825	1021	1174	1327	1460	SCN 1400
	R: kN	805	894	1174	1453	1671	1889	2078	
SCK 1700H	*E*: kJ	678	753	989	1225	1408	1592	1751	0.746
	R: kN	908	1009	1325	1641	1886	2132	2345	
SCK 2000H	*E*: kJ	1104	1227	1610	1994	2293	2592	2851	SCN 1400
	R: kN	1258	1397	1833	2268	2605	2942	3236	
SCK 2250H	*E*: kJ	1854	2060	2606	3151	3624	4096	4506	0.988
	R: kN	1876	2085	2637	3189	3668	4146	4561	
SCK 2500H	*E*: kJ	2544	2826	3575	4323	4971	5619	6181	1.098
	R: kN	2317	2574	3256	3937	4528	5119	5631	
SCK 3000H	*E*: kJ	3795	4217	5394	6571	7525	8479	9327	1.152
	R: kN	3310	3678	4683	5688	6526	7363	8099	

[a]Reference testing to the *Design and Testing Manual* (PIANC, 2002). Tested at 2–8 cm/min speed. Tolerance ±10%

Figure 14.44 Unit element fenders

Figure 14.45 Arch fenders

Table 14.6 Energy absorption and reaction forces on different sizes of unit 1000-mm long element fenders

Unit element

Energy index	Fender size	UE 250	UE 300	UE 400	UE 500	UE 550	UE 600	UE 700	UE 750	UE 800	UE 900	UE 1000	UE 1200	UE 1400	UE 1600
E0.9	*E*: kNm	8.1	11.7	21	32.4	40	47	63	73	84	106	131	186	257	337
	R: kN	79	95	113	142	157	171	199	214	228	256	284	340	398	455
E1.0	*E*: kNm	9.0	13.0	23	36.0	44	52	70	81	93	118	146	207	286	374
	R: kN	88	105	126	158	174	190	221	238	253	284	316	378	442	506
E1.1	*E*: kNm	9.3	13.4	24	37.1	45	54	72	84	96	122	150	213	294	385
	R: kN	90	108	130	163	179	196	228	245	261	293	326	389	455	521
E1.2	*E*: kNm	9.6	13.8	24	38.2	47	55	74	86	99	125	155	220	303	396
	R: kN	93	111	134	167	184	201	234	252	268	301	335	401	469	536
E1.3	*E*: kNm	9.9	14.2	25	39.3	48	57	77	89	101	129	159	226	311	407
	R: kN	95	114	137	172	190	207	241	259	276	310	345	412	482	552
E1.4	*E*: kNm	10.2	14.6	26	40.4	49	58	79	91	104	132	163	232	320	418
	R: kN	98	117	141	177	195	212	247	266	283	318	354	424	495	567
E1.5	*E*: kNm	10.5	15.0	27	41.5	51	60	81	94	107	136	168	239	328	429
	R: kN	100	121	145	182	200	218	254	274	291	327	364	435	509	582
E1.6	*E*: kNm	10.8	15.4	27	42.6	52	62	83	96	110	139	172	245	336	440
	R: kN	103	124	149	186	205	224	261	281	299	336	373	446	522	597
E1.7	*E*: kNm	11.1	15.8	28	43.7	53	63	85	99	113	143	176	251	345	451
	R: kN	106	127	153	191	210	229	267	288	306	344	383	458	535	612
E1.8	*E*: kNm	11.4	16.2	29	44.8	54	65	88	101	115	146	180	257	353	462
	R: kN	108	130	156	196	216	235	274	295	314	353	392	469	548	628
E1.9	*E*: kNm	11.7	16.6	29	45.9	56	66	90	104	118	150	185	264	362	473
	R: kN	111	133	160	200	221	240	280	302	321	361	402	481	562	643
E2.0	*E*: kNm	12.0	17.0	30	47.0	57	68	92	106	121	153	189	270	370	484
	R: kN	113	136	164	205	226	246	287	309	329	370	411	492	575	658
E2.1	*E*: kNm	12.3	17.5	31	48.5	59	70	95	109	125	158	195	278	381	499
	R: kN	117	140	169	211	233	253	296	318	339	381	423	507	592	678
E2.2	*E*: kNm	12.6	18.0	32	50.0	61	72	98	112	128	162	200	286	392	513
	R: kN	120	144	174	217	240	261	305	328	349	392	436	522	610	697
E2.3	*E*: kNm	12.9	18.5	33	51.5	62	74	100	115	132	167	206	294	404	528
	R: kN	124	149	179	224	246	268	313	337	358	403	448	537	627	717
E2.4	*E*: kNm	13.2	19.0	34	53.0	64	76	103	118	135	171	212	302	415	542
	R: kN	127	153	184	230	253	276	322	347	368	414	460	552	644	736
E2.5	*E*: kNm	13.5	19.5	35	54.5	66	79	106	122	139	176	218	311	426	557
	R: kN	131	157	189	236	260	283	331	356	378	426	473	567	662	756
E2.6	*E*: kNm	13.8	20.0	35	56.0	68	81	109	125	143	181	223	319	437	572
	R: kN	134	161	194	242	267	290	340	365	388	437	485	582	679	776
E2.7	*E*: kNm	14.1	20.5	36	57.5	70	83	112	128	146	185	229	327	448	586
	R: kN	138	165	199	248	274	298	349	375	398	448	497	597	696	795
E2.8	*E*: kNm	14.4	21.0	37	59.0	71	85	114	131	150	190	235	335	460	601
	R: kN	141	170	204	255	280	305	357	384	407	459	509	612	713	815
E2.9	*E*: kNm	14.7	21.5	38	60.5	73	87	117	134	153	194	240	343	471	615
	R: kN	145	174	209	261	287	313	366	394	417	470	522	627	731	834
E3.0	*E*: kNm	15.0	22.0	39	62.0	75	89	120	137	157	199	246	351	482	630
	R: kN	148	178	214	267	294	320	375	403	427	481	534	642	748	854
E3.1	*E*: kNm	16.5	24.2	43	68.2	83	98	132	151	173	219	271	386	530	693
	R: kN	163	196	235	294	323	352	413	443	470	529	587	706	823	939
Efficiency ratio (ε)		0.103	0.124	0.183	0.230	0.254	0.276	0.319	0.341	0.368	0.414	0.461	0.548	0.645	0.737

Table 14.7 Energy absorption and reaction forces on different sizes of AN and ANP fenders

AN arch fender

AN	E1		E2		E3		Efficiency ratio *e*
	E: kNm	*R*: kN	*E*: kNm	*R*: kN	*E*: kNm	*R*: kN	
150	4.3	74	5.6	96.2	7.4	127	0.058
200	7.6	98.6	10	128	13.1	169	0.078
250	11.9	123	15.6	160	20.5	211	0.097
300	17.1	148	22.5	192	29.5	253	0.117
400	30.5	197	40	256	52.5	338	0.155
500	47.6	247	62.4	321	82	422	0.194
600	68.6	296	89.9	385	116	507	0.231
800	122	394	160	513	210	675	0.311
1000	191	493	250	641	328	844	0.389

ANP arch fender

ANP	E1		E2		E3		Efficiency ratio *e*
	E: kNm	*R*: kN	*E*: kNm	*R*: kN	*E*: kNm	*R*: kN	
150	5.6	88.8	7.3	115	9.5	150	0.063
200	9.9	118	12.9	154	16.8	200	0.084
250	15.6	148	20.2	192	26.3	250	0.105
300	22.4	178	29.1	231	37.8	300	0.126
400	39.8	237	51.7	308	67.2	400	0.168
500	62.1	296	80.8	385	105	500	0.210
600	89.3	355	116	462	151	600	0.251
800	159	473	207	615	269	800	0.336
1000	249	592	323	769	420	1000	0.420

Performance values are for a fender 1000 mm long

Figure 14.46 Differing constructions for the Trelleborg MV and MI elements

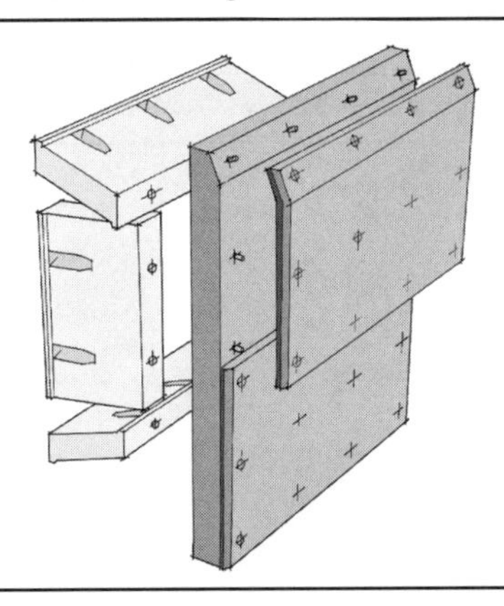

Figure 14.47 Differing constructions for the Trelleborg MV and MI elements

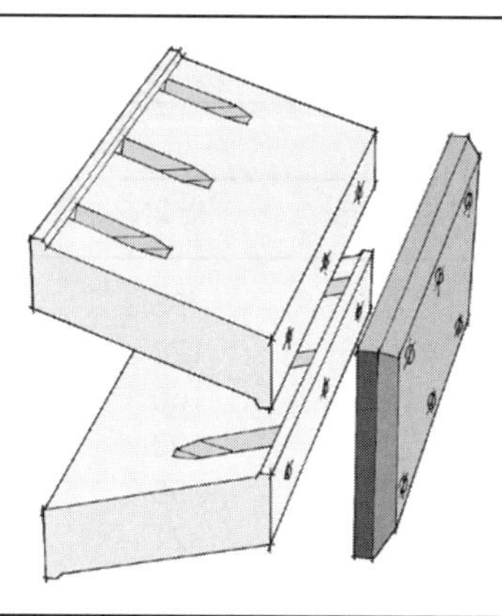

Figure 14.48 The Trelleborg element fender system

Table 14.8 The rated performance for one MV fender element

Element size, $H \times L$ Compound A or B	Rated performance for one element					
	E: t/m	R: t	E: kNm	R: kN	E: ft/kips	R: kips
MV 250 × 890 B	1.0	8.5	9.8	83.4	7.2	18.7
MV 250 × 890 A	1.4	12.1	13.7	118.7	10.1	26.7
MV 250 × 1000 B	1.1	9.5	10.8	93.2	8.0	20.9
MV 250 × 1000 A	1.6	13.6	15.7	133.4	11.6	30.0
MV 300 × 600 B	0.9	6.8	9	66	6	15
MV 300 × 600 A	1.3	9.8	13	96	9	21
MV 300 × 900 B	1.4	10.3	14	101	10	22
MV 300 × 900 A	2.0	14.7	20	144	14	32
MV 300 × 1200 B	1.8	13.7	18	134	13	30
MV 300 × 1200 A	2.6	19.6	26	192	19	43
MV 300 × 1500 B	2.3	17.2	22	168	16	38
MV 300 × 1500 A	3.3	24.5	32	240	24	54
MV 400 × 1000 B	2.8	15.3	27	150	20	34
MV 400 × 1000 A	4.0	21.8	39	214	29	48
MV 400 × 1500 B	4.2	22.9	41	224	30	50
MV 400 × 1500 A	6.0	32.7	59	321	43	72
MV 400 × 2000 B	5.6	30.6	55	300	41	67
MV 400 × 2000 A	8.0	43.6	78	428	58	96
MV 400 × 2500 B	7.0	38.2	68	375	51	84
MV 400 × 2500 A	10.0	54.5	98	535	72	120
MV 400 × 3000 B	8.4	45.8	83	449	61	101
MV 400 × 3000 A	12.0	65.4	117	642	87	144
MV 500 × 1000 B	4.3	19.0	43	187	32	42
MV 500 × 1000 A	6.2	27.2	61	267	45	60
MV 500 × 1500 B	6.5	28.6	64	280	47	63
MV 500 × 1500 A	9.3	40.8	91	400	67	90
MV 500 × 2000 B	8.7	38.2	85	374	63	84
MV 500 × 2000 A	12.4	54.4	122	534	90	120
MV 550 × 1000 B	5.3	21.0	52	206	38	46
MV 550 × 1000 A	7.6	30.0	75	294	55	66
MV 550 × 1500 B	8.0	31.5	78	309	58	69
MV 550 × 1500 A	11.4	45.0	112	441	82	99
MV 600 × 1000 B	6.3	22.8	62	224	46	50
MV 600 × 1000 A	9.0	32.6	88	320	65	72
MV 600 × 1500 B	9.5	34.2	93	336	69	76
MV 600 × 1500 A	13.5	48.9	132	480	98	108
MV 750 × 1000 B	9.8	28.7	96	282	71	63
MV 750 × 1000 A	14.0	41.0	137	402	101	90
MV 750 × 1500 B	14.7	43.1	144	423	106	95
MV 750 × 1500 A	21.0	61.5	206	603	152	135
MV 800 × 1000 B	11.2	30.5	110	299	81	67
MV 800 × 1000 A	16.0	43.6	157	428	116	96

Table 14.8 Continued

Element size, $H \times L$ Compound A or B	Rated performance for one element					
	E: t/m	R: t	E: kNm	R: kN	E: ft/kips	R: kips
MV 800 × 1500 B	16.8	45.8	165	449	122	101
MV 800 × 1500 A	24.0	65.4	235	642	174	144
MV 800 × 2000 B	22.4	61.0	220	599	162	134
MV 800 × 2000 A	32.0	87.2	314	856	232	192
MV 1000 × 900 B	15.8	34.3	155	337	113	76
MV 1000 × 900 A	22.5	49.0	221	481	162	108
MV 1000 × 1000 B	17.5	38.1	172	374	126	84
MV 1000 × 1000 A	25.0	54.4	245	534	180	120
MV 1000 × 1500 B	26.3	57.1	258	560	189	126
MV 1000 × 1500 A	37.5	81.6	368	800	270	180
MV 1000 × 2000 B	35.0	76.2	343	748	252	168
MV 1000 × 2000 A	50.0	108.8	490	1068	360	240
MV 1250 × 900 B	24.6	42.8	241	420	177	95
MV 1250 × 900 A	35.1	61.2	344	600	253	135
MV 1250 × 1000 B	27.3	47.6	268	467	197	105
MV 1250 × 1000 A	39.0	68.0	383	667	282	150
MV 1250 × 1500 B	41.0	71.4	402	701	296	158
MV 1250 × 1500 A	58.5	102.0	574	1001	423	225
MV 1250 × 2000 B	54.6	95.2	536	934	395	210
MV 1250 × 2000 A	78.0	136.0	766	1334	564	300
MV 1450 × 1000 B	36.8	55.3	361	543	266	122
MV 1450 × 1000 A	52.6	79.0	516	775	380	174
MV 1450 × 1500 B	55.2	83.0	542	813	399	183
MV 1450 × 1500 A	78.9	118.5	774	1162	570	261
MV 1450 × 2000 B	73.6	110.6	722	1085	532	244
MV 1450 × 2000 A	105.2	158.0	1032	1550	760	348
MV 1600 × 1000 B	44.8	61.0	440	599	323	135
MV 1600 × 1000 A	64.0	87.2	628	855	462	192
MV 1600 × 1500 B	67.2	91.6	659	898	485	202
MV 1600 × 1500 A	96.0	130.8	942	1283	693	288
MV 1600 × 2000 B	89.6	122.1	879	1197	647	269
MV 1600 × 2000 A	128.0	174.4	1256	1710	924	384

Table 14.9 The rated performance for one MI fender element

Element size, $H \times L$ Compound A or B	Rated performance for one element					
	E: t/m	R: t	E: kNm	R: kN	E: ft/kips	R: kips
MI 2000 × 1000 B	54.8	54.8	538	538	397	121
MI 2000 × 1000 A	89.7	89.7	880	880	649	198
MI 2000 × 1050 B	57.6	57.6	565	565	417	127
MI 2000 × 1050 A	94.1	94.1	923	923	681	208
MI 2000 × 1100 B	60.3	60.3	592	592	437	133
MI 2000 × 1100 A	98.6	98.6	967	967	714	217
MI 2000 × 1150 B	63.1	63.1	619	619	457	139
MI 2000 × 1150 A	103.2	103.2	1012	1012	747	228
MI 2000 × 1200 B	65.8	65.8	645	645	476	145
MI 2000 × 1200 A	107.6	107.6	1056	1056	779	237
MI 2000 × 1250 B	68.6	68.6	673	673	497	151
MI 2000 × 1250 A	112.1	112.1	1100	1100	812	247
MI 2000 × 1300 B	71.2	71.2	699	699	515	157
MI 2000 × 1300 A	116.6	116.6	1144	1144	844	257
MI 2000 × 1350 B	74.0	74.0	726	726	536	163
MI 2000 × 1350 A	121.1	121.1	1188	1188	877	267
MI 2000 × 1400 B	76.7	76.7	752	752	555	169
MI 2000 × 1400 A	125.6	125.6	1232	1232	909	277

REFERENCES AND FURTHER READING

BSI (British Standards Institution) (1994) BS 6349-4: 1994. Code of practice for design of fendering and mooring systems. BSI, London.

EAU (Empfehlungen des Arbeitsausschusses für Ufereinfassungen) (2004) *Recommendations of the Committee for Waterfront Structures, Harbours and Waterways*, 8th English edn. Ernst, Berlin.

Ministerio de Obras Públicas y Transportes (1990) ROM 0.2-90. Maritime works recommendations. Actions in the design of maritime and harbour works. Ministerio de Obras Públicas y Transportes, Madrid.

PIANC (International Navigation Association) (1984) *Report of the International Commission for Improving the Design of Fender Systems*. PIANC, Brussels.

PIANC (1990) *The Damage Inflicted by Ships with Bulbous Bows on Underwater Structures*. PIANC, Brussels.

PIANC (2002) *Guidelines for the Design of Fender Systems. Report of Working Group 33*. PIANC, Brussels.

PIANC (2014) *Berthing Velocity and Fender Design. Report of Working Group 145*. PIANC, Brussels.

Port Designer's Handbook
ISBN 978-0-7277-6004-3

http://dx.doi.org/10.1680/pdh.60043.403

Chapter 15
Erosion protection

15.1. General

The erosion of the sea bottom in front of a berth structure and of the filling under an open berth structure will generally be due to the wave action at the upper part of the filling and to the current from the main ship propellers and/or the bow and stern thrusters at the lower part of the filling and of the sea bottom (Figure 15.1).

The erosion problem in solid berth structures is limited to erosion of the bed material in front of the berth structure, whereas erosion in open berth structures is more complex and can include:

(a) erosion around the piles, in particular those near the berthing face
(b) erosion of the slope underneath the berth structure, even up to the top.

The most severe erosion effects on under-berth slopes or the sea bottom under a ship arise from the action of the main propeller and bow thrusters of container ships, roll-on/roll-off ships and ferries.

Although erosion can also occur near berth structures due to natural currents, berth structures are specifically vulnerable to erosion caused by ships' propeller action. Especially during berthing and unberthing, eroding forces on the seabed in front of the berth or on the slope underneath the berth can be substantial. The action of the ship's propeller is a main eroding factor due to the resulting current velocities, which can reach up to 8 m/s near the bottom, compared to, for example, the tidal current, which is typically limited to around 1 or 2 m/s crosswise to the longitudinal axis of the ship.

The propeller currents are due to the following:

(a) The **stern propeller** or screw will cause an induced jet current speed directly behind the propeller.
(b) The **bow thrusters** consist of a propeller, which works in a pipe and is located crosswise to the longitudinal axis of the ship. A typical bow thruster has a diameter of 1.5–2.5 m. It is used for manoeuvring out from the berth line. Current velocities of up to 7.0 m/s can be expected for bow thrusters of large, for example, container ships having a propeller output of up to about 1700 kW and a propeller diameter of 2.5 m (see Table 15.1). The thrust of the bow thrusters is in the range of about 3–15 t.

Impacts such as bottom erosion from the ship's main propeller depend on many factors, and these may be different in almost every situation. Typically, for design purposes the governing condition will be when the propeller is closest to the bottom, i.e. when the vessel is in the loaded condition and when

Figure 15.1 Erosion due to wave action and the current from ship propellers

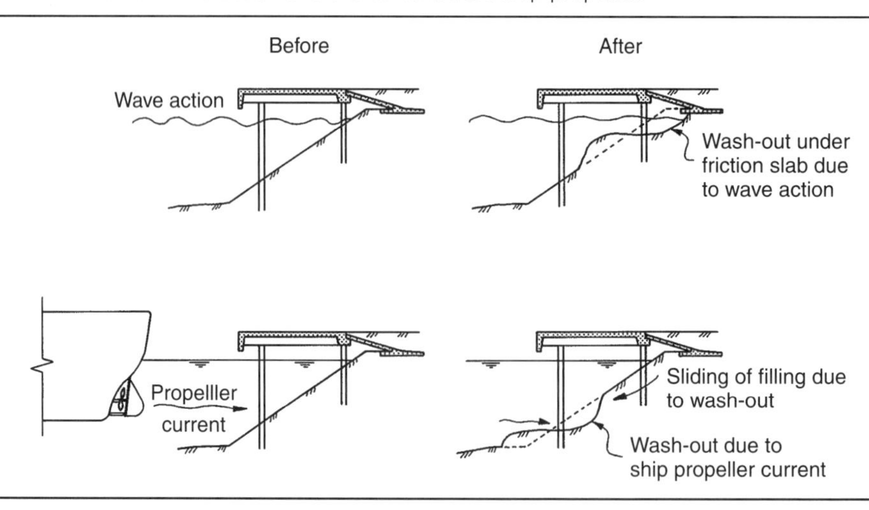

the tide is at its lowest. The use of the ship's main engine and propeller may be quite different, however, between berthing and unberthing operations.

The introduction in around 1960 of ship's bow and stern thrusters was due to the need to increase manoeuvrability and thereby minimise the manoeuvring time in the port. New and powerful ships with modern propeller systems, frequently combined with aggressive manoeuvring, can cause severe erosion in ports that otherwise would have remained stable for decades. The larger passenger ships can have two, three or four bow thrusters, container ships usually have one, and roll-on/roll-off ships and ferries may have one bow and one stern thruster and two main propellers.

Thrusters are propellers placed in a tube inside the ship's hull. Sometimes main propellers are placed in a nozzle, in particular in inland navigational vessels because of the restricted water depth that they operate in. The current from the thrusters against a vertical berth or quay front wall is shown in Figure 15.2.

Larger ships, such as ferries, use their propellers to manoeuvre in harbours. The function of the main propeller at the rear of the ship is predominantly forward thrust, but it can aid a manoeuvre by

Table 15.1 Diameter and power of propeller for container ships

Ship size: DWT	Main propeller		Bow thrusters	
	Power: kW	Propeller diameter: mm	Power: kW	Propeller diameter: mm
10 000	8 000	4500	500	1700
20 000	15 000	5500	750	1800
30 000	20 000	6500	1000	2000
40 000	26 000	7000	1200	2200
50 000	33 000	7400	1400	2300
60 000	40 000	7500	1700	2500

Figure 15.2 Illustration of the flow from thrusters

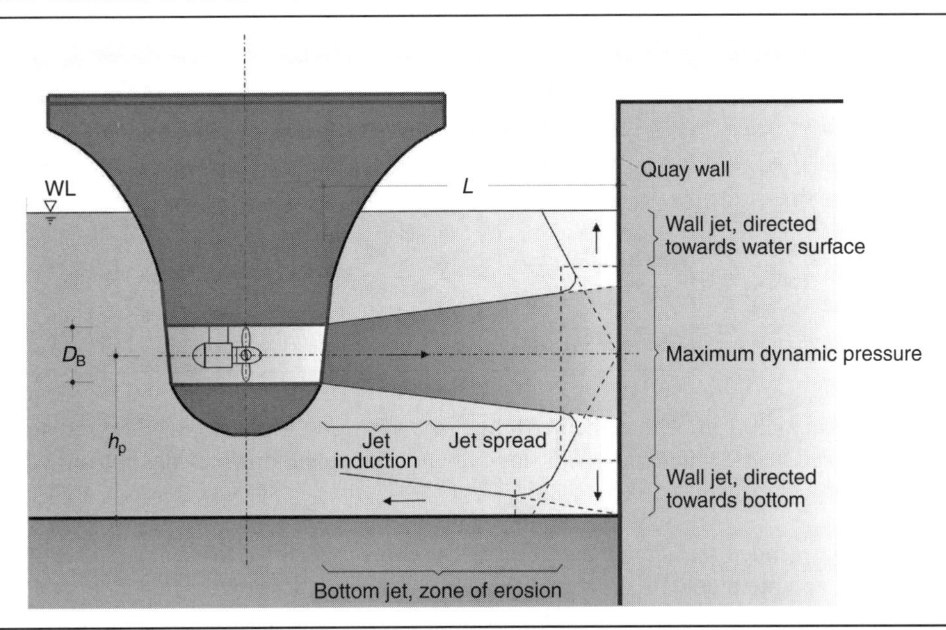

changing the direction of the rudders. The classic forces of the main propeller and stern thruster can be combined by a rotatable thruster, such as an azipod or a hydrojet type.

The availability and use of thrusters, and their characteristics vary largely with ship type. While bow thrusters are standard equipment on cruise, car and container vessels, they are less common on bulk carriers.

Bow thrusters are situated at the front of the ship. As a result of the increase in the size and engine power of bow thrusters, the flow velocities against the berth front and at the harbour basin bottom in front of the berth have increased considerably over the years. Because harbour bottoms are often not designed for these extreme velocities, they can lead to increased bottom erosion and possibly to berth failure.

The bow thrusters on large cruise vessels in particular can have a large damaging effect on old block wall berths which have no erosion protection in front of the wall (Figure 15.3). The potential failure mechanism for a gravity structure would be the loss of the passive soil pressure in front of the berth structure due to the erosion due to thruster action if there is no erosion protection. This bottom erosion can be minimised by placing well-dimensioned harbour bottom protection. The bow and stern thrusters on a very modern large cruise ferry are shown in Figure 15.4.

Modern ships are often equipped with azipods and Voith Schneider propellers. These systems are different from regular propellers in that there is no rudder. Azipods can turn around 360° and therefore are particularly implicated in erosion at quay walls. In a water jet system, the water jets are steerable, mainly to a maximum of 30° to starboard or port but also towards the bottom, and this is sufficient to create a strong water jet with a flow velocity up to 20 m/s in the direction of the quay wall or the seabed. The intersection point between the toe of the slope and the sea bottom should preferably

Figure 15.3 Erosion in front of an old block wall

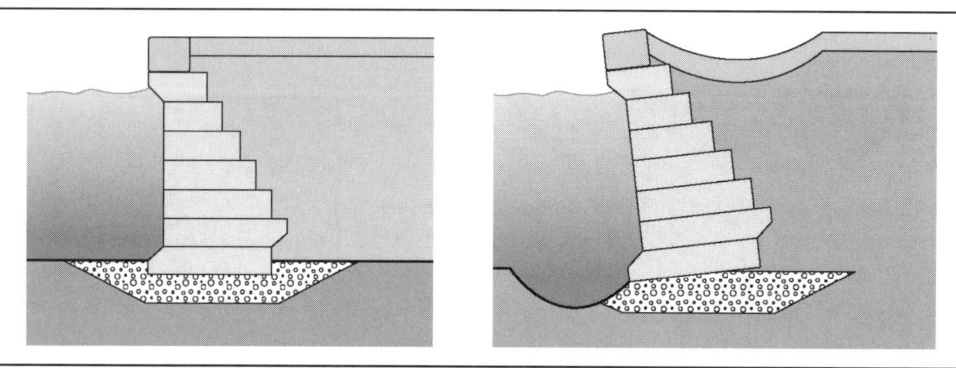

be set back approximately 1 m behind the berth line (see Figure 11.1 in Chapter 11) to ensure that no stone has been placed outside the theoretical slope line, or has fallen down to the bottom where it can damage the hull of a ship.

It is generally recommended that, when designing erosion protection for the seabed, it will be cheaper in the long run to accept some damage to the protection during its lifetime. But, due to difficulty of access under an open berth after the completion of the deck structure, it is recommended that for under-berth slope protection this should be designed to be maintenance free.

The erosion of the sea bottom from the main propeller will depend on whether the berthing or unberthing operations take place at low or high tide, but in the design of the protection one should always assume low tide. There is thus a practical need to lay the bottom protection below the lowest maintained level. Therefore, in terms of the erosion effect, the greater the underkeel clearance, the lower the water velocity at the level of the protection will be, and the lower the cost of construction of the erosion protection. For berth structures that are subjected to maintenance dredging, it is recommended that the bottom erosion protection should be placed at least 0.75 m below the lowest permitted dredge level.

Many factors must be taken into consideration during the design process. Examples of variables are type of seabed, depth and slope, type of berth construction, characteristics of vessels (type of propeller, engine size), frequency of arrivals and departures, angle of approach, etc.

Figure 15.4 The thrusters on the large cruise ferry *Color Fantasy*. (Courtesy of Color Line, Norway)

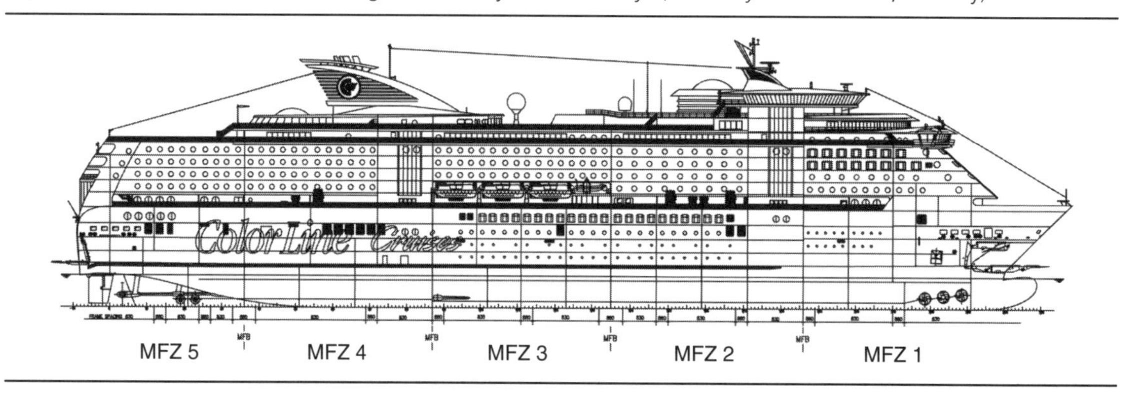

Table 15.2 The diameter and power of the propellers and bow thrusters of some cruise ferries

Name	Ship size: DWT	L_{OA}: m	Main propeller		Bow thrusters	
			Power: kW	Propeller diameter: mm	Power: kW	Propeller diameter: mm
Color Superspeed	1840	212	2 of 18 432	5250	2 of 2400	2650
Color Fantasy	5616	224	4 of 7800	5200	3 of 2200	2700

Berth pockets are used in harbours that have significant tidal water level variations and where ships enter and depart during the high-tide windows, so that deep-drafted vessels can stay along the berth during low tide. When designing such berth pockets, the underkeel clearance should be selected carefully. The underkeel clearance required during manoeuvring may be more than the minimum required while at berth. In situations where berth pockets are used, the impacts of the main propeller and thrusters are likely to extend to lower elevations of the berth structure, and this aspect will have to be taken into account when analysing and designing the structure. Depending on the depth of the berth pocket below grade, the transition slopes to the regular bottom elevation may also need to be a specific item of attention. The most important parameters for bed erosion are the underkeel clearance and the particle size of the bed material.

Seldom does the designer know the characteristics of the ships that will use the berth structure during its service life. Table 15.1 shows some figures for the diameter and power of the main propeller and bow thrusters for container ships which will give an approximate accuracy in calculations of $\pm 10\%$.

Nowadays the installed engine power as well as the number of thrusters in ships is increasing. For the new-generation container ships currently being built and intended to carry 18 000 TEU or more, the installed power for the main propulsion system is about 90–100 000 kW, while the installed power for each thruster is up to 3000 kW. Table 15.2 gives the diameter and power of the propellers and bow thrusters of the large cruise ferries *Color Fantasy* and *Color Superspeed.*

15.2. Erosion due to wave action

The design of protection against erosion due to wave action at the front of the stone fill is generally based on the Hudson formula:

$$W_{50} = \frac{\rho_s \times H_{des}^3}{K_D \times \left(\dfrac{\rho_s}{\rho_w} - 1\right)^3 \times \cot\alpha}$$

where

W_{50} = average block weight in kN
H_{des} = design wave height, H_s to $1.4H_s$
ρ_s = specific gravity of a block unit of quarry stone, 26 kN/m^3
ρ_w = specific gravity of seawater, 10.26 kN/m^3
α = slope angle of the cover layer

K_D = shape and stability coefficient, which for the berth front is 3.2, and for the berth end or the end of the fill under the quay is 2.3. For quarry stone and breaking waves the berth front value is 2.7

The block weight W_{max} should be less than (3.6 – 4.0) × W_{50}, and W_{min} should be greater than (0.2 – 0.22) × W_{50}.

If the berth structure is exposed to extreme wave conditions, the slope protection should be checked for stability against wave action down to at least a depth of 2 × H_s.

The equivalent rock or stone diameter will be:

$$d_{equ} = \sqrt[3]{\frac{6 \times W}{\pi \times \rho_s}}$$

where

d_{equ} = the equivalent rock or stone diameter in m
W = the bock weight in kN
ρ_s = the specific gravity of a block unit of quarry stone, 26 kN/m^3

15.3. Erosion due to the main propeller action

The exact design of the erosion protection against the actions of main ship propellers and bow and stern thrusters is, as the evaluation and comparison between EAU (2004) and the PIANC Working Group 48 (PIANC, 2010) shows, very difficult. This is due to the fact that the size and type of the protection will depend on the velocity of the propeller current, which will itself depend on the ship engine power, the speed, and the shape and the diameter of the propeller. To obtain all this information for all the different ships calling at a berth will be impossible. However, experience has shown that if the erosion protection against the propeller action has the same stone size and filter as those used for a design wave height of about 1.5–2.0 m, this will in most cases give sufficient erosion protection.

The erosion due to propeller action on the bottom seabed with stone filling is illustrated in Figure 15.5.

Figure 15.5 Erosion of the sea bottom due to the action of a main propeller

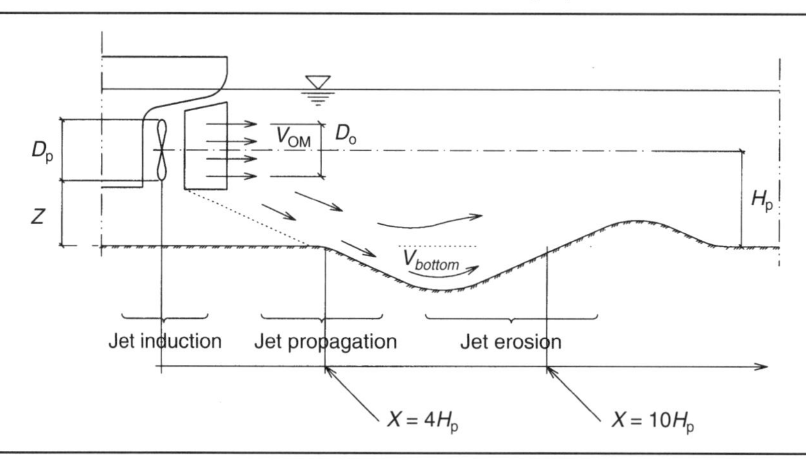

Figure 15.6 The velocity distribution of the current induced by a propeller

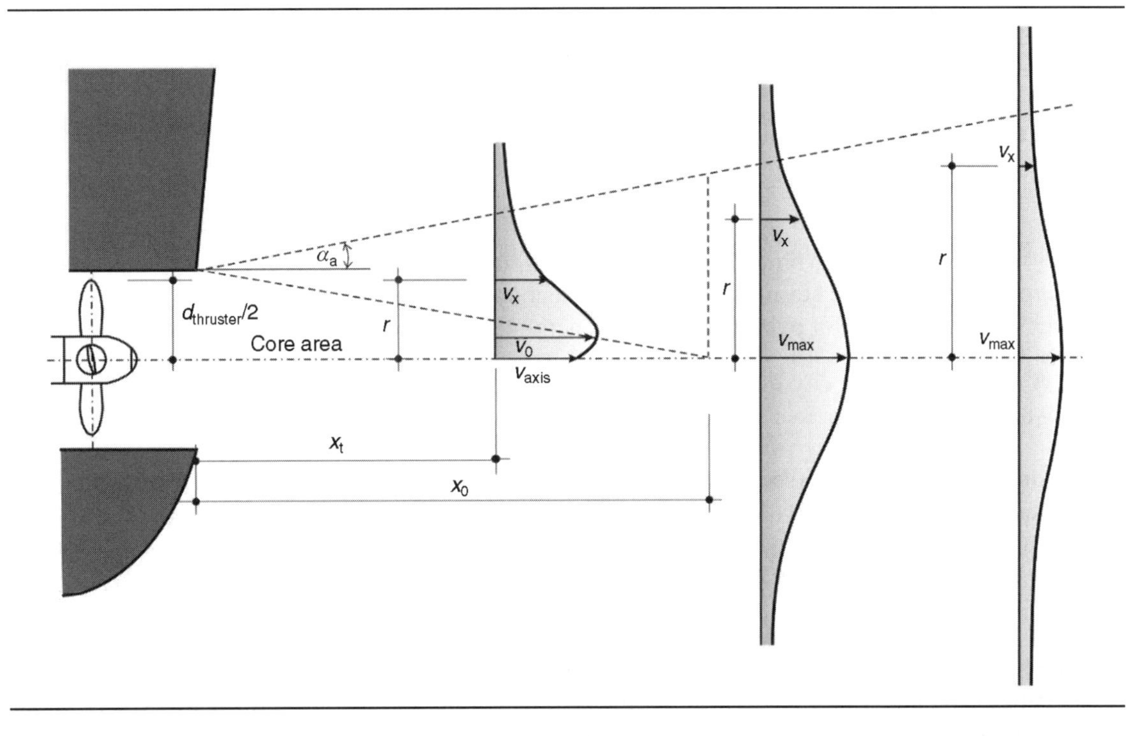

The velocity of the current produced by the propeller expands in a cone shape away from the propeller and decreases with increasing distance from the propeller. The zone of maximum seabed velocity V_{bottom}, which is essentially responsible for the erosion, lies approximately at a distance $4 \times H_{\text{p}}$ to $10 \times H_{\text{p}}$ from the propeller (Figures 15.5 and 15.6). Therefore the most important parameters for the erosion of the seabed are the underkeel clearance and the particle size of the seabed materials.

The jet velocity caused by the rotating main propeller is called the 'induced jet velocity', and occurs directly behind the main propeller. It is recommended by both the EAU (2004) and PIANC Working Groups 22 and 48 (PIANC, 1997, 2010) that this is calculated using the simplified formula:

$$V_{\text{OM}} = 0.95 \times n \times D_{\text{p}}$$

where

V_{OM} = initial centreline jet velocity from the main propeller
n = number of propeller revolutions per second
D_{p} = propeller diameter

If the output of the main propeller is known, rather than the velocity, the induced jet velocity can be calculated as follows:

$$V_{\text{OM}} = c \times \left[\frac{P}{\rho_{\text{O}} \times D_{\text{p}}^2}\right]^{1/3}$$

where

V_{OM} = initial centreline jet velocity from the main propeller
c = 1.48 for a free propeller or a non-ducted propeller, and 1.17 for a propeller in a nozzle or a ducted propeller
P = engine output power in kW
ρ_O = density of seawater, 1.03 t/m^3
D_p = propeller diameter

It is very unusual for the main propeller to be used at full power during berthing and unberthing, except in the case of ferries or in very strong wind and/or current conditions. Practical experience has shown that the machine power output for port manoeuvring generally lies between the following approximate values:

(a) 30% of the rated velocity for slow ahead
(b) 65–80% of the rated velocity for half-speed ahead.

It is recommended that a speed corresponding to 75% of the rated velocity be used for the design of the bottom erosion protection, but for particularly critical conditions with high wind and/or current forces acting on the ship, the rated velocity or increased velocity at maximum power output must be assumed to be up to 100% of the rated velocity.

It is important to realise that the installed power will be used by the captain if it is needed in difficult manoeuvring when berthing and unberthing. In other words, applying 100% of the installed power is a conservative estimate. This is underlined by the results of a questionnaire conducted by the Harbour Authorities of Antwerp, which showed that the applied power of bow thrusters was 75% of the installed power. For inland ships it is also recommended to apply 100% of the bow thruster power, although also in the field of inland navigation the applied power decreases with increasing installed power.

The EAU (2004) recommends that the seabed velocity be calculated using the following formula:

$$V_{\text{bottom}} = V_{OM} \times E \times \left[\frac{H_p}{D_p}\right]^a$$

where

V_{bottom} = bottom velocity due to the main propeller in m/s
V_{OM} = initial centreline jet velocity from the main propeller
E = 0.71 for a single-propeller ship with a central rudder, and 0.42 for a twin-propeller ship with a middle rudder
H_p = height of the propeller shaft over the bottom
D_p = propeller diameter
a = −1.00 for a single-propeller ship and –0.28 for a twin-propeller ship

Figure 15.7 shows the PIANC Working Group 22 (PIANC, 1997) recommendations for the bottom velocity. For the calculation of the initial jet diameter D_o, the PIANC recommends the following relationships:

(a) non-ducted propeller, $D_o = 0.71 \times D_p$
(b) ducted propeller, $D_o = D_p$.

Figure 15.7 The bottom velocity as per the PIANC recommendations

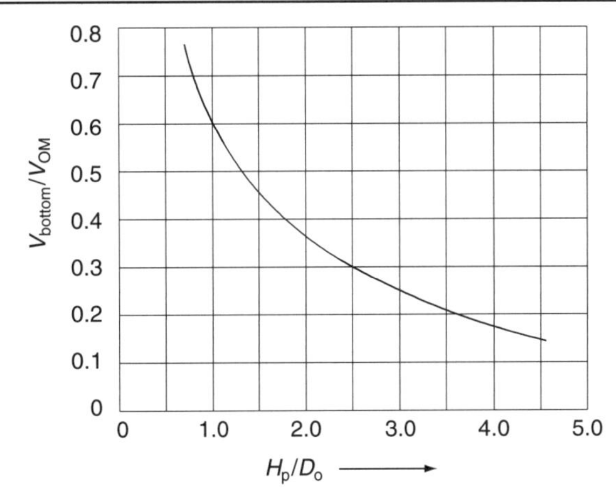

Water jet thrusters are a modern type of thruster that are commonly installed on fast ferries. Characteristic of these systems are the very high water outflow velocities of up to 25 m/s (Figures 15.8 and 15.9). The conventional propeller usually has a maximum water flow of about 8–10 m/s. The direction of these water jets can be towards the bottom of the harbour. For example, the Kamewa type of water jet has an angle of about 45°, but the catamaran between Lanzarote and Fuerteventura is equipped with water jets that can direct the water jet both horizontally when the catamaran is sailing from the ferry ramp and backwards 45° towards the seabed when the catamaran is sailing backwards to the ferry ramp.

15.4. Erosion due to thrusters

The bow and stern thrusters are used for easier manoeuvring inside a narrow port area and/or during berthing and unberthing operations. The thrusters consist of a propeller installed in a tube, and are

Figure 15.8 Scour caused by the water jet thrusters of fast ferries

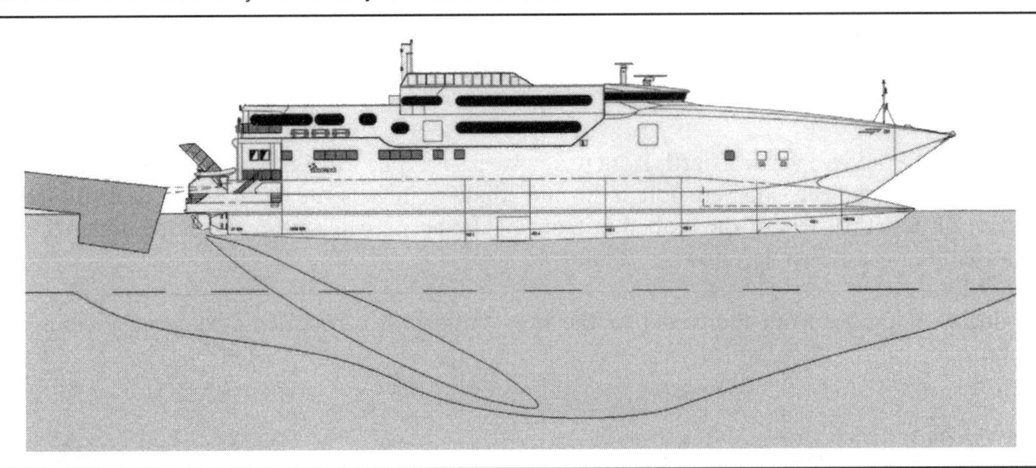

Figure 15.9 An example of the velocity of the jet produced by a water jet thrusters

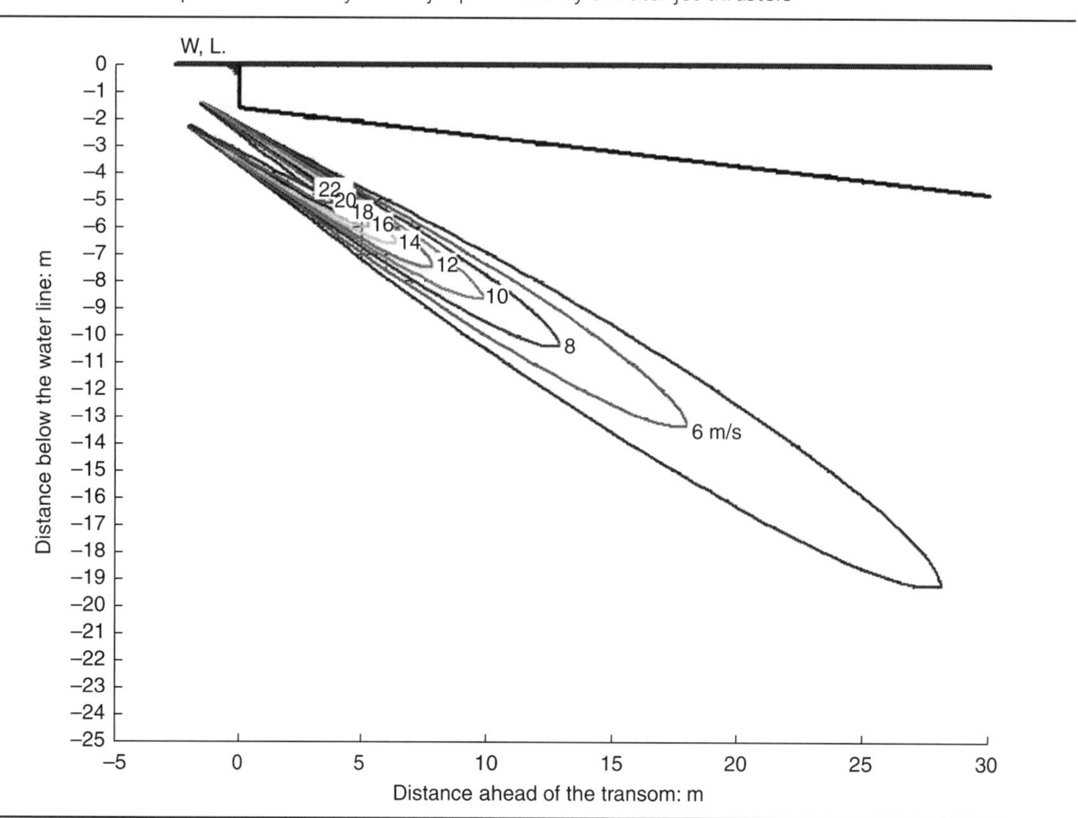

located crosswise to the longitudinal axis of the ship. They are always installed near the bow, and sometimes also at the stern. When the thrusters are used in the berthing operation, they generate a current that will hit directly the quay front or the slope below an open berth structure and be diverted to all sides from there.

Because the slopes under an open berth structure are more vulnerable to erosion than is a horizontal surface, it is recommended that the estimated D_{50} of the armour or rip-rap calculated for horizontal surfaces be increased by 50%.

If the water current hits a vertical berth front (e.g. a steel sheet pile wall), a part of the water current will hit the sea bottom and can cause erosion in the immediate vicinity of the berth wall. The erosion due to the action of bow thrusters of the seabed in front of the berth wall and/or the slope under the open berth structure is, in principle, shown in Figure 15.10.

The velocity of the jet from the outlet of the bow thruster can be calculated using the simplified formula:

$$V_{OB} = 1.04 \times \left[\frac{P}{\rho_o \times D_B^2}\right]^{1/3}$$

Figure 15.10 Erosion of the berth structure due to the action of bow thrusters

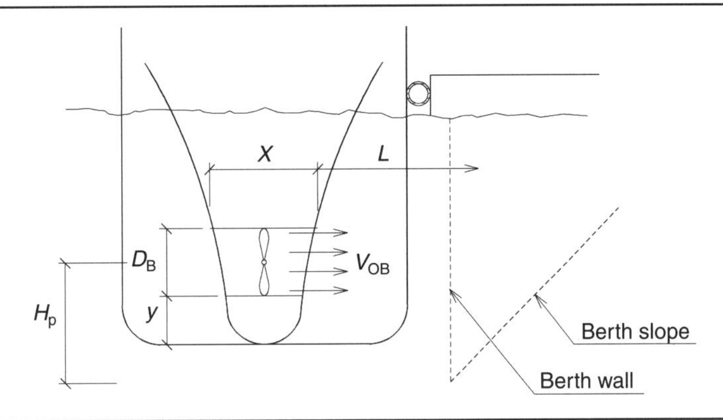

where

V_{OB} = initial centreline jet velocity from the bow propeller in m/s
P = bow engine output power in kW
ρ_o = density of water, 1.03 t/m
D_B = inner diameter of the bow thruster opening in m.

For combined combi-wall berth structures (Figure 15.11), the scour or erosion effect in front of the wall due to the action of thrusters can undermine the steel sheet piles between, for example, the tubular piles or H piles (see Section 10.1 in Chapter 10). Therefore, the effect of erosion in front of the sheet pile berth structure must be investigated very seriously.

For the design of the erosion protection due to bow thrusters one should assume that the ship uses the full power of the bow thruster when berthing and unberthing. For a large container ship at full power the velocity of the jet produced by the bow thruster can be assumed to be about 6.0–7.0 m/s. PIANC

Figure 15.11 Erosion in front of a sheet pile berth structure

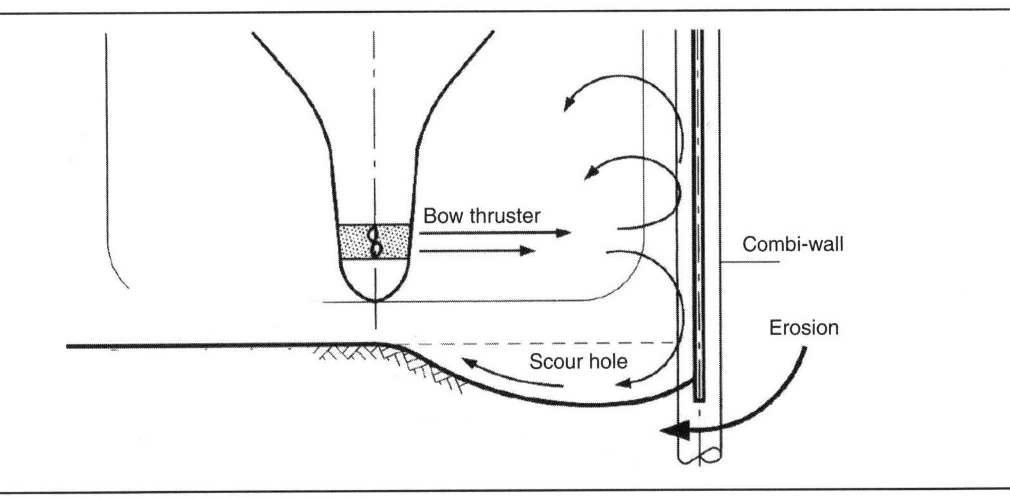

Figure 15.12 The velocity of the jet produced by bow thrusters

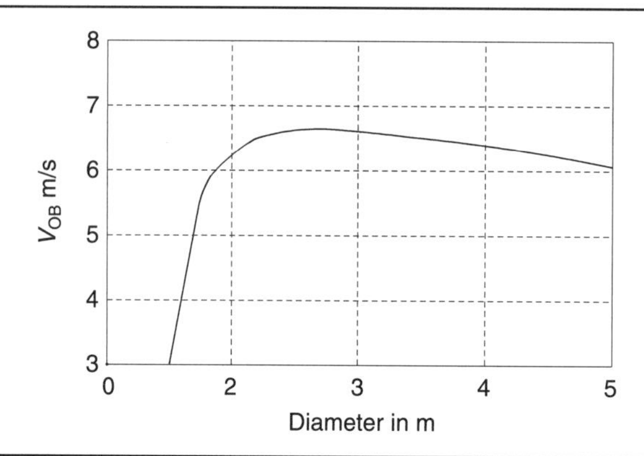

Working Group 22 (PIANC, 1997) gives the initial jet velocity under full power for various thruster diameters, as shown in Figure 15.12.

The water jet velocity that will hit the bottom or the slope below the berth can be assumed to be:

$$V_{\text{bottom}} = V_{\text{OB}} \times 2.0\left[\frac{D_{\text{B}}}{L}\right]$$

V_{bottom} = bottom velocity due to the bow thrusters in m/s
V_{OB} = initial centreline jet velocity from the bow propeller in m/s
D_{B} = inner diameter of the bow thruster opening in m
L = distance from the opening of the bow thruster to the berth wall in m

For design purposes the length x of the thruster tube, as shown in Figure 15.10, can be taken as approximately 30% of the beam of the ship, and the height y of the bow thruster above the keel can be taken as approximately equal to D_{B}. The distance L from the opening of the bow thruster of the ship should be to the berth wall or the slope.

15.5. The required stone protection layer

EAU (2004) recommends the following formula for the necessary stone diameter to ensure stability against the propeller current:

$$d_{\text{req}} \geq \frac{V_{\text{bottom}}^2}{B^2 \times g \times \dfrac{(\rho_{\text{s}} - \rho_{\text{o}})}{\rho_{\text{o}}}}$$

where

d_{req} = required diameter of the stones in m
V_{bottom} = bottom velocity in m/s
B = stability coefficient: 0.90 for ships without a central rudder, 1.25 for ships with a central rudder
g = acceleration due to gravity, 9.81 m/s^2

Figure 15.13 The required stone size for a given bed velocity

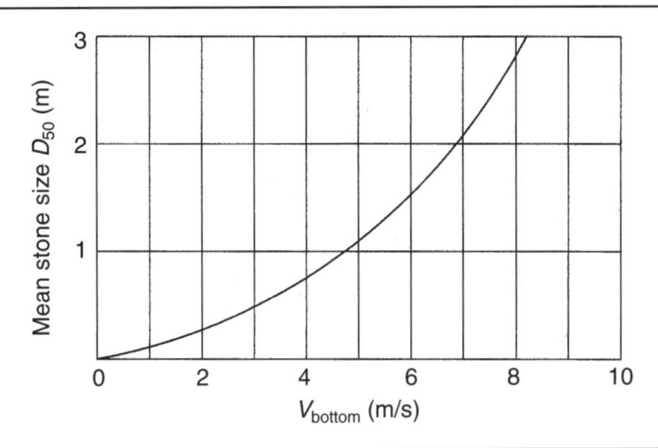

r_s = density of stone, 2.65 t/m^3
r_o = density of water, 1.03 t/m^3

Figure 15.13 shows the PIANC Working Group 22 (PIANC, 1997) recommended mean stone size D_{50} for there to be no erosion and for a given bed velocity.

The equivalent rock weight will be:

$$W = \frac{d_{equ}^3 \times \pi \times \rho_s}{6}$$

W = rock weight in kN
d_{equ} = equivalent rock or stone diameter in m
r_s = specific gravity of a block unit of quarry stone, 26 kN/m^3

15.6. Erosion protection systems

The following protection systems are used to protect the sea bottom and the fill from erosion:

(a) rock blocks or stones and rip-rap placed on a filter layer of gravels and/or a filter fabric
(b) filling loose stone with grouting
(c) covering with reinforced concrete slabs
(d) covering with flexible composite systems.

The most commonly used erosion protection systems are (a) and (d).

Erosion protection using **rock blocks** or **stones** as fill under an open pile berth structure is illustrated in Figure 15.14. The composition of the erosion protection layer and its thickness will depend on the current velocity, the angle of the slope and the coarseness of the materials in front of the fill. This protection system is one of the most frequently used, and the requirements shown in Figure 15.14 must be met.

The natural inclination of the rock fill is approximately 1 : 1.2 when constructed. The final slope of the fill is usually recommended not to be steeper than 1 : 1.5 when constructed under water, or the slope should not be steeper than the acceptable safe geotechnical stability of the slope. This condition can be achieved by reworking the slope using excavating equipment or by dumping material from a barge.

The thickness of the erosion protection layer should be more than 3 × d_{50} or 1.5 × d_{max}, and not less than about 1.0–1.5 m. A thickness of two layers of rock is recommended. The smallest rock size to be used for the primary rock layer should be approximately 500–1000 kg. In addition to quarry stone, other materials can also be used as protection, for example, reinforced concrete units such as tetrapods and dolosse. A filter layer should be constructed between the stone filling and the protection layer.

The difference between rock and stone armour and rip-rap is that rock armour comes in a narrow range of sizes and units must be placed individually, whereas rip-rap comes in a large or wide range of sizes and is placed by dumping. The rock layer derives its stability due to the interlock achieved through the method of placing it unit by unit, whereas rip-rap gets its stability from the packing effect due to its wide range of sizes. When the stones are larger approximately 500 kg they can only be placed properly by using a grabber.

The thickness and size of the stones in the filter layer will depend on the materials comprising the core stone filling. The detail of rock or stone erosion protection is shown in Figures 15.14, 15.15 and 15.16. A general empirical rule is that the weight of the rock in the second layer should not be less than 1/10 to 1/15 of the weight of the rock in the primary armour layer. A geotextile mat can be used between the filter layer and the stone fill to ensure that there is no migration of finer particles from the stone fill.

In the case of a soft seabed in front of the stone fill slope under an open pile berth structure, the erosion protection layer for the slope should be extended at least 3–5 m out in front of the berth line (Figure 15.17). As the slope under an open berth is more exposed to erosion than is the horizontal sea bottom, it is recommended that the estimated D_{50} of the quarry stone or rip-rap is increased by 50%.

Backfilling against or behind the berth structure itself should be done using either crushed or excavated rock having a maximum stone size of 300 mm. The fill above sea level should be compacted by performing at least four passes over the area with a 600 kg vibrating plate.

Figure 15.14 Erosion protection using a stone fill

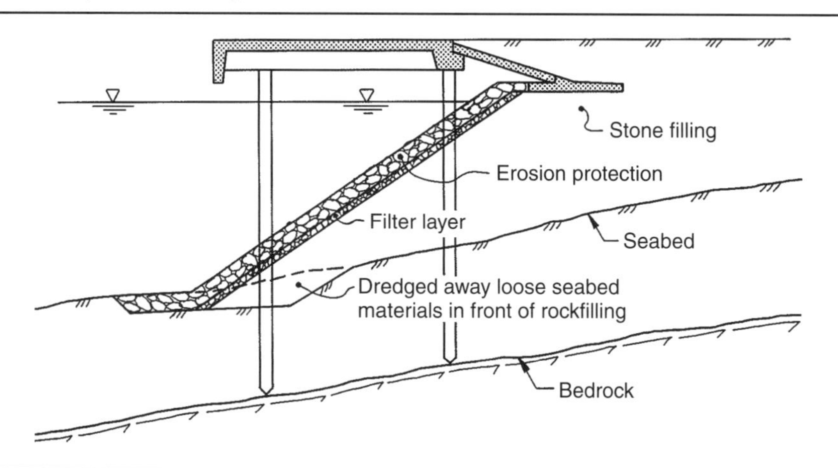

Figure 15.15 Detail of rock protection before construction of the berth itself

Figure 15.16 Detail of rock or stone erosion protection under a berth slab

Figure 15.17 Detail of rock or stone erosion protection

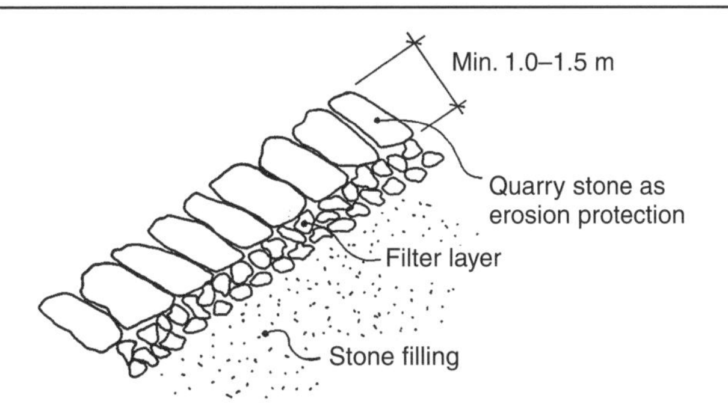

For bottom erosion protection using **loose stone cover**, the following requirements must be met:

(a) The installation of the stone should be in at least two layers.
(b) The stone cover should be stable against the velocity due to propeller action.
(c) A filter layer of grain or a textile filter should be installed between the subsoil and the stone layers.

If quarried rock of the required weight for rip-rap is not available, lighter rock **grouted** with concrete such that the porosity of the protection is retained can be used. It is recommended that the following aspects must be taken into consideration if the bottom protection of stone fill is grouted using, for example, concrete to make the erosion protection more stable against erosion:

(a) Depending on the grouting area, a minimum pore volume of about 15–20%, which should be continuous from the bottom to the top surface of the fill to compensate for any hydrostatic pressure, if necessary.
(b) Grouted stone fill can be stable up to very high bottom velocity of approximately 7 m/s.
(c) As the grouted stone fill forms a stable but rigid unit, erosion can occur at the edges due to underwashing, because the grouting cannot react flexibly to this action.

An ideal form of erosion protection is an underwater in situ reinforced **concrete slab**, because the thickness of the protection can be constructed with greater accuracy than with a stone fill layer. The underwater concrete slab can be constructed in thicknesses from approximately 30 cm up to 80 cm or more, depending on the concreting technique. The advantage of this system over a stone fill is that, in the latter, the stones can be dislodged by propeller action or by an anchor. The disadvantage of the concrete slab system is the need to install the concrete under water, which can be a complicated and very costly process.

Erosion protection comprising flexible composite systems, for example mattresses filled with concrete, such as the **FlexiTex** system from Norway or equivalent, should be placed on the prepared slopes and sea bottom when empty, joined together and then pumped full of concrete. The mattresses, which are made from double-weave mattress, are woven together at regular points that act as filters to even out water pressure. When using double-weave mattresses filled with concrete it is important to level the supporting bed. Mattresses can tolerate unevenness of up to ± 0.15 m/m^2.

The mattresses, which are generally supplied in widths of about 3.75 m and lengths of up to about 100 m, can be attached to panels in advance for rapid installation in the area to be protected. When calculating the required mattress size, it should be remembered that when the mattresses are filled with concrete they will shrink by approximately 10–15% in both directions. The thickness of the mattresses can be 7.5 cm up to 60 cm, with area weights, when filled with concrete of approximately 150–1200 kg/m^2. The mattresses used for berth protection are usually delivered with a thickness of at least 20 cm, to give an approximate weight of around 500 kg/m^2.

The filling of mattresses is shown in Figures 15.18 and 15.19. The concrete used to fill the mattresses must be designed for pumping. When the concrete is pumped into the mattresses, it will cause the water to evacuate through the fabric. The fabric acts like a sieve, and is designed to prevent concrete particles from passing through it. The filling of a mattress should always start from the lowest part of the mattress. The individual mattresses should be connected by zippers.

Figure 15.18 Filling of mattresses

The design of the concrete should be suitable for pumping through a pipeline of 50–75 mm diameter. From experience, the proportions of concrete constituent material per m^3 of concrete should be approximately as follows:

Cement = 350–500 kg of standard Portland
Silica = 30–50 kg

Figure 15.19 Filling of mattresses. (Courtesy of EB Marine, Norway)

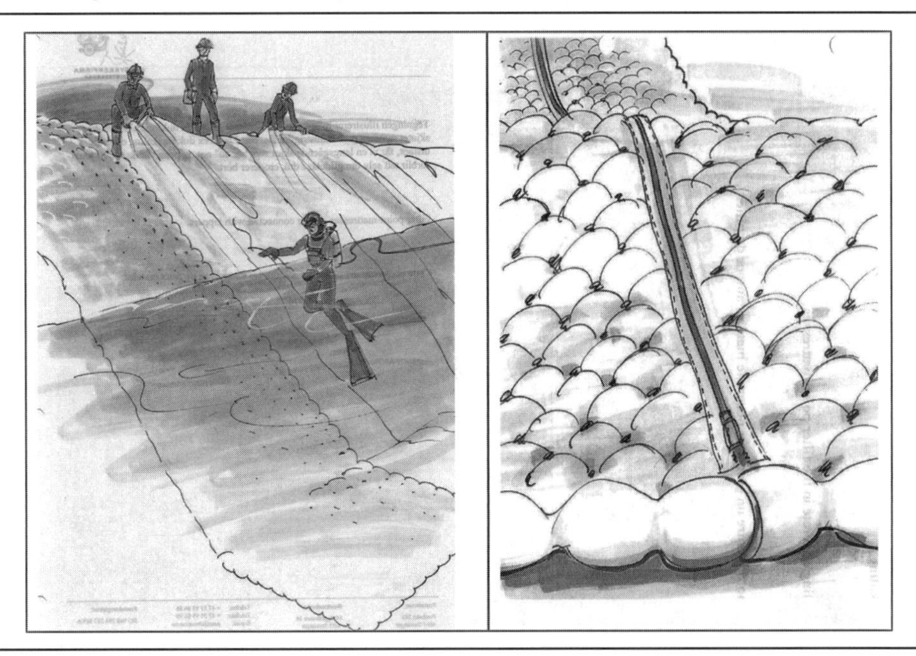

Plasticiser = 5–15 kg
Sand = 1400–1600 kg, with grain size 0–8 mm. It is recommended that the sand is very fine in order to facilitate pumping
w/c ratio = 0.4–0.5

The mattresses are usually fitted with industrial-quality zip fasteners for easy joining below water level. The result is a mass of pillow-like concrete units (Figures 15.20 and 15.21). The method is fast to construct; the cost could be lower than for a rock protection system and they are considered durable. Figure 15.22 shows the erosion protection of a ferry berth using concrete-filled mattresses.

Figure 15.23 shows the repair of an old quay damaged by propeller erosion. The repair is done using textile bags filled with anti-washout (AWO) concrete, and filling behind with AWO concrete.

A different flexible system is the wire box-like gabion system filled with small rock stones. This system could be acceptable if the current speed is very slow. The disadvantage of wire gabion mats is that the wire boxes or mesh are liable to corrosion, and the wire can break due to movement between mats.

When the sea bottom needs to be repaired or protected against erosion, an excellent choice can be to use a concrete mattress to cover the area. It is essential that the repair work is extended well outside the point where the erosion starts. As the weakest point, i.e. where the erosion can start, is at the edge of the mattress, it is very important that the end of the outer edge of the mattress assembly is secured down in a trench covered with concrete-filled bags or rocks or gravel (Figure 15.24).

Where the current due to propeller action may act against a solid berth wall, resulting in a strong downward current, the erosion protection layer should be extended for some distance beyond the berth wall (Figure 15.25). Depending on the seabed, the type of ship, etc., this distance should be approximately

Figure 15.20 A mattress filled with concrete

Figure 15.21 A mattress filled with concrete

Figure 15.22 Protection provided by mattresses. (Courtesy of EB Marine, Norway)

Figure 15.23 Repair of a quay using textile bags

Figure 15.24 A typical section of concrete mattress protection in front of a sheet pile

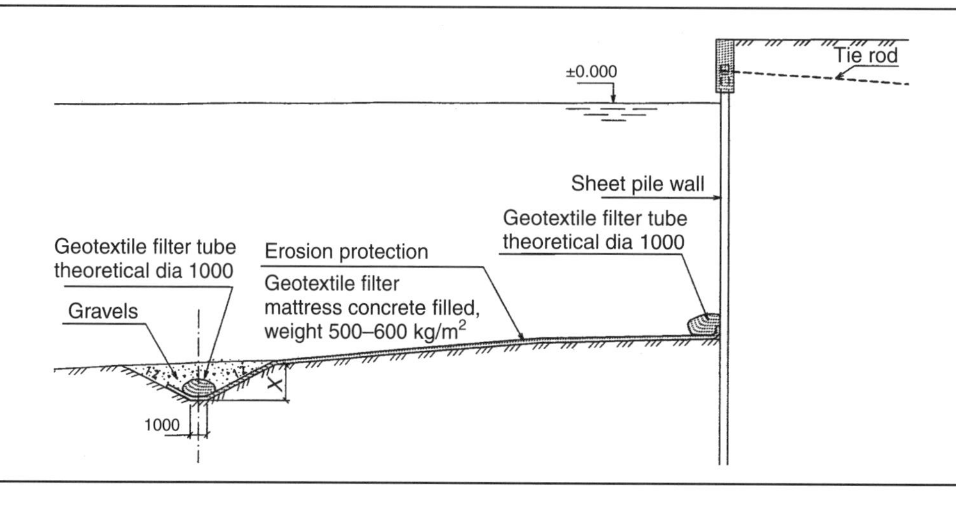

Figure 15.25 A container ship moored alongside a sheet pile berth with a protected bottom

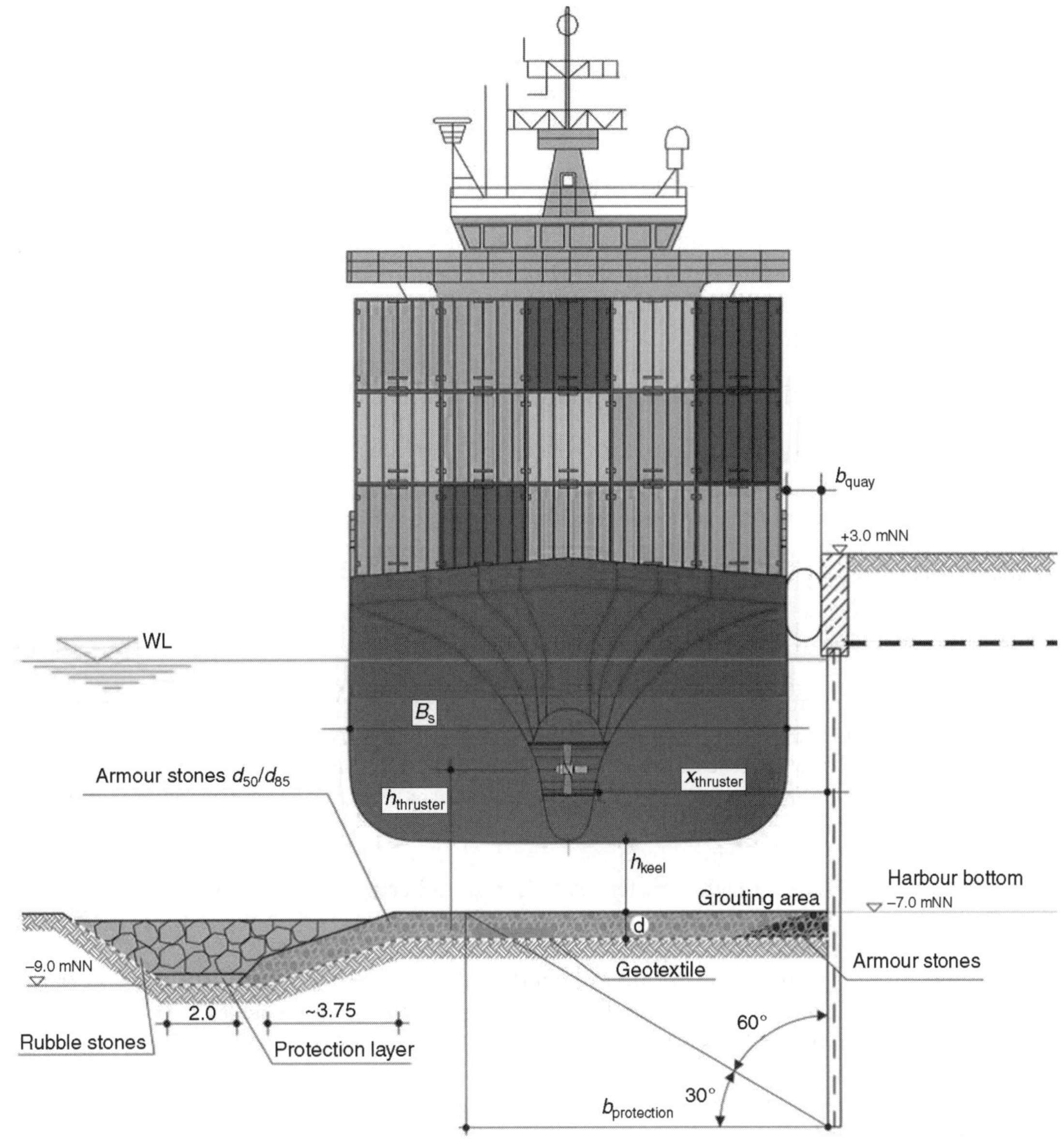

the width of the largest ship using the berth, or at least 5 m beyond the longitudinal axis of the design ship. The depth of the protection should be approximately 1 m more than the expected erosion.

Delft Hydraulics uses the following equation to determine the width to be protected, as measured from the quay wall:

$$b_{\text{protection}} = b_{\text{kade}} + 0.5B_{\text{s}} + 0.5H_{\text{oh}} + 0.5D_{\text{p}} + 5 \text{ (m)}$$

where

b_{kade} = distance between ship and quay wall in m
B_s = ship width in m
H_{oh} = distance between main propeller shafts in m
D_p = propeller diameter

The extra width of 5 m is a value based on experience.

For most conditions this will result in a width of protection that is smaller than the ship's width, and therefore it is recommended that the width of the protection provided should be at least equal to the ship width.

The protected length along the quay is a least equal to the ship's length and extended 50 m in front of the bow and 50 m behind the stern. However, this length depends on the berthing procedures used.

15.7. Operational guidelines

In addition to the option of designing either for the berth structure to withstand erosion in front of it or for the bed material in front of the structure to withstand erosion forces from thrusters, in certain cases there may also be the option to avoid or minimise erosion forces. This could be relevant both for designers of new structures and for the operators of existing structures, and it could be especially relevant for existing structures that are being exposed to higher forces and loads than were anticipated during their design, due to developments in thruster power and use.

In addition to structural measures to avoid or minimise erosion forces resulting from, for example, bow thrusters, such as installation of current deflectors or energy dissipation features at the face of the dock, operational measures can be considered as well. Factors to include when considering whether operational measures are a suitable solution or alternative to other means of erosion protection are:

(a) Frequency of exposure to high thruster loads (e.g. for a berth that is only occasionally used by ships having high thruster forces, it may be more effective to provide a tug in those cases than to protect the berth against scouring).
(b) Variability in the location of exposure to high thruster loads (e.g. if exposure to erosion is always at the same location of the berth, it may be easier to install dedicated very localised erosion protection than it would be when the location of erosion exposure varies).
(c) Level of control over users of the berth and the use of bow thrusters (e.g. in the case of a single-user berth it is easier to optimise between the cost of the structure and the impact to the ships using the berth than it is in case of a multi-user berth).
(d) Impacts of operational measures to shipping (e.g. if availability of tugs is problematic, the requirement to use tugs could be unacceptable for ships using the berth).
(e) Temporary implementation of operational measures (e.g. operational measures could provide temporary relief or solutions while permanent protection is being considered, evaluated or prepared).

Either way, one needs to realise that a ship operator who has made an investment in equipping a ship with bow thrusters in order not to rely on tugs (or to rely on tugs to a lesser degree) will probably not look favourably upon restrictions on the use of thrusters and/or a requirement to use more tugs.

An example of an operational measure is a requirement to use tug assistance in certain specific situations when a ship's own bow thruster forces could otherwise be expected to be particularly high. Depending on the situation, a requirement for additional tug assistance could be limited to certain circumstances such as:

(a) when keel clearance is low (only deeply loaded ships and/or only during low tide)
(b) when wind forces are high (only empty ships and/or during high wind events)
(c) for ships over a certain tonnage or weight
(d) for ships with the potentially most damaging thrusters.

Other even simpler operational measures could be effective as well. A good example is to have periodic communications with pilots and ship operators to ensure that they understand the issue. Sharing hydro survey information showing scour can help them understand the relationship between the use of thrusters and erosion effects, and the impact that certain procedures or the modification thereof can have.

Another option is some sort of signal system, simply calling attention to high flow from bow thrusters. It is natural for pilots and masters to have the perception that a solid wall berthing structure is sturdier and thus less vulnerable than a pile-supported structure. Not being aware that erosion damage can pose a significant risk to either type of structure, they feel more naturally inclined to limit use of thrusters near piles, where they can see actual movement or see and sense the structure's vulnerability. Simply calling attention to high flow velocities generated by bow thrusters by installing sirens and/or flashing lights that would be triggered by flow sensors near the berth could further increase awareness of operators controlling how thruster power is applied.

REFERENCES AND FURTHER READING

Breusers HNC and Raudkivi AJ (1991) *Scouring*. Balkema, Rotterdam.

BSI (British Standards Institution) (2000) BS 6349-1: 2000. Maritime structures. Code of practice for general criteria. BSI, London.

CUR (Centre for Civil Engineering Research and Codes) (2005) *Handbook of Quay Walls*. CUR, Gouda.

EAU (Empfehlungen des Arbeitsausschusses für Ufereinfassungen) (2004) *Recommendations of the Committee for Waterfront Structures, Harbours and Waterways*, 8th English edn. Ernst, Berlin.

McConnell K (1998) *Revetment Systems against Wave Attack. A Design Manual.* Thomas Telford, London.

PIANC (International Navigation Association) (1992) *Guidelines for the Design and Construction of Flexible Revetments Incorporating Geotextiles in Marine Environment. Report of Working Group 21.* PIANC, Brussels.

PIANC (1996) *Reinforced Vegetative Bank Protection Utilising Geotextiles. Report of Working Group 12.* PIANC, Brussels.

PIANC (1997) *Guidelines for the Design of Armoured Slopes under Open Piled Quay Walls. Report of Working Group 22.* PIANC, Brussels.

PIANC (2010) *Guidelines for Port Construction, Related to Thrusters. Report of Working Group 48.* PIANC, Brussels.

Romisch K (1993) Propellerstraahlinduzierte Erosionserscheinungenin Hafen. *HANSA 130th Annual Set*, No. 8.

Romisch K (1994) Spezielle Probleme. *HANSA 131st Annual Set*, No. 9.

Port Designer's Handbook
ISBN 978-0-7277-6004-3

http://dx.doi.org/10.1680/pdh.60043.427

Chapter 16
Steel corrosion

16.1. General

The corrosion of steel sheet piles varies in differing conditions of sea air and seawater exposure. Experience has shown that severe corrosion occurs in saline water and under marine growth, especially in the splash zone and the lower tidal zone with alternate wetting and drying. Another type of corrosion by sulphate-reducing bacteria has been found in the sea bottom zone. These bacteria are active in waters containing nearly no oxygen, such as the conditions found in some very polluted harbour basins. Figure 16.1 shows the general pattern of steel corrosion in the marine environment.

Steel will be subjected to a natural corrosion process when it comes into contact with water while at the same time in the presence of oxygen. The material abrasion from corrosion will depend on the local hydrological conditions and on the local vertical position regarding the water line, which means that there will be different zones where corrosion forms. The degree of corrosion (the rusting rate and the intensity) decreases with an increasing layer of rust thickness, unless the covering rust layer is continuously being destroyed by, for example, wave-washing actions against the steel face.

Under aquatic conditions, the corrosion rate is directly proportional to the electrical conductivity of the water. The conductivity of seawater is high, resulting in a higher corrosion rate than for freshwater. The corrosion protection of steel in seawater has to be evaluated separately in different zones. The corrosion rate is at its highest in the splash zone and immediately below the water level.

For reasons related to corrosion, circular steel sheet pile cell constructions can be a better solution than traditional sheet pile walls. Circular steel sheet pile cells are heavily strained, a little over the sea bottom, where the maximum tensile forces from the fill are acting, whereas in a sheet pile wall the profiles have to resist large moments due to the loads in the tidal zone where the anchors are connected to the wall.

Whether the corrosion is acting on both sides or only on the outside of the steel profiles depends very much on the kind of fill used in the steel pile cells or behind the sheet pile wall. If dense material such as sand and gravel is used, corrosion on the outside can only be assumed, whereas rock fill leaving water pockets behind the sheet piles implies the danger of corrosion on both sides. One should therefore specify and check that a fill of at least 1 m thickness close up to the sheet piles is sand or gravel.

The rate of corrosion will depend on the following:

(a) Atmospheric conditions of the environment.
(b) Seawater salinity. Normally, the dissolved salt concentration of seawater lies between 3.2% and 3.6%. The maximum corrosion rate occurs when the salt concentration is a little lower, at

Figure 16.1 Corrosion of a steel sheet pile

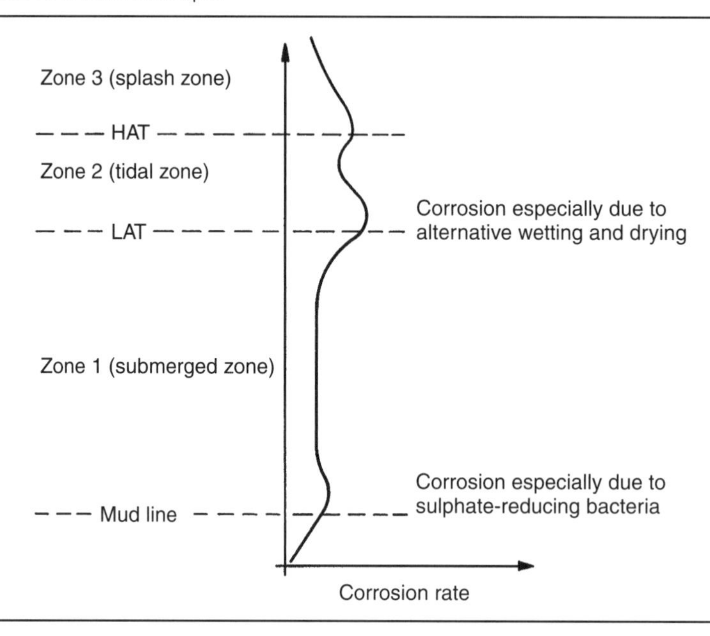

around 2.5–3.0%. This is typically found in estuarine locations, where freshwater from a river mixes with the seawater.

(c) The pH value of the seawater. If the pH value is less than 4, the rate of corrosion will increase dramatically.
(d) Dissolved oxygen. If the dissolved oxygen content in the seawater increases, the rate of corrosion will also increase.
(e) Temperature. The rate of corrosion increases in direct relation to the increase in temperature.
(f) Waves and the current. The rate of corrosion increases in direct relation to the wave action against the steel structure and the current speed.
(g) Chemical composition of the stratum into which the steel is embedded.

The corrosion protection of steel piles will vary according to their ambient conditions. The corrosion is generally taken into account as a corrosion allowance for the material thickness. The extent of the corrosion allowance depends on the planned design working life of the structure and on the estimated corrosion rate. The corrosion in soil is usually so low and uniform in the different soil layers that the protection of the steel is achieved simply by slightly over-dimensioning the steel thickness.

16.2. Corrosion rate

In the absence of accurate corrosion recordings, it can be assumed that, as a rule of thumb, the average corrosion of steel structural elements in berths amounts to about 0.10–0.15 mm/year per waterside of the steel sheet pile. The pitting corrosion rate in the tidal zone can in Scandinavian harbours be up to 0.5 mm/year, with an average of about 0.3 mm. In tropical waters, the rate of corrosion is usually higher. Eurocode 3 recommends the corrosion allowances under normal conditions as shown in Table 16.1.

Table 16.1 Recommended corrosion allowances (mm) under normal conditions

Soil conditions	Design working life: years				
	5	25	50	75	100
Undisturbed natural soils (sand, silt, clay, schist, etc.)	0.00	0.30	0.60	0.90	1.20
Polluted natural soils and industrial grounds	0.15	0.75	1.50	2.25	3.00
Aggressive natural soils (swamp, marsh, peat, etc.)	0.20	1.00	1.75	2.50	3.25
Non-compacted and non-aggressive fills (clay, schist, sand, silt, etc.)	0.18	0.70	1.20	1.70	2.20
Non-compacted and aggressive fills (ashes, slag, etc.)	0.50	2.00	3.25	4.50	5.75

Source: EN 1993-5: 2007, 'Eurocode 3: Design of steel structures. Part 5: Piling'
The corrosion rate in compacted landfill is slower than in uncompacted landfill. The design for compacted landfill can be made using the corrosion allowances for non-compacted landfill divided by 2. The values given are for guidance only. Local conditions should be taken into consideration. The values given for 5 and 25 years are based on measurements. The other values have been obtained by linear extrapolation and are therefore on the safe side

16.3. Corrosion protection systems

The reason why metal corrodes in the seawater tidal zone is due to the fact that parts of the metal surface act as anodes and other parts act as cathodes. Where the electrical current leaves the metal surface, the corrosion attack will begin. Pitting corrosion can be dangerous if a pit has been formed, because the chemical composition of the electrolyte in the pit can accelerate the corrosion in the pit.

Where there is great uncertainty about the rate of corrosion in the environment of the berth structure, preparations should be made during construction for the later installation of cathodic protection, which is an electrochemical method of corrosion control. By installation of cathodic protection, the corrosion of steel completely immersed under water (zone 1, see Figure 16.1) can be substantially eliminated, and corrosion of steel alternatively exposed to wet and dry conditions (i.e. in the tidal zone (zone 2) and the splash zone (zone 3)) can be significantly protected with an impressed current system in the tidal and splash zone. This installation must be carried out by companies specialising in corrosion protection, for example Corroteam A/S, Norway.

Marine steel structures can be protected by the following two main types of cathodic protection systems (Figure 16.2):

(a) The **sacrificial anode** or **passive protective system**, which consists of a sacrificial anode immersed in the seawater (the electrolyte) and electrically connected to the marine steel structure (e.g. the berth structure). The protected surface of the marine steel structure will now act as a cathode. The sacrificial anode system requires no external source of electrical power, and is relatively easy to install and maintain, and it is an attractive system if the required protective current is not large. The system will, when properly installed, require very little attention and maintenance during its design service life. The anodes that are used nowadays have a design life of about 15–20 years.
When two different metals are coupled together in an electrolyte (e.g. seawater), one of the metals will act as a sacrificial anode and corrode, and the other will act as a cathode. The one that acting as a cathode will be the most noble metal of the two. For example, if steel and zinc are connected in seawater, the zinc will act as the anode because the current will flow from the

Figure 16.2 Cathodic protection systems

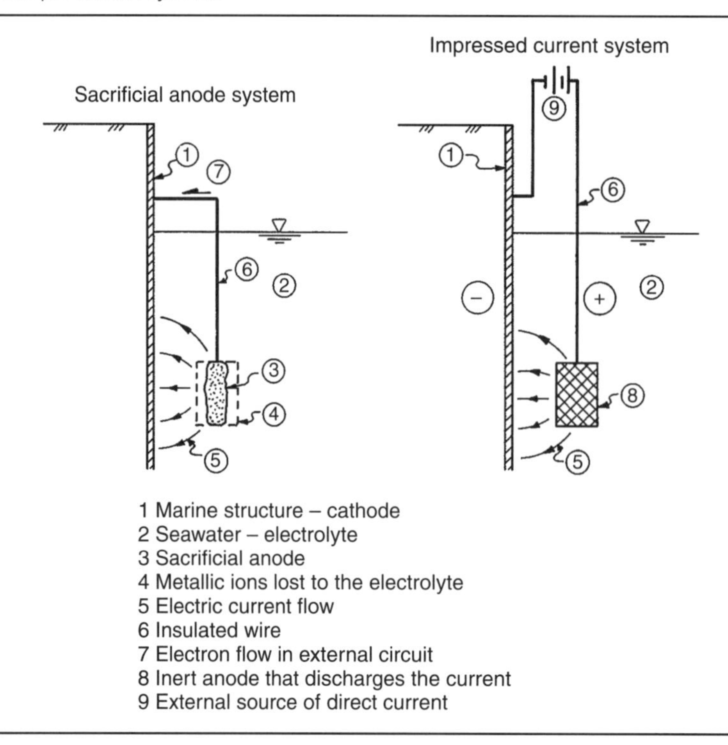

zinc to the steel. This reaction can only take place when there is an electrolyte present between the two dissimilar metals.
The list below shows some metals' relationships to each other:

Protected end – cathodic or noble
 Gold, platinum
 Nickel (passive)
 Copper
 Nickel (active)
 Steel, iron, cast iron
 Aluminium
 Zinc
 Magnesium
Corroded end – anodic

The anodes that are used nowadays are mostly anodes of aluminium alloy. The aluminium anodes give longer lifetimes with less anode weight compared with zinc anodes.

(b) **The impressed current** or **active corrosion system** is used to protect marine steel structures that require a high current. As the name indicates, a protective direct current is impressed into the cathode surface by external means. A rectifier consisting of a step-down transformer and a rectifier stack converts alternating current to direct current.
The anodes used are inert anodes of platinum, platinised titanium, lead alloys, magnetite or other suitable materials.

The difference between the impressed current system compared with the sacrificial anode system is that in the sacrificial system the anodes corrode because the current is leaving their surface, while in the impressed current system the anodes can be made of non-corroding anode materials, which enables the anode to last much longer. For both the sacrificial and the impressed current systems, the following must be considered to give the marine structure the desired service life:

(a) determine the required protective current
(b) determine the most suitable number, location, size and type of anode
(c) develop specifications for suitable mounting of the equipment to the structure
(d) develop specifications for proper maintenance inspections.

Instead of using a cathodic protection system, it is possible to **paint the steel** elements with anti-corrosion compositions or protective coatings, to form a barrier to environmental exposure and thereby delay the corrosion. The usefulness of this is often questionable because these barriers invariably break down after a number of years. Important factors in ensuring optimum performance of the protective coatings are the choice of coatings, the method of application and the thickness of coats. In arctic harbours, a coating system is generally not recommended because floating ice can destroy the coating.

The ideal and optimum protective system for steel in a marine environment could theoretically be a combination of different protective systems, because one system that is economical and effective in one zone might not be suitable for another zone. For example, some coatings are effective and economical in the splash zone but less attractive for the submerged part of the structures due to high maintenance costs. In the submerged part, the cathodic protection systems would be the most suitable. Therefore, if combinations of selected coatings and an impressed current system are compatible, they can be an economical solution to the corrosion problem.

Where protective coatings or cathodic protection are not practical or their maintenance is doubtful, increased section thickness or an **extra thickness of steel** equal to the amount of corrosion expected for the lifetime of the berth structure may be economically justified and a technically better solution.

As a rough guide, the thickness of steel used in marine structures should be a minimum of 10 mm where cathodic protection is not used and 6 mm where cathodic protection is used.

Figure 16.3 shows the corrosion of a steel sheet on a cell berth structure where the corrosion has gone too far prior to any corrosion protection being applied to the structure. Figure 16.4 shows steel corrosion beyond repair.

16.4. Astronomical low water corrosion

The phenomenon commonly described as astronomical low water corrosion (ALWC) has become increasingly prevalent in recent years. Although it is thought to be caused by sulphate-reducing bacteria, the origin of these bacteria is not always clear, as they occur in both non-polluted marine environments and polluted waterways.

It is characterised by red oxide staining on steel piles or by holes in more advanced cases just on or below the lowest astronomical tide level. The rate of corrosion is much higher than within the splash zone, and because of its location can be difficult to detect.

Figure 16.3 Corrosion of the steel sheet on a cell berth structure

Figure 16.4 Steel corrosion beyond repair

Figure 16.5 The principle of stray current corrosion

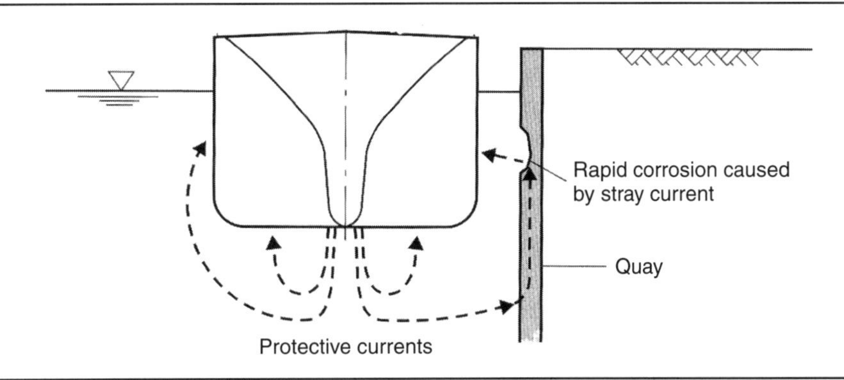

On cleaning the affected area with a high-pressure water jet, beneath the red oxide outer layer lies a layer of black sulphide. Once the sulphide is removed, the revealed steel has a shiny pitted appearance with a considerable reduction from its original thickness.

16.5. Stray current corrosion

Various types of steel corrosion in seawater can occur (e.g. stray current corrosion), due to the following:

(a) The ship's own cathodic protection system can partly incorporate the steel sheet pile berth structure (Figure 16.5). This can be the case in, for example, a ferry terminal where the same ferry uses the same berth every time.
(b) An earth leakage from an electrical installation at the berth can escape from the steel sheet pile wall to the seawater, and generate a corrosion system.

REFERENCES AND FURTHER READING

BSI (British Standards Institution) (2000) BS 6349-1: 2000. Maritime structures. Code of practice for general criteria. BSI, London.

CUR (Centre for Civil Engineering Research and Codes) (2005) *Handbook of Quay Walls*. CUR, Gouda.

EAU (Empfehlungen des Arbeitsausschusses für Ufereinfassungen) (2004) *Recommendations of the Committee for Waterfront Structures, Harbours and Waterways*, 8th English edn. Ernst, Berlin.

PIANC (International Navigation Association) (2004) *Inspection, Maintenance and Repair of Maritime Structures Exposed to Damage and Material Degradation Caused by the Salt Water Environment. Report of Working Group 17*. PIANC, Brussels.

PIANC (2005) *Low Water Corrosion. Report of Working Group 44*. PIANC, Brussels.

Port Designer's Handbook
ISBN 978-0-7277-6004-3

http://dx.doi.org/10.1680/pdh.60043.435

Chapter 17
Underwater concreting

17.1. General

Placing of concrete in water is a very difficult operation. All aspects of the procedure, from mixing, transportation, placing and control of the work, have to be carefully evaluated, and should only be performed by very experienced engineers and workers. The aim when placing concrete under water is to keep the fresh concrete and water apart as much as possible during the process, and to avoid a rapid flow of either of the water or the contact when they come in contact, so that the cement will not be washed out. For these reasons, selection of the correct placing method is the most important factor with respect to final quality.

Underwater concreting is not a new technique: it has been experimented with since about 1850. In 1910, the Norwegian, August Gundersen, took out a Norwegian patent on a 'Method of underwater casting for concrete columns and the like'. In the same year, the method was tried for the first time in Norway for underwater concreting of a reinforced structure. This method is, nowadays, the main underwater concreting method, and is known as the tremie pipe method.

Since the 1980s, admixtures have been developed that increase the cohesion of the concrete and make direct contact with water possible without significantly changing the properties of the concrete, and these are widely used. The anti-washout (AWO) admixtures, e.g. ResconMapei from Norway and similar products, have certain properties that influence the fresh concrete, and its setting and hardening. Knowledge about these properties is crucial for all parties involved in the process.

17.2. Different methods of underwater concreting

A short summary of the most common methods used for underwater concreting is given in the following sections.

17.2.1 Bucket concreting

The simplest way of placing concrete in a formwork under water is to lower the concrete through the water in an open bucket to a diver, who the carefully places the concrete in the formwork. Bucket concreting should only be used for very minor and temporary work.

17.2.2 Sack concreting

This method is used in minor permanent works and repair works. The concrete is placed in porous sacks of woven materials and lowered down through the water to a diver. As the sacks are only 50–70% full, the diver can mould the sacks into suitable shapes to give them a good contact area with each other, either side by side and/or on top of each other. As the cement paste will squeeze out through the woven sacks, a certain amount of cementation will occur between the sacks. The opening of one sack should always be placed against another sack. To provide a stronger and better result, the diver can drive reinforced steel bars through the sacks. The sacks are usually laid in bond, similar to block walling.

17.2.3 Container concreting

The concrete is lowered down through the water in a closed bag or skip in one of the following ways:

(a) *The bag method.* Where small amounts of concrete are required, for example in repair work, a canvas bag, about 2 m long and about 0.5 m in diameter, is a useful means of placing concrete under water. The canvas bag, which is reusable, is filled with concrete and closed at both ends, and then lowered to the specified location. Just above the casting spot, the bottom of the bag is slowly opened, letting the concrete flow out of the bag into the form.

(b) *The steel container or skip method.* In this method a cylindrical steel container or skip is used, with a top and bottom lid. This method is more effective than the bag method, as it is possible to bury the bottom or the mouth of the skip in previously laid concrete, and in this way prevent or reduce the possibility of washout of cement. When loaded, the skip should be full, and a flexible cover or lid should be placed over the top opening. This will reduce the washout of cement during lowering during discharging of the contents. The flexible cover will follow the top of the concrete down during pouring. To allow free flow of the concrete through the skip, the skip should always be vertical during discharge of the concrete. The weight of the skip together with the concrete will be sufficient to ensure that it sinks into the concrete surface. To reduce the possibility of washout, the skip should be provided with a skirt. During pouring the skip should be raised slowly.

For concreting of small foundations under water, a concrete with a cohesion-increasing admixture or AWO admixture will diminish the risk of washout of cement. In this case, the skip concreting method would be a better alternative than the tremie pipe method.

17.2.4 Injection (prepact method)

In this method the formwork is first filled with specially washed coarse graded aggregate. The voids in the aggregate are then filled by injection with a mortar or grout consisting of cement, sand, and expanding and stabilising material. This method can be especially useful in flowing water and in areas inaccessible to skips, tremie, hydrovalve or pump concreting, such as undercuts, for example, under a foundation.

17.2.5 Hydrovalve concreting

This method is a refinement of the tremie pipe method, or it can be said to be a cross between the skip method and the tremie pipe method. Instead of using a rigid pipe, the concrete slides down a collapsible tube, which is kept closed by water pressure until the weight of the concrete in the tube overcomes the hydrostatic pressure and the tube skin friction. The concrete plug will then slide slowly through the tube, and the tube is sealed behind each plug by the water pressure. A valve at the bottom end of the tube controls the concrete discharge.

17.2.6 Tremie pipe concreting

The concrete is transported and poured through the water by means of a rigid pipe that dips into the fresh concrete already placed. When the concreting starts, the first batch is passed through the pipe under the control of a sliding valve. The method is described in detail later in this chapter, as it contains the fundamental principle of nearly all underwater concreting.

17.2.7 Pump concreting

The pump concreting method can also be said to be an extension or a variation of the tremie pipe method. Instead of delivering the concrete into the formwork by the pressure created by the concrete's own weight, the concrete is placed into the formwork by hydraulic pumps, which pump the concrete through the pipe. Pump concreting is nowadays generally considered superior to other methods, especially when concreting large volumes. If the pipeline is equipped with an outlet valve, the concrete pump method is both versatile and safe for many applications.

Generally, placement by pumping is emphasised as a more reliable method than the conventional tremie method, because the concrete can be subjected to greater forces in the pipe than those of gravity alone. The placement pipe can then be used at greater depths, which gives a more favourable flow profile and thus less risk of sedimentation.

17.3. The tremie pipe method

17.3.1 General

The general and the fundamental principle of pouring concrete under water is generally best explained and understood if one fully understands the principle of the tremie pipe method. As shown in Figure 17.1, the concrete is poured down a ridged pipe, usually of steel or plastic, from a hopper above the surface, and pressed into the mass of concrete in the formwork by the weight of the concrete in the pipe. If plastic pipes are used, it must be ensured that they are sufficiently strong to be adequate for use at the water depth they will be placed at. The pipe and hopper are suspended from a staging, and mounted such that the steel pipe and hopper can be smoothly lifted and lowered vertically, and independently of waves and tidal variations.

The height of the hopper above the water will depend on the required casting pressure and the length of the tremie pipe. The pipe diameter should be between 15 and 30 cm. Nowadays 20 cm is the most common.

Reusable steel pipes should be built up of lengths of 1–2 m joined by watertight joints, such as bolted flanges with rubber gaskets. The pipes must be watertight and well cleaned. The lowest section of the pipe should have no flange at its lower end. Each pipe length should be easy to unscrew and remove. Figure 17.2 shows a tremie pipe being lifted into and down the column formwork. Figure 17.3 shows a tremie pipe in the centre of the column formwork, together with the column reinforcement, and Figure 17.4 shows the complete arrangement just before the start of concreting work.

17.3.2 Formwork

The formwork should be watertight to prevent water flow through the formwork and thus the possibility of washing the cement out of the fresh concrete. When wooden formwork is used, the boards should be of the tongue and groove type. Ordinary shuttering boards should be used only in massive concrete structures in water without current. An overflow should be provided just above the waterline for the water displaced by the concrete.

The formwork should be either adjusted to the shape of the rock footing or sealed by other means (see Figure 17.1). Before the concrete work starts, a diver should check for and seal possible leaks between the formwork and the rock. When wooden formwork is used, it must be weighted or anchored, and attention should be paid to vertical uplifting forces against the formwork surfaces, which are not vertical. Column bases on rock should be enlarged by at least 10 cm in all directions. The enlargement of the column base will result in an increase in buoyancy of the formwork, which must be taken into account.

Figure 17.1 The tremie pipe method

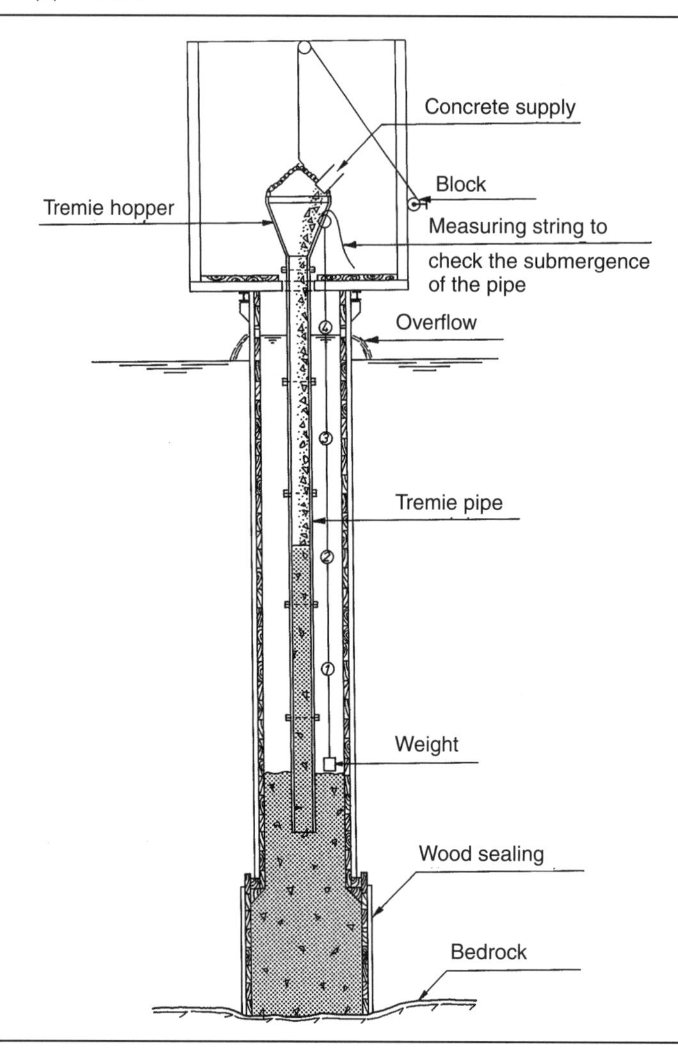

The formwork must generally be robust, simple and easy to assemble or dismantle by divers or frogmen. Tie rods must be placed where they will not obstruct the movement of the tremie pipes or the flow of the concrete into the formwork.

Generally all formwork, except that placed in the tidal zone, should be removed in order to facilitate detailed inspection and control of the concrete. In particular, the foundations of columns, walls, casting joints and expansion joints should be carefully examined for any defects.

17.3.3 Spacing of tremie pipes

The horizontal distance to be poured from one pipe with a diameter of about 20 cm should not exceed approximately 2.5 m. The supply of concrete must be regular in order to ensure a satisfactory form-filling rate. If these requirements cannot be satisfied, the area should be sectioned by means of partition

Figure 17.2 A tremie pipe being lifted into formwork

walls, or, if the capacity of the batching plant is large enough, two or more pipes can be used simultaneously. Alternatively, AWO concrete having a retarded setting could be used.

The number of pipes or sectioning will also depend on the vertical tolerance required of the finished top surface. Generally, the spacing between the pipes will be about 4–5 m. The concrete will flow about 2.5 m horizontally when using rounded gravel, and about 2.0 m horizontally when using suitable crushed stone. When using AWO concrete the spacing can be greater. The slope of the concrete surface is likely to be in the range 1 : 6–1 : 9, unless the concreting rate is very high or AWO concrete is used. A closer spacing will give a more level top surface.

17.3.4 Pouring of concrete

At the start of the concreting operation, the pipe will be full of water. The pipe is lowered to the bottom of the formwork and a plug is then placed just above the water level. The pipe and the hopper are then filled (Figure 17.5).

The controlled lowering of the concrete down the pipe is achieved by suspending the plug from a wire. Figure 17.6 shows different types of plug hanging from a wire. The rod and plate plug is the most

Figure 17.3 A tremie pipe in the centre of formwork, together with the reinforcement

Figure 17.4 The complete tremie pipe arrangement

Figure 17.5 The start of the tremie pipe method

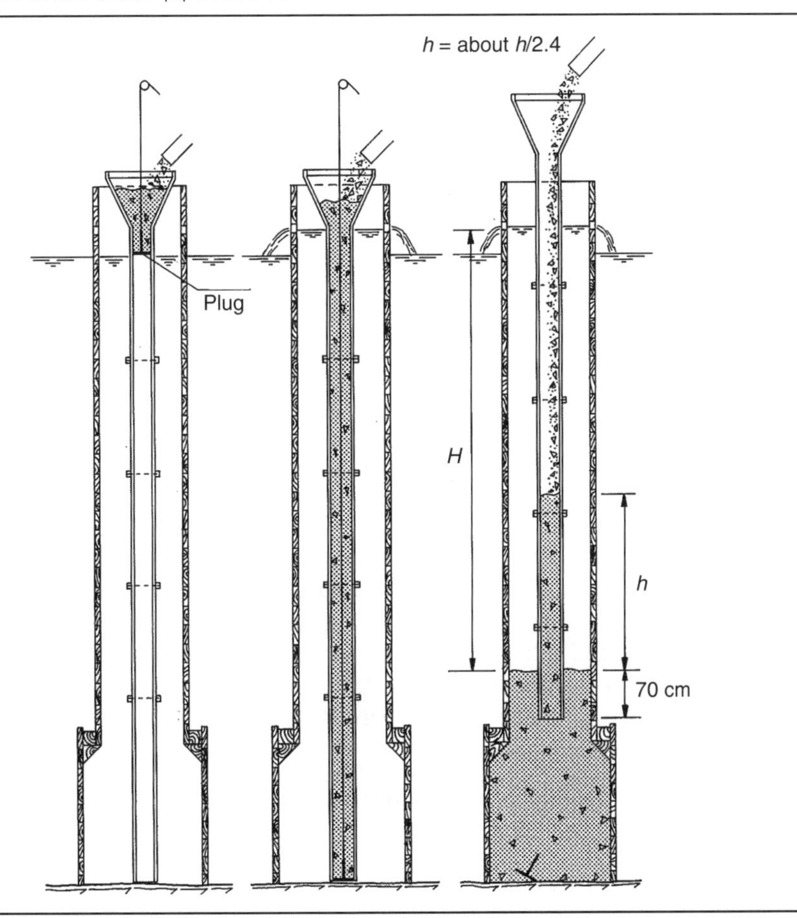

Figure 17.6 Different types of plug

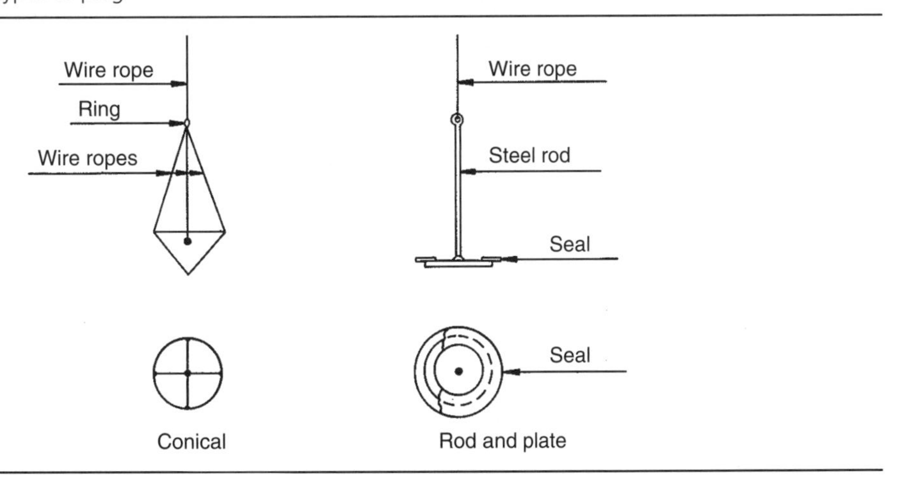

Figure 17.7 Detail of the rod and plate sections of a tremie pipe

commonly used type. In the rod and plate plug, the length of the rod should be a minimum of 5 times the diameter of the plate. The plug should be passed down the pipe, prior to pouring, in order to check that there is sufficient clearance. Plugs made of rubber balls should not be used because the water pressure will decrease the ball diameter, and therefore not prevent the water and concrete from mixing. Figure 17.7 shows the detail of the rod and plate section of a tremie pipe, and Figure 17.8 shows a pneumatic valve on the end of a concreting flexible tube.

Figure 17.8 The pneumatic valve in the end of a flexible tube for concreting

Figure 17.9 The concrete flow from the pipe into the column

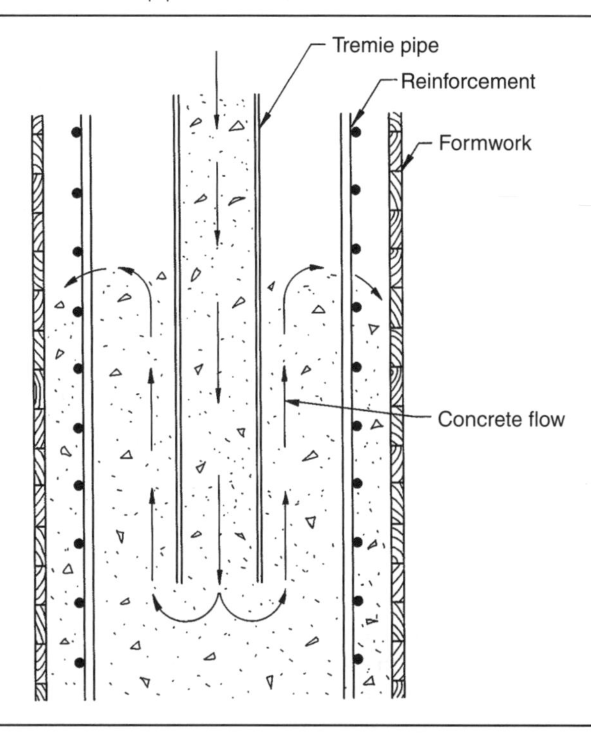

The first concrete batch should always be an oversanded and cement-rich mix. The plug is then slowly lowered down the pipe, while the pipe is continuously being filled with concrete. When the plug reaches the bottom of the pipe, the pipe should be filled to the top. The tremie hopper should also be filled completely, and additional concrete should be kept ready in a hopper above the tremie hopper. The wire for the plug should then be cut, and the pipe lifted slowly, whereupon the concrete will start to flow into the formwork. The pipe should then be lowered to reduce the speed of the concrete leaving the pipe, and simultaneously the pipe should be filled with more concrete. With constant refilling of the pipe, the concrete will be pushed upwards and outwards in the formwork (Figure 17.9).

The concrete will flow from the pipe into the poured mass of concrete, as shown in Figure 17.9. The concrete moves down the pipe and will flow the easiest route after leaving the pipe, i.e. it will flow up along the outside of the pipe, due to the friction of the reinforcement to the surface of the concrete, and roll over to the formwork (Figure 17.10). This means that nearly all the concrete will come in contact with water. It is therefore very important that the pipe outlet is submerged at least a minimum of 70 cm into the concrete. The reason for this is to slow down the flow speed of the concrete as it comes up along the pipe and rolls out on the top of the concrete (see Figure 17.9).

If the flow rate is too high, the cement in the concrete might be washed out. This will be noticeable, as the cement will discolour the water and white foam may float to the top of the water. To obtain successful underwater concrete pouring, the flow of the concrete must be correct in both the pipe and the formwork.

Figure 17.10 The concrete flow from the tremie pipe

During pouring, the pipe outlet should be submerged at least a minimum of 70 cm into the concrete. If the immersion depth is too small, a breakthrough of water from the outside into the pipe can occur, or the concrete flow up the tremie pipe and to the surface can be too fast, with the result that the cement is washed out. If the immersion depth is too large, the concreting speed is reduced to virtually zero.

A main advantage of using a concrete pump for underwater concreting is that the immersion depth can be kept at a safe level without reducing the casting speed. In addition, the speed pattern is changed from transport along the pipe to a volume increase inside the fresh concrete.

Accurate measurements must be made continually to check that the immersion depth is kept at a safe level as the concrete rises in the formwork and the pipe is withdrawn upwards. The difference in level between the concrete surfaces inside and outside the pipe is usually about one-quarter of the water depth in the formwork (see Figure 17.5). This also favours use of the pumping method in shallow water.

The following two levels must be checked continuously:

(a) the level of the outlet of the tremie pipe
(b) the level of the concrete surface in the formwork.

When the concrete level approaches the level of the water surface, a reasonable overpressure in the supply pipe requires that a 'tower' is built (see Figure 17.4); if concrete pumping is applied, no 'tower' is necessary.

If the ground or sea bottom is loose or soft below the pipe, for instance clay or sand, a metal sheet, a concrete tile or the like is laid under the outlet of the pipe to avoid disturbance of the loose material.

Once the underwater concreting has started, the process should and must proceed continuously without breaks or other interruptions until the concrete reaches a predetermined level above the water level.

Generally, concreting under water should proceed as quickly as the formwork pressure allows. For columns with a cross-section of about 0.5–2 m^2, a concrete filling rate of about 2–4 m of column length per hour is usual, the actual rate depending on the strength of the formwork. For walls and structures with larger cross-sections, the filling rate may be reduced to about 60 cm/h. If, for any reason, the filling rate falls below 60 cm/h, the flow of concrete is normally unsatisfactory. If the pipe outlet is embedded too deep in the concrete, or the filling rate has been too low, a 'plug' may build-up in the pipe. By using a rod or a special vibrator inside the pipe, the concrete flow may be restarted without letting water into the pipe, which may occur as a result of lifting the pipe too high. Again, the use of a pump has advantages, as plugs that are troublesome in ordinary tremie pipes are no problem when using a pump. Vibration of the concrete should generally not be allowed in the formwork itself due to the risk of washing out the cement.

For more extensive works, where an interruption has far-reaching consequences, the pouring speed may, in extraordinary cases, be reduced temporarily. If the pouring speed is reduced, problems are not usually experienced immediately. Problems due to poor concrete flow are usually experienced 2–6 h later. The pouring must be kept at high speed, and the concrete, the temperature and other relevant conditions must be met to suit underwater concreting. In addition, qualified personnel must be present on site. In such cases, testing and sampling, in order to confirm that the structures are being satisfactorily poured, must be undertaken.

When the area to be poured is large and the bottom is uneven and sloping with depressions, several pipes may be used, and the pouring should start in the pipe that is deepest. The other pipes are brought successively into use when their outlets are about to be covered by the concrete poured from the lower pipes. Just prior to reaching these levels, the corresponding pipe must be filled with concrete, as shown in Figure 17.11.

When new concrete is poured from the hopper, a characteristic hollow thump is heard in the pipe. Shock feeding of the pipe must be avoided, as this may lead to pockets of air in the pipe. To a certain extent, air locks may be prevented by hanging plastic hoses or a ventilation tube down through the pipe, as shown in Figure 17.12. The concrete should always be placed on the tremie hopper wall and not directly into the pipe.

Figure 17.11 Tremie pipes on a sloping bottom

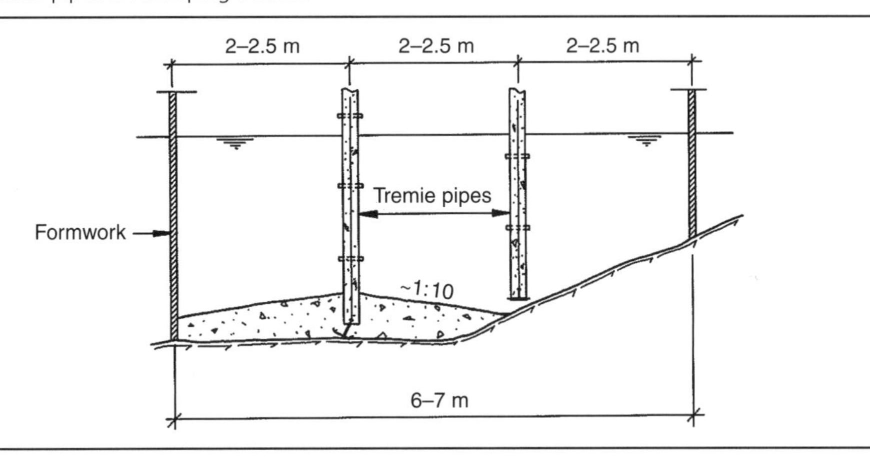

Figure 17.12 An air pocket in a tremie pipe

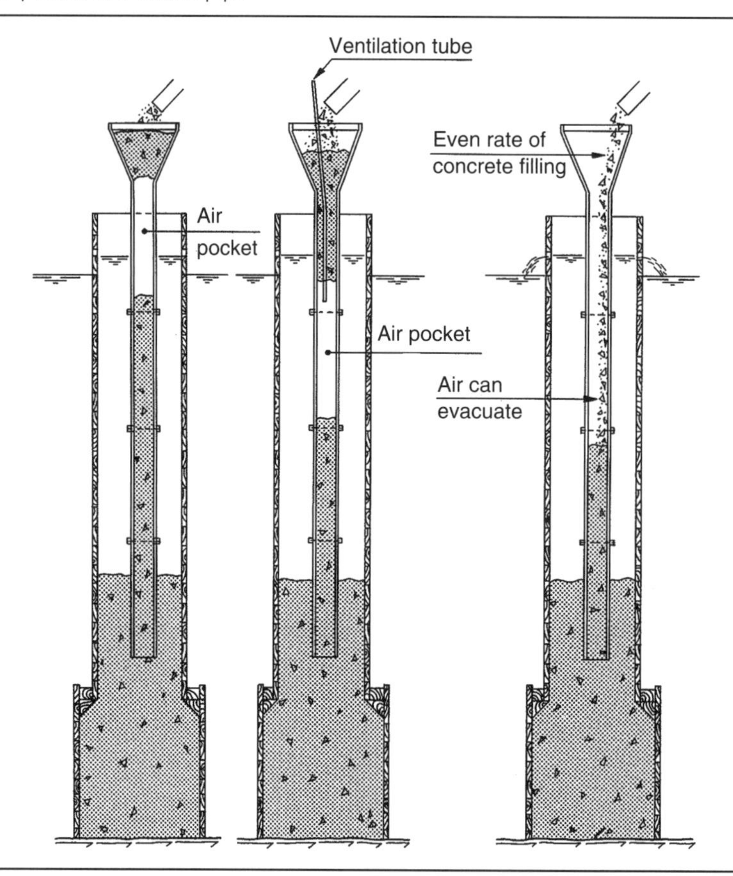

Figure 17.13 shows how the concrete will flow during the concreting of an underwater slab, and what can happen if the pipe, by mistake, is lifted up too much. The concrete will flow to the surface at too high a speed, resulting in a 'blow-out' and in washed-out concrete. As is also shown in Figure 17.13, due to the water pressure, the water can penetrate into the pipe, which will also result in washed-out concrete.

Completion of a foundation under water can be done as shown in Figure 17.14. When the pouring is almost complete, the concrete will stand in the tremie pipe at an equilibrium height, as shown. The pipe is then slowly filled with water to reduce the possibility of a breakthrough of water from the outside through the concrete and into the pipe. The pipe is then slowly withdrawn from the foundation. When a concrete structure has to be levelled off under water, experience has shown that the tremie pipe method will give a good result when carried out carefully. The concrete is poured some centimetres higher than the required level. Then, before the concrete has set and hardened, the excess concrete is removed by scraping in one direction across the whole surface in a single operation. If a sound surface with the greatest possible compressive strength is required, the upper 2–3 cm of the finished surface must be removed, and cleaned by water jetting once the concrete has hardened.

Figure 17.13 Concreting of an underwater slab

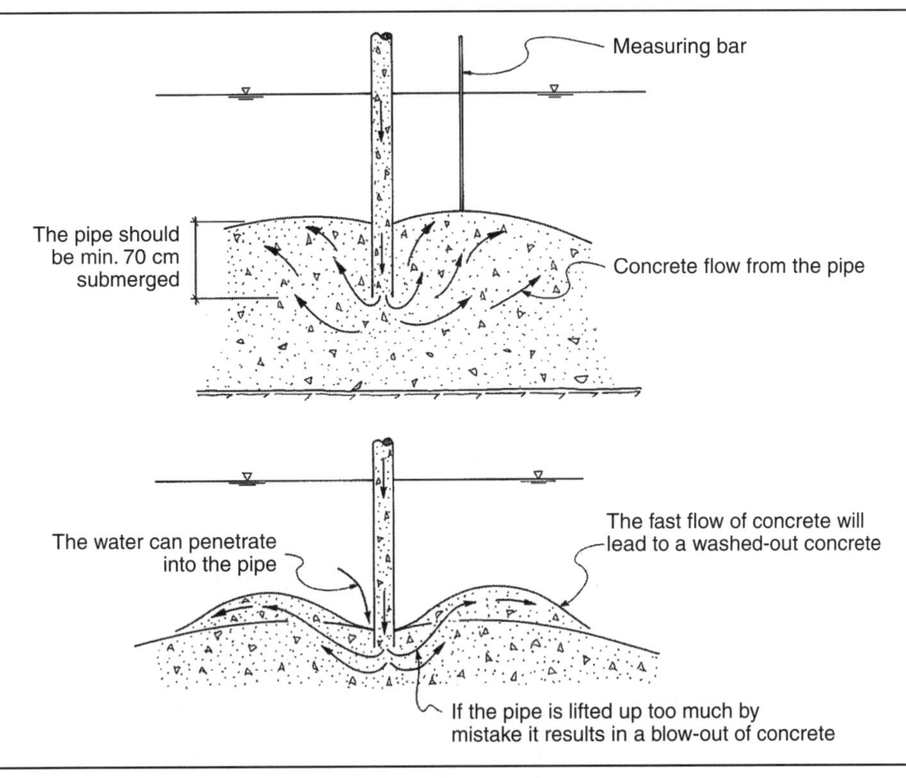

When concreting temporary structures, a larger area may be covered by carefully moving the tremie pipe sideways. This should not be allowed in the case of permanent structures, as washing out of cement cannot be avoided if a closable valve is not used or if the submerged depth is more than 70 cm.

17.3.5 Structural aspects

Underwater concrete structures should be designed and constructed in accordance with accepted international codes and regulations. This chapter is based on common and proven Norwegian practice and guidelines.

When designing reinforced concrete structures that have to be poured under water, the method of concreting has to be taken into account during the design phase. For the tremie pipe method, the least horizontal dimension of a structural cross-section is governed by the size of the tremie pipe flange, which is normally about 35 cm for a 20 cm pipe. A reinforced column must, therefore, not be less than 70 cm in diameter, and a reinforced wall must not be thinner than 60 cm. In shallow water, where a pipe without flanges can be used, the minimum thickness can be reduced.

Sufficient spacing between the reinforcing bars must be provided for the tremie pipe. Figure 17.15 shows the most usual arrangement of the reinforcement stirrups in a rectangular column. The stirrups should be made of steel of not less than 10 mm diameter. Formwork and reinforcement baskets for concrete columns are often prefabricated and mounted ashore. The reinforcement in the baskets is usually welded together. The formwork with the reinforcement must, therefore, be sufficiently strong

Figure 17.14 Completion of concrete pouring under water

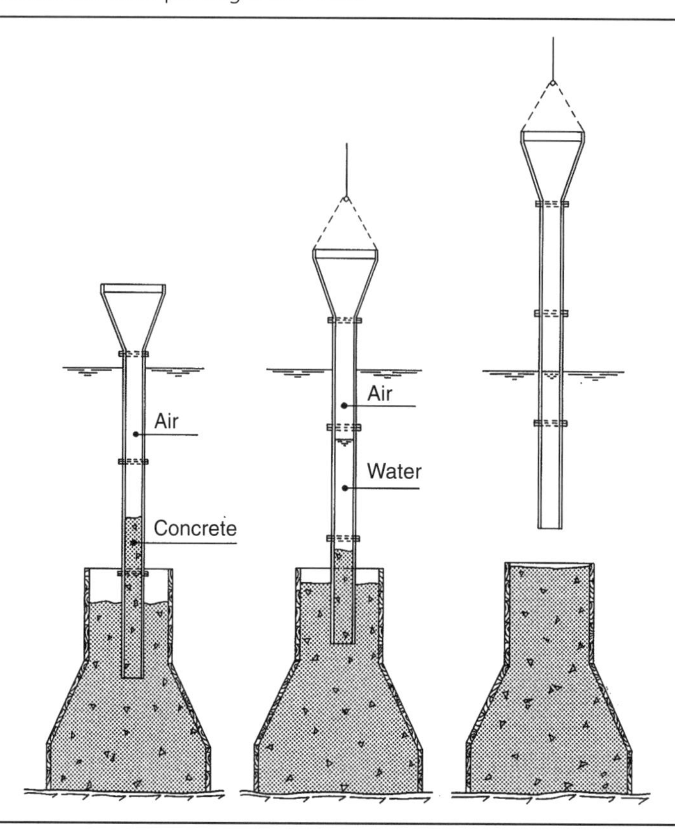

to be lifted into the water. From a construction and maintenance point of view, a circular cross-section of the columns is best.

Where the capacity of a concrete section is fully utilised, the base should be increased in size by at least 10 cm, or preferably 15 cm, in all directions, in order to compensate for possible concrete washout at the start of pouring (Figure 17.16). Reductions in column or wall cross-sections must not occur suddenly. Cross-sections should be formed in such a way that concrete can easily flow and fill the formwork. Any reduction in cross-sections or other changes in structural shape should be formed with slopes of not less than 45°, and preferably 60°.

Figure 17.15 The complete arrangement of reinforcement stirrups in a rectangular column

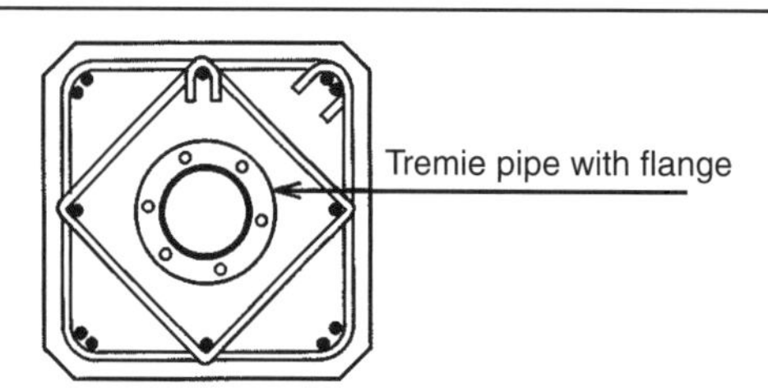

17.4.4 Water/cement ratio

The water/cement (w/c) ratio with Portland cement, or the water/binder content where silica fume is used, should not exceed 0.45. For silica fume, an efficiency factor of 2.0 is used. A w/c ratio factor below 0.4 is not recommended because the concrete flow may be reduced.

17.4.5 Aggregates

The aggregates used should be non-reactive. The use of low-alkali cements as well as blended cements will improve the resistance of the concrete to alkali–aggregate reactions.

The aggregates used for tremie pipe concrete should be well graded, with an excess of fine gravel. The total amount of aggregate passing a 0.25 mm sieve should be 8–15%. The maximum size of the aggregate should normally not exceed 22 mm, and under no circumstances should aggregates larger than 32 mm be used. Natural river gravel should preferably be used. If this is not available, crushed rock with a cubic shape may be used. Sea sand or sea aggregates contaminated with salt, etc., should never be used, even if there is sufficient freshwater available for washing.

The aggregates for tremie pipe concrete should contain less coarse material than is used in normal vibrated concrete due to the flow in the tremie pipe. The percentage by weight of coarse aggregates should be about 45–48%, and the fine aggregate should be about 55–52%. Usually, about 45% of coarse aggregates larger than 8 mm and 55% of fine aggregates smaller than 8 mm are used. The first batch of concrete used to start the concreting procedure should always be even sandier. In Norway, materials $\leq$8 mm in size are regarded as fine aggregates.

17.4.6 Workability

The workability must be adequate to ensure satisfactory placing of the concrete in the formwork. For tremie pipe concrete this will correspond to a slump of 18–22 cm.

17.4.7 Admixtures

Plasticiser should be used in underwater concrete, and its use is mandatory if silica fume is used. When pouring slabs, or where there are large cross-sections and low pouring rates, retarders should also be used. When pouring in the tidal zone, where freezing temperatures can occur, the inclusion of air-entraining agents in addition to plasticising agents is mandatory.

During the past decade, admixtures that increase the cohesion of the fresh concrete mixture and AWO admixtures have been introduced. These admixtures give the fresh concrete a high resistance to washout and segregation when it is being placed under water. As these types of admixture also make the concrete highly fluid and self-compacting, they should be used at the start of underwater pouring and in when pouring underwater concrete slabs having a thickness of less than 80–100 cm. The disadvantage of using these admixtures is that they make the concrete very sticky, and therefore make it more difficult to clean the mixing and transport equipment, as compared with ordinary concrete. The concrete also moves considerably slower in the tremie pipe.

17.4.8 Concrete compressive strength

The compressive strength of concrete placed under water should be a minimum of C35 (cube strength). The maximum prescribed strength for ordinary concrete should be C55 and that for AWO concrete should be C45 (see below).

The compressive strength class may be different for different structural parts, but as a minimum a strength class of C40/50 is recommended in BS EN 206: 2013 (BSI, 2013). This means a minimum characteristic cylinder strength of 40 N/mm^2.

17.5. Anti-washout (AWO) concrete

17.5.1 General

AWO concrete is a special concrete for use underwater. It contains an AWO agent, such as Rescon-Mapei T from Norway, or equivalent products. The AWO agent is a formulation made up of a stabiliser, a high-range water reducer (superplasticiser), special fillers and additives. The superplasticiser ensures that the cement flocks are adequately dispersed. The stabiliser encapsulates the cement grains, which prevents the cement from being washed out, even when in close contact with water. This is illustrated in Figure 17.17. In the left-hand tube in the figure, concrete containing an AWO agent is falling freely through water, without the cement being washed out. In the right-hand tube, which shows concrete without an AWO agent, the cement is washed out.

When AWO agents added to concrete its properties are altered radically compared with those of ordinary tremie pipe concrete. AWO concrete makes the diver's work more efficient, as the good visibility makes it possible to achieve better control and correction during casting.

With a correct mix design, AWO concrete flows easily. The yield stress is extremely low, allowing the concrete to flow to give an almost even surface (self-levelling). The concrete can pass obstacles, surround any reinforcement and fill the form completely. In situ core tests reveal perfect self-compacting abilities. The flow of AWO concrete through water is relatively slow, due to its high viscosity. AWO concrete retains its slump for a substantial period of time, this being longer the greater the proportion of AW agent in the concrete. This allows for longer transport and casting times, which

Figure 17.17 Concrete with (left) and without (right) anti-washout agent added

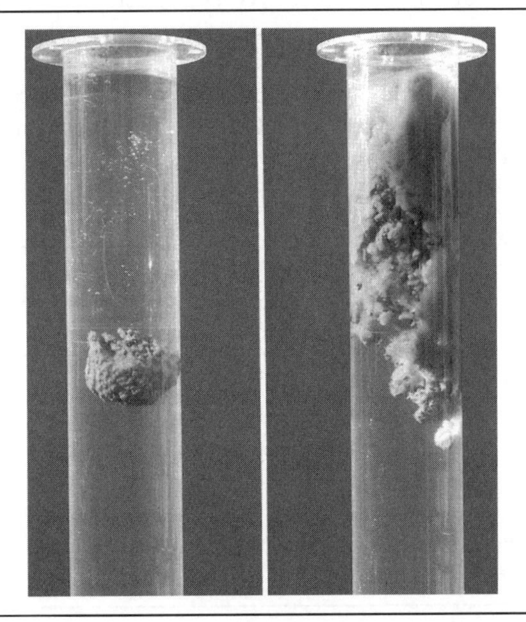

can extend to over several hours. Adjustment of slump flow can be done on site using a sufficiently efficient truck mixer.

17.5.2 Mix design consequences

It is the dose of admixture rather than the sand/stone ratio, or the cement content, that determines the amount of water needed for AWO concrete: the greater the amount of AWO agent, the more mixing water is needed. Compared with ordinary underwater concrete, an increase of 30–50 l of water is not unusual.

Unlike traditional underwater concrete or tremie pipe concrete, the amount of coarse aggregates must not be reduced in AWO concrete. Equal amounts of coarse (8 mm) and fine (8 mm) aggregates can be chosen. The aggregates should be well graded, with a maximum size of 16–26 mm. Rounded particles are always preferable, but crushed stones that are not too flaky or elongated are highly acceptable. As with normal underwater concrete, the first batch of AWO concrete to be poured should contain more sand.

Depending on the casting method and type of cement used, the addition of 5–10% of condensed silica fume, by weight of cement, is advantageous. The particle shape, the fineness and the pozzolanic efficiency of silica fume all improve the flow and internal cohesion of fresh concrete, as well as improving the long-term compressive strength, permeability and durability of the hardened concrete. The addition of silica also reduces the retarding effect of the AWO agent.

As in ordinary concrete, the final strength of AWO concrete is determined by the water/cement ratio, and the addition of silica contributes positively in this respect. The influence of the cement type is as for concrete in general, with the normal differences in early strength development. AWO agents have a noticeable retarding effect, which becomes significant at lower temperatures (e.g. below 10°C). Using cement having a higher Blaine fineness and adding silica fume reduces this retardation of setting.

Table 17.1 gives two examples of mix designs for AWO concrete.

17.5.3 Mixing procedure

AWO agents can be added into a central mixer or directly into a truck mixer, providing this has a sufficiently efficient mixing capacity. When added into the mixer of a ready-mix plant, the AWO agent can

Table 17.1 Two mix designs for AWO concrete (per m^3)

Concrete type	Water/cement ratio	
	0.42	0.50
ResconMapei T	4	4
Total water	220	225
CEM I-42.5R	450	430
Condensed silica fume (CSF)	38	20
Sand 0–8 mm	840	850
Stones 8–22/26 mm	840	850
Superplasticiser	2–3	2–3

be added either on the aggregate scales or at the same time as the aggregates are put into the mixer, but it can also be added after the ordinary concrete has been mixed. The correct amount of water should be added immediately, as adjusting the concrete flow by adding water after the introduction of the AWO agent is both time-consuming and extremely difficult because of the AWO properties. Furthermore, some of the concrete will stick to the sides of the mixer, especially if the water content is too low, thus making it more difficult to empty the mixer.

The best method is to add the AWO powder directly into the truck mixer. The efficiency of the mixer must be verified beforehand, but most automixers are perfectly capable of mixing AWO concrete. The addition can be done in two ways: either by adding the powder into the concrete flow as the batch is poured into the truck mixer, or by placing the powder in the truck prior to the fresh concrete. In both cases, rapid rotation of the mixer is essential. Both methods require a thorough mixing at full rotation speed of the mixer. The minimum time of mixing at full speed, while the truck is at a standstill, is 15 minutes for a 6 m^3 load. During transport, the mixer is normally rotating slowly, but the general stability of AWO concrete even makes it possible to transport concrete of this type without rotation.

If, on arrival at the construction site, it is found that the concrete flow is not high enough, it is possible to adjust the flow by adding a high-range water reducer/superplasticiser. Superplasticisers based on both melamine and the new co-polymers can be used. Sulphonated naphthalene must not be used. To be safe, the compatibility of the admixture should always be checked with the supplier. It is essential that this adjustment be done in cooperation with the mixer diver, who can verify the actual flow into the form. In this way, the corrections can be communicated to the ready-mix plant, to ensure correct mixing of ensuing batches.

17.5.4 Casting procedures

Many problems that result in damaged structures in ordinary underwater concrete constructions occur as a result of bad craftsmanship, lack of knowledge or bad planning. The introduction of an AWO washout agent into a sound concrete does not make these factors irrelevant. It is still essential to execute the casting according to codes and good craftsmanship, and, indeed, it is absolutely necessary to undertake a significant amount of planning prior to the concreting process. A deviation from the ideal plan can lead to unforeseen problems, which must be handled accordingly. The better the planning, the more adequately any problems can be solved. For a discussion of proper planning, see Section 17.8.

17.5.5 Placing and casting methods

AWO concrete can be cast in most of the usual ways, and often the need for costly and complicated rigs is reduced. Both new constructions and repair works at smaller depths have been successfully carried out with a crane and skip. If the concrete is required sustain a free fall through water, it is advisable to prescribe the maximum dosage of admixture (e.g. with ResconMapei T, up to 6 kg or one bag per m^3), but just the introduction of a slope to reduce the direct free fall reduces the risk of washout dramatically. For smaller works it is also possible to use buckets, and simply pour the concrete into the form.

When executing larger works and at larger sea depths, the use of pumping is normal. Even with doses as low as 3–4 kg/m^3, the AWO effect is high.

When pumping AWO concrete, it is essential to give the concrete time to flow slowly through the pipe. Trying to increase the speed will only result in a higher pumping pressure and an increased temperature of the hydraulic oil, and thus increased wear on the equipment. The high viscosity of AWO concrete prevents fast pumping, so the only way to increase the rate of casting is to increase the diameter of the pipes (a minimum of approximately 12.5 cm is recommended) or to use more pumps.

17.5.6 Formwork consequences

AWO concrete has extremely good flow properties, thus ensuring that the form is completely filled without the need for any additional compaction energy. The active flow is also followed by a penetrating quality, which enables the AWO concrete to find its way out through even small holes. It is therefore paramount to ensure that the forms used are absolutely tight. The transition zone between the rock/ground and the formwork is especially important, as a leak here can have serious consequences.

Because of the retarding effect of the AWO agent, the formwork must be dimensioned to tolerate the loads of a fresh concrete pillar from the bottom to the top of the form. Again the zone between the formwork and the foundation is important, and good anchoring either by bolting or by means of sand bags is absolutely necessary.

AWO concrete is virtually self-levelling if correctly designed and mixed, and it is therefore not possible to obtain slanting surfaces without the use of an over-formwork.

17.5.7 Combination

It is possible to combine AWO concrete and traditional underwater concrete in the same construction. If it is certain that the casting pipe is permanently submerged pipe, the start of the casting (i.e. when the concrete is most exposed to washing out) can be done with concrete to which an AWO agent has been added. The amount of this 'start' is relative to the possible exposure to water – slender structures with smaller cross-sections have a relatively small surface area as compared with massive constructions. The subsequent casting is continued with the pipe always submerged approximately 70 cm, using an ordinary well-designed underwater concrete having the appropriate water/cement ratio. The initial AWO concrete will function partly as a buffer for the primary contact with water. The possibility that the normal concrete will be exposed to water cannot be ruled out, especially when using complex forms and where there are currents in the sea, but, more often than not, the use of this method has radically improved the final result.

17.5.8 Hardened concrete

For traditional underwater concrete, the reduced quality of the concrete that is in contact with water means that the effective cross-section of the construction is reduced. The outer 10 cm in slender constructions cannot be considered as carrying loads, and for massive constructions the outer 20 cm must be excluded. In accordance with the improved performance of concrete containing AWO agents, the Norwegian Concrete Association's *Guidelines for the Design and Construction for Underwater Concrete Structures* (2003) allow the full cross-section to be taken as load carrying. This makes for slender constructions and a reduced quantity of concrete.

The exposed surfaces with traditional underwater concrete are often rather porous, often leaving larger depths of 'concrete' with an extremely high water/cement ratio. With AWO concrete, practical applications show no, or only very thin, layers having reduced quality, thus ensuring a high quality of the concrete cover, which is so essential for the durability of concrete.

All concrete follows the 'law' of slow starters: in the end the strength is higher than the 'false starters'. AWO concrete is normally a slow starter, and the surroundings of underwater constructions can also be unfriendly in terms of temperature. On the other hand, underwater concrete always has sufficient water to ensure complete hydration of the cement. It is therefore not surprising that the development of strength, measured at in situ constructions, shows a marked growth after 28 days.

Concrete that is totally submerged in water does not freeze, but in tidal zones the cycles of freezing and thawing can lead the concrete to deteriorate from within. AWO concrete is not susceptible to air-entraining agents, and is therefore not considered to be frost resistant. However, to date, frost damage of AWO concrete has not been observed on real structures. Nevertheless, the normal procedure is to stop using AWO concrete approximately 2 m below the tidal zone, and then continue the concreting with air-entrained frost-resistant concrete in the tidal zone and into the construction above sea level.

17.6. Damage during construction of new structures

Damage to newly poured concrete can be due to one or more of the following factors.

17.6.1 Unskilled labour

Perhaps the most common reason for damage during construction is the lack of skill and experience of the concrete workers. For instance, when starting the pouring of concrete through the immersed tremie pipe, a common error is that the lower end of the pipe is placed too high, or lifted too high, resulting in a washing out of cement and fine aggregates (Figure 17.18). Similarly, during the concreting of columns, the tremie pipe may be lifted out of the concrete and then replaced in it again, resulting in a layer of washed out concrete (Figure 17.19). Therefore a person experienced in underwater concreting work should always be in attendance during the pouring.

17.6.2 Unsatisfactory concreting equipment

The capacity of rigs and other equipment may be either insufficient to achieve the necessary working continuity, or otherwise unsatisfactorily adapted to the work at hand. For instance, if the jointing material between the sections of a tremie pipe is not tight, the result will be leakage and washed-out concrete. Alternatively, if the diameter of the tubes is too small, discontinuous pouring will occur; and,

Figure 17.18 The tremie pipe placed in too high a position

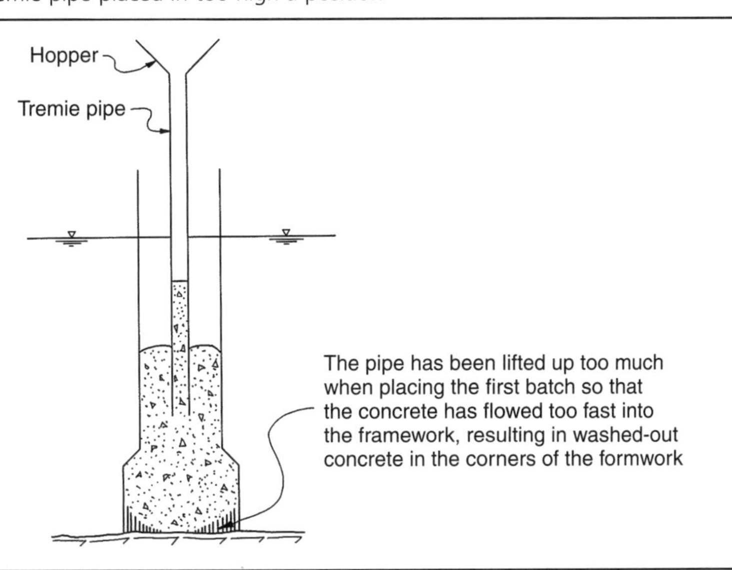

Figure 17.19 A washed-out concrete layer in the column

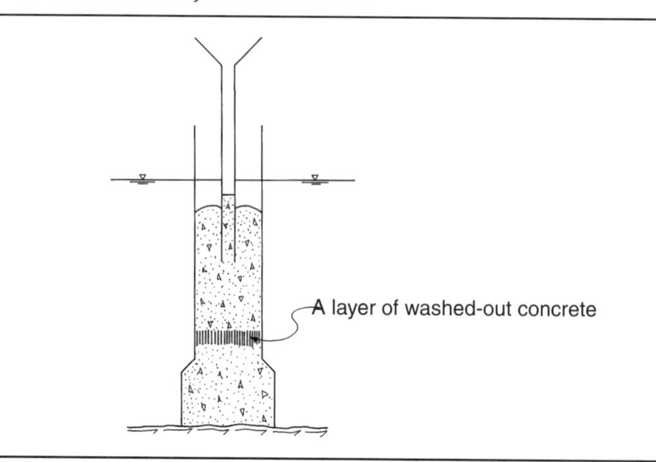

if the hopper is overfilled with concrete, some of the concrete will flow over the hopper edge and fall down freely through the water, and place itself on top of the already placed concrete (Figure 17.20).

17.6.3 Deficient delivery of concrete

It may be that continuous delivery of concrete of the prescribed consistency has not been planned or arranged in advance. A trial pour will show whether, for instance, the amount of retarding agent used per m^3 of concrete permits a satisfactory form filling rate.

Figure 17.20 Overflow of concrete when the hopper is overfilled

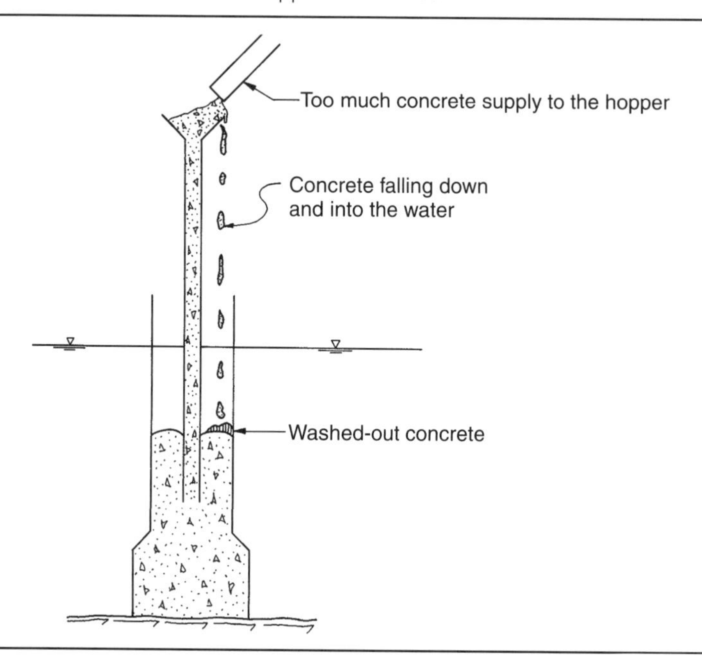

Figure 17.21 The production of an air pocket due to the formwork giving way

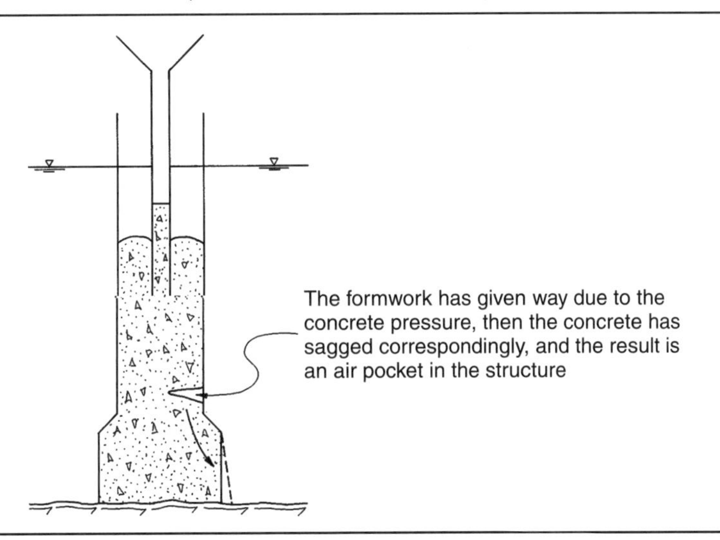

17.6.4 Faulty formwork

Incorrect design of the formwork can cause distortion under the pressure of wet concrete (Figure 17.21), formwork may not fit the underlying rock surface, so that some of the concrete leaks out (Figure 17.22), and if the form is not tight waves and current can wash out the concrete (or especially the cement in the concrete) (Figure 17.23).

17.6.5 Physical damage

Damage caused by severe impacts from ships, waves, ice, etc. are usually outside the owner's control, whereas damage from overloading (e.g. the formwork during construction due to application of too high live loads or construction cranes, etc.) can be avoided.

Figure 17.22 Concrete leaking out due to the formwork not fitting the underlying rock surface

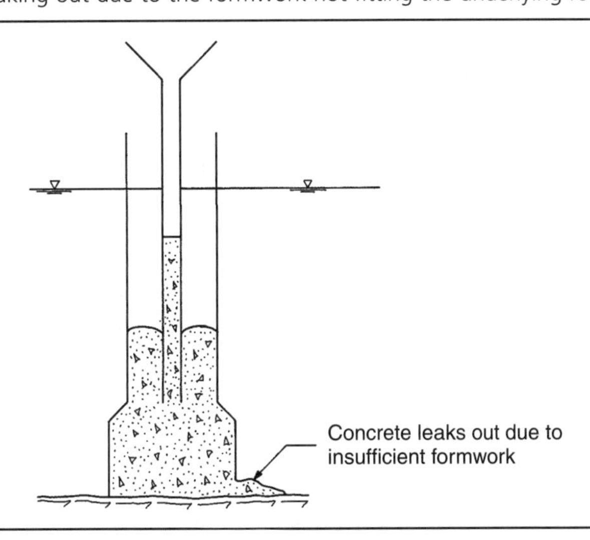

Figure 17.16 Enlargement of the column cross-section due to possible concrete washout

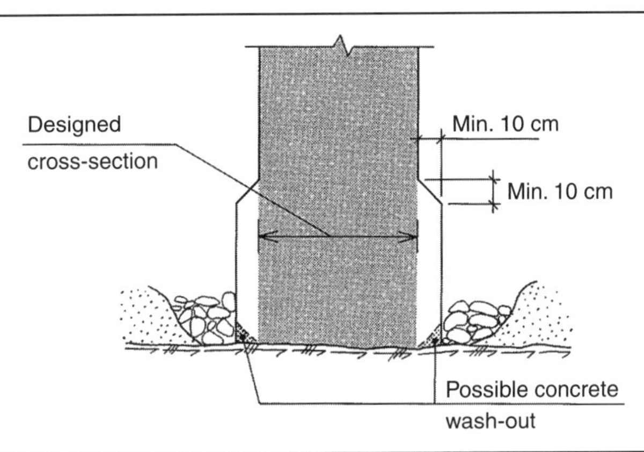

When concrete is poured towards an upper horizontal surface, it is possible that a washed out layer of about 1–4 cm can occur. The top reinforcement should, therefore, be given an extra 5 cm of cover.

Underwater concreting of slabs by the tremie pipe method is a difficult operation. Even under favourable conditions the concrete is likely to have a slope of not less than 1 : 10. In practical terms, this will determine both the minimum thickness and the horizontal dimensions of the structure. Usually, a minimum thickness of 80–100 cm of the slab represents a practical limit, unless a well-flowing AWO concrete is used. If the concrete has a good flow and a concrete pump is used to obtain a high pouring rate, the minimum thickness can be reduced.

17.4. The production of concrete for use tremie pipes

17.4.1 General

The component materials and strength of the concrete should be in accordance with accepted and relevant codes and regulations. The basic concrete mix should comply with the following.

17.4.2 Cement

The most suitable cement for berth structures is ordinary Portland cement. Sulphate-resisting Portland cement is not recommended. The importance of a moderate C_3A content of 5–8% has been clearly demonstrated by studies of the cement properties and their effect on concrete durability.

The content of cementitious material of concrete placed in the tidal zone (Zone 2) and the splash zone (Zone 3) should normally be not less than 400 kg/m^3 of compact concrete. If it can be shown that the workability can be kept unchanged by use of filler, the cementitious quantity may be reduced to 350 kg/m^3, but then the content of fine aggregates of size $\leq$0.25 mm, including cement, should exceed 400 kg/m^3.

Silica fume has performed favourably in underwater concrete to reduce the water sensibility, increase the flow and improve the durability of the concrete. The amount of silica fume should preferably be in the range of 5–10% of the cement content.

17.4.3 Water

For concrete production only fresh potable water should be used. Seawater may not be used as mixing water, or as curing water for the young concrete.

Figure 17.23 Leaking formwork

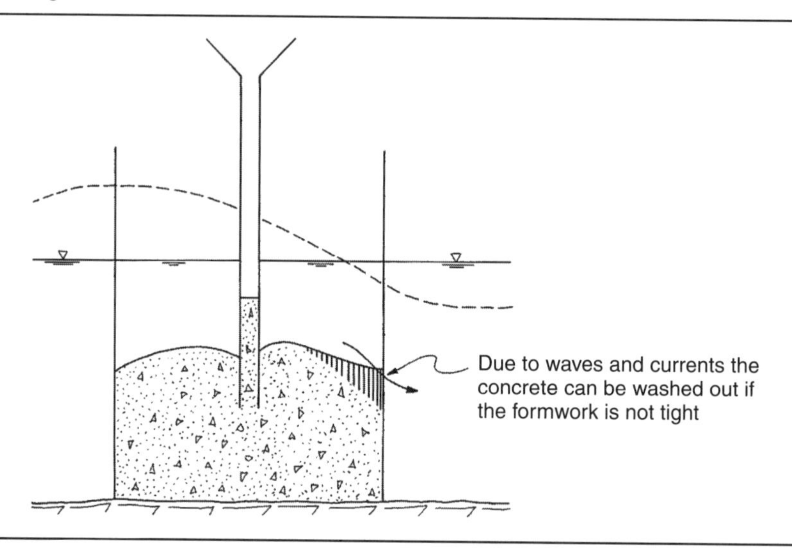

17.7. Repairs of new concrete

Generally repairs to underwater concrete are so costly that, if an error is discovered during the concreting, it pays to stop the concreting work and wash out or remove what has already been cast. Repairs to new concrete mean the repair immediately after the construction of the new concrete has finished. The supervising engineer must have sufficient knowledge of concrete technology, the theory of structures, the construction of underwater concrete structures and construction supervision. For instance, if anything irregular happens during the pouring of the concrete, the supervisor must be sufficiently competent to understand what the consequences can be, and what measures must immediately be taken.

The supervising engineer should, therefore, be properly informed in advance about the qualities required in the finished structure, so that he or she will be able to judge whether to:

(a) Stop the pouring and remove at once all the concrete already poured. For instance, if during concreting of an important column the tremie pipe is lifted into a too high position so that water penetrates into the tube, the formwork should be removed, the concrete washed away with high-pressure water, and the work started anew.
(b) Stop the pouring and continue the next day. This could be the correct measure if, for instance, the concreting of the column described above was nearly finished when the tremie pipe was lifted too high. The formwork could then be removed above the concrete surface and the washed-out concrete on top (at least 10 cm) removed by divers. Concreting may be continued after at least 12 hours. In most cases it will be necessary to strengthen the joint between the two parts of the column. A reinforced concrete mantle surrounding the column then increases the column cross-section at the joint.
(c) Continue pouring the concrete and make the repair afterwards. For instance, during the concreting of a long, high and thick concrete wall, using four or five tremie pipes at the same time, one of the pipes has been lifted too high so that a fault occurs. In less important structural elements, it is possible to put the pipe down again and continue the pouring, provided the contractor finds it more convenient to make a repair when the wall is finished. In

such cases, the contractor shall cover all the costs of repair and control, including those of a diamond drill test.

17.8. Concrete plant and supervision

17.8.1 General

As already mentioned, during the pouring of underwater concrete it is important that the work proceeds continuously without any breaks. The contractor must, therefore, carefully plan and organise the concrete work. The batching plant and the transportation system have to be dependable and must have sufficient capacity. For larger or more important jobs, an additional batching plant and a power generator must always be provided. The additional equipment must be capable of starting at short notice in order to avoid any interruption in the pouring of the concrete.

17.8.2 Construction supervision

Both the consulting engineer and the contractor should document that they have previous experience of underwater concrete work before they take or are given any responsibility for underwater concrete work.

Before the concreting starts, a check of the contractor's equipment and personnel should be made to ensure that all requirements are fulfilled.

17.8.3 Checklist for underwater concreting

The following checklist for the quality control of underwater concreting using the tremie pipe method is strongly recommended by the Norwegian Concrete Association.

Planning of the concreting operations

(a) Brief all the key personnel, e.g. inspectors, foremen and representatives of the contractor and others.
(b) Check that the concrete mix design is executed and that the mix is approved.
(c) Check that the pouring rig/plant is sufficiently designed.
(d) Check that the owner and the designer approve possible deviations from the specifications or regulations.
(e) If requested, test pours in order to check the workability of the concrete or the suitability of unusual rig arrangements.
(f) Pouring speed, submerged pipe length, sectioning of the area has to be planned.
(g) Sufficient divers for inspection and work operations.
(h) Has the contractor approved the specified method, or does he want changes to be made concerning equipment or methods?
(i) Consider possible tidal variations.

Before pouring: concrete delivery

The aggregates:

(a) sieve analysis of sand and gravel
(b) grain shape (for good workability and reduced segregation)
(c) impurities
(d) maximum size of aggregates.

The admixtures and their effects:

(a) air-entraining agents, specified air content
(b) plasticising agents, specified workability
(c) retarding agents, specified retardation
(d) sufficient amounts of admixtures available.

Cement:

(a) sufficient amount of specified cement available
(b) prescribed type.

Batching plant:

(a) sufficient capacity for continuous pouring
(b) scales in order
(c) reliable water supply and proportioning.

Transport arrangements:

(a) sufficient capacity
(b) reliability (weather changes, etc.)
(c) flexibility.

Transport distance (danger of segregation, etc.).

Sufficient capacity for filling the pipes when starting.

Plan of work for continuous pouring (no breaks, i.e. lunch).

Men available, shift work, one person responsible attending each shift.

Additional standby equipment:

(a) sufficient capacity
(b) operational.

At the site

Foundation:

(a) check cleaning
(b) the base shall be free from mud, fines, seaweed, etc.
(c) centre bolt shall be removed.

Formwork:

(a) suitable shape, good outlet of the concrete secured
(b) sealing of formwork and footing

(c) sufficient strength
(d) clean and free from pieces of wood and other debris
(e) dimensions in accordance with drawings, specifications, sufficient tie rods and/or anchors
(f) displacement of the formwork
(g) openings for water outlet at the top
(h) horizontal surfaces are avoided as far as possible
(i) at changes in cross-sections.

Reinforcement:

(a) correct placing
(b) sufficiently tied and stiffened
(c) cover, type, number of spacing

Tremie pipe:

(a) check dimensions, strength, waterproof joints and valves
(b) check hose quality for pumped concrete
(c) check of tremie pipe position in the mould, distance from the mould and area to be covered
(d) check pipe construction, the hopper and lifting arrangements.

Checks during concreting operations

Check that start-up is done according to plan:

(a) plug and plug suspension
(b) hoses for venting the tremie pipe
(c) protection against concrete overflow at the hopper
(d) start pouring in deepest pipe.

Quality control of concrete:

(a) receipt of dispatch notes, with indications of time
(b) concrete that shows signs of separation shall be rejected
(c) check slump
(d) check air content
(e) check compressive strength.

Ensure that filling of concrete into the tremie pipe is controlled by a plug. Check pipe is submerged by measurement.

Check that the height of the concrete inside and outside the tremie pipe is held above the critical value.

Check that no water enters the tremie pipe.

On restarting, ensure that proper cleaning of mud and starting procedure are performed. Check that the reinforcement does not move upwards. Ensure that there is sufficient over-height of concrete when pouring is completed.

Ensure that lateral displacement of pipes does not occur.

Ensure that pouring proceeds smoothly from all pipes.

After concreting

Check the removal of the upper layer (laitance) after the concrete has set.

Ensure thorough cleaning of pipes, etc. for later use.

After removal of formwork, test the surface with a sharp object.

Core drilling must be done in doubtful areas.

Make a report.

REFERENCES AND FURTHER READING

BSI (British Standards Institution) (2013) BS EN 206: 2013 Concrete. Specification, performance, production and conformity. BSI, London.

Gerwick BC, Holland TC and Komendant GJ (1981) *Tremie Concrete for Bridge Piers and Other Massive Underwater Placements*. University of California Press, Berkeley, CA.

Gjørv OE (1968) *Durability of Reinforced Concrete Wharves in Norwegian Harbours*. Ingeniørforlagct, Oslo.

Gjørv O (2009) *Durability Design of Concrete Structures in Severe Environments*. Taylor & Francis, London.

Ligtenberg FK, Dragosavic M, Loof HW, Strating J and Witteveen J (1973) Underwater Concrete, *Heron* **19**(3).

Marine Concrete (1986) *Papers for the International Conference on Concrete in the Marine Environment*. American Concrete Institute/Institution of Civil Engineers, London.

Norwegian Concrete Association (2003) *Guidelines for the Design and Construction for Underwater Concrete Structures*. Norwegian Concrete Association, Oslo. Publication No. 5 (in Norwegian).

Perkins PH (1978) *Concrete Structures: Repair, Waterproofing and Protection*. Applied Science, London.

Port Designer's Handbook
ISBN 978-0-7277-6004-3

http://dx.doi.org/10.1680/pdh.60043.463

Chapter 18
Concrete deterioration

18.1. General

Deterioration of concrete in marine environments can occur in different zones. The usual horizontal zonal division of structures in marine environments is as follows (Figure 18.1).

(a) **Zone 1**: the submerged zone, which is the area below the lowest astronomical tide (LAT) (i.e. the part of the structure that is always submerged in water).
(b) **Zone 2**: the tidal zone, which is the area between the LAT and the highest astronomical tide (HAT).
(c) **Zone 3**: the splash zone or the area above the HAT, which is periodically exposed to water from waves. Berth beams and bottoms of berth decks are normally included in this zone.
(d) **Zone 4**: the atmospheric zone or the areas that are only sporadically exposed to seawater due to splash from waves and spray from wind. The tops of berth decks, concrete walls on beaches, etc., are included in this zone.

These four zones can have different requirements on the composition of the concrete, the placing and covering of the reinforcement, the designed load coefficients, the materials coefficients, etc.

Experience has shown that any defect or weakness in a concrete structure will show up relatively quickly in a marine environment. It is therefore very important that anyone who designs structures for marine environments has a thorough knowledge of the potentially destructive mechanisms endangering the structures and how to repair the structure.

The reasons for the deterioration of concrete could be that the design engineer may have chosen unfavourable dimensions for the structural elements by prescribing, for instance, too high and narrow beams under the berth deck, incorrect cover to the reinforcement and unfavourable locations of casting joints, or that the contractor has not carried out a satisfactory concreting. It is important that the form has been cleaned of all debris before concreting, and that during concreting the reinforcement is not trodden down, leaving it insufficiently covered by concrete in the finished structure.

Likely reasons for the damage or deterioration of the concrete structures in a marine environment are:

(a) the poor quality of the concrete used
(b) the concrete has been poured without proper care
(c) the cover of the reinforcement bars was too small
(d) the surface drainage system has not been effective
(e) there has been no maintenance or service inspection.

Figure 18.1 Zonal division

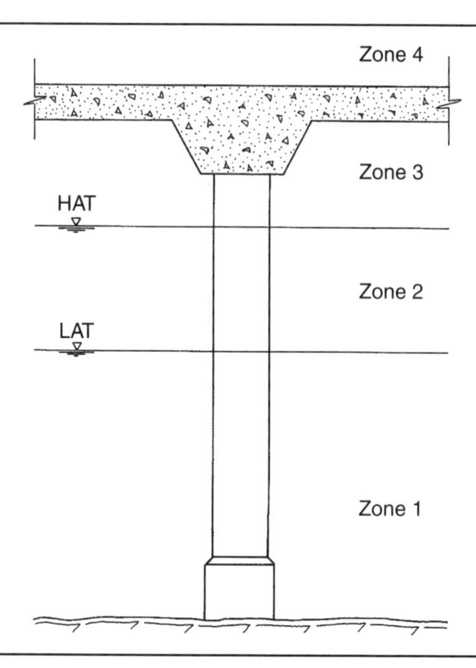

Under these circumstances, corrosion of the reinforcement steel may start very early and go on unhindered, leading to the concrete cracking and spalling off, further corrosion and breakdown of the structure. Once corrosion starts, it becomes progressively worse as the rust or the corrosion products spall and crack the concrete, thus admitting more oxygen and chlorides from seawater to the reinforcement.

A Swedish investigation (Ligtenberg *et al.*, 1986) on the causes of deterioration of berth structures showed that the frequency of damages can be divided into the following:

(a) environmental conditions (frost, corrosion, salt, ice, etc.), 45%
(b) excessive loading (ship collision, too heavy live load, etc.), 20%
(c) wrong design of the structure, 20%
(d) various other mistakes, 15%.

Deterioration can generally be divided into groups, namely those appearing during and immediately after pouring of the concrete, and those first arising after some years (Table 18.1).

18.2. Durability of concrete berth structures

To ensure satisfaction for the minimum durability of the berth structures, it is important to provide for exact first-class workmanship: concrete composition and production, execution of the concrete works and, most importantly, control and documentation of the workmanship. Research has shown that it is important to make a point of the early age control of the concrete cover. This means providing the concrete cover with good curing conditions, be it cold weather or warm weather conditions, and be it large beams or a thin wall.

Table 18.1 Deterioration to be expected in the different zones

Zone	Cause of deterioration	Deterioration occurring immediately	Deterioration occurring after some years
Zone 1	Faulty formwork	×	×
	Faulty pouring	×	×
	Corrosion		×
	Chemical reactions		×
	Erosion		
Zone 2	Faulty pouring	×	×
	Freezing and thawing	×	×
	Physical actions		×
	Corrosion		×
	Chemical reactions		×
	Erosion		×
Zone 3	Freezing and thawing		×
	Corrosion		×
	Chemical reactions		×
Zone 4	Corrosion		×
	Chemical reactions		×

Therefore, for concrete berth structures along the Norwegian coastline, extensive field investigations and research have been carried out on the durability and long-term performance of concrete structures in the marine environment. This work has revealed that an uncontrolled rate of chloride penetration and the corrosion of embedded steel have created a serious threat to the safety and economy of berth structures.

The research has shown that the minimum durability requirements in current concrete codes may not be satisfactory to ensure the good long-term performance of concrete structures in the marine environment. The experiences have shown that a high chloride penetration may already be reached at an early age during concrete construction before the concrete has gained sufficient maturity.

This may be especially true if the concrete construction work is carried out in rough and cold weather where the curing conditions during concrete construction can be poor and render the concrete more vulnerable to early chloride exposure compared with that under milder climatic conditions. Therefore, experience has shown that concrete in marine environments will show defects and deterioration relatively quickly if the composition of the concrete and the execution of the concreting work have been deficient. The deterioration of concrete in an aggressive or exposed environment can be due to the following.

18.3. Freezing and thawing

In concrete exposed to repeated cycles of freezing and thawing, for instance in the tidal zone, a suitable amount of air-entraining agent should be added. It is the smallest air pores of less than about 300 μm (0.3 mm) that determine the degree of resistance.

A low water/cement (w/c) ratio also improves the freezing and thawing resistance.

It is thus the spacing factor and the specific surface that we are interested in.

The spacing factor is defined as the maximum distance of any point in the cement paste from the periphery of any air void, calculated to the nearest 0.01 mm. The normal requirements are:

in freshwater $\leq$0.23 mm
in salt water $<$0.16 mm.

The specific surface is defined as the ratio of the surface of the air voids to their volumes, calculated to the nearest mm^{-1}. The normal requirements are:

in freshwater $>$23 mm^{-1} (mm^2/mm^3)
in saltwater $>$30 mm^{-1} (mm^2/mm^3).

The various air-entraining agents can give a different number and distribution of air pores, depending on the type of cement, aggregates, other admixtures and the type of mixers. It is necessary to do trial mixes and testing of the actual concrete composition and mixing procedure.

18.4. Erosion

Probably the most usual type of deterioration in zone 2 (the tidal zone) is mechanical erosion caused by waves, currents, ice action, etc. Unprotected concrete columns are too often found to have the characteristic shape of an hourglass with rusted-off reinforcement (see Figure 11.15 in Chapter 11). Therefore, concrete columns, seawalls, etc., that are subjected to severe erosion should be provided with up to 300 mm of cover to the reinforcement to obtain adequate life.

Due to the mechanical erosion caused by waves, ice action, etc., the minimum concrete strength should be 55 MPa. Adding reinforcing fibres to the concrete mix can substantially enhance the concrete resistance. Fibre reinforcement or fibremesh will provide an internal restraining mechanism, which will stabilise the intrinsic stresses, particularly during the first week, when the concrete is most vulnerable to cracking due to shrinkage.

18.5. Chemical deterioration

18.5.1 Sulphate reaction

Research has shown that the less resistant a type of cement is to chemical salt-water aggression, the more important it is that the permeability of the concrete is low. Standard Portland cement concrete can exhibit satisfactory resistance to salt-water attacks if made sufficiently impermeable.

Chemical attack by sulphates in the seawater on the calcium hydroxide ($Ca(OH)_2$) or the tricalcium aluminate hydrate (C_3A) components of the hardened cement can result in softening or disruption of the concrete. This problem is generally less severe in marine conditions than in sulphate-bearing groundwater.

18.5.2 Alkali–aggregate reaction

The alkali–aggregate reaction (AAR) is of the slow/late-expansive type developing over decades, and takes place in the aggregate fraction. The AAR is caused by a high content of alkalis in the cement, and occurs in structures situated in environments with high humidity. There are three requirements for the AAR (Figure 18.2):

(a) reactive aggregates

Figure 18.2 Requirements for the AAR

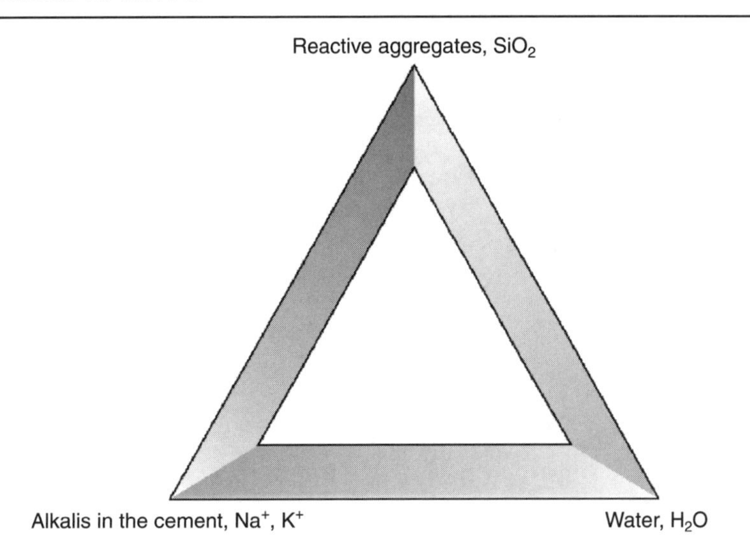

(b) alkalis in the cement
(c) relative humidity >80%.

The AAR appears as small cracks in the concrete construction (Figure 18.3).

18.6. Corrosion of reinforcement

Although deteriorating processes such as the AAR, freezing and thawing, and chemical seawater attack may also represent a potential problem for durability and have to be properly addressed, extensive experience has shown that it is the chloride penetration and corrosion of embedded steel that represents the major challenge to the durability and service life of concrete structures in a marine environment. The corrosion of reinforcement is usually the most serious problem related to the durability and safety of concrete structures in marine environments, particularly in zones 2 and 3. Generally, the Portland cement concrete is sufficiently alkaline (basic) to initially protect the reinforcing steel embedded in concrete (i.e. a thin passivating film of gamma ferric oxide is formed on the steel surface).

Even for what is stated as 'high-performance concrete' in the marine environment, current experience has shown that, for Portland cement types of concrete, it appears to be just a question of time before detrimental amounts of chloride will reach embedded steel. For cold and rough-weather conditions during concrete construction, rapid chloride penetration may also take place during concrete construction before the concrete has had sufficient curing and reached sufficient maturity. As soon as the chlorides have reached the embedded steel, it becomes both technically difficult and very expensive to get the corrosion of the embedded steel under control.

The passivity of the steel is broken as soon as the pH of the concrete component nearest to the steel surface is reduced due to carbonation, which involves interaction with atmospheric carbon dioxide (CO_2), or as soon as the concrete is polluted by chlorides (salts) from the seawater penetrating to the steel surface. A Portland cement concrete usually has a pH above 13.0, whereas the pH for salt water is about 8. The Portland cement protects the embedded steel with a thin layer of gamma ferric oxide on the steel surface (Figure 18.4).

Figure 18.3 Example of the AAR in a base plate

The reinforcement will not corrode before either the pH has become low, <9, or there are chlorides in contact with the reinforcement. The amount of chlorides must be above a critical level to start the corrosion process (Figure 18.5).

The pH will be reduced when CO_2 from the air penetrates the concrete (Figure 18.6).

Figure 18.4 Reinforcement protected in concrete

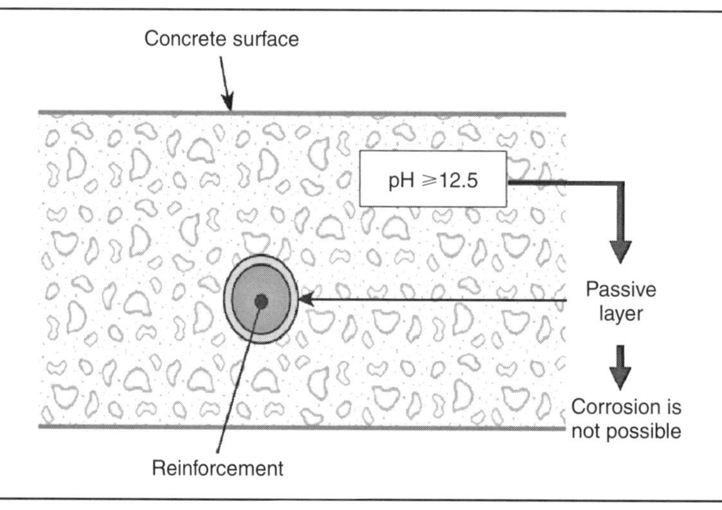

Figure 18.5 Possible corrosion in concrete

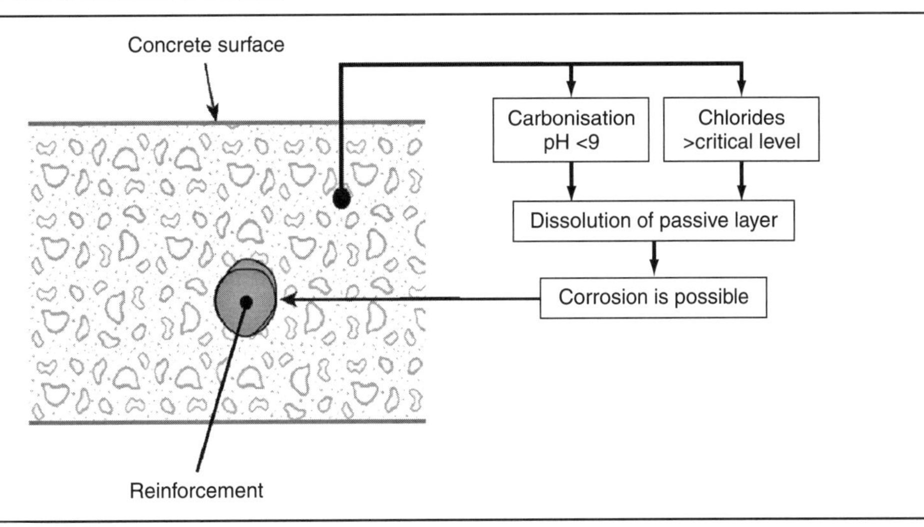

Carbonation ceases being a cause of corrosion only in zone 4, the atmospheric zone.

The most widespread cause of corrosion is penetration by chlorides. The chlorides penetrate the concrete and destroy the protective layer. The corrosion caused by chlorides is called pitting corrosion, and can totally break down the steel on a small anodic area while the cathode is intact (Figures 18.7 and 18.8).

Corrosion of the reinforcement in concrete is caused by either carbonation or chloride penetration, and the corrosion rate is controlled by humidity (resistivity) and oxygen.

When the chloride ions in seawater penetrate the concrete cover, the steel passivity resulting from a high alkaline concrete environment will be broken down. With sufficient oxygen and water present, this breakdown will cause a difference in the electrode potential between the exposed steel at the point at which depassivation has occurred in the oxide coated steel. The process is illustrated in Figure 18.2.

Generally, the service life of a concrete structure in a chloride-containing environment can be divided into two phases (Figure 18.9):

(a) The initiation period is the period when the reinforcement embedded in the concrete still remains passive, but environmental changes are taking place that may terminate the passivity. The initiation period is the time it takes for detrimental amounts of chloride to penetrate the concrete cover and depassivate the embedded steel.
(b) The corrosion period is the period that begins at the moment of depassivation, and involves the propagation or starting of corrosion of the reinforcement at a significant rate until a final stage, and the end of the service life, is reached, when the structure is no longer considered acceptable on the grounds of unacceptable structural integrity, appearance or serviceability. The propagation period is characterised by the development of electrochemical corrosion basically controlled by the availability of oxygen, electrical resistivity and temperature. Experience has shown that the initiation period can be as short as 5–10 years.

Figure 18.6 Corrosion caused by the penetration of carbonation

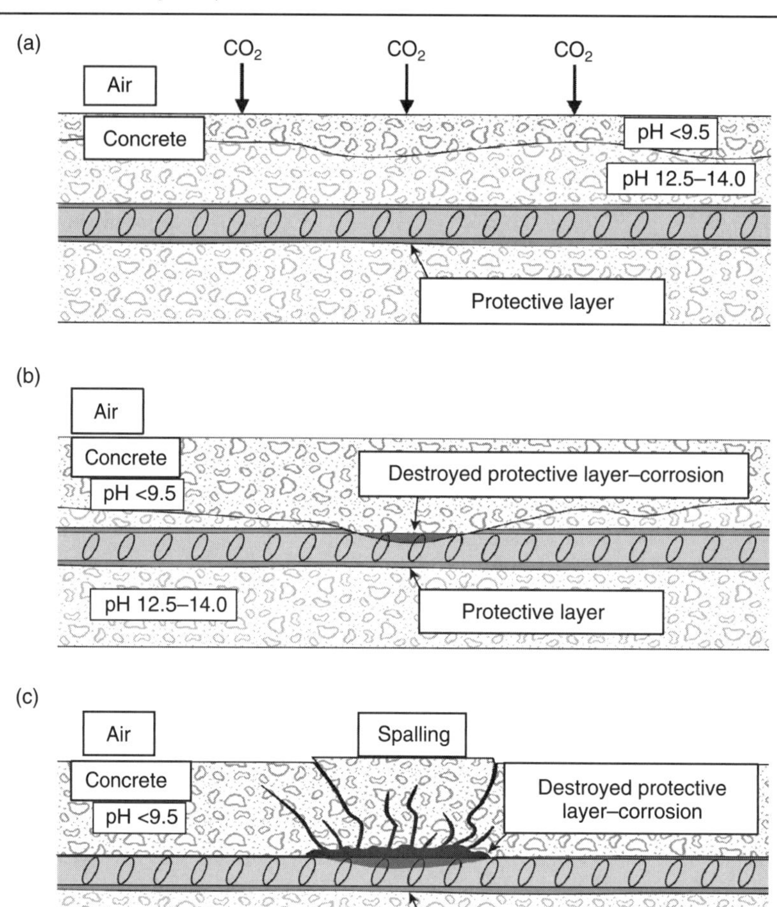

Figure 18.7 Corrosion caused by chlorides

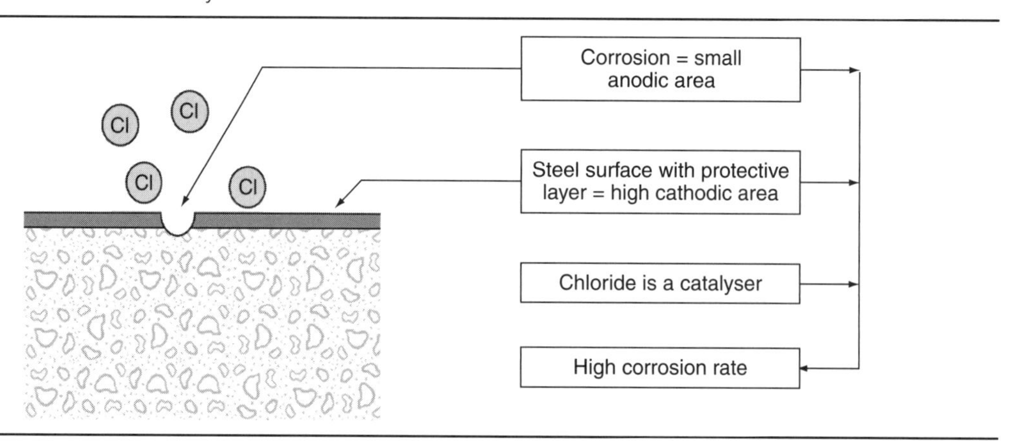

Figure 18.8 Pitting caused by chlorides

The difference in potential provides the basis for an electrolytic reaction between the exposed steel anode and the passivated steel cathode, resulting in the steel reinforcement corroding and starting an expansive reaction that will generate sufficient tensile stresses to crack and spall the concrete cover (Figure 18.10). These cracks can further provide easy access for oxygen, moisture and chlorides. For

Figure 18.9 Corrosion processes in the marine environment

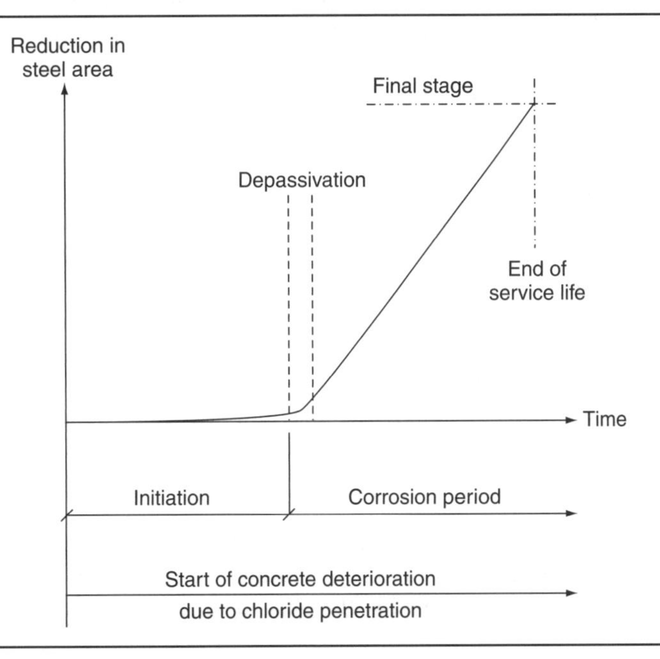

Figure 18.10 Spalling of concrete cover due to corrosion

concrete structures that are generally well below the water level and thus continuously submerged, few corrosion problems occur.

When the steel becomes depassivated by chloride penetration in the area above the water, the rate of corrosion will primarily be controlled by the electrical resistivity of the concrete in combination with the geometry and location of the anodic and cathodic areas forming on the steel surface of the reinforcement system. For a concrete with a given w/c ratio and maturity, the electrical resistivity is primarily controlled by the type of binder, the degree of water saturation and the temperature. For a concrete based on a binder with blastfurnace slag or pozzolanic materials, such as fly ash or silica fume, the electrical resistivity is much higher than that of a pure Portland cement, and hence the rate of corrosion is much lower.

In order to break down the passivity of the steel, fully or partly, the electrochemical potential has to become more negative locally (anodic areas), while other areas of the steel surface where the passivity is still intact can promote the entry potential of oxygen and form cathodic areas. Therefore, since wet concrete is a good electrolytic conductor, a complicated system of galvanic cells can come into existence in the concrete structure. Research indicates that when concrete structures pass through several environmental zones, the concrete in the splash and tidal zones, where there is a plentiful oxygen supply, can act as a cathode for corrosion under water. The intensity of the electromotive force in such a cell depends upon the pH and the chloride concentration in the water component nearest to the steel surface and upon the amount of dissolved oxygen penetrating the concrete cover.

Experience has indicated that during the propagation period it is very difficult to get the chloride-induced corrosion under control. When sufficient amounts of chloride have penetrated the concrete cover, a cathodic protection system is probably the only repair method that can, in principle, stop the corrosion.

18.7. Resistivity

The corrosion rate is dependent upon the electrical resistivity in the concrete. The electrical resistivity, ρ, of a material is given by

$$\rho = R \times A/l$$

where

ρ = the static resistivity (Ω m)
R = the electrical resistance of a uniform specimen of the material (Ω)
l = the length of the piece of material (m)
A = the cross-sectional area of the specimen (m^2)

The corrosion rate is dependent on the electrical resistivity of the concrete:

$\rho < 100\ \Omega$ m	possible high corrosion activity, but the resistivity is not decisive
$100\ \Omega$ m $< \rho < 500\ \Omega$ m	moderate to high activity
$500\ \Omega$ m $< \rho < 1000\ \Omega$ m	low activity
$\rho > 1000\ \Omega$ m	no, or very low activity

High electrical resistivity in concrete can be achieved by using pozzolans as fly ash. There is good experience from Norway in the use of 40–65% fly ash as a percentage of the cement weight (CEM I), and the resistivity measured after 2 years is 1200–1600 Ω m. Concrete based on CEM I has a low w/c ratio resistivity.

18.8. Condition survey

Physical testing should be carried out on suitable representative components and locations, both at the site and in the laboratory. The test programme should include a cover depth survey of the concrete cover to the steel reinforcement, and half-cell mapping to determine the steel potentials and contouring, as well as chloride profiles.

Determining the amount of chloride in the concrete is a very important part of the survey to assess the condition of deteriorating concrete. To chemically test for chloride penetration, a small sample of the concrete in the form of small-diameter cores and/or dust from drillings should be collected from the berth structure for chemical laboratory testing and analysis. The chemical testing should generally include: chloride ion depth profiling and testing, cement content and type, alkali–silica reaction testing, sulphate content and carbonation depth testing. Figure 18.11 shows a general profile of the chloride penetration, which has penetrated the hardened concrete, as a percentage of the concrete weight for different depths of the concrete from the surface.

Figure 18.12 indicates that, due to high tide and wind during the concrete construction of the beams and deck of an open berth structure, chloride penetration has occurred. Gjørv has shown that, during the construction stage, concrete is very sensitive to chloride, especially when the concrete construction work is carried out during cold and rough weather conditions.

A spray-on indicator with phenolphthalein is used for the determination of the carbonation depth in concrete. The test procedure is more complicated for the indication of the chloride levels: the concrete chloride samples are treated with acid to dissolve the cement, and the chloride content is determined by,

Figure 18.11 Chloride penetration of the concrete surface

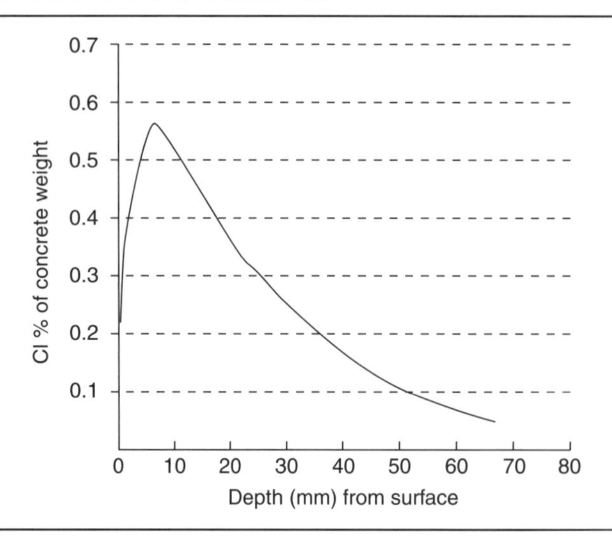

for instance, titration against silver nitrate. Chloride meters for rapid field tests are available (e.g. the Quantab and Hach methods). Table 18.2 shows the risk of corrosion due to chloride ions.

Chloride, in harmful amounts, can penetrate further into high-quality concrete than the practical limit of the concrete cover thickness. Research has shown that even if the cement content is increased to 500 kg/m^3, the penetration of chloride cannot be prevented. Increased thickness of the cover and/or increased cement content would only delay the penetration of chloride. General experience indicates that a concrete mixture with a w/c ratio of 0.40 or less may give a high resistance against chloride penetration (i.e. low chloride diffusivity). It should be noted, however, that a reduction of the w/c ratio from 0.45 to 0.35 for a concrete based on a pure Portland cement might reduce the chloride diffusivity by only a factor of 2, while replacement of the Portland cement by a blastfurnace slag cement may reduce the chloride diffusivity by a factor of 50. The utilisation of pozzolanic materials, such as silica fume, will also improve the chloride resistance. It should be noted, however, that utilisation of

Figure 18.12 Observed chloride penetration in the beams and deck structure of an open berth

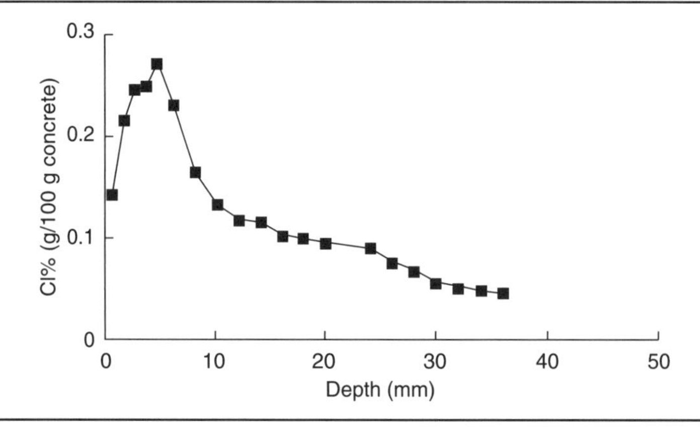

Table 18.2 Risk of corrosion due to chloride ions

Chloride ions: % of cement weight	Chloride ions by % of the concrete weight	Risk of corrosion
<0.4	<0.07	Negligible
0.4–1.0	0.07–0.17	Moderate
1.0–2.0	0.17–0.33	High
>2.0	>0.33	Very high

blastfurnace slag cements or pozzolanic materials will generally render the concrete more sensitive to good curing conditions and, hence, the execution of the concrete construction becomes more important. As a practical guideline, the minimum cement content should be at least 350–370 kg/m^3. The w/c ratio should not exceed a value between 0.40 and 0.45.

Altogether, the best method of ensuring that the natural alkaline protective mechanism is maintained is by providing or mixing a concrete that has the lowest possible permeability.

18.9. Concrete cover

The most critical construction area is zone 2: the tidal zone. Eurocode EN 1992-1-1: 2004 (BSI, 2004) on the design of concrete structures designates zones 2 and 3 as exposure class XS3, which is defined as the tidal, splash and spray zones. The minimum concrete cover for reinforcement steel is 55 mm with a design working life of 100 years. In Norway, there is the recommendation $C_{nom} = C_{min} + C_{dev}$, where C_{dev} is the allowed tolerance, C_{minus} and C_{plus}. The reinforcement should be installed in accordance with C_{nom} (Figure 18.13).

If an owner wants increased security against chloride penetration, it is possible to increase the concrete cover. In berth structures in Norway, C_{nom} is often taken as 90 ± 15 mm, which is $C_{min} = 75$ mm.

The concrete cover to the reinforcement in maritime structures should not, for the various zones, be less than the following:

zone 4: above the berth slab = 50 mm
zone 3: the splash zone = 100 mm
zone 2: the tidal zone = 120 mm
zone 1: the submerged zone = 100 mm

Figure 18.13 Reinforcement and concrete cover

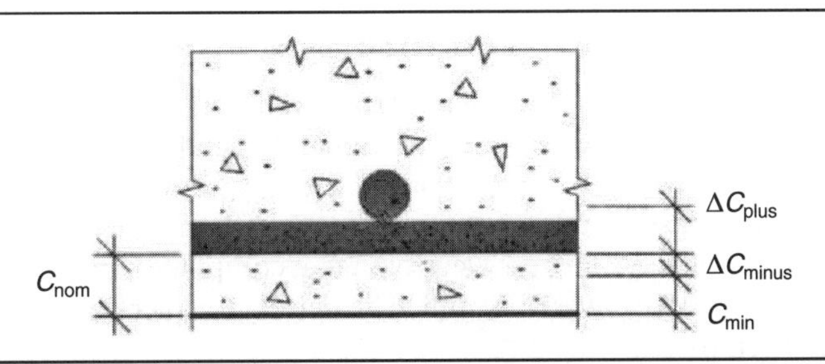

Research indicates that the cover thickness needed to prevent a reduction in the passivation due to chloride penetration should be more than twice the cover thickness needed to prevent carbonation.

To estimate the thickness of the concrete cover to the reinforcement, an electromagnetic cover-meter is a helpful instrument. The depth to the reinforcement can be difficult to estimate if the bar diameter is unknown, because small-diameter reinforcement near the surface can give the same depth readings as larger-diameter reinforcement with larger concrete cover. If the bar diameter is known, the cover-meter readings can estimate the depth to the reinforcement with an accuracy of ± 5 mm. Before reading, the cover-meter should always first be calibrated at the site by testing the meter to locate a reinforcement bar and then doing a control drilling to control its depth.

18.10. Surface treatments

Prevention against chloride penetrating into the concrete can also be achieved by using different types of products on the concrete surface that is exposed to chloride. Over recent years, a number of different surface protective systems have been developed for the prevention of chloride penetration or for retarding the rate of chloride ingress into concrete. If possible, a protective system should be applied immediately after concreting and before any exposure to chlorides.

Solid coating materials have shown that there may, after some years, be problems due to de-bonding and peeling off of the coating. Therefore, during recent years, impregnation of the concrete surface in the form of a hydrophobic treatment has been more widely adopted. Products for this treatment are available in the form of either a pure liquid, a paste or a gel. Experience has indicated that the efficiency of penetration depends not only on the time of action of the hydrophobic agent but also on the porosity and the moisture conditions of the concrete at the time of application: these are very important for the obtained depth of penetration.

Most protective treatment or coating products require regular maintenance throughout the life of the structure. Since a concrete based on slag cement is more sensitive to curing, surface protection of this kind of concrete will be even more important. For the concrete platforms in the North Sea that were given a protective coating of the concrete surface during construction, little chloride penetration has been observed after 20–25 years of exposure. Therefore, in spite of little information being available on the long-term efficiency, current experience indicates that the proper application of surface protective systems may provide valuable retardation of chloride penetration.

18.11. Condition survey

For berth structures, a condition survey or service life design should be carried out, and appropriate programmes for the life-cycle management and for later regular monitoring of the chloride penetration should be established.

The quality of the construction works is important to ensure that the owner gets what is expected, with a view to durability. The building contractor has a responsibility to prove that all materials used are fit for purpose, and to prove the actual performance. This means controlling concrete production, from developing the prescription, documentation of properties and the production control. And the building contractor must undertake stepwise documentation during their work. It is important to measure the concrete cover before casting the concrete, because it is very difficult – almost impossible – to control >60 mm concrete cover after casting. There is electromagnet equipment that can measure the cover, but its accuracy is very low for concrete cover >60 mm. The inaccuracy increases with increasing cover and an increasing amount of reinforcement.

An increase in the service life might be achieved by prolonging the initiation period or decreasing the rate of propagation once the damage has started to occur. Therefore, it is important to establish a concrete and a cover depth that provides a durability level that will match the required service life. Furthermore, it is important to make sure that the level of concrete quality based on laboratory testing is reached at the construction site. The most important factors for durability will be chloride diffusivity and concrete cover. The strategy for the durability design would basically involve the following:

(a) An appropriate concrete mixture in combination with sufficient reinforcement covers should be addressed, and the appropriate curing conditions also have to be addressed in order to meet the required quality level. Necessary back-up also has to be addressed if deviations occur at the construction site. Adequate concrete mix will most probably include blastfurnace slag cements or pozzolans (fly ash, silica fume, etc.). Back-up systems could be coating or back-up cathodic protection, if this is acceptable.
(b) Reinforcing steel should conform to an international standard for reinforcing steel (e.g. EN 10080 or similar). The surface of the reinforcement should be free from loose rust and deleterious substances, which may adversely affect the steel or the concrete, or the bond between them.
(c) Stainless steel reinforcement should be used in combination with a high-quality concrete mix based on Portland cements. For concrete structures in a chloride-containing environment, experience has shown that a partial replacement of the traditional reinforcing steel in slabs and beams with stainless steel can be advantageous from a technical and economic point of view. For such applications, steel of type 1.4436 (EU 10088-3) or 316 (ASTM) has been shown to be the most effective, but a simpler type of steel such as 1.4301 or 304 has also been shown to be beneficial.

For chloride-induced corrosion, repairs are both technically more difficult and disproportionately more expensive compared with regular monitoring of the chloride penetration in combination with a protective coating and/or cathodic protection at a suitable stage. Therefore, the following should be emphasised so as to obtain an appropriate design for the durability of a new berth structure:

(a) Cathodic protection: control of reinforcement corrosion based on cathodic protection and prevention.
(b) Blastfurnace slag cements: those countries having extensive experience claim that a high-performance concrete based on a blastfurnace slag type of cement will give a much higher durability than that of a Portland cement type of concrete.
(c) Prefabricated concrete elements: for the construction of a berth structure in a rough marine environment, reinforced non-prestressed concrete elements could be economical due to both the construction time and the design life. Prefabricated elements constructed under protected and controlled conditions will reduce or avoid problems due to early exposure from splashing and spraying of seawater. Appropriate protection systems for the concrete surface can be applied under controlled and optimum conditions.
(d) Concrete based on Portland cement will increase the durability, the resistance against chloride penetration, and the resistivity, by using large amounts of fly ash instead of cement.

18.12. Overloading of the berth structure

The application of excessively heavy live loads on the berth structure and/or heavy impacts from ships have caused cracks and damage to concrete, which in turn lead to corrosion of the reinforcement and, finally, to breakdown of the structure if not repaired in time.

18.13. In-situ quality control

Generally, during concrete construction, variations in the construction quality such as concrete production, the curing conditions and workmanship may produce variations in the concrete quality. As a result, the in situ properties may be different from those specified or produced in the laboratory. For all concrete structures, where durability and long-term performance are of great importance, the documentation of the most critical in-situ properties is crucial.

REFERENCES AND FURTHER READING

BSI (British Standards Institution) (2000) BS 6349-1: 2000. Maritime structures. Code of practice for general criteria. BSI, London.

BSI (2004) BS EN 1992-1-1: 2004. Design of concrete structures. General rules and rules for buildings. BSI, London.

Concrete Society (1986) *Marine Concrete: Papers for the International Conference on Concrete in the Marine Environment*. Concrete Society, London.

Gjørv O (1968) *Durability of Reinforced Concrete Wharves in Norwegian Harbours*. Ingeniør-forlaget, Oslo.

Gjørv O (1994) Steel corrosion in concrete structures exposed to Norwegian marine environment. *Concrete International* **16(4)**: 35–39.

Gjørv O (2009) *Durability Design of Concrete Structures in Severe Environments*. Taylor & Francis, London.

Gjørv O, Sakai K and Banthia N (eds) (1998) *Proceedings of the 2nd International Conference on Concrete under Severe Conditions, Environment and Loadings*. Spon, London.

Lahus O (1999) An analysis of the condition and condition development of concrete wharves in Norwegian fishing harbours. Dr Ing thesis, Department of Building Materials. Norwegian University of Science and Technology, Trondheim (in Norwegian).

Lahus O, Gjørv O and Johansen R (1998) *Durability of Reinforced Concrete Wharves in Norwegian Harbours*. PIANC, The Hague.

Norwegian Concrete Association (2003) *Guidelines for the Design and Construction for Underwater Concrete Structures*. Norwegian Concrete Association, Oslo. Publication No. 5 (in Norwegian).

Perkins PH (1978) *Concrete Structures: Repair, Waterproofing and Protection*. Applied Science, London.

Pullar-Strecker P (1987) *Corrosion Damaged Concrete*. CIRIA, London.

Schiessel P (1988) *Corrosion of Steel in Concrete. RILEM Report*. Chapman and Hall, London.

Taylor D and Davies K (2002) Engineering the rehabilitation of reinforced concrete port structures. *Proceedings of PIANC 2002: 30th International Navigation Congress*, Sydney.

Våg-och Vattenbyggaren No. 9 (1986) Svenska, Våg-och vannenbyggaren Riksförbund, Stockholm.

Port Designer's Handbook
ISBN 978-0-7277-6004-3

http://dx.doi.org/10.1680/pdh.60043.479

Chapter 19
Concrete repair

19.1. General

The durability and lifetime of concrete structures are becoming increasingly important. In all repair work, an evaluation of the remaining service-life of the berth structure should be part of the repair design evaluation.

Damage has been discovered on many old and new berth structures in the past few decades and, unfortunately, also on a number of relatively new ones. The main problem found has been chloride-induced corrosion of berth structure reinforcement. Contributing factors to these unfortunate findings were that designers had a relaxed view about the problem and a misguided belief that concrete structures could last forever. Poor concrete quality, insufficient concrete cover and casting errors seem to be the most important reasons for the many cases of damage and the need for repair.

Deterioration of structural elements located in the tidal zone and above the water level is nearly always caused by chloride-induced corrosion of the reinforcement or freeze–thaw bursting. Deterioration of this type will inevitability occur to some extent on old structures, and is difficult and expensive to repair.

Generally, the repair work should be defined according to the condition of the concrete and the reinforcement, the remaining design life of the structure and economic considerations. When confronted concrete structure that has deteriorated, it is often the case that a method of repair is chosen that involves the lowest cost at that particular time, without considering the lifetime of the structure. Unfortunately, this will usually imply repeated repairs, and is an altogether wasteful solution.

However, under-water damage often occurs during the construction period, and reinforcement corrosion is seldom a serious problem, provided that the cover is adequate and the concrete has not been washed out during casting. This means that the structure under water should be investigated after construction and before it is taken over by the owner.

19.2. Assessment

The most important step in the design of repair and/or strengthening work is a careful assessment of the existing structure. The purpose of this assessment is to identify all defects and damages, to diagnose their cause, and hence to assess the present and future likely adequacy of the structure. The information obtained from the structural assessment can then be used to determine whether or not corrective work is required or is economical (when compared against the cost of demolition and replacement) and, if so, how it can best be accomplished. Without prior planning and proper assessment, any programme of corrective work is likely to prove ineffective.

Owing to the safety consequences that have to be considered, an engineer with a broad knowledge of and experience in materials technology, deterioration mechanisms, structural behaviour, repair techniques and construction procedures should perform the assessment.

The assessment can be carried out in several stages. In most cases it is useful first to carry out a general in situ survey, to gain an estimate of safety hazards and to give an indication of whether immediate safety precautions are needed. This first-stage survey may help to plan the next stage of the survey, by using the information obtained to choose the required type, number and location of future investigations and measurements to be carried out.

The final report, based on the assessment, the laboratory tests, the information from the owner and the structural analyses, should contain information on the following:

(a) structural design data
(b) environmental conditions
(c) information on future use: expected service lifetime of the repair and required load-carrying capacity
(d) data from visual inspection
    (i) state of corrosion
    (ii) amount of spalling, cracking and patches
(e) data from in situ and laboratory investigation
    (i) concrete strength
    (ii) concrete cover
    (iii) chloride profiles
(f) load-carrying capacity of the structure
(g) description of each structural element and the cause of the deterioration
(h) evaluation of different repair techniques
(i) need for strengthening
(j) cost evaluation of different repair strategies
(k) recommendations.

19.3. Maintenance manual and service inspection

For all concrete berth structures and equipment, both periodic and regular inspections are necessary so that any damage may be detected in good time in order that costly repair and renewal of the structure can be avoided.

The owner should perform a service inspection of the completed berth structure at least every 5 years. The inspection should both provide an evaluation of the structural condition of the berth and detail a maintenance procedure if necessary. The areas of special interest for inspection will normally be the splash zone, construction joints, previously repaired areas and areas of vital load transfer. The basis for the inspection and observations should be the 'as built' drawing of the construction and, if performed, notes from the latest service inspection.

Usually, the first inspection of any structure would be at the takeover or acceptance of the structure, and the second inspection should be at the end of the guarantee period. All results of inspections should be recorded for later necessary evaluations of maintenance measures. The service reports from the inspection should also be sent to the designer and the contractor for their information.

The maintenance manual and service inspection should contain the following.

The first chapter should describe the construction, and include a description of its original purpose and drawings. Design loadings and special materials also form a natural part of this chapter, as does a list of the key personnel involved in the design and erection. Then the geotechnical solution should be described in detail, including soil and rock conditions, the foundation method used, problems encountered during construction and equipment used during erection. Next is a description of the materials used and the service lifetime. A detailed description of the chosen materials and concrete cover must be provided. When there are differences between the design concrete cover and the actual concrete cover at site for parts of or the whole construction, these must be described in detail. All material parameters should be provided in this chapter. It is common that some or all equipment has a lower design lifetime than the basic construction. A description of all equipment, including surface treatment and expected lifetime, concludes this chapter.

The second chapter contains the planning of future inspections. This is divided into yearly inspections, the inspection at the end of the guarantee period (e.g. 3 years) and the main inspections every 5 years. Yearly inspections should contain a visual check of the reinforcement corrosion and a visual check of all equipment. This inspection can be done by the owner without further assistance. The inspection done at the end of the guarantee period should cover the same aspects as a yearly inspection, but a design engineer and contractor should participate. The main inspection done every 5 years should cover the same aspects as the yearly inspections, but chloride profiles should be taken and the design engineer should participate on site. Main inspections should also include a verification of maintenance performed and deviations, including treatments.

The third chapter covers the planning of maintenance, which is a yearly event. Yearly maintenance should contain a hose-down of the whole construction, but this can be done more often. Equipment should be checked for corrosion, and repainted if necessary. It is also very important to check all safety equipment for malfunctions. Malfunctioning equipment must be replaced. The planning includes the preparation of planned maintenance and a register of which personnel and working teams are designated to carry out the maintenance. At the end of the chapter there should be a register of completed maintenance and, where there have been deviations, a list of treatments completed.

The fourth chapter covers future budget planning to avoid unexpected costs. It is important that all necessary recourses are available ahead of execution.

19.4. Condition of a structure

To give the owner of a berth structure a very rough idea of the condition of the structure before the repair work starts, the classification system shown in Table 19.1 can be used. The system depends on a thorough inspection of the columns, beams and slabs, and gives each part a score according to its condition. The corrosion process due to chloride-induced deterioration of reinforced concrete over time is shown in Figure 19.1. The score allocated will also indicate the expected lifetime and live loading of the structure (see Table 19.1).

Based on this system, each of the columns, beams, slabs, etc. can be assigned an average score, and the berth itself can be given an average total score. This will give the owner a very rough indication of the condition of the berth structure. If the damage is so severe that the capacity of the structure is below the ultimate loading condition and the cost of repair is too expensive, it might be possible to reduce the design loading and recalculate the score for the structure using effective reinforcement in line with the

Table 19.1 Expected lifetime and live loadings

Marks	Definition	Remaining lifetime of the structure	Live loading
1	Satisfactory	40–50 years	As designed
4	Damaged but worth repairing	10–20 years	Nearly as designed
7	Costs more to repair than to rebuild	5–10 years	Has to be reduced substantially
10	Damaged and cannot be repaired	0–1 year	No live load

correction. This might give the structure a few more years of remaining service life. However, it is important to be aware that actual loading must be below the new design loading, and that these calculations and verified assumptions have to be done by a very skilled and experienced engineer.

Figure 19.2 shows a berth slab and beam that are beyond repair because no maintenance or repair work has been carried out during the structure's service life.

19.5. Repairs of concrete

The choice of method for repairing deteriorated concrete will, in each case, depend on the zone in which the deterioration is found and the cause of the deterioration. It is therefore essential to establish what has gone wrong before a repair procedure is implemented. If there is careful evaluation of the extent and the cause of the concrete deterioration, the procedures for the repair work will also accomplish one or more of the following:

(a) restore the structural strength
(b) increase the structural strength
(c) improve the appearance of the concrete surface.

Figure 19.1 Chloride-induced determination of reinforced concrete over time

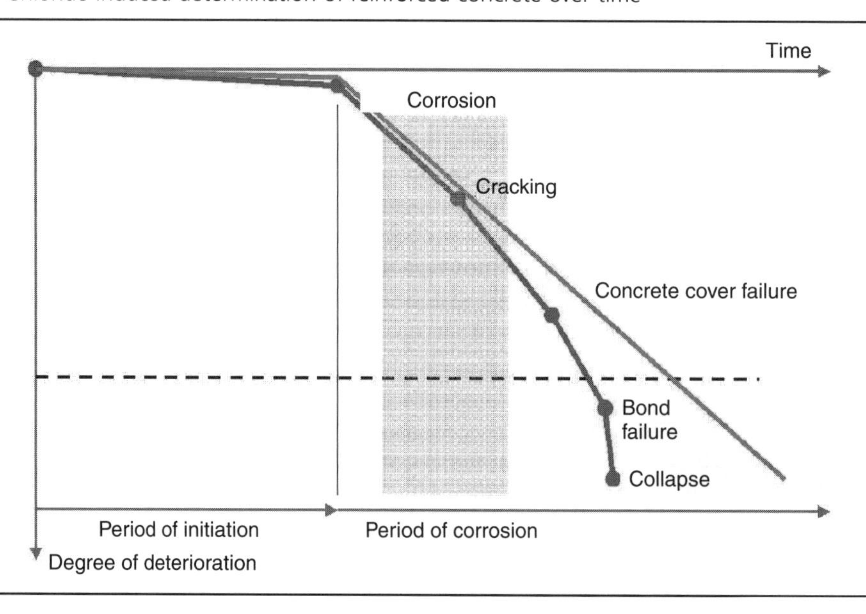

Figure 19.2 A berth slab and beam that are beyond repair

In connection with repairs to old concrete, the electrolytic conditions in the structure should be altered as little as possible. Much research has been undertaken to find ways to seal the concrete surface against salt water (chlorides) penetrating into fresh concrete and reaching the reinforcement. Means of protection, such as bitumen, epoxy, etc., applied on beams and slabs have been shown not to last for more than 5–10 years. The reason for this is probably that the coating is too tight, leading to condensation inside and subsequent freezing and spalling. To obtain a satisfactory coating, it must be sprayed onto a clean, not too smooth surface so that it has a good overall adhesion. However, the coating itself must not be so impermeable that condensation water cannot escape (i.e. the coating must be able to 'breathe').

Galvanised reinforcement has also been tried against corrosion, but the galvanisation is expensive and galvanised bars are not produced in long lengths. Therefore, cathodic protection of the reinforcement should be the alternative, but in practice care should be taken to keep the potential value constant throughout the reinforcement.

In the tidal zone (zone 2) the formwork should normally not be removed but remain as an additional protection. Research has clearly shown that such 'permanent' formwork is an efficient measure against freeze–thaw attacks. Concrete that is permanently submerged (zone 1) needs no such special protection, and all formwork should be removed so that the concrete surface can be inspected.

Routine control should always be established so that faults can be detected and repaired as soon as possible. In the long run, the most economical forms of 'protection' are correctly composed concrete and the cast concrete having a correct cover thickness outside the reinforcing bars. The methods of repair suitable for the various zones are given in Table 19.2.

Regardless of the zone in which it is located, all poor concrete must be cut or chiselled off and the concrete surfaces cleaned. Chloride-infected concrete approximately 3 cm behind the main reinforcement in zones 2 and 3 must be removed before further repair. Rust must be carefully removed,

Table 19.2 Methods of concrete repair

Methods of repair	Zone 1	Zone 2	Zone 3
Tremie pipe concreting	×	×	
AWO concrete	×	×	
Injection	×	×	×
Micro-concrete	×	×	×
Special epoxy	×	×	×
Shotcrete		×	×
Cathodic protection embedded in shotcrete			×

preferably by sandblasting, before fresh concrete or epoxy is applied. It is quite possible to carry out compressed air chiselling, sandblasting and pressure water jet washing under water to a depth of about 20 m.

In order to avoid the growth of algae, etc. on the cleaned concrete surfaces below the water, and thereby to reduce the adherence between old and fresh concrete, the placing of fresh concrete must be carried out immediately after the cleaning has been done (Figure 19.3).

19.6. Repairs in zone 1 (permanently submerged)

No matter whether the deterioration has taken place on old or fresh concrete, the choice of repair method in zone 1 will be the same. Repairs in this zone are made by tremie pipe concreting, injection, micro-concrete or special epoxy. As these methods involve diving, and the work must be planned in such a way that it can be carried out in a simple and straightforward manner under water.

Figure 19.3 A cleaned concrete surface before repair

19.6.1 Tremie pipe concreting

Berth columns and structural elements of similar dimensions being repaired by the tremie pipe method should be provided with a mantle all around in order to give the newer part of the element a good bond with the old. This applies even if deterioration is found only on one side. Additional reinforcement must be provided or, if this is not needed for structural strength reasons, a reinforcement net should be applied to keep the concrete in place.

When deterioration has occurred at the joint between the concrete and the bedrock, the latter must be cleaned of old concrete, and it should be carefully considered whether additional bolts into the rock ought to be provided together with the concrete mantle. The formwork for the mantle should be carefully tailored to the rock surface in advance, so that the diver does not have to pack between formwork and rock during the concrete pouring.

The thickness of the mantle must be sufficient to provide room for the tremie pipe between the additional reinforcement and the formwork. For the concreting of such a mantle, two tubes should be used, placed on opposite sides of the column. Figure 19.4 illustrates the repair of a column base on rock.

If the part to be repaired is not very far below the berth slab, flexible textile bags can be mounted around the damaged area. After all loose old concrete has been removed, the new reinforcement is mounted around the damaged area, and concrete is pumped into the textile bag (Figure 19.5). The tremie pipe concrete is placed in the usual manner, as tremie pipe concrete can be replaced by tremie pipe of steel and holes can be made for them in the slab.

If the AWO concreting technique is used, the procedure will be the same.

19.6.2 Injection

This method is usually carried out by specialist contractors who have the necessary equipment for injection concreting. The method is normally used for older structures undergoing comprehensive repair

Figure 19.4 Repair of a column base rock using the tremie pipe method

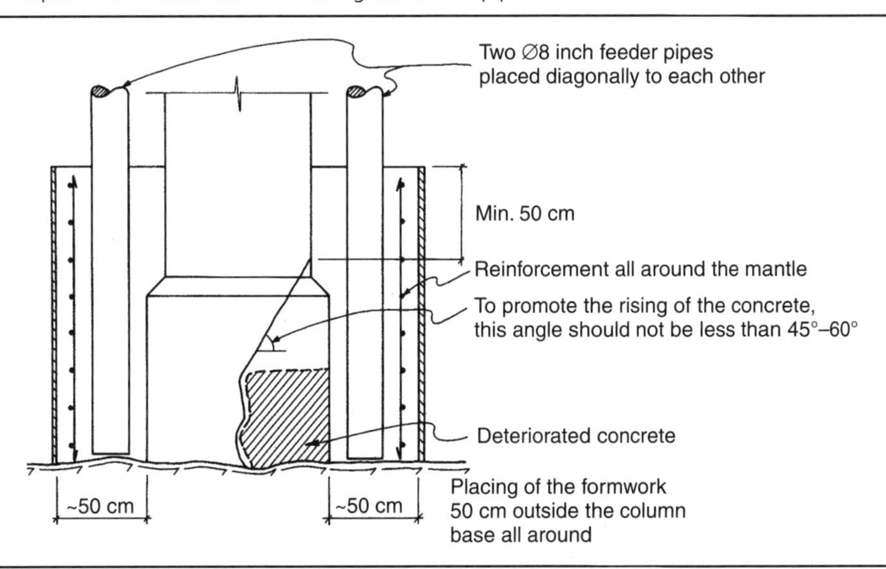

Figure 19.5 Repairing a damaged column using a textile bag. (Courtesy of EB Marine, Norway)

and maintenance works. One reason for this is that. in columns that have been recently concreted, deteriorations are usually discovered before the slab is concreted, and so there is access for tremie pipe equipment to be used from above. However, in new heavily reinforced structures the injection method can be a good alternative.

When using injected concrete, the thickness of the concrete mantle can be reduced compared with the thickness required in tremie pipe concreting. After placing the formwork outside the reinforcement, the injection pipe is put in place and, finally, the aggregates. Figure 19.6 illustrates a repair of this type just above a column base.

The coarse aggregates should consist of cleaned and sieved gravel and crushed stone, with a minimum size 8–10 times the maximum grain size of the injection mortar. The specified minimum size of coarse

Figure 19.6 A repair just above column base done using injected concrete

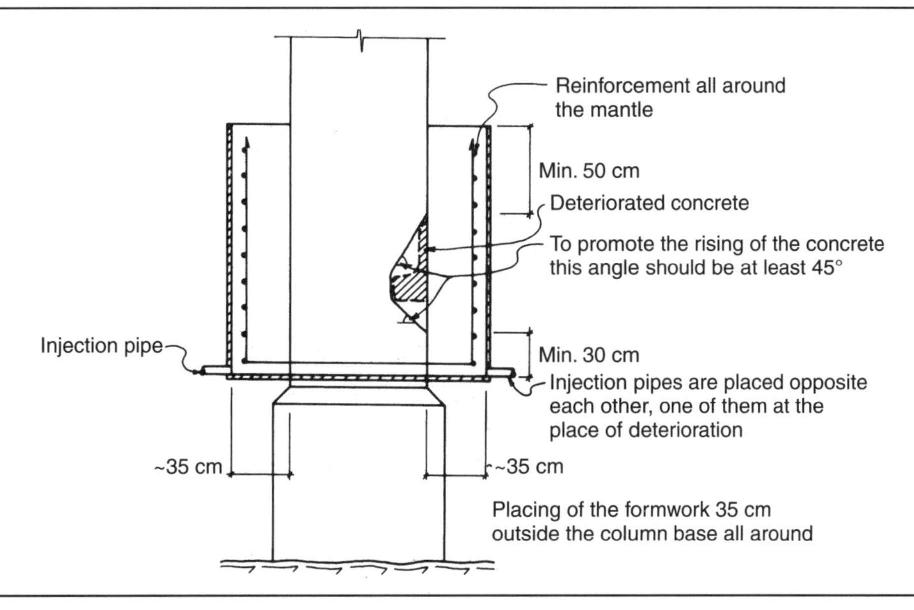

aggregates is 2 cm, and the maximum size of aggregates should be 5 cm. The injection mortar should have a cement/sand ratio of 1 : 1 by weight, plus expanding and stabilising agents.

19.6.3 Micro-concrete

The micro-concrete method represents a simplification of the injection method. The procedure is the same, except that the formwork is not filled with aggregates but only with injection mortar. The micro-concrete method lends itself to smaller repairs in particular. Figure 19.7 shows a micro-concrete repair of a wall.

If the part to be repaired is so long that two injection pipes above each other are needed, the procedure is as follows: first, the mortar is injected via of the lowest pipe until the concrete level reaches the next injection pipe. The lower pipe is then closed and the injection hose is disconnected. While the hose is still full of mortar, it is moved up to the next injection pipe and the process is continued. When starting the injection, the pressure must be adjusted with a view to avoiding a too-fast flow of mortar, and thereby to avoid washing-out its finer components.

19.6.4 Special epoxy

For smaller repairs of volume up to, say, 0.01 m^3, the use of epoxy may prove successful, provided that the curing temperature is above 5°C. If the volume to be repaired is approximately 0.25 litres, the diver will be able to carry out the repair with his hands (using disposable gloves) and using a plastic epoxy smooth material. The pot life of this material is about 20 minutes at 20°C depending on the type of epoxy.

19.6.5 ResconMapei method

Cement and epoxy resin products have been used in underwater repair, but used alone both have drawbacks. For example, in contact with water, cement can be leached out from the repair grout, leaving a low-strength product. Also, the use of epoxy resin mortars is limited due to their high cost and their

Figure 19.7 Repair of a wall using the micro-concrete method

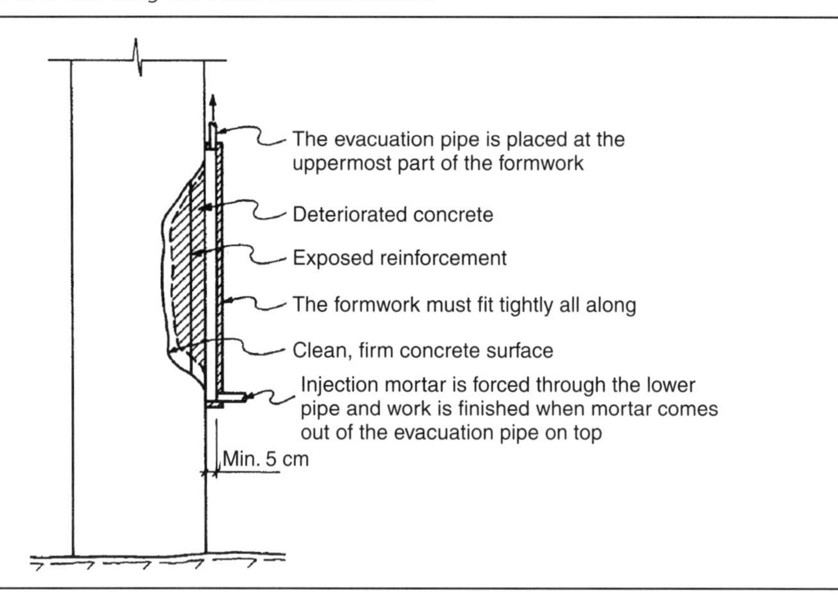

mechanical properties, which differ substantially from those of concrete. For example, the Norwegian ResconMapei method combines both repair materials and utilises their best properties:

(a) The *E* modulus and thermal coefficient of expansion of the cement grout is similar to that of concrete.
(b) The repair is more economical in material costs compared with an equivalent volume of epoxy mortar.
(c) Cement washout is eliminated by the epoxy resin in the ResconMapei method.
(d) The epoxy resin ensures good adhesion between the repair grout and the parent concrete (up to 2.5 MPa).

Procedure

(a) Remove the damaged or eroded area, leaving only sound material, and prepare a suitable bonding surface using grit blasting or water jetting.
(b) Construct shuttering using smooth, preferably transparent, sheets. Make tight at the base and side, using foam strips, mechanical fixing and underwater putty. The shuttering is positioned slanting away from the structure at the top to allow access for the hose down which the repair materials are pumped from the surface.
(c) The epoxy resin is mixed at the surface and pumped into the base of the mould to an approximate depth of 10–20 cm down the hose, which reaches into the lowest part of the mould.
(d) The epoxy is immediately followed by an expanding cement grout, which displaces the epoxy resin from the base of the mould. This action coats the structure with epoxy, improving the adhesion, coating the shuttering and ultimately giving a protective epoxy coating to the grout while maintaining a layer of epoxy on the surface of the rising grout, preventing cement leaching.

(e) When the epoxy and grout have finally cured (the increase in temperature of the curing cement also helps to cure the epoxy), the shuttering is stripped for reuse.

19.7. Repairs in zone 2 (tidal zone)

The above descriptions of repairs in zone 1 are valid also for repairs in zone 2. In addition, the shotcrete method can be used in zone 2.

19.7.1 Shotcrete or gunite

A repair under dry conditions can be achieved by installing a waterproof box or 'cofferdam', which serves as a working platform around the column or alongside the wall to be repaired. The cofferdam provides a dry working space for chiselling off the deteriorated concrete and repairing it with shotcrete. The methods are illustrated in Figures 19.8 and 19.9. For repair of a column, two steel box halves are lowered down, one on each side of the column, and then put together and fastened underneath the berth slab. The box must be high enough to cover an ample space below the damaged area and also above the tidal zone. Soft rubber is used as a seal between the two parts of the box and between the box and the column. The box is then pumped empty.

If more than one column is to be repaired, it pays to put some effort into the design of the box system, with a view to making it easy to install and remove the boxes for reapplication at several columns.

Figure 19.8 A waterproof box or 'cofferdam'

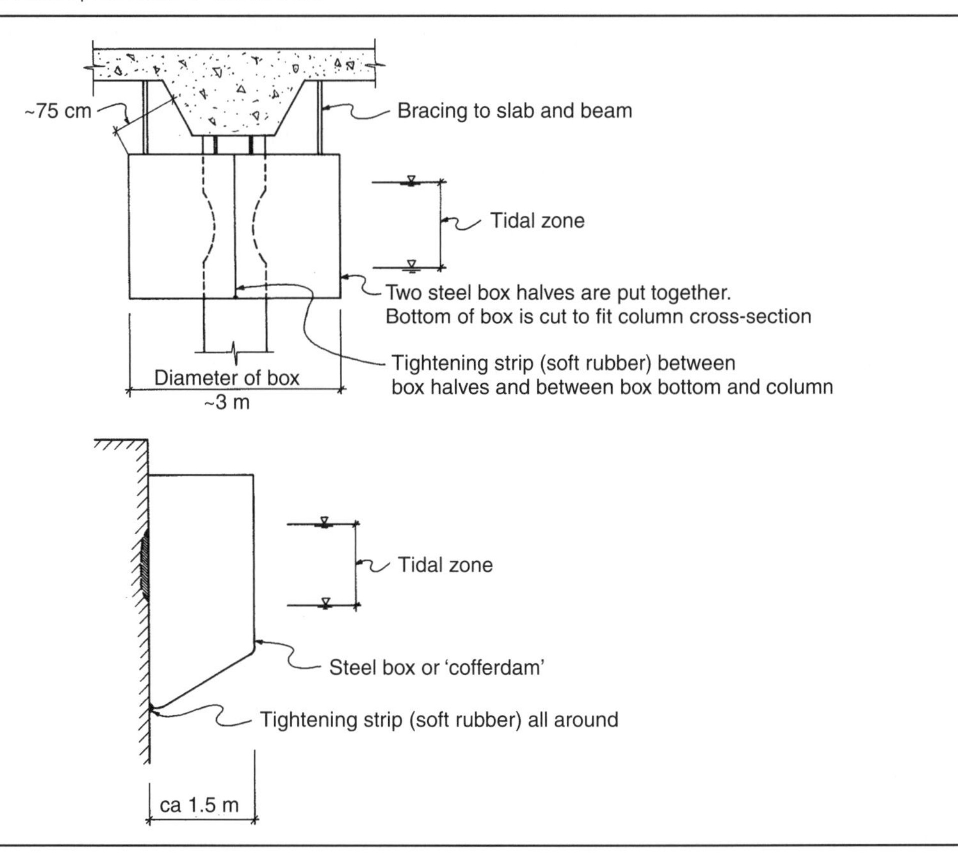

Figure 19.9 A steel cofferdam. (Courtesy of AF Gruppen ASA, Norway)

When the concrete surface has been carefully chiselled and sandblasted, the reinforcement, if required, is put in place, and shotcrete is applied until the column has regained its cross-sectional dimensions. Where the deteriorated column has become hour-glass shaped and additional reinforcement has to be applied, concrete should be sprayed onto the old concrete before the new reinforcement is installed.

It is better to spray two layers of 2–3 cm thickness of concrete than to add an accelerating agent to the concrete and spray a single 4–5 cm layer in one operation. Due to material loss, however, the latter method is more usual, notwithstanding the fact that accelerating agents tend to reduce the compressive strength of the concrete. The extent to which such reduction takes place depends on the type and make of the agent used. A strength reduction of 50% has been recorded in concrete having an agent content of 5%. Generally, concrete of high compressive strength also has better adhesion properties than low compressive strength concrete.

19.7.2 Tremie pipe concreting

Figure 19.10 illustrates a tremie pipe concreting repair, where the formwork is hanging on bolts drilled into the column higher up. The bottom of the formwork consists of two semicircular formwork plates, with a cut-out for the column, hanging on bars from the bolts above. The sides of the form consist of two semi-cylinders of glass-fibre-reinforced polyester. These are later used as permanent formwork.

The concrete is known as 'pea-shingle concrete', and has a slump of 16–18 cm. The concrete is placed on the bottom of the form through a 200 mm diameter pipe from a concrete feeder screw. During the concreting operation, the pipe is not moved before the concrete overflows the top of the form. A hammer or similar implement is used to knock on the sides of the form in order to compact the concrete

Figure 19.10 A feeder tube

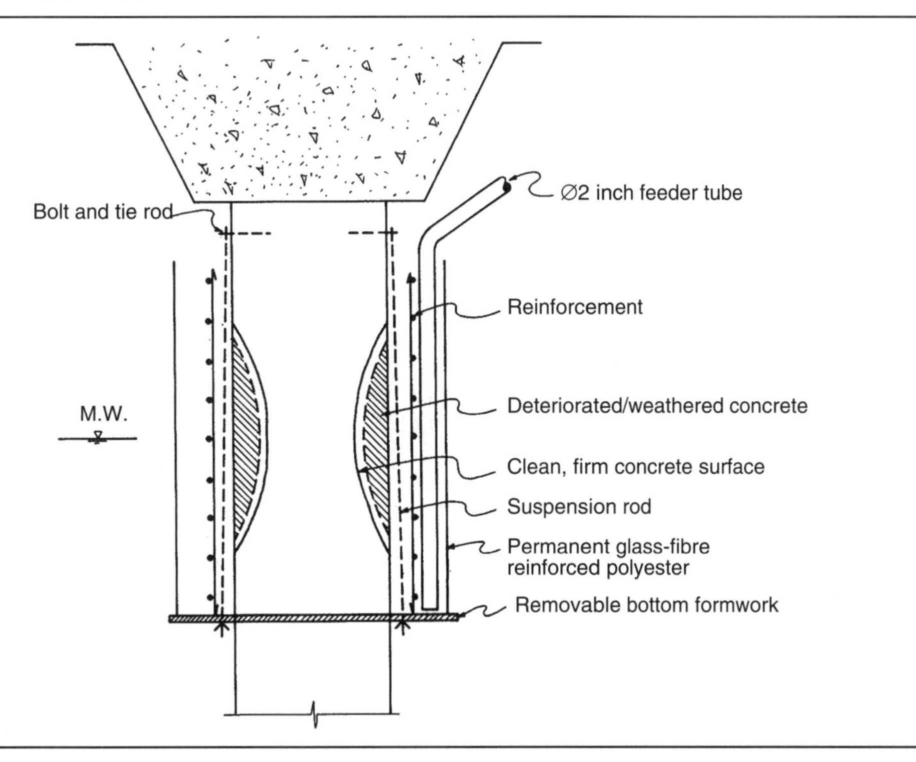

in the formwork. The amount of new reinforcement that needs to be used depends on how much steel has rusted away, but a steel mesh should be applied in any case. Figures 19.11 and 19.12 show a glass-fibre formwork before and after being filled with concrete.

19.8. Repairs in zone 3 (the splash zone or the area above HAT)

In order to successfully repair a berth structure, particularly in zone 3, it is important to have a full understanding of the deterioration process that is occurring in the structure. This requires full investigation of the concrete, with mapping of the condition to gain a full understanding, before selecting the most appropriate technical repair method.

An understanding of the deterioration mechanisms and the structural behaviour is a prerequisite for developing rational and sound repair strategies. The only way of influencing a deterioration process is to influence the parameters that govern the mechanism of deterioration.

In the evaluation to decide on the repair type for the structure, the following strategies should always be considered:

(a) postpone the repair and monitor the structure
(b) recalculate the structure and reduce the load-bearing capacity
(c) repair the structure to increase the service life
(d) strengthen the structure
(e) rebuild parts of or the whole structure
(f) demolish.

Figure 19.11 Glass-fibre formwork before being filled with concrete

As repairs are, in general, very expensive, the strategy or combination of strategies that is both technically and economically favourable should be chosen. In addition, after the repair the structure should have the lowest possible life-cycle cost.

Marine structures that have deteriorated as a result of chloride- induced reinforcement corrosion can be repaired using one of the following methods. The most common and most proven methods that have been adopted over recent decades are:

(a) concrete restoration or patch repairs

Figure 19.12 The column after repair

(b) cathodic protection
(c) chloride extraction.

Experience from repair work done in the past shows that the commonest mistake is to wait too long before the repair work is carried out. This is a very expensive way to operate the structure, and a lot of money could be saved by providing a correct assessment of the structure and starting the repair at the optimum time. This expense of undertaking a later than optimum repair is mainly explained by the high cost of removing and replacing damaged concrete.

19.8.1 Concrete restorations or patch repair

The patch repair method has been used for many years, but does not represent a successful long-term repair method. In many cases it has to be accepted that the repaired berth structure will eventually need a new repair, because the patch repair method is neither technically nor economically viable. This often leads in the end to new and severe deterioration, and later perhaps to demolition and replacement of the structure.

The traditional way of repairing spalled and cracked concrete, due to chloride-induced corrosion, in zone 3 is to use the patch repair procedure, as illustrated in Figure 19.13.

For instance, in a concrete beam located in zone 3 that has deteriorated due to corrosion of the reinforcing steel, it is not possible to simply replace the poor or deteriorated concrete with fresh concrete. The fresh concrete is not chloride-ionised, and will probably cause an accelerated corrosion when coming into contact with older chloride-ionised concrete. Therefore the patch repair method should be applied to new structures only, where chlorides have not generally reached the level of the reinforcement. Such a repair will not last long, and does not take into account available knowledge on the mechanism of corrosion due to chloride.

Generally, the patch repair should be carried out in such a way that all concrete containing chlorides above a certain content is removed around and to 3 cm behind the reinforcement. This is, in fact, perhaps the most expensive way of repairing, and it is important to be aware that new damage may occur within 5–10 years.

Figure 19.13 The patch repair method

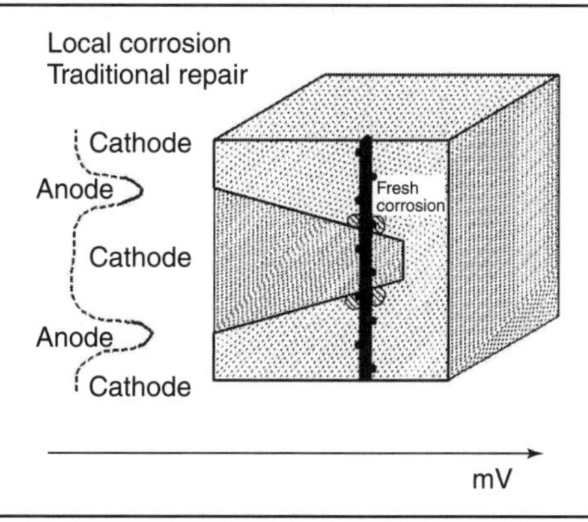

Figure 19.14 A water jet for removing the loose concrete. (Courtesy of Betongfornyelse AS, Norway)

The patch repair method may be chosen when structure needs to be strengthened temporarily for safety reasons, or where the remaining service life of the structure is short. Either way, the repair should be carried out by proper removal of damaged concrete, cleaning of the reinforcement and using a repair cement mortar of proven good quality.

Deterioration of old concrete in zone 3 is usually due to corrosion of the reinforcement. Longitudinal cracks and wounds in connection with corroded steel bars can often be seen on the bottom parts of older rectangular high beams. Figure 19.14 shows a water jet arrangement for removing all loose concrete around the reinforcement.

When all loose concrete and chloride-infected concrete to approximately 3 cm behind the main reinforcement has been removed (Figures 19.15 and 19.16) and the concrete and the reinforcement have

Figure 19.15 A slab with all the loose concrete removed. (Courtesy of Betongfornyelse AS, Norway)

Figure 19.16 A beam with all the loose concrete removed

been sandblasted, the decision must be taken as to whether additional reinforcement should be installed before the existing reinforcement is covered by shotcrete.

The injection method is used only for the repair of beams. Figure 19.17 shows a typical example of such a repair. The sloping sides of the beam will ensure a better and easier filling of the form. The whole length of the beam ought to have this new cross-section, not just the deteriorated parts. This applies

Figure 19.17 Repair of a beam

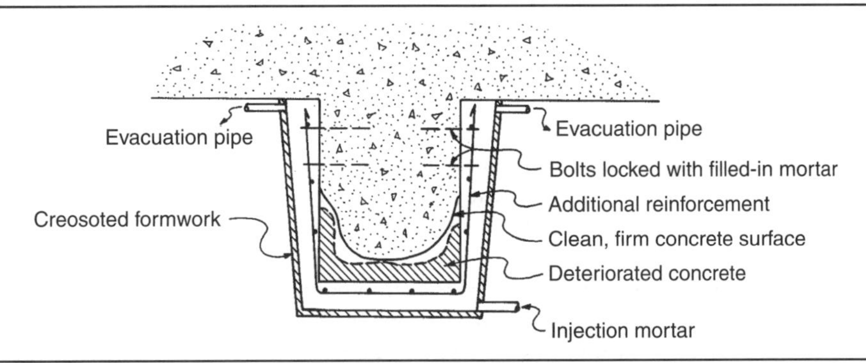

also when the shotcrete method is used. The stone aggregates and injection mortar used are as described for repairs in zone 1. Micro-concrete can also be used as described for zone 1.

If the cross-sectional area of reinforcement has been reduced by corrosion to such an extent that additional reinforcement has to be installed, or if it has been necessary to remove deteriorated concrete behind existing reinforcement, the spraying of concrete behind the bars can prove difficult. In order to avoid cavities in this concrete, the bar diameter should not be more than 12 mm. However, in beams, 25 mm diameter bars can hardly be avoided, and therefore micro-concrete should be used.

19.9. Cathodic protection

Although the principle of cathodic protection has been around since the 1820s, its use in increasing the service lifetime of reinforced concrete did not commence until the late 1950s.

Chloride corrosion is a very serious type of deterioration and, because a patch repair is not expected to last long, the most reliable method of repair is cathodic protection. This method is the only way to reduce the corrosion rate to a minimum and experience over recent decades has shown it to be very effective.

In principle there are two types of cathodic protection system available on the market today:

(a) The **sacrificial anode system**, which is generally used only below the mean water level, and is almost maintenance free. For example, zinc alloy sacrificial anode bracelets fixed to the berth structure at approximately LAT.
(b) The **impressed current system** using various types of anodes, is usually used above the mean water level, and requires regular monitoring and a degree of expertise to operate it with maximum efficiency.

A sketch of a cathodic protection system is shown in Figure 19.18.

Impressed current cathodic protection works by passing a small direct current (DC) from a permanent anode, fixed on top of the surface or into the concrete, to the reinforcement. The power supply passes sufficient current from the anode to the reinforcing steel to force the anode reaction to stop and make a cathodic reaction the only one to occur on the steel surface.

The Pourbaix diagram explains the cathodic protection system. By applying a small impressed current, the potential will move, for example, from A to B in the negative direction, as shown in Figure 19.19. The steel will then become immune and corrosion will stop due to the cathodic protection system.

There are different methods of installing cathodic protection, as shown in Table 19.3. The selection of the most appropriate type of cathodic protection system will depend on the nature of the structure, financial considerations, the anticipated service life of the berth structure, environmental conditions and the maintenance capability of the port owner. There are a lot of different types of anode available on the market and new ones are constantly being developed. For marine structures one should choose robust systems. A sketch of a titanium mesh is shown in Figure 19.20.

To minimise the maintenance and repair cost it is recommended that repairs are carried out before corrosion and spalling occur. With cathodic protection, the requirement for breaking out sound but chloride-contaminated concrete is unnecessary. Only delaminated concrete and rust product has to

Figure 19.18 A cathodic protection system

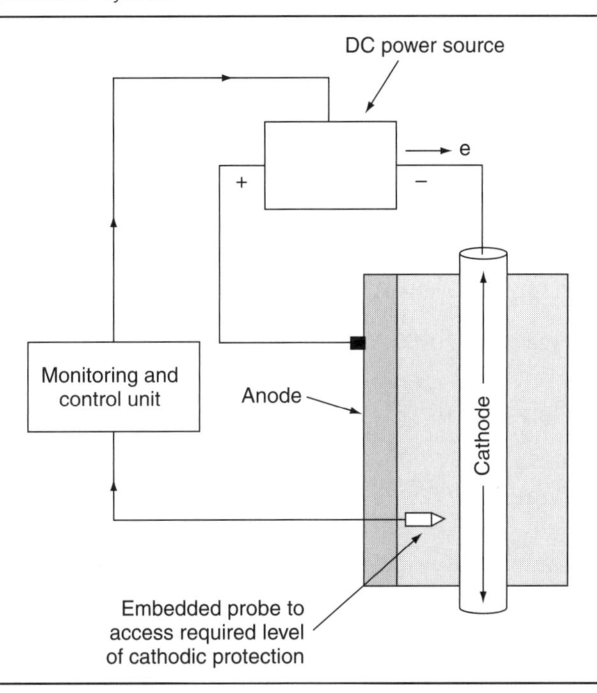

be removed, to ensure a homogeneous concrete material (Figure 19.21). The new concrete used for replacement should generally closely match the original concrete.

Titanium mesh and ribbon systems are mechanically fixed to the old concrete surface (Figures 19.22 and 19.23) and covered by a 15–20 mm thick layer of sprayed concrete (Figure 19.24).

Figure 19.19 The Pourbaix diagram

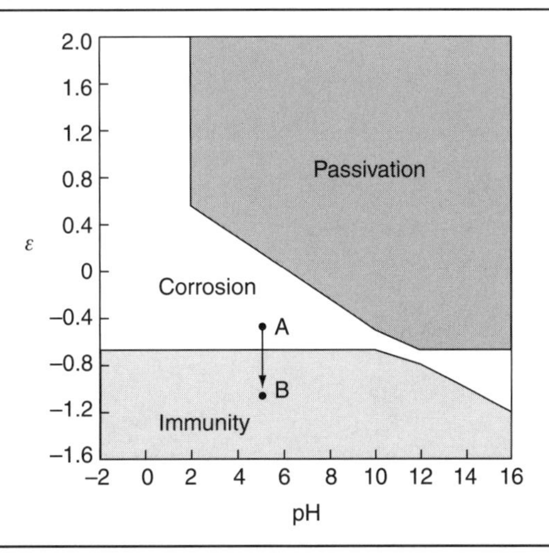

Table 19.3 Methods of installing cathodic protection

Anode material	Expected service life	Current density: mA/m^2	Comments
Titanium mesh	+25 years	10–20	Durable established system. Main problem may be overlay application and increased weight
Carbon mesh	+25 years		A relatively new product with possible great potential but limited experience
Titanium ribbon	+25 years	10–20	Main problem may be overlay application
Conductive mortars	+25 years	20–50	The system will secure an even distribution of current, but it is important to avoid short-circuiting
Discrete anodes (embedded probes)	+20 years	10–20	Durability of the backfill may be a problem, but is relatively easy to maintain
Sprayed zinc	+15 years		An offer anode system
Sprayed electrical leading mortar	–		Polymer-modified cement mortar with nickel-covered graffiti fibres

A conductive mortar system has been developed in recent years, and this could be particularly economical where large areas have to be repaired. The conductive mortar system uses nickel-coated carbon fibres to provide conductivity. It is applied by using similar equipment and methods as for sprayed gunite concrete.

Figure 19.20 Titan mesh

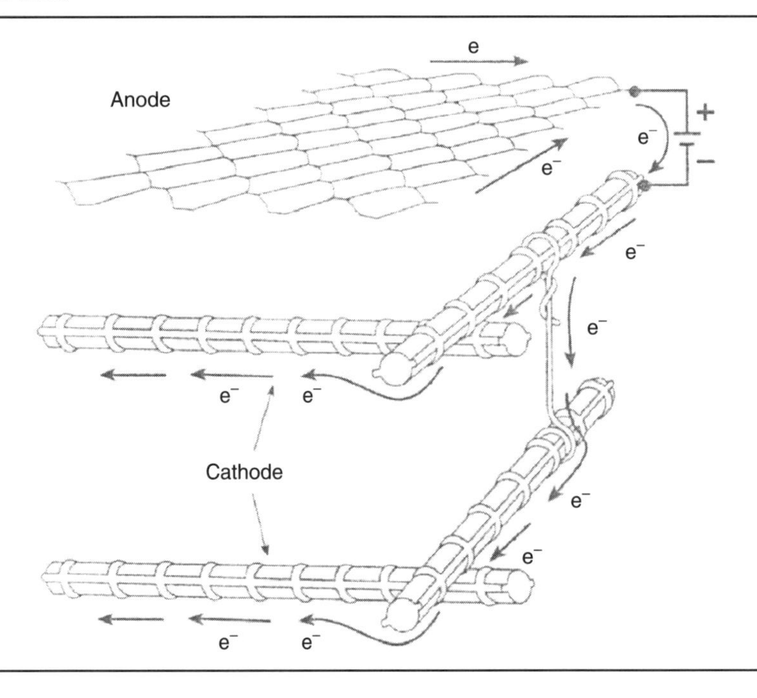

Figure 19.21 Removing delaminated concrete and rust from the beam. (Courtesy of Norconsult AS, Norway)

If the cathodic protection system is designed, installed and maintained according to well-established knowledge, it is the best repair system on the market to increase the service life of a structure and the only system that can reduce corrosion to a minimum.

The design, preparation of tender documents and installation of such a system require specialists. Experienced personnel should monitor the system. It is therefore recommended that the owner sign a maintenance agreement with the supplier or another experienced consultant company to look after, adjust and run the system for its life span.

Figure 19.22 Installed titanium mesh. (Courtesy of Norconsult AS, Norway)

Figure 19.23 Installation of titanium ribbon. (Courtesy of Norconsult AS, Norway)

19.10. Chloride extraction

Desalination is a method that may remove a major part of the chlorides from the concrete cover by electrochemical means. The principle of this method is the same as for chloride protection except that requirement for electrical power is higher and the time of execution is 6–7 weeks compared to cathodic

Figure 19.24 An anode system covered with dry-sprayed concrete. (Courtesy of Norconsult AS, Norway)

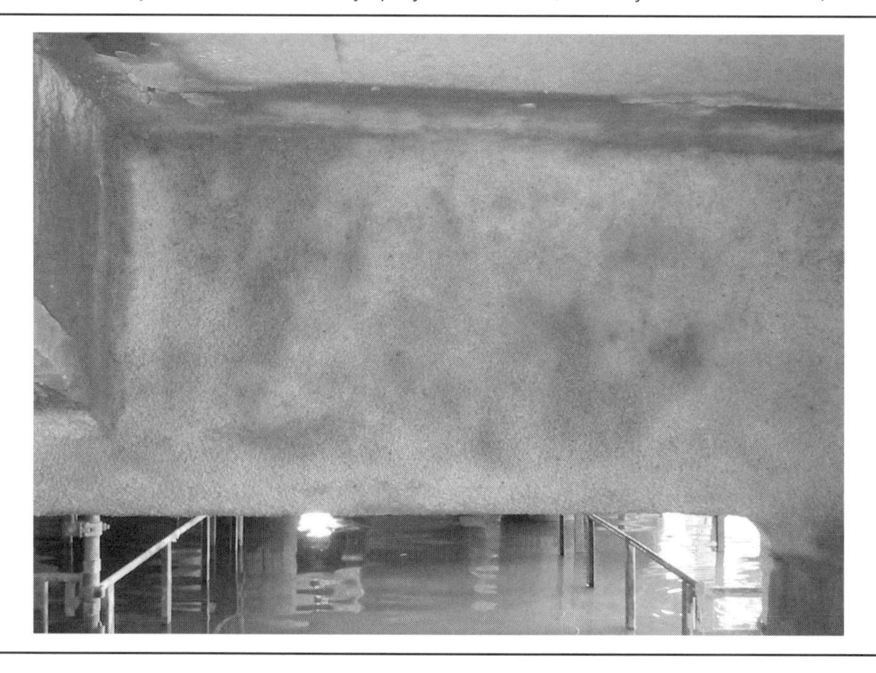

protection which must be in service for the remaining service lifetime of the construction. This method has been reported as being successfully used in some projects. Nevertheless, one should bear in mind that the concrete has to be properly protected from further chloride ingress after removal. This method may be the right choice for some projects, but the life span of the repair must be expected to be lower than for the well-documented cathodic protection systems.

19.11. Costs of repairs

Owners tend, unfortunately, to choose a method of repair that involves the lowest cost at that moment in time, without considering the expected lifetime of the structure. Irrespective of whether the berth structure to be repaired is new or 40–50 years old, in both cases the cost of repair and the assumed remaining life of the berth structure should be considered as a whole.

To estimate the costs of repair on the basis of a visual inspection is difficult, because the extent of deterioration first becomes apparent when all deteriorated concrete has been removed. The costs, therefore, tend to be higher than expected.

Usually, repair work is paid for on account, i.e. the contractor gets all direct expenses reimbursed, plus a fee. The fee can be a percentage of the direct expenses or a fixed sum, or a combination of these, according to the advance agreement between the client and the contractor. However, the cost and the quality of the work depend first of all on the expertise and management employed by the contractor.

REFERENCES AND FURTHER READING

Broomfield JP (1997) *Corrosion of Steel in Concrete. Understanding, Investigation and Repair*. Spon, London.

FIP (Fédération Internationale de la Précontrainte) (1991) *Repair and Strengthening of Concrete Structures*. Thomas Telford, London.

Gjørv O (1968) *Durability of Reinforced Concrete Wharves in Norwegian Harbours*. Ingeniør-forlaget, Oslo.

Gjørv O (1994) Steel corrosion in concrete structures exposed to a Norwegian marine environment. *Concrete International* **16(4)**: 35–9.

Gjørv O (2009) *Durability Design of Concrete Structures in Severe Environments*. Taylor & Francis, London.

Lahus O, Gjørv O and Johansen R (1998) *Durability of Reinforced Concrete Wharves in Norwegian Harbours*. PIANC, The Hague.

Mallett GP (1996) *Repair of Concrete Bridges*. Thomas Telford, London.

Norwegian Concrete Association (2003) *Guidelines for the Design and Construction for Underwater Concrete Structures*. Norwegian Concrete Association, Oslo. Publication No. 5 (in Norwegian).

Perkins PH (1978) *Concrete Structures: Repair, Waterproofing and Protection*. Applied Science, London.

Pullar-Strecker P (1987) *Corrosion Damaged Concrete*. CIRIA, London.

Schiessel P (1988) *Corrosion of Steel in Concrete*. Chapman and Hall, London.

Port Designer's Handbook
ISBN 978-0-7277-6004-3

http://dx.doi.org/10.1680/pdh.60043.503

Chapter 20
Port maintenance

20.1. Responsibility for maintenance

Responsibility for the maintenance of plant and equipment must be controlled by properly experienced personnel under the control of a qualified engineering maintenance manager and supervisory staff.

In addition to the manager, the maintenance management team typically involves engineering inspection and supervisory staff. Often these are the same individuals. The inspectors collect the required information in the field and produce subsequent reports while the supervisory staff prepare plans and specifications for the repair and maintenance of facilities, plant and equipment.

Responsibility for the infrastructure should be under the control of a civil engineer and the facilities manager, who will be responsible for the long-term planned maintenance, replacement of items prone to degradation through wear and tear, and for responding to daily emergencies.

Minor and planned civil, building, mechanical and electrical maintenance is generally carried out by in-house personnel who will be based in dedicated workshops located within the port facility.

Due to its variability and scope, major maintenance is best carried out by suitably qualified contractors or, in respect of plant and equipment under service agreements, with the manufacturers of the equipment. This is particularly applicable to quayside gantry cranes, mobile cranes, specialised loading/unloading equipment, rail-mounted gantries, automated stacking cranes, reach stackers and empty container handlers. Other equipment such as tractor units, trailers, fork-lift trucks and other smaller pieces of equipment is best maintained by qualified in-house tradespeople.

20.2. Spares

It will also be necessary to provide a stores section within the maintenance function, with the responsibility of holding necessary spares and materials. This is particularly important for spares that may have to come from overseas.

20.3. Management information

The effectiveness of maintenance is normally measured by reference to budget expenditure and the availability of plant and equipment or other aspects of the infrastructure.

For modern terminals, the latter aspect is most important, and it will be necessary to determine acceptable levels of plant and equipment downtime in order to provide the engineering maintenance function with workable and acceptable targets.

It will be necessary to set up a system of inspection schedules through which planned preventative maintenance expenditure and work executed can be budgeted and monitored on a monthly basis.

Any unscheduled repairs (breakdowns and accident damage) will also need to be recorded for both the work carried out and the associated costs.

There are a number of proprietary maintenance IT systems available for both the recording of work carried out and the capture of costs. The final choice of system will be dependent on the actual method of operation chosen and the details of financial and task interrogation required.

20.4. Maintenance personnel

A typical grouping of in-house engineering management personnel is:

(a) management
(b) senior management
(c) junior and middle management
(d) blue-collar labour
(e) skilled
(f) semi-skilled
(g) labour.

Many terminals within ports operate on a 24/7 basis, and in this case it will be necessary to have appropriate maintenance staff available on a 24 h basis to cover breakdowns and emergency repairs. This is normally achieved by operating a two- or three-shift system.

A typical maintenance organisation is shown in Figure 20.1.

Apart from the managing director or the chief executive and the operations director, who are the direct line managers for the engineering manager, all maintenance personnel should be housed in a maintenance building located close to the operational part of the port. This building should accommodate the administrative, management, supervisory and trades personnel together with the actual workshop and stores facilities themselves.

20.5. Plant and equipment

Following the initial commissioning of plant and equipment, and the establishment of an efficient engineering maintenance workshop, it may take time to provide appropriately trained maintenance personnel.

Planned preventative maintenance, major maintenance works and statutory inspections of plant and equipment are normally carried out during the day shift when all specialist trades, both in-house and from contractors, are available. Outside of the day shift, minimum manning levels are normally retained to cover breakdowns and emergency repairs only.

For other specialist areas such as IT and electronics, it is usual to retain in-house personnel, due to the specific needs of such systems and equipment.

Mechanical and electrical engineering and IT personnel will be responsible for the daily maintenance of cargo-handling equipment and other aspects of the facility that require these skills and for specific IT operating hardware and software such as the terminal operating system.

High-voltage electrical cables and switchgear will have to be maintained by specialist contractors, while maintenance of low- and medium-voltage cables and domestic electrics can be undertaken by fully qualified in-house electrical tradespeople.

However, in order to ensure the infrastructure maintains its original design life, it is essential that the inspection and maintenance of all elements of the infrastructure is carried out on a predetermined regular basis.

In the case of berths, pavements and buildings, maintenance work is usually carried out during the normal working week. It is likely that most of this work will be carried out by outside contractors, although a small in-house team may be retained for minor or emergency repairs.

The civil engineer and the facilities manager will be responsible for the maintenance of these facilities, and will have to consider preventative maintenance strategies as well as having to react to emergencies and the replacement of elements of the infrastructure.

Records of previous maintenance work carried out, including the frequency and nature of the work together with the commensurate level of expenditure, will be necessary.

A typical example of how this might be achieved for berth structures and yard areas is summarised in Table 20.1.

20.7. Optimisation of design to reduce future maintenance costs

In keeping with the principles of life-cycle management, it is important to give consideration to the optimisation of designs to ensure that future maintenance costs are kept to a minimum. It is recommended that whole-life costing is considered as an effective tool to evaluate important alternatives for major elements of the infrastructure that will not only affect the capital cost of the project but can also help to reduce future maintenance costs.

A number of these major elements of the proposed port facility are briefly discussed below.

20.7.1 Earthworks

The main earthworks activity for many port projects includes reclamation and the potential for ground improvement to reduce settlement.

From an economic perspective, ideally the reclamation material should be sea-won dredged material, but this may not be possible.

In any event, ground improvement is an effective means of reducing long-term settlements and subsequent maintenance, and will be determined by the operator's settlement criteria in respect of pavements and foundations.

However, as ground improvement can add significant costs to the project, careful consideration of the type and the effect on the whole-life costs will have to be made.

Types of ground improvement include:

(a) surcharging
(b) vertical (wick) drains
(c) horizontal drains
(d) vibro-compaction
(e) stone columns.

Table 20.1 Examples of typical maintenance strategies

Item	Daily	Periodic	Replacement	Emergency
Area: berth structures				
Fenders	Check for damage	Fender panels	Complete unit	Spare units in store
Bollards	Check for damage			Spare units in store
Cathodic protection	Check after reported incident	Check 3-monthly	Anodes or cables	
Steel sheet piles	Check after reported incident	Annual walkover survey		Plan for ship impact damage
Anchor wall		Check front nuts	Tighten or renew nuts	
Concrete capping beam	Check after reported incident	Check for damage	Replace concrete and/or reinforcement	
Area: yard				
Surfacing and drainage	Records from daily sweep	Check annually by walk over for damage and settlement		Local settlement repairs if affecting equipment or stacking stability. Have materials in store
Line markings	Review operational records	Check annually by walkover survey	Annually	
Crane rails	Check for damage	Monthly or as reported	Sections as required	Spares in store
Cable slot and turnover pits	Check for damage	Monthly or as reported	Sections as required	Spares in store
Fuel, water, sewage and electrical cables		As reported	As required	Spares in store
Ladders and quay furniture	Check for damage	As reported	As required	Spares in store

Figure 20.2 Damage to asphalt pavement

20.7.2 Pavements

The choice of pavement will have a significant impact on the maintenance costs. The loading criteria for each area will vary for different locations. For example, container-stacking areas have relatively high point loads from container corner castings, while roadways have lower loads with high levels of repeat loadings, and will therefore require a heavy duty pavement design – preferably not asphalt pavement (Figure 20.2).

It will therefore be uneconomic to design all areas to the highest pavement loading, and it will be important to identify each area in order that the appropriate type of paving is used, bearing in mind the need to retain a degree of flexibility to accommodate potential future changes in use and the likely level of future maintenance required.

Types of paving for consideration in heavy duty loading areas include:

(a) concrete block pavers
(b) asphalt
(c) grouted asphalt (e.g. Densiphalt)
(d) reinforced concrete slabs
(e) gravel beds and ground beams in container-stacking areas.

20.7.3 Steelwork

It is advised that all general steelwork surfaces are protected against corrosion with appropriate paint systems or galvanising as appropriate, to provide the required durability and design life and in order to minimise maintenance costs.

It is recommended that all steel piling is provided with a cathodic protection system, and from a maintenance perspective the use of sacrificial anodes is more straightforward than an impressed current system, which requires constant monitoring.

All steelwork and piles should be designed without taking account of corrosion protection.

20.7.4 Concrete

For all cast-in-situ or precast concrete, it is recommended that a thickness cover for reinforcement steel appropriate for the marine environment is adopted to provide the necessary durability and to minimise maintenance costs.

20.7.5 Utilities

Requirements for the following utilities will have to be agreed with potential operator(s), bearing in mind any long-term maintenance implications:

(a) water – potable mains distribution and hydrants and the collection, storage and use of collected rainwater/grey water for non-potable use.
(b) electrical mains distribution.
(c) foul sewage – mains distribution and treatment facilities
(d) surface water mains distribution – need for oil interceptors
(e) area lighting – choice of high mast (lattice or column)
(f) ducting – telecommunications and IT cabling
(g) fuelling – location and type of station(s), including bunding
(h) fire – mains distribution, fire hydrants and ships' bunkering, including the choice, location and capacity of the pumps.

20.7.6 Fenders

It is important to understand the type and size of fenders required for the range of ships anticipated to use the facility. Fenders are prone to ship damage, and often require repair or replacement. Ease of repair can be crucial to operational downtime due to berthing implications, and consideration must be given to the ability to carry out rapid and effective maintenance in all states of the tidal range. Figure 20.3 shows a complete lack of maintenance of the fender system on an industrial open berth structure.

20.7.7 Bollards

It will be necessary to determine the size and capacity of bollards, bearing in mind future ship size developments. Generally, there are few day-to-day maintenance issues with bollards, although through-deck fixings should be considered for ease of repair in the event of major damage.

20.7.8 Pumps

Pumps will be required for the foul sewage and fire main system, and their choice should reflect not only their capacity but also the ease of maintenance and repair.

20.7.9 Fencing

The choice and type of perimeter and other fencing should reflect both the required security protection and the ease of maintenance.

Figure 20.3 Complete lack of maintenance of a fender system

20.8. Maintenance management

20.8.1 General

The lifetime costs of maintaining the infrastructure of a port can be in the range 10–25% of the cost of the original investment. For plant and equipment, the maintenance costs over the lifetime of the asset will generally exceed the original purchase price.

It is of course possible to overmaintain an asset, and in some circumstances the cost to the business of the occasional breakdown of plant or equipment may be less than the cost of the maintenance input to prevent the breakdown in the first place. Planned preventative maintenance at predetermined intervals including statutory inspections is usually kept separate from the cost of corrective maintenance caused by a breakdown, damage or general deterioration.

20.8.2 Operational records

Records of work undertaken on structures, or on plant and equipment, form part of the continuing evaluation of the performance of port facilities, and include the purchase and storage of spares. This information is used to evaluate the actual and relative performance of structures and plant and equipment and for the pre-ordering and stocking of spare parts.

20.9. Maintenance strategy

It is important to provide a maintenance strategy for the infrastructure plant and equipment, and once adopted will require any maintenance undertaken to be continuously monitored. In this way, feedback from the monitoring will enable the strategy to be reviewed on a regular basis to establish if, for example, patterns of similar repairs appear, and where it would be beneficial to the productivity and cost-effectiveness of the port to amend the strategy put in place.

Therefore, continuous review of planned maintenance against what is actually happening will identify areas that can be modified to improve the overall maintenance regime.

20.9.1 Maintenance costs

The costs of all the maintenance undertaken must also be recorded. Again, comparison of actual costs against the budgeted cost for items of work will identify areas that may give cause for concern. Reasons for cost overruns and underruns may be established to ascertain if improvements can be made in the original assumptions made for the execution of items of work.

20.9.2 Operation and maintenance cost planning

Identify the likely total operation and maintenance expenditures that can be expected. This methodology will only account for the regular wear and tear that a facility undergoes. It will not account for accidents that are unpredictable by nature generally. For most port facilities, elements of structures and plant and equipment that get the most use and/or abuse can be easily identified.

Fenders, for example, are sacrificial elements that are installed to protect both the quay structure and vessels. In the normal functional life of a facility (e.g. 30 years), fender units will require replacement and/or major repairs at least once within the functional life. Asphalt surfaces usually last only 10–15 years due to degradation from use by heavy equipment and exposure to the sun. Sacrificial anodes on steel piles are usually good for up to 5 years.

By the same rationale, nothing or very little happens to the fill materials behind a sheet pile wall or buried tie rods and dead man anchors. If the various components of a facility are considered together with the number of times that these components have to be replaced in the functional life of the structure, the total operational and maintenance costs can be predicted. By dividing the total operation and maintenance costs by the functional life, the 'annual' operation and maintenance budget can be established.

Although the results obtained will not necessarily be exact, they will be useful in establishing operation and maintenance budgets. Of course, the actual year-to-year operation and maintenance cost will vary, but such a prediction model will at least give some indication of the cost.

20.9.3 Structures and facilities

Most items of plant and equipment will be subject to planned preventative maintenance programmes as set out by manufacturers and statutory inspections laid down by the statutory authorities/insurance companies.

However, for structures and other facilities, it will be necessary to develop maintenance management programmes for the various different types of structures and facilities.

20.10. Inspections

Inspections should be conducted and ratings assigned against distinct structural units. For example, a steel sheet pile and reinforced concrete deck should be divided into at least two distinct structures for purposes of inspecting and assigning condition ratings. Structural units should typically be of uniform construction type and material and, in the case of pile-supported structures, should be in a continuous bent numbering sequence.

The boundaries of structures must be clearly defined at the outset of the work. Common boundaries include expansion joints, configuration changes and changes in the age of construction, in direction or in the bent numbering sequence.

Consistent with the American Society of Civil Engineers' Manual 101, *Underwater Investigations: Standard Practice Manual* (Childs, 2001), six inspection types may be considered in maintenance management, with types (d) to (f) being part of routine maintenance inspections:

(a) new construction inspection
(b) baseline inspection
(c) routine inspection
(d) repair design inspection
(e) special inspection
(f) post-event inspection.

New construction inspections are conducted only in association with newly constructed structures/ components to ensure proper quality control.

Baseline inspections for new structures serve to verify that construction plans have been followed and to ensure that construction is free from significant defects prior to owner acceptance. For existing structures, this inspection serves to verify dimensions and construction configuration details. Baseline inspections are typically conducted close to completion of new construction, prior to owner acceptance. On existing structures, they should be coincident with the first routine inspection

Routine inspections are intended to assess the general overall condition of the structure, assign a condition assessment rating, and recommend actions for future maintenance activities. The inspection should be conducted to the level of detail required to evaluate the overall condition of the structure. Documentation of inspection results should therefore be limited to the collection of the data necessary to support these objectives, in order to minimise the expenditure of maintenance resources.

Repair design inspections serve to record the relevant attributes of each defect to be repaired such that a schedule of repairs may be generated. Contrary to routine inspections, repair design inspections are conducted only when repairs have been performed, as determined from the routine inspection. Repair design inspections may take considerably longer to execute than routine inspections because they require the detailed documentation of all defects that have been repaired.

By using this two-tiered approach for the inspection process, resources can be utilised in a very efficient manner. It is not always required that a routine inspection is performed prior to a repair design inspection. In situations where the need for repairs is known or is obvious, or for small facilities, it may be advantageous to conduct the routine inspection and the repair design inspection simultaneously.

Special inspections are intended to perform detailed design testing or investigation of a structure, required to understand the nature and/or extent of the deterioration, prior to determining the need for and type of repairs necessary. It may involve various types of in situ and/or laboratory testing. This type of inspection is conducted only when deemed necessary as a result of a routine or repair design inspection.

Finally, **post-event inspections** are conducted in emergency or damage situations to perform a rapid evaluation of a structure, following a storm, vessel impact, fire or similar event, in order to determine if further attention to the structure is necessary as a result of the event. Such inspections are conducted only in response to a significant loading or event having the potential to cause (severe) damage.

20.11. Rating and prioritisation

Ratings are assigned to each structure upon completion of routine inspections and post-event inspections. The ratings are important in establishing the priority of follow-up actions to be taken. This is particularly true when many structures are included in an inspection programme and follow-up activities must be ranked or prioritised due to limited resources.

The rating system used for post-event inspections differs from that used for routine inspections because post-event inspection ratings must focus on event-induced damage only, excluding long-term defects such as corrosion deterioration. An alphabetical scale is used for post-event inspections, to distinguish from the numerical condition assessment scale use for routine inspections.

20.12. Condition assessment ratings

Condition assessment ratings should be assigned upon completion of the routine inspection, and remain associated with the structural unit until the structure is re-rated following a quantitative engineering evaluation, repairs or upon completion of the next scheduled routine inspection.

A scale of 1 to 6 is used for the rating system (Table 20.2). A rating of 6 represents a structure in good condition, while a rating of 1 represents a structure in critical condition. Other suitable rating systems may be substituted for a particular owner's purpose as appropriate.

Table 20.2 Routine condition assessment ratings

Rating	Description
6 Good	No visible damage or only minor damage noted. Structural elements may show very minor deterioration, but no overstressing observed. No repairs required
5 Satisfactory	Limited minor to moderate defects or deterioration observed, but no overstressing observed. No repairs required
4 Fair	All primary structural elements are sound, but minor to moderate defects or deterioration observed. Localised areas of moderate to advanced deterioration may be present, but do not significantly reduce the load-bearing capacity of the structure. Repairs recommended, but the priority of the recommended repairs is low
3 Poor	Advanced, deterioration or overstressing observed on widespread portions of the structure, but does not significantly reduce the load-bearing capacity of the structure. Repairs may need to be carried out with moderate urgency
2 Serious	Advanced, deterioration or overstressing observed on widespread portions of the structure, but does not significantly reduce the load-bearing capacity of the structure. Repairs may need to be carried out with moderate urgency
1 Critical	Very advanced deterioration, overstressing or breakage has resulted in localised failure(s) of primary structural components. More widespread failures are possible or likely to occur and load restrictions should be implemented as necessary. Repairs may need to be carried out on a very high priority basis with high urgency

It is important to understand that ratings are used to describe the structure in its current condition relative to its condition when newly built. The fact that the structure was designed for loads that are lower than the current standards for design should have no influence upon the ratings.

It is equally important to understand that the correct assignment of ratings requires both experience and an understanding of the structural concept of the structure to be rated. Judgement must be applied considering:

(a) the scope of the damage (the total number of defects)
(b) the severity of the damage (the type and size of defects)
(c) the distribution of the damage (local versus general)
(d) the types of components affected (their structural 'sensitivity')
(e) the location of a defect on a component (relative to the point of maximum moment/shear).

The experience of individuals assigning ratings is important in ensuring that the ratings are assigned consistently and uniformly in accordance with sound engineering principles and the guidelines provided in this chapter.

20.13. Post-event condition ratings

The post-event condition rating should be assigned upon completion of the post-event inspection, preferably prior to leaving the site. The rating should be used to reflect whether additional attention is necessary and, if so, at what priority level. Table 20.3 lists the four post-event condition ratings. A rating of 'A' indicates no further action is required, while a rating of 'D' indicates major structural damage requiring urgent attention.

The following guiding principles should be followed when assigning post-event condition ratings:

(a) Ratings should reflect only the damage that was likely to have been caused by the event. Long-term or pre-existing deterioration such as corrosion damage should be ignored unless the structural integrity of the structure is immediately threatened.
(b) Ratings are used to describe the existing in-place structure as compared with the structure when new. The fact that the structure was designed for loads that are lower than the current standards for design should have no influence upon the ratings.
(c) Assignment of ratings should reflect an overall characterisation of the entire structure being rated. The correct assignment of a rating should consider both the severity of the deterioration and the extent to which it is widespread throughout the structure.

Table 20.3 Post-event condition ratings

Rating	Description
A	No significant event-induced damage observed. No further action required
B	Minor to moderate event-induced damage observed but all primary structural elements are sound. Repairs may be required but the priority of repairs is low
C	Moderate to major event-induced damage observed that might have significantly affected the load-bearing capacity of primary structural elements. Repairs necessary on a priority basis
D	Major event-induced damage has resulted in localised or widespread failure of primary structural components. Additional failures are possible or likely to occur. Urgent remedial attention necessary

(d) It should be recognised that the assignment of rating codes will require judgement. The use of standard rating guidelines is intended to make assignment of these ratings uniform among inspection personnel.

20.14. Recommendations and follow-up actions

Whereas condition assessment and post-event condition ratings describe the urgency with which, or when, follow-up action should be taken, the recommended actions describe what specific actions should be taken. Recommended actions are assigned upon completion of each inspection type described above, with the exception that new construction and repair construction inspections are in-process activities that typically require immediate follow-up action in the event of non-conformities.

A description of typical recommended action choices is provided in Table 20.4.

Multiple recommended actions may be assigned upon completion of each inspection; however, guidance should be provided to indicate the order in which the recommended actions should be carried out. For example, consider a structure that has received a routine inspection, and has been assigned recommended actions of emergency action (due to broken piles), repair design inspection (due to deteriorated and broken piles) and special inspection (because the cause of deteriorated piles is not known, and coring, testing and analysis is required). In this example, the guidance in the report should state that the emergency action should be taken first (erect barricades/close portion of the structure);

Table 20.4 Recommendations and/or follow-up actions

Recommended action	Description
Emergency action	Recommended whenever an unsafe condition is observed. If the situation is life threatening, significant property damage may occur or significant environmental damage may occur, appropriate owner representatives should be contacted immediately. Emergency actions may consist of barricading or closing all or portions of the structure, placing load restrictions or unloading portions of a structure
Engineering valuation	Recommended whenever significant damage or defects are encountered that require a structural investigation or evaluation or to determine what method of repair is appropriate. While the scope of the routine inspections should include the structural assessment of the damage or defects on the capacity of typical structures, consider the actual/anticipated loads that are or will be imposed on the structure
Repair design inspection	Recommended whenever repairs are required, typically as a result of a routine inspection, but may also result from a special inspection or post-event inspection
Special interpretation	Typically recommended to determine the cause of significance of non-typical deterioration, usually prior to designing repairs. Special testing, analysis, monitoring or investigation using non-standard equipment/techniques is typically required
Develop repair plans	Recommended when the repair design inspection has been completed and any special inspections recommended have been completed. Indicates that the field data have been collected and the structure is ready to have repair documents prepared
No action	Recommended when no further action is necessary on the structure until the next scheduled routine inspection

then the special inspections should be executed to determine the cause of the deterioration; then the repair design inspection should follow.

20.15. Repair prioritisation

Repair prioritisation must also be defined at the defect level. Not all defects need to be repaired with the same urgency. For example, corrosion cracking on a reinforced concrete element should be addressed with some urgency, whereas a corrosion crack of the same dimensions on a non-reinforced concrete component may be of less concern. Knowledgeable inspectors while recording the defects during field inspections should ideally assign prioritisation of defects for repair. Guidelines should be established for the inspectors to follow, allowing the inspectors' judgement to interpret the guidelines.

The guidelines must address the following defect attributes:

(a) construction materials
(b) component type
(c) structure type and function
(d) location of the component on the structure
(e) location of the defect on the component
(f) defect type
(g) defect dimensions
(h) accessibility for repair
(i) feasibility of repair
(j) structural redundancy within the design
(k) severity of defects on adjacent components
(l) presence or absence of anticipated loading on the component prior to repair execution.

20.16. Maintenance data management

Maintenance management activities generate significant data and also require data feedback to make informed decisions. It is useful to establish a database to manage these data and to facilitate access to the data to all appropriate levels of management.

The database should ideally manage the following:

(a) inventory database – may contain information on the location, dimensions, design criteria, designer, constructor, modification history, upgrade history, etc.
(b) environmental – may contain information as to the wind, waves, currents, tidal conditions, etc.
(c) maintenance database – may contain information on past maintenance activities and current condition assessment ratings
(d) operational – may contain information on operational restrictions, load restrictions, etc.
(e) financial data – may contain information on original construction costs, maintenance and upgrade costs, historic demolition costs, and historic unit bid prices for repair work.

REFERENCES AND FURTHER READING

Childs Jr KM (2001) *Underwater Investigations: Standard Practice Manual*. American Society of Civil Engineers, Reston, VA. ASCE Manual 101.

PIANC (International Navigation Association) (2008) *Life Cycle Management of Port Structures: Recommended Practice for Implementation*. PIANC, Brussels. Report No. 103.

Port Designer's Handbook
ISBN 978-0-7277-6004-3

http://dx.doi.org/10.1680/pdh.60043.517

Chapter 21
Ship dimensions

21.1. General

The following ship definitions are used in the design of ports and harbours:

- **Deadweight tonnage** (DWT): the carrying capacity of the ship, namely the total weight of cargo, fuels, freshwater, etc.
- **Gross registered tonnage** (GRT): the total internal capacity of the ship divided by 100 ft^3 or 2.83 m^3, depending on the application of relevant laws and regulations.
- **Displacement tonnage** (DT): the total weight or mass of the ship is obtained by multiplying the volume of water displaced by the ship by the density of the water. The DWT for mixed-cargo and bulk-cargo ships is roughly equal to 1.2–1.4 times the DWT and equal to 2.0 times the gross registered tonnage. For passenger ships, the displacement tonnage is roughly equal to 1.0 times the gross registered tonnage.
  The displacement of a ship is therefore the product of the length between perpendiculars (perps), the beam, the draft, the block coefficient and the density of the water. The block coefficient is the ratio between the volume of the wetted portion of the ship's hull (the displacement) and the volume of the enclosing block (length between the perps × the beam × the draft).
  The displacement light varies from about 15–25% of the displacement fully loaded. The displacement light (i.e. the ship without ballast or any load) should for safety reasons only occur when the ship is moored in a dock or at a shipyard for fitting-out or repair.
  The ballast condition is the minimum weight or ballast that a ship has to carry for safe manoeuvring stability. For example, having discharged all oil cargo, the oil tanker has to increase its weight by taking in seawater as ballast to increase the draft in order to obtain the necessary safe manoeuvring stability. The ballast displacement is about 30–50% of the displacement fully loaded, depending on the weather conditions.
- **Air draught**: the maximum distance from the water level to the highest point of the ship at the prevailing draft (Figure 21.1).
- **Scantling draught**: the draft for which the structural strength of the ship has been designed.
- **Designed draught**: the draft on which the fundamental design parameters of the ship are based.
- **Trim**: the difference between the aft and the forward draft.
- **Bow to centre manifold/stern to centre manifold**: the distance from the extreme point of the bow or stern to the manifold centreline for tankers.

The definitions of a ship's overall dimensions (length, draught, etc.) are illustrated in Figure 21.2.

Figure 21.1 The air draught

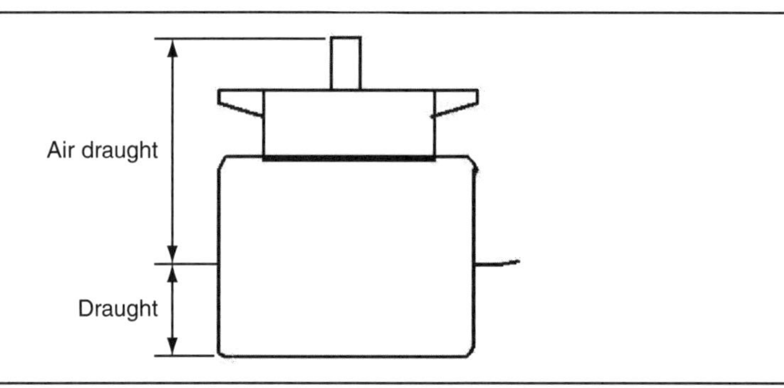

21.2. Ship dimensions

When the dimensions of ships are not clearly known, the average ship dimensions shown in the tables in this chapter may be used in the design of berths, dolphins and fenders. The following tables show the general average dimensions for the beam, the overall length and the fully loaded draft for general cargo ships, oil tankers and container ships. The length between perps is roughly 95% of the length overall.

Figure 21.2 Definitions of ship dimensions

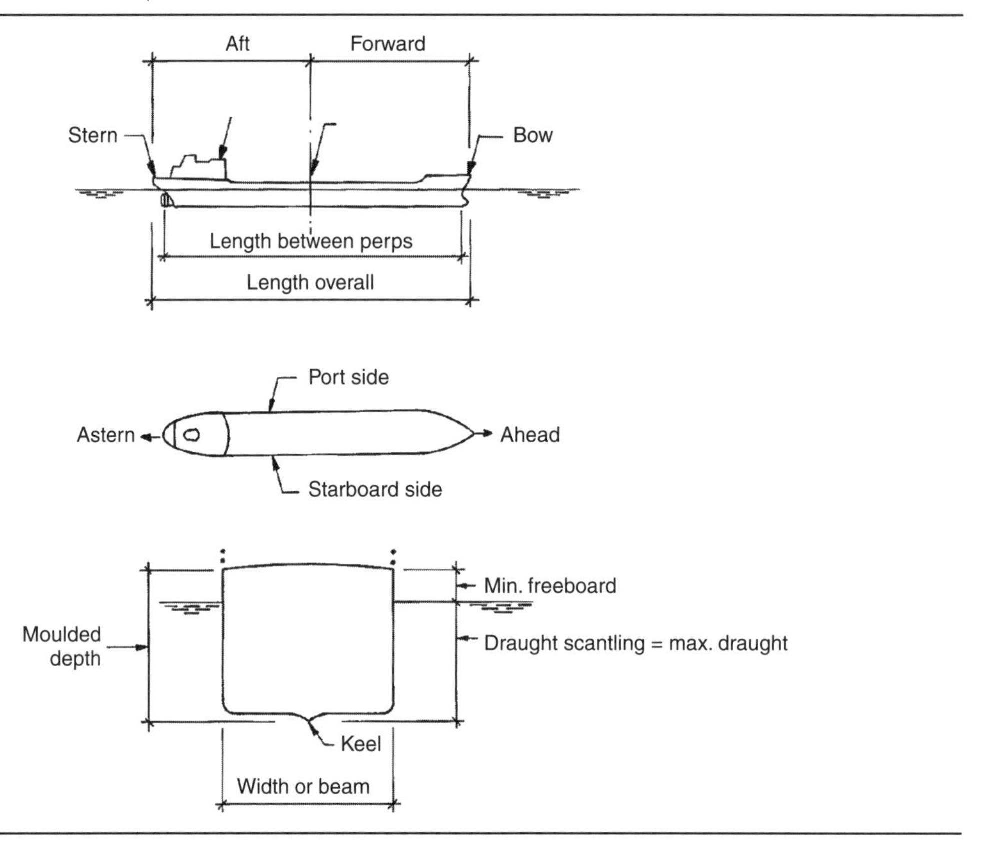

Passenger ships

Passenger ships are classified as shown below.

GRT: t	DWT t	Displacement t	Overall length m	Length between perps m	Beam m	Max. draught: m
70 000	–	38 000	260	220	33.0	7.6
50 000	–	28 000	230	200	31.0	7.6
30 000	–	18 000	195	165	27.0	7.6
10 000	–	6 700	135	120	20.5	5.0

The two largest passenger ships today are the *Oasis of the Seas* and *Allure of the Seas*, and their main dimensions are shown in the table below.

GRT: t	Displacement: t	Overall length: m	Length between perps: m	Beam water-line: m	Beam deck 15 and 17: m	Max. draught: m	Air draught: m	Lateral wind area: m^2	Front wind area: m^2	Passengers
225 282	105 983	361	300	47.0	65.7	9.30	72.0	15 764	3354	5400

Figure 21.3 compares the *Oasis of the Seas* and the *Titanic*.

Figure 21.3 The *Oasis of the Seas* and the *Titanic*

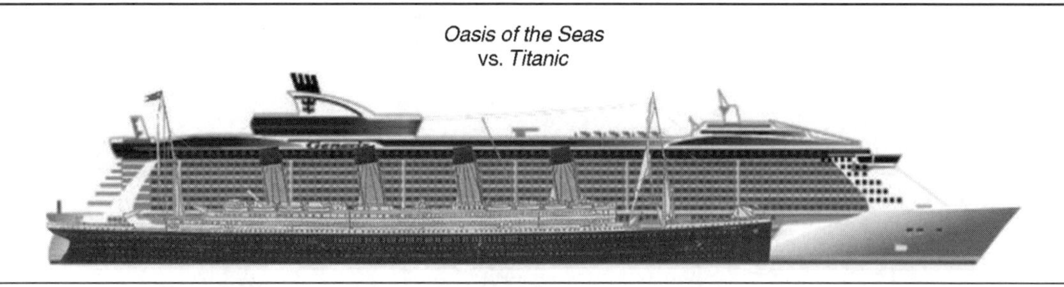

The table below lists the dimensions of other large passenger ships.

Name	Overall length: m	Beam: m	Max. draught: m	Passengers
Vision of the Seas	279	32.2	8.00	2435
Majesty of the Seas	279	32.2	7.65	2435
Emerald Princess	290	36.0	8.00	3114
Brilliance of the Seas	293	32.2	8.50	2501
Adventure of the Seas	312	48.0	8.85	3114
Freedom of the Seas	339	38.6	8.50	4370
Queen Mary 2	345	41.0	10.10	2620

Mixed cargo ships (full deck construction)

GRT: t	DWT: t	Displacement: t	Overall length: m	Length between perps: m	Beam: m	Moulded depth: m	Draught ballast loaded: m	Max. draught: m
10 000	15 000	20 000	165	155	21.5	12.0	4.9	9.5
7 000	10 000	14 000	145	135	20.0	11.5	4.4	8.5
5 500	8 000	11 000	135	125	18.0	10.5	4.1	8.0
4 000	6 000	8 000	125	115	16.5	9.5	3.8	7.5
3 500	5 000	7 000	105	100	15.0	8.5	3.6	7.0
2 000	3 000	4 000	90	85	13.0	7.3	3.0	6.0
1 300	2 000	3 000	80	75	12.0	6.5	2.6	5.3
1 000	1 500	2 000	70	65	10.0	5.1	2.2	4.3
500	700	1 000	55	50	8.5	4.5	1.9	3.8

Bulk cargo (oil tankers, bulk carriers, etc.)

Tankers are usually classified as shown below.

Type of tanker	DWT: t
General-purpose/product carrier	<25 000
Super tankers and large tankers	25 000–150 000
VLCC	150 000–300 000
ULCC (Figure 21.4)	>300 000

Figure 21.4 Ultra-large crude carrier. (Photograph courtesy of Bergesen DY, Norway)

Bulk carriers are usually classified as shown below.

Type of bulk carrier	DWT: t
Mini bulk carrier	12 000
Small handy-sized	15 000–25 000
Handy-sized	25 000–50 000
Handy max	35 000–50 000
Panamax	50 000–80 000
Cape-sized	100 000–180 000
Very large bulk carrier	>180 000

DWT: t	Displacement: t	Overall length: m	Length between perps: m	Beam: m	Moulded depth: m	Draught ballast loaded: m	Max. draught: m
400 000	460 000	412	392	66.0	29.0	12.5	24.0
300 000	340 000	360	340	57.0	27.0	12.0	21.0
250 000	300 000	330	315	55.5	26.0	11.5	20.0
200 000	240 000	310	300	55.0	25.0	10.5	19.0
150 000	180 000	290	270	47.0	23.0	9.0	17.0
100 000	125 000	250	240	42.0	21.0	8.0	15.0
70 000	90 000	230	215	38.0	18.5	7.2	13.5

Common sizes of crude oil tankers are given in Figure 21.5.

Figure 21.5 Common sizes of crude oil tankers. (Data source courtesy of Lloyd's Register-Fairplay)

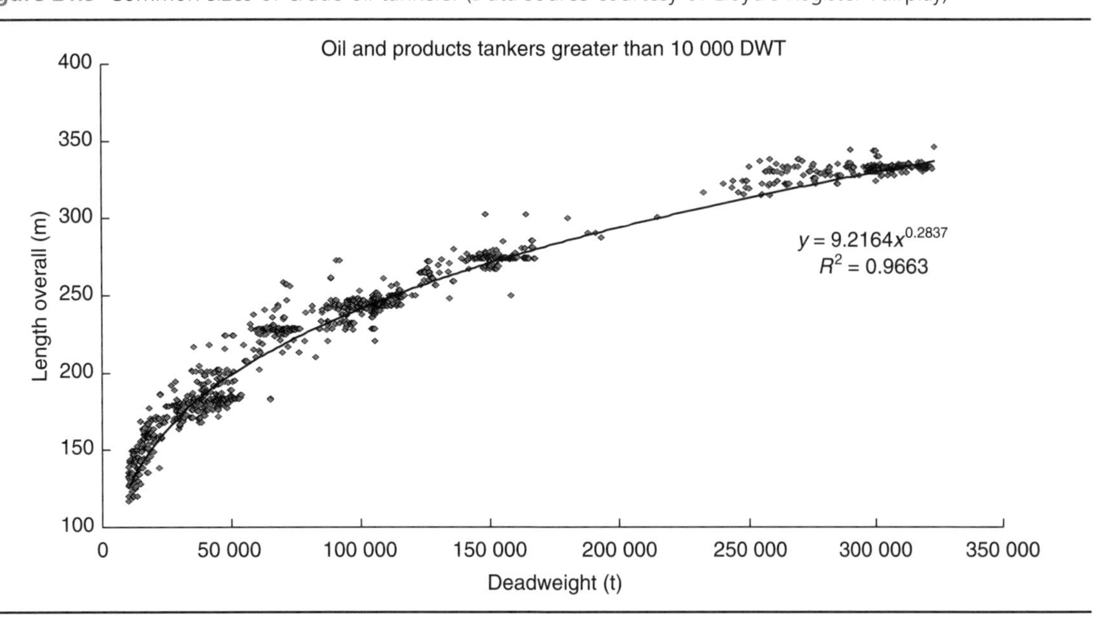

Figure 20.1 Typical maintenance organisation flowchart

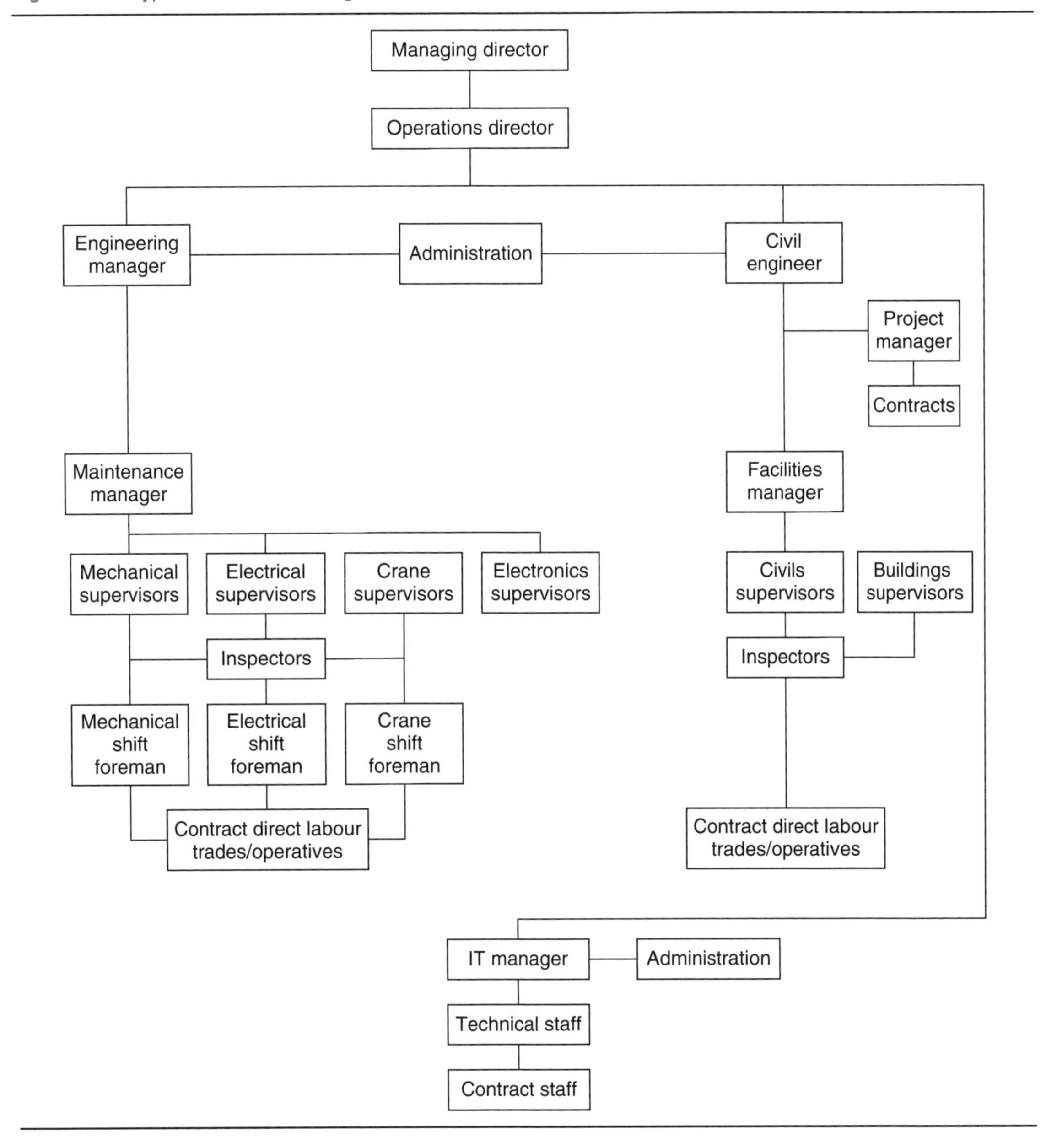

20.6. Infrastructure

Generally, the planned preventative maintenance of terminal plant and equipment is taken for granted. This is usually not the case for the maintenance of the terminal infrastructure. For example, a berth structure, with the exception of the fenders and berth hardware (bollards, ladders, etc.), will be assumed to perform throughout its design life with little or no maintenance. For berth structures, it is often a case of 'out of sight, out of mind'.

Container ships

DWT: t	Displacement: t	Overall length: m	Length between perps: m	Beam: m	Max. draught: m	Number of containers (approximate)	Generation
165 000	NA	400	NA	59.0	16.0	18 000	Triple E class
156 900	NA	397	NA	56.4	15.5	15 500	PS class
Post-Panamax							
104 000	143 000	340	330	42.8	14.5	8 000	7th
85 000	119 000	320	300	42.0	14.2	6 400	6th
70 000	95 000	285	270	41.8	14.0	5 000	6th
60 000	82 000	260	252	38.0	13.4	4 400	5th
Panamax							
55 000	75 000	270	260	32.2	12.5	3 900	4th
50 000	68 000	260	250	32.2	12.5	3 500	3rd
40 000	54 000	235	225	32.2	11.8	2 800	3rd
30 000	41 000	210	200	30.5	10.8	2 100	2nd
20 000	27 000	175	165	26.0	10.0	1 700	2nd
15 000	20 000	152	145	23.5	8.7	1 000	1st
10 000	13 500	130	125	21.0	7.6	600	1st
7 000	10 300	120	110	20.0	6.8	400	1st

Containership characteristics are listed in the following table.

TEU class[a]	Description	Number of ships	Average draught: m	Average beam: m	Average L_{OA}: m
≤500	Small feeder	429	6.01	17.00	104.22
501–1000	Feeder	819	7.69	21.03	133.04
1001–2000	Small containership	1258	9.49	25.84	168.92
2001–3500	Medium-sized containership	895	11.50	31.04	217.67
3501–5000	Panamax/post-Panamax	669	12.70	32.63	272.26
5001–7500	Post-Panamax	514	13.80	39.09	293.30
7501–9500	Very large post-Panamax	207	14.43	43.56	334.37
≥9501	Ultra-large post-Panamax	49	15.19	47.90	359.25

Source: Clarkson Research Services (2010), *Report on Container Vessel Fleet*
[a]TEU, 20 ft equivalent units

Gas tankers

Capacity: m³	DWT: t	Displacement: t	Overall length: m	Length between perps: m	Beam: m	Moulded depth: m	Max. draught: m
LPG tankers							
75 000	46 900	75 000	229	218	36.0	21.0	12.1
52 000	38 500	53 200	206	196	31.4	18.6	11.3
35 000	36 200	7 500	185	176	27.8	18.0	12.5
24 000	18 100	32 800	157	149	25.3	16.0	10.1
15 000	16 200	25 000	151	140	25.0	14.3	9.6
8 300	9 800	15 500	128	116	20.0	12.1	9.4
5 000	5 400	9 000	106	98	17.0	10.0	7.4
2 500	2 800	5 000	75	70	14.0	7.9	6.8
LNG tankers							
250 000	122 500	177 000	369	354	55.7	31.2	12.8
220 000	108 000	158 000	365	341	53.8	30.5	12.5
200 000	100 000	146 000	340	325	51.3	28.0	12.0
168 000	84 500	125 000	298	285	48.7	28.0	11.9
163 700	84 000	125 000	292	280	45.2	27.5	11.6
145 000	74 400	110 000	288	274	49.0	26.8	12.3
137 000	71 500	100 000	290	275	48.1	28.0	11.3
125 000	66 800	102 000	272	259	47.2	26.5	11.4
87 600	53 600	74 000	250	237	40.0	23.0	10.6
65 000	36 400	52 000	214	204	37.8	21.5	9.8
29 000	22 100	32 600	182	171	29.0	16.5	9.0
1 000		1 600	65		12.0	6.0	3.5

Today, the largest LNG spherical tanker is 177 000 m^3, and the largest LNG membrane tanker is 266 000 m^3. Figures 21.6 and 21.7 show common sizes for LNG and LPG tankers.

Figure 21.6 Common sizes of LNG tankers. (Data source courtesy of Lloyd's Register-Fairplay)

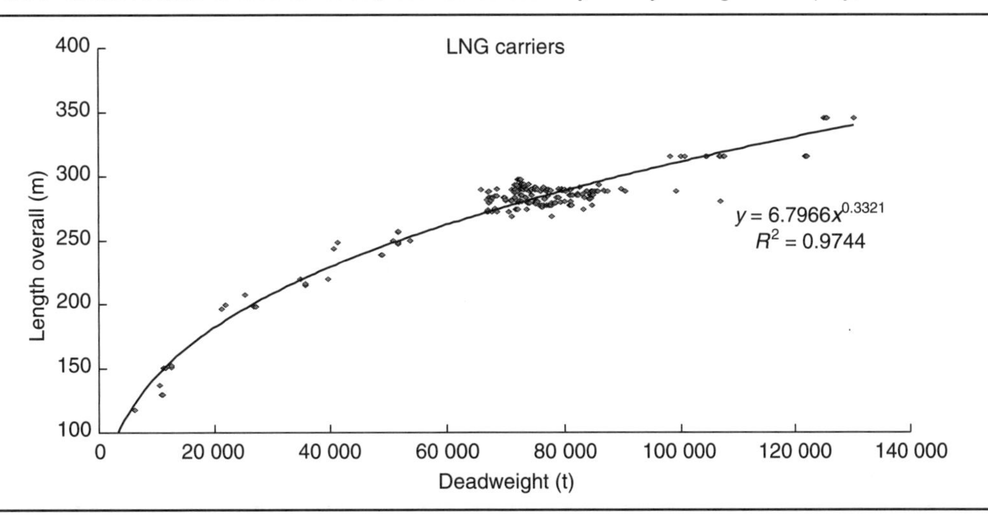

Figure 21.7 Common sizes of LPG tankers. (Data source courtesy of Lloyd's Register-Fairplay)

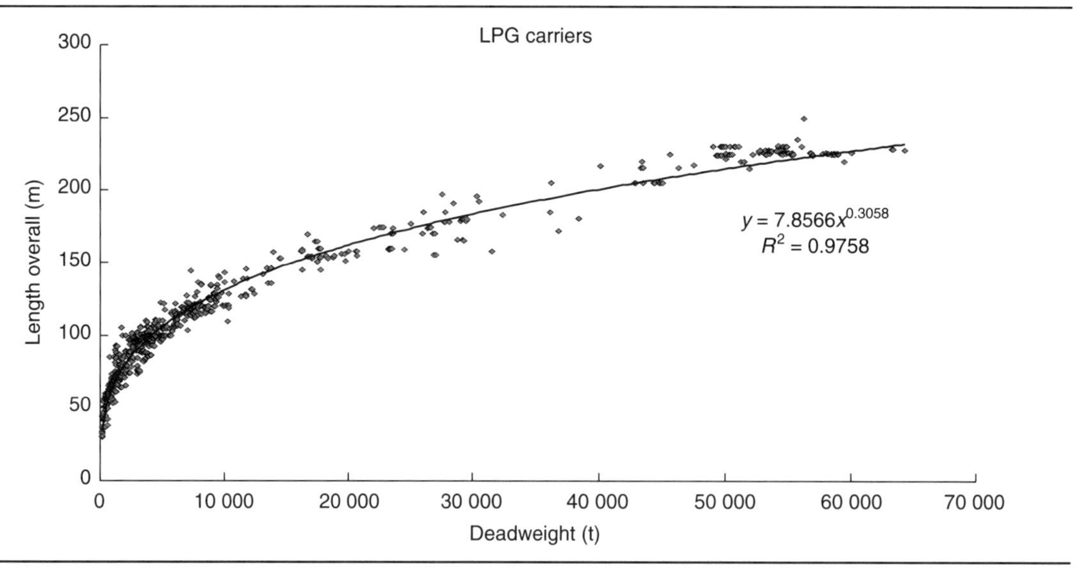

Ferries

DWT: t	Displacement: t	Overall length: m	Length between perps: m	Beam: m	Max. draught: m
50 000	40 000	230	215	33.0	12.5
40 000	30 500	225	210	32.0	9.0
30 000	23 000	205	190	30.0	7.5
20 000	15 500	175	160	27.0	6.7
10 000	7 900	140	125	23.0	5.5
5 000	4 000	110	100	19.0	4.5

The Color Line (Norway) cruise ferry *Color Fantasy*, shown in Figure 15.4 in Chapter 15, has the following dimensions.

DWT: t	Overall length: m	Beam: m	Max. draught: m	Wind lateral area: m^2
5616	223.72	35.0	7.0	7410

Ro/ro ships

DWT: t	Displacement: t	Overall length: m	Length between perps: m	Beam: m	Max. draught: m
30 000	45 000	230	215	30.5	11.3
20 000	32 000	200	185	27.5	10.0
15 000	25 000	180	165	25.8	8.8
10 000	17 000	155	140	23.3	7.6
5 000	8 800	120	110	19.5	5.9

Car transport ships

DWT: t	Displacement: t	Overall length: m	Length between perps: m	Beam: m	Max. draught: m	Number of cars (approximate)
28 000	45 000	198	183	32.3	11.8	6200
26 300	42 000	213	198	32.3	10.5	6000
17 900	33 000	195	180	32.3	9.7	5600

Fishing boats

GRT: t	DWT: t	Displacement: t	Overall length: m	Length between perps: m	Beam: m	Moulded depth: m	Draught ballast loaded: m	Max. draught: m
2500	–	2800	90	80	14.0	–	–	5.9
2000	–	2500	85	75	13.0	–	–	5.6
1500	–	2100	80	70	12.0	–	–	5.3
1000	–	1750	75	65	11.0	–	–	5.0
800	–	1550	70	60	10.5	–	–	4.8
600	–	1200	65	55	10.0	–	–	4.5
400	–	800	55	45	8.5	–	–	4.0
200	–	400	40	35	7.0	–	–	3.5

General wind and current areas for different types of ship

Figure 21.8 shows the approximate laterally projected areas perpendicular to the wind direction above water of typical oil tankers, bulk and ore carriers, and cargo ships in ballasted and loaded conditions relative to the ship's displacement. For comparison, the data shown in Figures 21.9 and 21.10 are taken

Figure 21.8 Lateral projected wind area of a ship relative to the displacement

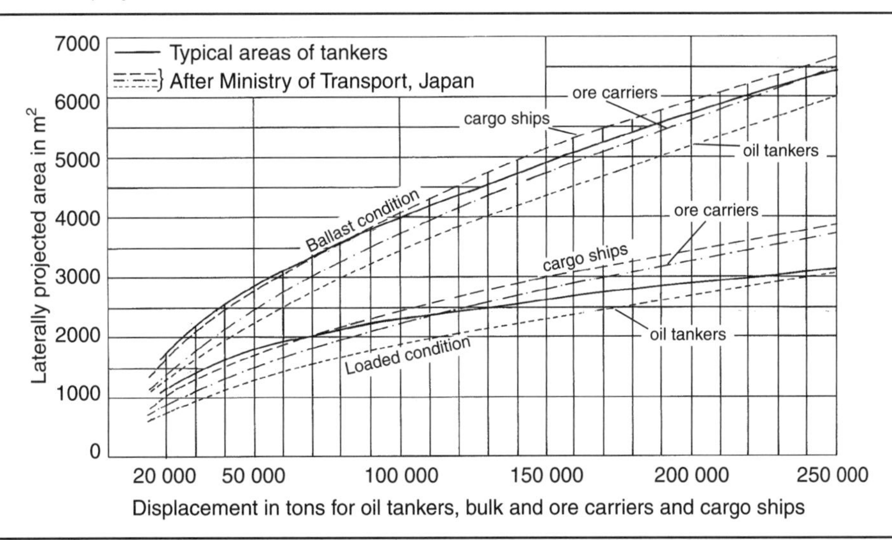

from the British standard BS 6349-1 and from research done by the Port and Harbour Research Institute, Ministry of Transport, Japan. The values given in these figures must only be used as a rough indicator for a ship's dimensions. For important calculations, the actual dimensions of the ships that will call at the berth or harbour should be used.

Figure 21.9 Typical longitudinal projected areas of tankers relative to the DWT

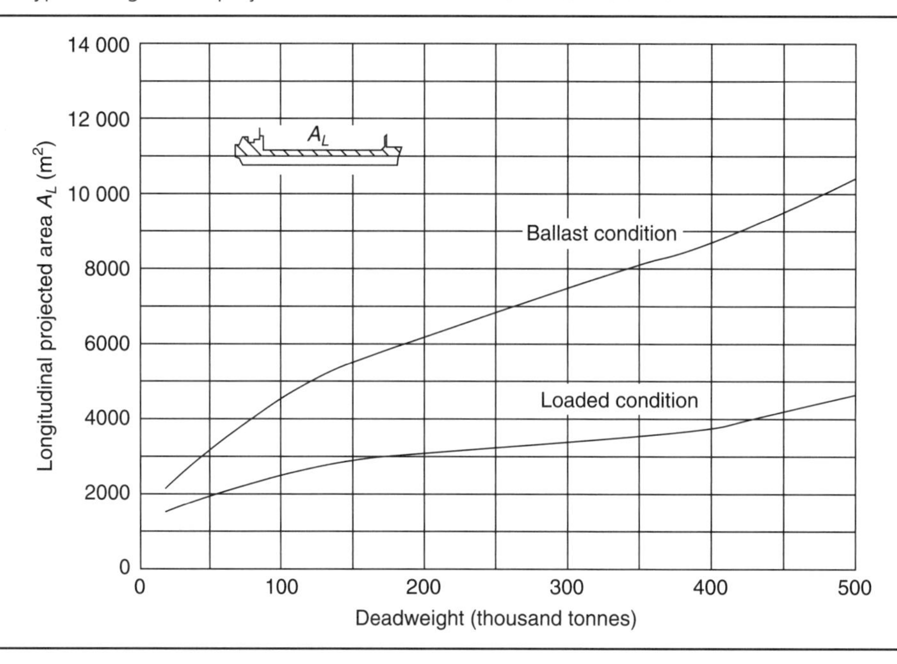

Figure 21.10 Length and longitudinal projected areas of container ships relative to the DWT

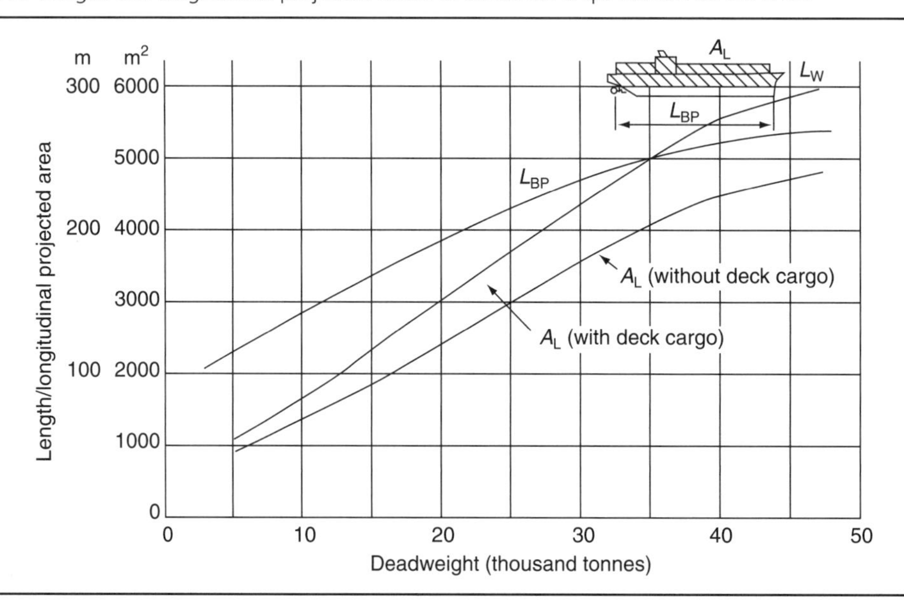

Deadweight	Container ship wind area in m^3	
	Without containers on deck	With containers on deck
50 000	4900	6100
42 000	4600	5700
30 000	3600	4400
25 000	3000	3600
20 000	2400	3100
15 000	1900	2300
10 000	1400	1700

Figure 21.10 shows data from the British standard BS 6349-1 for the length and the longitudinal projected areas of container ships in loaded and ballasted conditions, compared with general approximately lateral projected wind area data for container ships with and without containers on the deck shown in the table below the figure.

The approximately lateral projected wind and current areas above and below sea level for some LPG and LNG tankers in fully loaded and ballast-loaded conditions are shown in the following table.

	Above sea level				Below sea level			
	Fully loaded: m^2		Ballast loaded: m^2		Fully loaded: m^2		Ballast loaded: m^2	
	Lateral	Front	Lateral	Front	Lateral	Front	Lateral	Front
LPG tankers								
75 000 m^3	2 400	760	3 640	960	2970	490	1760	280
52 000 m^3	2 100	630	3 000	770	2450	370	1560	240
24 000 m^3	1 000	320	1 850	460	1790	300	920	160
LNG tankers								
220 000 m^3	11 500	2200	13 000	2400	4400	670	2800	430
200 000 m^3	9 700	2000	11 200	2300	4000	610	2500	390
168 000 m^3	8 700	2300	9 800	2500	3300	600	2400	400
145 000 m^3	7 600	1950	8 400	2100	3400	600	2500	450
137 000 m^3	7 000	1800	8 400	2100	2300	550	2100	360
125 000 m^3	7 320	1600	8 430	2000	3190	530	2120	350
87 600 m^3	5 900	1200	6 780	1400	2490	420	1660	280
65 000 m^3	4 300	1100	4 900	1200	2000	370	1380	250
29 000 m^3	3 160	900	3 700	990	1540	260	1090	180

LNG carriers with a spherical containment system (Moss-Rosenberg) will have proportionately higher windage than those with a prismatic/membrane containment system, and vessels in ballast have relatively more windage than laden vessels because of the reduction in draught. Consequently, while all ships will be affected by the wind conditions, LNG carriers with spherical tanks and vessels in ballast will be more sensitive to the wind than laden vessels or those with membrane tanks. Figures 21.11 and 21.12 show the different tankers in question.

Figure 21.11 A spherical LNG tanker

Figure 21.12 A membrane LNG tanker

Some of the largest LNG membranes are shown in detail in the following table.

Vessel size reference	QFlex	QFlex	QFlex	QMax	Max
Vessel capacity: m^3	210 000	216 200	217 300	263 000	266 000
Type of cargo containment system	Membrane GTNO96	Membrane Mark III	Membrane Mark III	Membrane GTNO96	Membrane Mark III
Main particulars:					
Length overall (L_{OA}): m	315	315	315	345	345
Length between perpendiculars (L_{BP}): m	303	303	302	333	332
Breadth moulded: m	50	50	50	55.0	53.8
Depth moulded to upper deck at midship: m	27	27	27	27.0	27
Depth moulded to freeboard deck: m	21.92	21.82	21.2	22.16	2.5
Design draught moulded: m	12	12	12	12	12
Scantling draught moulded: m	13.60	13	13	13.7	13
Displacement – ballast draught	109 000 t	112 060 mt/ 108 988 m^3	111 900 mt/ 108 900 m^3	125 436	141 990 mt/ 138 000 m^3
Bow type	Bulbous bow	Bulbous bow	Bulbous bow	Bulbous bow	Bulbous bow
Parallel mid-body (PMD) length at ballast draft					
Total: m	142.81	143.3	145.2	167	178
Wind areas – ballast draught:					
Transverse (head-on) wind area: m^2	1650	1660	1650	1850	18 794
Longitudinal (broadside) wind area: m^2	8300	8030	8000	9300	8895
Wind areas – loaded draught:					
Transverse (head-on) wind area: m^2	1550	1540	1510	1730	1668
Longitudinal (broadside) wind area: m^2	7500	7320	7130	8200	8065

Figure 21.13 Laterally projected wind area of ships relative to their displacement

Type of ship				General cargos		Oil tankers		Ore carriers	
Range of tonnage in DWT				500–140 000		500–320 000		500–200 000	
Coefficients				α	β	α	β	α	β
Displacement fully loaded = DT				2.463	0.936	2.028	0.954	1.687	0.969
Laterally projected area	Above sea level	Fully loaded	$= \alpha(\mathrm{DWT})^{\beta}$	8.770	0.496	4.964	0.522	4.390	0.548
		Ballast loaded		9.641	0.533	5.943	0.562	5.171	0.580
	Below sea level	Fully loaded		3.495	0.608	3.198	0.611	2.723	0.625
		Ballast loaded		1.404	0.627	1.629	0.610	1.351	0.633
Front area	Above sea level	Fully loaded		2.763	0.490	2.666	0.478	1.971	0.510
		Ballast loaded		3.017	0.510	2.485	0.517	1.967	0.538
Total surface area below sea level		Fully loaded		9.260	0.639	6.162	0.673	4.576	0.702
		Ballast loaded		4.637	0.669	3.865	0.686	3.471	0.701
Displacement ballast loaded			$= \alpha(\mathrm{DT})^{\beta}$	0.199	1.084	0.383	1.018	0.385	1.023
Draft ballast loaded			$= \alpha(\mathrm{draft}_{max})^{\beta}$	0.352	1.172	0.548	0.966	0.551	0.993

21.3. Recommended design dimensions

The Port and Harbour Research Institute, Ministry of Transport, Japan, has given the following formulas (Figure 21.13) for the correlation between the displacement tonnage (DT), the laterally projected area, the front area, the total surface area below sea level, the displacement ballast loaded, the draft ballast loaded, etc., for general cargo, oil tankers and ore carriers.

As a useful rule of thumb, the following approximate formula gives the fully loaded draught of cargo ships and bulk carriers in meters:

$$\text{Fully loaded draught} = \sqrt{\frac{\mathrm{DWT}}{1000}} + 5$$

The formula will give the fully loaded draught to within 1 m for general cargo ships and dry- and liquid-bulk carriers over the range 5000–400 000 DWT.

As an **example** of the use of Figure 21.13, the following approximate values provide a rough indicator for a 50 000 DWT oil tanker:

$$\text{Displacement fully loaded} = 2.028 \times 50\,000^{0.954} = 61\,643\ \text{t}$$

$$\text{Laterally projected area above sea level fully loaded} = 4.964 \times 50\,000^{0.522} = 1408\ \text{m}^2$$

$$\text{Displacement ballast loaded} = 0.383 \times 61\,643^{1.018} = 28\,792\ \text{t}$$

Maximum draught using Figure 21.13 $= 12.0$ m

$$\text{Maximum draught} = \sqrt{\frac{50\,000}{1000}} + 5 = 12.07\,\text{m}$$

Draught ballast loaded $= 0.548 \times 12.07^{0.966} = 6.08$ m

Draught minimum in accordance with the International Maritime Organisation

$$= 0.02 \times L_{BP} + 2.0 = 6.3\ \text{m}$$

Draught ballast using Figure 21.13 $= 6.4$ m

When designing harbours and berth facilities including the fender design, the design vessel should be the largest ship expected to berth. The data in the tables below are from Akakura and Takahashi (1998), *Technical Note of the Ports and Harbour Research Institute*, No. 911, and the Port and Harbour Bureau of Ministry of Transport, Japan. The tables show the 50, 75, 90 and 95% confidence limits. L_{OA} is the ship's overall length, L_{BP} is the length between perps, B is the beam and D is the moulded depth.

Confidence limit: 50%

Type	DWT: t	Displacement: t	L_{OA}: m	L_{BP}: m	*B*: m	*D*: m	Max. draught: m	Wind lateral area: m^2		Wind front area: m^2	
								Full-load condition	Ballast condition	Full-load condition	Ballast condition
General cargo ship	1 000	1 580	63	58	10.3	5.2	3.6	227	292	59	88
	2 000	3 040	78	72	12.4	6.4	4.5	348	463	94	134
	3 000	4 460	88	82	13.9	7.2	5.1	447	605	123	172
	5 000	7 210	104	96	16.0	8.4	6.1	612	849	173	236
	7 000	9 900	115	107	17.6	9.3	6.8	754	1060	216	290
	10 000	13 900	128	120	19.5	10.3	7.6	940	1340	274	361
	15 000	20 300	146	136	21.8	11.7	8.7	1210	1760	359	463
	20 000	26 600	159	149	23.6	12.7	9.6	1440	2130	435	552
	30 000	39 000	181	170	26.4	14.4	10.9	1850	2780	569	709
	40 000	51 100	197	186	28.6	15.7	12.0	2210	3370	690	846
Bulk carrier[a]	5 000	6 740	106	98	15.0	8.4	6.1	615	850	205	231
	7 000	9 270	116	108	16.6	9.3	6.7	710	1010	232	271
	10 000	13 000	129	120	18.5	10.4	7.5	830	1230	264	320
	15 000	19 100	145	135	21.0	11.7	8.4	980	1520	307	387
	20 000	25 000	157	148	23.0	12.8	9.2	1110	1770	341	443
	30 000	36 700	176	167	26.1	14.4	10.3	1320	2190	397	536
	50 000	59 600	204	194	32.3	16.8	12.0	1640	2870	479	682
	70 000	81 900	224	215	32.3	18.6	13.3	1890	3440	542	798
	100 000	115 000	248	239	37.9	20.7	14.8	2200	4150	619	940
	150 000	168 000	279	270	43.0	23.3	16.7	2610	5140	719	1140
	200 000	221 000	303	294	47.0	25.4	18.2	2950	5990	800	1310
	250 000	273 000	322	314	50.4	27.2	19.4	3240	6740	868	1450
Container ship[b]	7 000	10 200	116	108	19.6	9.3	6.9	1320	1360	300	396
	10 000	14 300	134	125	21.6	10.7	7.7	1690	1700	373	477
	15 000	21 100	157	147	24.1	12.6	8.7	2250	2190	478	591
	20 000	27 800	176	165	26.1	14.1	9.5	2750	2620	569	687

	25 000	34 300	192	180	27.7	15.4	10.2	3220	3010	652	770
	30 000	40 800	206	194	29.1	16.5	10.7	3660	3370	729	850
	40 000	53 700	231	218	32.3	18.5	11.7	4480	4040	870	990
	50 000	66 500	252	238	32.3	20.2	12.5	5230	4640	990	1110
	60 000	79 100	271	256	35.2	21.7	13.2	5950	5200	1110	1220
Oil tanker	1 000	1 450	59	54	9.7	4.3	3.8	170	266	78	80
	2 000	2 810	73	68	12.1	5.4	4.7	251	401	108	117
	3 000	4 140	83	77	13.7	6.3	5.3	315	509	131	146
	5 000	6 740	97	91	16.0	7.5	6.1	419	689	167	194
	7 000	9 300	108	102	17.8	8.4	6.7	505	841	196	233
	10 000	13 100	121	114	19.9	9.5	7.5	617	1040	232	284
	15 000	19 200	138	130	22.5	11.0	8.4	770	1320	281	355
	20 000	25 300	151	143	24.6	12.2	9.1	910	1560	322	416
	30 000	37 300	171	163	27.9	14.0	10.3	1140	1990	390	520
	50 000	60 800	201	192	32.3	16.8	11.9	1510	2690	497	689
	70 000	83 900	224	214	36.3	18.9	13.2	1830	3280	583	829
	100 000	118 000	250	240	40.6	21.4	14.6	2230	4050	690	1010
	150 000	174 000	284	273	46.0	24.7	16.4	2800	5150	840	1260
	200 000	229 000	311	300	50.3	27.3	17.9	3290	6110	960	1480
	300 000	337 000	354	342	57.0	31.5	20.1	4120	7770	1160	1850
Ro/ro ship	1 000	1 970	66	60	13.2	5.2	3.2	700	810	216	217
	2 000	3 730	85	78	15.6	7.0	4.1	970	1110	292	301
	3 000	5 430	99	90	17.2	8.4	4.8	1170	1340	348	364
	5 000	8 710	119	109	19.5	10.5	5.8	1480	1690	435	464
	7 000	11 900	135	123	21.2	12.1	6.6	1730	1970	503	544
	10 000	16 500	153	141	23.1	14.2	7.5	2040	2320	587	643
	15 000	24 000	178	163	25.6	16.9	8.7	2460	2790	701	779
	20 000	31 300	198	182	27.4	19.2	9.7	2810	3180	794	890
	30 000	45 600	229	211	30.3	23.0	11.3	3400	3820	950	1080

[a]Full-load condition of wind lateral/front areas of log carrier do not include the areas of logs on deck
[b]Full-load condition of wind lateral/front areas of container ships include the areas of containers on deck

Type	Gross tonnage: t	Displacement: t	L_{OA}: m	L_{BP}: m	B: m	D: m	Max. draught: m	Wind lateral area: m^2		Wind front area: m^2	
								Full-load condition	Ballast condition	Full-load condition	Ballast condition
Passenger ship	1 000	850	60	54	11.4	4.1	1.9	426	452	167	175
	2 000	1 580	76	68	13.6	5.3	2.5	683	717	225	234
	3 000	2 270	87	78	15.1	6.2	3.0	900	940	267	277
	5 000	3 580	104	92	17.1	7.5	3.6	1270	1320	332	344
	7 000	4 830	117	103	18.6	8.6	4.1	1600	1650	383	396
	10 000	6 640	133	116	20.4	9.8	4.8	2040	2090	446	459
	15 000	9 530	153	132	22.5	11.5	5.6	2690	2740	530	545
	20 000	12 300	169	146	24.2	12.8	7.6	3270	3320	599	614
	30 000	17 700	194	166	26.8	14.9	7.6	4310	4350	712	728
	50 000	27 900	231	197	30.5	18.2	7.6	6090	6120	880	900
	70 000	37 600	260	220	33.1	20.7	7.6	7660	7660	1020	1040
Ferry	1 000	810	59	54	12.7	4.6	2.7	387	404	141	145
	2 000	1 600	76	69	15.1	5.8	3.3	617	646	196	203
	3 000	2 390	88	80	16.7	6.5	3.7	811	851	237	247
	5 000	3 940	106	97	19.0	7.6	4.3	1150	1200	302	316
	7 000	5 480	119	110	20.6	8.5	4.8	1440	1510	354	372

	10 000	7 770	135	125	22.6	9.5	5.3	1830	1930	419	442
	15 000	11 600	157	145	25.0	10.7	6.0	2400	2540	508	537
	20 000	15 300	174	162	26.8	11.7	6.5	2920	3090	582	618
	30 000	22 800	201	188	29.7	13.3	7.4	3830	4070	705	752
	40 000	30 300	223	209	31.9	14.5	8.0	4660	4940	810	860
Gas carrier	1 000	2 210	68	63	11.1	5.3	4.3	350	436	121	139
	2 000	4 080	84	78	13.7	6.8	5.2	535	662	177	203
	3 000	5 830	95	89	15.4	7.8	5.8	686	846	222	254
	5 000	9 100	112	104	17.9	9.4	6.7	940	1150	295	335
	7 000	12 300	124	116	19.8	10.6	7.4	1150	1410	355	403
	10 000	16 900	138	130	22.0	12.0	8.2	1430	1750	432	490
	15 000	24 100	157	147	24.8	13.9	9.3	1840	2240	541	612
	20 000	31 100	171	161	27.1	15.4	10.0	2190	2660	634	716
	30 000	44 400	194	183	30.5	17.8	11.7	2810	3400	794	894
	50 000	69 700	227	216	35.5	21.3	11.7	3850	4630	1050	1180
	70 000	94 000	252	240	39.3	24.0	11.7	4730	5670	1270	1420
	100 000	128 000	282	268	43.7	27.3	11.7	5880	7030	1550	1730

Confidence limit: 75%

Type	DWT: t	Displacement: t	L_{OA}: m	L_{BP}: m	*B*: m	*D*: m	Max. draught: m	Wind lateral area: m^2		Wind front area: m^2	
								Full-load condition	Ballast condition	Full-load condition	Ballast condition
General cargo ship	1 000	1 690	67	62	10.8	5.8	3.9	278	342	63	93
	2 000	3 250	83	77	13.1	7.2	4.9	426	541	101	142
	3 000	4 750	95	88	14.7	8.1	5.6	547	708	132	182
	5 000	7 690	111	104	16.9	9.4	6.6	750	993	185	249
	7 000	10 600	123	115	18.6	10.4	7.4	922	1240	232	307
	10 000	14 800	137	129	20.5	11.6	8.3	1150	1570	294	382
	15 000	21 600	156	147	23.0	13.1	9.5	1480	2060	385	490
	20 000	28 400	170	161	24.9	14.3	10.4	1760	2490	466	585
	30 000	41 600	193	183	27.8	16.2	11.9	2260	3250	611	750
	40 000	54 500	211	200	30.2	17.6	13.0	2700	3940	740	895
Bulk carrier[a]	5 000	6 920	109	101	15.5	8.6	6.2	689	910	221	245
	7 000	9 520	120	111	17.2	9.5	6.9	795	1090	250	287
	10 000	13 300	132	124	19.2	10.6	7.7	930	1320	286	340
	15 000	19 600	149	140	21.8	11.9	8.6	1100	1630	332	411
	20 000	25 700	161	152	23.8	13.0	9.4	1240	1900	369	470
	30 000	37 700	181	172	27.0	14.7	10.6	1480	2360	428	569
	50 000	61 100	209	200	32.3	17.1	12.4	1830	3090	518	723
	70 000	84 000	231	221	32.3	18.9	13.7	2110	3690	586	846
	100 000	118 000	255	246	39.2	21.1	15.2	2460	4460	669	1000
	150 000	173 000	287	278	44.5	23.8	17.1	2920	5520	777	1210
	200 000	227 000	311	303	48.7	25.9	18.6	3300	6430	864	1380
	250 000	280 000	332	324	52.2	27.7	19.9	3630	7240	938	1540
Container ship[b]	7 000	10 700	123	115	20.3	9.8	7.2	1460	1590	330	444
	10 000	15 100	141	132	22.4	11.3	8.0	1880	1990	410	535
	15 000	22 200	166	156	25.0	13.3	9.0	2490	2560	524	663
	20 000	29 200	186	175	27.1	14.9	9.9	3050	3070	625	771

	25 000	36 100	203	191	28.8	16.3	10.6	3570	3520	716	870
	30 000	43 000	218	205	30.2	17.5	11.1	4060	3950	800	950
	40 000	56 500	244	231	32.3	19.6	12.2	4970	4730	950	1110
	50 000	69 900	266	252	32.3	21.4	13.0	5810	5430	1090	1250
	60 000	83 200	286	271	36.5	23.0	13.8	6610	6090	1220	1370
Oil tanker	1 000	1 580	61	58	10.2	4.5	4.0	190	280	86	85
	2 000	3 070	76	72	12.6	5.7	4.9	280	422	119	125
	3 000	4 520	87	82	14.3	6.6	5.5	351	536	144	156
	5 000	7 360	102	97	16.8	7.9	6.4	467	726	184	207
	7 000	10 200	114	108	18.6	8.9	7.1	564	885	216	249
	10 000	14 300	127	121	20.8	10.0	7.9	688	1090	255	303
	15 000	21 000	144	138	23.6	11.6	8.9	860	1390	309	378
	20 000	27 700	158	151	25.8	12.8	9.6	1010	1650	355	443
	30 000	40 800	180	173	29.2	14.8	10.9	1270	2090	430	554
	50 000	66 400	211	204	32.3	17.6	12.6	1690	2830	548	734
	70 000	91 600	235	227	38.0	19.9	13.9	2040	3460	642	884
	100 000	129 000	263	254	42.5	22.5	15.4	2490	4270	761	1080
	150 000	190 000	298	290	48.1	25.9	17.4	3120	5430	920	1340
	200 000	250 000	327	318	52.6	28.7	18.9	3670	6430	1060	1570
	300 000	368 000	371	363	59.7	33.1	21.2	4600	8180	1280	1970
Ro/ro ship	1 000	2 190	73	66	14.0	6.2	3.5	880	970	232	232
	2 000	4 150	94	86	16.6	8.4	4.5	1210	1320	314	323
	3 000	6 030	109	99	18.3	10.0	5.3	1460	1590	374	391
	5 000	9 670	131	120	20.7	12.5	6.4	1850	2010	467	497
	7 000	13 200	148	136	22.5	14.5	7.2	2170	2350	541	583
	10 000	18 300	169	155	24.6	17.0	8.2	2560	2760	632	690
	15 000	26 700	196	180	27.2	20.3	9.6	3090	3320	754	836
	20 000	34 800	218	201	29.1	23.1	10.7	3530	3780	854	960
	30 000	50 600	252	233	32.2	27.6	12.4	4260	4550	1020	1160

[a]Full-load condition of wind lateral/front areas of log carrier do not include the areas of logs on deck
[b]Full-load condition of wind lateral/front areas of container ships include the areas of containers on deck

Type	Gross tonnage: t	Displacement: t	L_{OA}: m	L_{BP}: m	B: m	D: m	Max. draught: m	Wind lateral area: m^2		Wind front area: m^2	
								Full-load condition	Ballast condition	Full-load condition	Ballast condition
Passenger ship	1 000	1 030	64	60	12.1	4.9	2.6	464	486	187	197
	2 000	1 910	81	75	14.4	6.3	3.4	744	770	251	263
	3 000	2 740	93	86	16.0	7.4	4.0	980	1010	298	311
	5 000	4 320	112	102	18.2	9.0	4.8	1390	1420	371	386
	7 000	5 830	125	114	19.8	10.2	5.5	1740	1780	428	444
	10 000	8 010	142	128	21.6	11.7	6.4	2220	2250	498	516
	15 000	11 500	163	146	23.9	13.7	7.5	2930	2950	592	611
	20 000	14 900	180	160	25.7	15.3	8.0	3560	3570	669	690
	30 000	21 300	207	183	28.4	17.8	8.0	4690	4680	795	818
	50 000	33 600	248	217	32.3	21.7	8.0	6640	6580	990	1010
	70 000	45 300	278	243	35.2	24.6	8.0	8350	8230	1140	1170
Ferry	1 000	1 230		61	14.3	5.5	3.4	411	428	154	158
	2 000	2 430	86	78	17.0	6.8	4.2	656	685	214	221
	3 000	3 620	99	91	18.8	7.7	4.8	862	903	259	269
	5 000	5 970	119	110	21.4	9.0	5.5	1220	1280	330	344
	7 000	8 310	134	124	23.2	10.0	6.1	1530	1600	387	405
	10 000	11 800	153	142	25.4	11.1	6.8	1940	2040	458	482

	15 000	17 500	177	164	28.1	12.6	7.6	2550	2690	555	586
	20 000	23 300	196	183	30.2	13.8	8.3	3100	3270	636	673
	30 000	34 600	227	212	33.4	15.6	9.4	4070	4310	771	819
	40 000	45 900	252	236	35.9	17.1	10.2	4950	5240	880	940
Gas carrier	1 000	2 480	71	66	11.7	5.7	4.6	390	465	133	150
	2 000	4 560	88	82	14.3	7.2	5.7	597	707	195	219
	3 000	6 530	100	93	16.1	8.4	6.4	765	903	244	273
	5 000	10 200	117	109	18.8	10.0	7.4	1050	1230	323	361
	7 000	13 800	129	121	20.8	11.3	8.1	1290	1510	389	434
	10 000	18 900	144	136	23.1	12.9	9.0	1600	1870	474	527
	15 000	27 000	164	154	26.0	14.9	10.1	2050	2390	593	658
	20 000	34 800	179	169	28.4	16.5	11.0	2450	2840	696	770
	30 000	49 700	203	192	32.0	19.0	12.3	3140	3630	870	961
	50 000	78 000	237	226	37.2	22.8	12.3	4290	4940	1150	1270
	70 000	105 000	263	251	41.2	25.7	12.3	5270	6050	1390	1530
	100 000	144 000	294	281	45.8	29.2	12.3	6560	7510	1690	1860

Confidence limit: 90%

Type	DWT: t	Displacement: t	L_{OA}: m	L_{BP}: m	B: m	D: m	Max. draught: m	Wind lateral area: m^2		Wind front area: m^2	
								Full-load condition	Ballast condition	Full-load condition	Ballast condition
General cargo ship	1 000	1 790	72	66	11.4	6.5	4.2	333	394	67	98
	2 000	3 440	89	83	13.8	8.0	5.3	511	623	107	149
	3 000	5 040	101	94	15.4	9.0	6.0	656	815	140	192
	5 000	8 150	118	111	17.8	10.5	7.1	899	1143	197	262
	7 000	11 200	131	123	19.5	11.6	8.0	1106	1430	247	323
	10 000	15 700	146	138	21.5	12.9	8.9	1380	1810	313	402
	15 000	22 900	166	157	24.1	14.6	10.2	1770	2370	410	516
	20 000	30 100	181	172	26.1	15.9	11.2	2110	2860	496	615
	30 000	44 000	205	195	29.2	18.0	12.8	2710	3740	650	789
	40 000	57 700	224	214	31.6	19.6	14.0	3240	4530	788	942
Bulk carrier[a]	5 000	7 090	111	103	16.0	8.7	6.4	763	970	237	259
	7 000	9 740	123	114	17.7	9.7	7.1	880	1160	268	303
	10 000	13 700	136	127	19.8	10.8	7.9	1020	1400	306	358
	15 000	20 000	152	143	22.5	12.1	8.9	1220	1740	356	433
	20 000	26 300	165	156	24.6	13.2	9.6	1370	2030	395	495
	30 000	38 600	186	176	27.9	14.9	10.9	1630	2510	459	599
	50 000	62 600	215	206	32.3	17.4	12.7	2030	3290	555	761
	70 000	86 000	236	227	32.3	19.3	14.0	2340	3930	628	892
	100 000	121 000	262	253	40.5	21.4	15.5	2720	4750	717	1050
	150 000	177 000	294	286	45.9	24.2	17.5	3240	5890	833	1280
	200 000	232 000	319	311	50.2	26.4	19.1	3660	6860	926	1460
	250 000	287 000	340	333	53.8	28.2	20.4	4020	7720	1006	1620
Container ship[b]	7 000	11 200	129	121	21.1	10.3	7.4	1600	1830	358	492
	10 000	15 800	148	139	23.2	11.9	8.3	2060	2290	445	594
	15 000	23 200	174	164	25.9	14.0	9.3	2740	2950	570	735
	20 000	30 500	195	184	28.0	15.7	10.2	3360	3530	679	855

	25 000	37 800	213	201	29.8	17.1	10.9	3930	4060	778	960
	30 000	45 000	229	216	31.3	18.4	11.5	4460	4550	869	1060
	40 000	59 100	256	243	32.3	20.6	12.6	5460	5450	1040	1232
	50 000	73 200	280	266	32.3	22.5	13.5	6390	6260	1190	1380
	60 000	87 100	301	286	37.8	24.2	14.2	7260	7020	1330	1520
Oil tanker	1 000	1 710	64	61	10.6	4.7	4.2	210	293	94	90
	2 000	3 320	80	76	13.1	6.0	5.2	309	442	130	132
	3 000	4 890	91	87	14.9	6.9	5.8	388	562	158	165
	5 000	7 970	107	102	17.5	8.2	6.8	516	760	201	219
	7 000	11 000	119	114	19.4	9.3	7.5	623	928	235	263
	10 000	15 500	133	128	21.6	10.5	8.3	760	1150	279	320
	15 000	22 800	151	146	24.5	12.1	9.3	950	1460	338	401
	20 000	30 000	165	160	26.8	13.4	10.1	1120	1730	387	469
	30 000	44 200	188	182	30.4	15.4	11.4	1400	2190	469	587
	50 000	72 000	220	215	32.3	18.5	13.2	1870	2970	598	777
	70 000	99 200	245	239	39.6	20.8	14.6	2250	3620	701	935
	100 000	140 000	274	268	44.2	23.5	16.2	2750	4470	830	1140
	150 000	206 000	312	306	50.2	27.1	18.2	3450	5690	1010	1420
	200 000	271 000	341	336	54.8	30.0	19.8	4050	6740	1150	1670
	300 000	399 000	388	382	62.2	34.6	22.3	5080	8570	1400	2080
Ro/ro ship	1 000	2 400	79	72	14.8	7.3	3.8	1080	1130	248	248
	2 000	4 560	102	94	17.5	9.9	4.9	1480	1550	335	344
	3 000	6 630	118	109	19.3	11.8	5.8	1790	1860	400	416
	5 000	10 620	143	131	21.9	14.8	7.0	2270	2350	499	530
	7 000	14 500	161	149	23.8	17.1	7.9	2650	2740	578	621
	10 000	20 200	184	170	26.0	20.0	9.0	3130	3230	675	736
	15 000	29 300	213	197	28.7	23.9	10.5	3780	3880	805	891
	20 000	38 200	237	219	30.8	27.2	11.6	4320	4430	912	1020
	30 000	55 600	275	255	34.0	32.5	13.5	5210	5330	1090	1240

[a]Full-load condition of wind lateral/front areas of log carrier do not include the areas of logs on deck

[b]Full-load condition of wind lateral/front areas of container ships include the areas of containers on deck

Type	Gross tonnage: t	Displacement: t	L_{OA}: m	L_{BP}: m	*B*: m	*D*: m	Max. draught: m	Wind lateral area: m^2		Wind front area: m^2	
								Full-load condition	Ballast condition	Full-load condition	Ballast condition
Passenger ship	1 000	1 220	68	65	12.8	5.7	3.3	502	518	207	218
	2 000	2 260	86	82	15.2	7.4	4.4	804	822	278	292
	3 000	3 240	99	94	16.9	8.7	5.1	1060	1080	330	346
	5 000	5 110	119	111	19.2	10.5	6.3	1500	1510	410	428
	7 000	6 900	133	124	20.8	12.0	7.2	1880	1890	473	493
	10 000	9 480	151	139	22.8	13.7	8.2	2400	2400	551	573
	15 000	13 600	173	159	25.2	16.0	8.4	3160	3150	654	679
	20 000	17 600	192	175	27.1	17.9	8.4	3850	3810	740	766
	30 000	25 200	220	200	30.0	20.9	8.4	5070	4990	879	907
	50 000	39 700	263	237	34.1	25.4	8.4	7170	7020	1090	1120
	70 000	53 600	296	265	37.1	28.9	8.4	9020	8780	1260	1290
Ferry	1 000	1 790	74	68	15.9	6.3	4.3	434	451	167	171
	2 000	3 540	95	87	18.9	7.8	5.3	693	722	232	239
	3 000	5 260	110	101	20.9	8.8	5.9	911	951	281	291
	5 000	8 690	133	122	23.8	10.4	6.9	1290	1350	358	372
	7 000	12 100	150	139	25.9	11.5	7.6	1610	1690	420	438
	10 000	17 100	170	158	28.3	12.8	8.4	2050	2150	497	521

	15 000	25 500	197	184	31.3	14.5	9.5	2700	2840	602	633
	20 000	33 800	219	204	33.6	15.9	10.3	3270	3450	690	728
	30 000	50 300	253	237	37.2	18.0	11.7	4300	4540	836	886
	40 000	66 800	281	264	39.9	19.7	12.7	5230	5520	960	1020
Gas carrier	1 000	2 740	74	68	12.2	6.0	5.0	431	493	144	160
	2 000	5 050	91	85	14.9	7.7	6.1	659	750	211	233
	3 000	7 230	104	97	16.8	8.9	6.9	845	958	265	291
	5 000	11 300	121	114	19.6	10.7	8.0	1160	1300	351	385
	7 000	15 300	135	126	21.7	12.0	8.8	1420	1600	423	463
	10 000	20 900	150	141	24.1	13.7	9.8	1770	1980	515	563
	15 000	29 900	170	161	27.2	15.8	11.0	2260	2530	645	702
	20 000	38 500	186	176	29.6	17.5	11.9	2700	3010	756	822
	30 000	55 100	211	200	33.4	20.2	12.8	3460	3850	946	1026
	50 000	86 400	247	235	38.9	24.3	12.8	4740	5240	1250	1360
	70 000	116 000	274	262	42.9	27.4	12.8	5820	6420	1510	1630
	100 000	159 000	306	293	47.7	31.1	12.8	7240	7960	1840	1980

Confidence limit: 95%

Type	DWT: t	Displacement: t	L_{OA}: m	L_{BP}: m	B: m	D: m	Max. draught: m	Wind lateral area: m^2		Wind front area: m^2	
								Full-load condition	Ballast condition	Full-load condition	Ballast condition
General cargo ship	1 000	1 850	74	69	11.7	6.9	4.4	372	428	70	101
	2 000	3 560	92	86	14.2	8.5	5.5	570	678	111	154
	3 000	5 210	104	98	15.9	9.6	6.3	732	887	146	198
	5 000	8 440	122	115	18.3	11.2	7.5	1003	1243	205	271
	7 000	11 600	136	128	20.1	12.4	8.3	1234	1550	256	333
	10 000	16 200	151	143	22.2	13.8	9.3	1540	1970	325	414
	15 000	23 700	172	163	24.8	15.6	10.7	1970	2570	426	532
	20 000	31 100	188	179	26.9	17.0	11.7	2360	3110	516	634
	30 000	45 600	213	203	30.1	19.2	13.4	3030	4070	675	814
	40 000	59 800	233	223	32.6	20.9	14.7	3610	4930	818	971
Bulk carrier[a]	5 000	7 190	113	105	16.3	8.8	6.5	811	1010	247	267
	7 000	9 880	124	116	18.1	9.8	7.2	936	1210	280	312
	10 000	13 800	138	129	20.2	10.9	8.0	1090	1460	319	369
	15 000	20 300	155	146	22.9	12.3	9.0	1290	1810	371	447
	20 000	26 700	168	159	25.0	13.4	9.8	1460	2110	412	511
	30 000	39 100	188	179	28.4	15.1	11.0	1740	2610	479	618
	50 000	63 500	218	209	32.3	17.6	12.8	2160	3420	578	786
	70 000	87 200	240	231	32.3	19.5	14.2	2490	4090	655	920
	100 000	122 000	266	257	41.2	21.6	15.8	2890	4940	747	1090
	150 000	179 000	298	290	46.8	24.4	17.8	3440	6120	868	1320
	200 000	236 000	324	316	51.1	26.6	19.4	3890	7130	965	1510
	250 000	291 000	345	338	54.8	28.5	20.7	4270	8020	1048	1670
Container ship[b]	7 000	11 500	133	125	21.5	10.6	7.6	1700	2000	377	524
	10 000	16 200	153	144	23.7	12.3	8.4	2180	2490	468	632
	15 000	23 900	179	169	26.4	14.4	9.5	2900	3210	599	782
	20 000	31 400	201	190	28.6	16.1	10.4	3550	3850	714	910

	25 000	38 800	219	208	30.4	17.6	11.1	4150	4420	818	1020
	30 000	46 200	236	223	31.9	18.9	11.8	4720	4950	914	1130
	40 000	60 800	264	251	32.3	21.2	12.8	5780	5930	1090	1310
	50 000	75 200	288	274	32.3	23.2	13.7	6760	6820	1250	1470
	60 000	89 400	310	295	38.5	24.9	14.5	7680	7640	1390	1620
Oil tanker	1 000	1 800	66	63	10.9	4.8	4.4	223	302	99	93
	2 000	3 480	82	78	13.5	6.1	5.3	328	455	137	137
	3 000	5 130	93	89	15.3	7.1	6.0	412	578	166	171
	5 000	8 360	109	105	17.9	8.5	7.0	548	782	211	226
	7 000	11 500	122	118	19.9	9.5	7.7	661	954	248	272
	10 000	16 200	136	132	22.2	10.8	8.5	806	1180	294	332
	15 000	23 900	155	150	25.2	12.4	9.6	1010	1500	356	414
	20 000	31 400	169	165	27.5	13.7	10.4	1190	1770	408	486
	30 000	46 300	192	188	31.2	15.8	11.7	1490	2260	494	607
	50 000	75 500	226	222	32.3	19.0	13.6	1980	3050	630	804
	70 000	104 000	251	247	40.6	21.3	15.0	2390	3720	739	968
	100 000	146 000	281	277	45.3	24.2	16.7	2920	4600	875	1180
	150 000	216 000	320	316	51.4	27.9	18.8	3660	5850	1060	1470
	200 000	284 000	350	346	56.2	30.8	20.4	4300	6930	1210	1730
	300 000	418 000	398	395	63.7	35.5	23.0	5390	8810	1470	2160
Ro/ro ship	1 000	2 540	83	76	15.2	8.0	4.0	1210	1240	258	257
	2 000	4 820	107	99	18.1	10.9	5.2	1680	1700	348	357
	3 000	7 010	125	115	20.0	13.0	6.1	2020	2050	416	432
	5 000	11 200	150	139	22.6	16.3	7.3	2560	2590	519	551
	7 000	15 300	170	157	24.6	18.9	8.3	3000	3010	601	645
	10 000	21 300	194	179	26.8	22.1	9.5	3540	3550	702	764
	15 000	31 000	225	208	29.6	26.4	11.0	4270	4270	837	925
	20 000	40 400	250	231	31.8	30.0	12.3	4880	4860	949	1060
	30 000	58 800	290	269	35.1	35.8	14.3	5890	5850	1130	1280

[a]Full-load condition of wind lateral/front areas of log carrier do not include the areas of logs on deck
[b]Full-load condition of wind lateral/front areas of container ships include the areas of containers on deck

Type	Gross tonnage: t	Displacement: t	L_{OA}: m	L_{BP}: m	B: m	D: m	Max. draught: m	Wind lateral area: m^2		Wind front area: m^2	
								Full-load condition	Ballast condition	Full-load condition	Ballast condition
Passenger ship	1 000	1 350	70	69	13.2	6.3	3.9	525	539	219	232
	2 000	2 500	90	86	15.7	8.2	5.1	842	855	295	310
	3 000	3 590	103	99	17.4	9.5	6.0	1110	1120	350	368
	5 000	5 650	123	117	19.8	11.6	7.3	1570	1570	435	456
	7 000	7 630	138	131	21.5	13.2	8.4	1970	1970	502	525
	10 000	10 500	156	147	23.5	15.1	8.7	2510	2500	585	609
	15 000	15 000	180	168	26.0	17.6	8.7	3310	3270	695	722
	20 000	19 400	199	185	28.0	19.7	8.7	4030	3960	785	815
	30 000	27 900	229	211	31.0	23.0	8.7	5310	5190	933	966
	50 000	44 000	273	250	35.2	27.9	8.7	7510	7300	1160	1200
	70 000	59 300	307	279	38.3	31.8	8.7	9440	9130	1340	1380
Ferry	1 000	2 240	79	72	17.0	6.9	4.9	449	466	175	179
	2 000	4 430	102	93	20.2	8.5	6.0	716	746	243	250
	3 000	6 590	118	108	22.3	9.6	6.7	941	982	295	305
	5 000	10 900	142	131	25.4	11.3	7.8	1330	1390	376	390
	7 000	15 100	160	148	27.6	12.5	8.7	1670	1750	441	459
	10 000	21 500	182	169	30.1	14.0	9.6	2120	2220	522	545

	15 000	31 900	210	196	33.4	15.8	10.8	2790	2930	632	664
	20 000	42 300	233	218	35.8	17.3	11.8	3380	3560	724	763
	30 000	63 000	270	253	39.6	19.6	13.3	4450	4690	877	928
	40 000	83 500	300	282	42.6	21.5	14.5	5400	5700	1010	1070
Gas carrier	1 000	2 910	75	70	12.5	6.2	5.3	457	511	151	166
	2 000	5 370	94	87	15.3	8.0	6.4	699	777	222	243
	3 000	7 680	106	99	17.3	9.2	7.2	896	992	278	303
	5 000	12 000	124	116	20.1	11.1	8.4	1230	1350	369	401
	7 000	16 200	138	129	22.2	12.5	9.2	1510	1660	444	481
	10 000	22 200	154	145	24.7	14.2	10.2	1870	2050	541	585
	15 000	31 700	174	165	27.9	16.4	11.5	2400	2620	677	730
	20 000	40 900	190	180	30.4	18.2	12.5	2870	3120	794	855
	30 000	58 500	216	205	34.2	21.0	13.1	3670	3990	994	1067
	50 000	91 800	253	241	39.9	25.2	13.1	5030	5430	1320	1410
	70 000	124 000	280	268	44.0	28.4	13.1	6180	6650	1590	1700
	100 000	169 000	313	300	49.0	32.3	13.1	7680	8250	1930	2060

21.4. Recommendations

As can be seen from the different tables in Section 21.2, the beam, the overall length and the draft may vary, depending on the country of origin of the ship and the ship's construction. The ship's dimensions may vary by as much as ±10–15% between different sources. For important calculations and if the actual design ship's dimensions are not known, use of the information given in Section 21.3 from Japan is recommended.

REFERENCES AND FURTHER READING

Akakura Y and Takahashi H (1998) *Technical Note of the Ports and Harbour Research Institute*, No. 911. Ports and Harbour Research Institute, Ministry of Transport, Tokyo.

BSI (British Standards Institution) (2000) BS 6349-1: 2000. Maritime structures. Code of practice for general criteria. BSI, London.

EAU (Empfehlungen des Arbeitsausschusses für Ufereinfassungen) (2004) *Recommendations of the Committee for Waterfront Structures, Harbours and Waterways*, 8th English edn. Ernst, Berlin.

Ministerio de Obras Públicas y Transportes (1990) ROM 0.2-90. Maritime works recommendations. Actions in the design of maritime and harbour works. Ministerio de Obras Públicas y Transportes, Madrid.

Ministerio de Obras Públicas y Transportes (2007) ROM 3.1-99. Recommendations for the design of the maritime configuration of ports, approach channels and harbour basins. English version. Puertos del Estado, Madrid.

PIANC (International Navigation Association) (2002) *Guidelines for the Design of Fender System. Report of Working Group 33*. PIANC, Brussels.

PIANC (2012) *The Safety Aspect Affecting the Berthing Operations of Tankers to Oil and Gas Terminals. Report of Working Group 55*. PIANC, Brussels.

PIANC (2014) *Design Principles for Container Terminals in Small and Medium Ports. Report of Working Group 135*. PIANC, Brussels.

Port and Harbour Research Institute (1999) *Technical Standards for Port and Harbour Facilities in Japan*. Port and Harbour Research Institute, Ministry of Transport, Tokyo.

Port Designer's Handbook
ISBN 978-0-7277-6004-3

http://dx.doi.org/10.1680/pdh.60043.549

Chapter 22
Definitions

The following main definitions, some of which are illustrated in Figure 22.1, are frequently used in port design.

Figure 22.1 Definitions

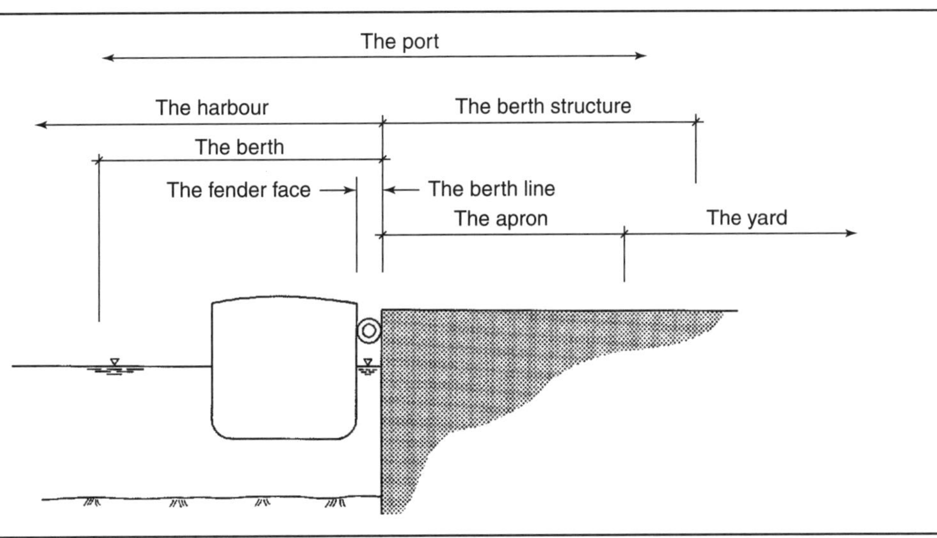

There are currently numerous dictionaries and lexicons available covering almost all facets of the port, environment and navigation fields. In this book the following main terms and definitions, which are considered to be presently in common use in the field, especially by the International Navigation Association (PIANC), have been used.

Accretion	Natural accretion – the build-up of land, solely by the action of the forces of nature, on a beach by deposition of water or airborne material. Artificial accretion – similar build-up of land by reason of an act of man, such as the accretion formed by a groin, breakwater or beach fill deposited by mechanical/hydraulic means.
Active tugboat escort (ATE)	Escort with at least one tugboat connected to the ship with at least one towing line. The tug will only follow the ship, which will be going ahead under its own power, and only assist the ship in situations when necessary.
Actual time of arrival (ATA)	The time that the ship passes or arrives at a reference point (pilot station, terminal, etc.).

Actual time of departure (ATD)	The time that the ship leaves the berth.
Aerobic bacteria	Bacteria that oxidise a substrate (feed) by oxygen respiration and live on the energy generated in the process. Aerobic bacteria are the opposite of anaerobic bacteria, which need no oxygen gas. Aerobic bacteria play an important role in the natural purification process of water bodies, the activated sludge process, the sprinkling filter method and other water quality preservation processes.
Air draught (AD)	The minimum free height from sea level to obstacles above sea level outside the berth or in the approach channel at high tide.
Anaerobic	Oxidation occurring in the absence of free or dissolved oxygen often facilitated by specific bacterial strains, such as methane-producing bacteria present during the anaerobic digestion of sewage sludge.
Anaerobic bacteria	Bacteria that grow without requiring dissolved oxygen. The decomposition of organic substances by anaerobic bacteria produces hydrogen sulphide, ammonia, methane and low molecular weight fatty acids. The process of digestion by anaerobic bacteria is used for the treatment of human waste and sewage sludge.
Anchorage area	That proportion of a harbour area or designated areas outside of the harbour in which ships are permitted to lie at anchor.
Approach route (AR)	The defined route where the ship is entering the terminal from the arrival point.
Apron	The area between the berth line and the transit shed or the storage area for the loading and unloading of cargo.
Arrival point (AP)	The appointed place where the pilot or the first tugboat meets the ship. The AP should be clearly defined by the terminal, including the longitude and latitude.
Artesian water level	A hydrostatic pressure level higher than the ground level.
Astronomical tide	Tide due to the gravitational attraction of the sun, moon and other astronomical bodies.
ATT	Admiralty Tide Table.
Ballast water	Water taken on board a vessel to ensure stability while navigating in an unladen or partially laden state.
Basin	Tidal – a dock or basin without water gates, in which the water level changes. (a) Turning – an area of water or enlargement of a channel used for the turning around of ships. (b) Wet – an area of impounded water within which ships can remain afloat at a uniform level, independent of external tidal action.
Bathymetry	The physical configuration of the seabed, the measurement of depths of water in the ocean, etc., and also information derived from such measurement.
Belting	Substantially horizontal continuous narrow rigid steel fender along the ship's side above the water line.

Term	Definition
Berth	A place where the ship can moor. In the case of a quay or jetty structure, it will include the section of the structure where labour, equipment and cargo move to and from the ship.
Berth departure time (BDT)	The total time from the ship's departure from the berth to arrival at the AP.
Berth draught (BD)	The minimum depth outside the berth and under the ship at the shallowest point at low tide.
Berth line	The line along the outermost part of the superstructure. Removable equipment such as fenders will be on the outside of the berth line.
Berth structure	An artificial landing place for the loading and unloading of ships. Berth structures can be subdivided into: (a) Quay or wharf: a berth structure that generally is aligned parallel to the shoreline. (b) Jetty or pier: a berth structure that projects out into the water from the shore. (c) Dolphin: a berth structure for mooring the ship on the open sea.
Berthing arrival time (BAT)	The transit time from the ship's arrival at the AP to when the ship is moored at the berth.
Berthing operation arrival (BOA)	The time period of arrival of the non-moored ship near the berth, when the speed of the ship is so low that the ship is dependent on external aids such as tugs for steering.
Best available techniques (BAT)	The latest stage in the development of activities, processes and their methods of operation that indicate the practical suitability of particular techniques as the basis of preventing or minimising emissions to the environment.
Bollard	A vertical post to which the eye of a mooring line can be attached.
Breakwater	A rubble mound and/or a concrete structure to protect the harbour area from wave action.
Breakwater berth	A berth structure on the lee side of the breakwater.
Breasting dolphin	A dolphin structure designed to take the impact from a berthing ship.
Bulkhead	A structure for retaining or to prevent earth or fill from sliding into water.
Bulkhead line	The farthest line offshore to which a fill or solid structure may be constructed.
Capital dredging	Dredging carried out to create new channels, etc., as distinct from maintenance dredging. This is also called new work dredging.
Channel	A dredged waterway through which ships proceed from the sea to the berth or from one berth to another within the harbour.
Chloride ion	Refers to ionised chloride (Cl^-), which forms ionised compounds with various metals. The chloride ion is also found in natural water, measuring several ppm in river water and 1.9% in seawater.
Chock	A guide for a mooring line, which enables the line to be passed through a ship's bulwark or other barrier.

Coastal berth	A berth fully exposure to wind, waves and current.
Compressed natural gas (CNG)	Natural gas pressurised and stored at pressures up to 25 MPa. Typically, it is the same composition as the local 'pipeline' gas.
Confined dredged material placement	Placement of dredged material within dikes near shore or upland confined placement facilities that enclose and isolate the dredged material from adjacent waters during placement. Confined dredged material placement does not refer to sub-aqueous capping or contained aquatic dredged material placement.
Containment (retention) basin	Containment constructed of dikes for the purpose of retaining dredge spoil until much of the suspended material has settled out when the water itself is released.
Contaminated sediment or contaminated dredged material	Sediment or material that has an unacceptable level of contaminant(s) that have been demonstrated to cause an unacceptable adverse effect on human health or the environment.
Convention on the Prevention of Marine Pollution by the Dumping of Wastes and Other Matter	This Convention was prepared at the United Nations Conference on the Human Environment (Stockholm) in June 1972, adopted on 13 November in the same year in London and took effect on 30 August 1975. The Convention mainly contained (1) an absolute ban on the ocean dumping of organic halogen compounds and mercury; (2) the issuance of a special licence for the ocean dumping of lead, copper and zinc; and (3) the issuance of a general licence for the ocean dumping of substances other than the above. Upon the issuance of either special or general licences, the following must be taken into consideration in establishing criteria: the shape and characteristics of the substances to be dumped; the location, water depth, and distance from land of dumping sites; and pre-dumping treatment methods.
Crude oil	Also called petroleum. This is a naturally occurring liquid consisting of a complex mixture of hydrocarbons. In its naturally occurring form it may contain other non-metallic elements such as sulphur and oxygen. It occurs in porous rock formations in the upper strata of some areas of the Earth's crust. There is also petroleum in oil sands (tar sands).
Deadweight (DWT)	The carrying capacity of a ship, including cargo, bunkers and stores, in metric tonnes.
Debris	Wastes or remains of something broken down or destroyed. Also, any oversized material adversely affecting the hydraulic transport system.
Density	The total weight of all materials/unit of volume.
Deposit (geology)	The matter made of crushed pieces of disintegrated rocks, carcasses, volcanic ejecta, etc., and physically and chemically settled and accumulated in water or on the land. In other words, it is clay, sand or gravel formed by weathering or erosion and transported by rivers or sea currents to sink and settle on the bottom of a sea or lake.
Design life	The period of time from the beginning of the construction of a structure until it is dismantled, put out of service or used for another purpose.

Diffraction of water waves When a part of a train of waves is interrupted by a barrier (e.g. a breakwater), the effect of diffraction is manifested by the propagation of waves into the sheltered region within the geometric shadow of the barrier.

Displacement The mass of water in tonnes displaced by a ship at a given draught.

Dock A harbour basin where the basin is cut off from the tides by dock gates.

Dolphin An isolated piled or gravity structure used either to manoeuvre a ship or to facilitate holding it in position at its berth.

Dredged material Material excavated from inland or ocean waters. Dredged material refers to material that has been dredged usually from the bed of a water body, while 'sediment' refers to material in a water body prior to the dredging process.

Dredging Refers to loosening and lifting earth and sand from the bottom of water bodies. Dredging is often carried out to widen the stream of a river, deepen a harbour or navigational channel, or collect earth and sand for landfill; it is also carried out to remove contaminated bottom deposit or sludge to improve water quality.

Emergency towing arrangement Equipment and fittings provided at both ends of a tanker to facilitate towing in the case of an emergency.

Environmental assessment (EA) A written environmental analysis that is conducted to determine whether a proposed undertaking would significantly affect the environment. The conducting of an environmental assessment for a proposed project is usually a mandatory requirement of various jurisdictional authorities. (See also *environmental impact assessment*, EIA.)

Environmental impact assessment (EIA) A system involving the investigation, estimation and evaluation of the effect of a project or activity on the environment, and is usually conducted by the proponent for the proposed undertaking in the process of planning. (See also *environmental assessment*, EA.)

Environmental management plan A plan that is required to reflect socioeconomic trends and the general public needs, with air, water, forests, soil and other natural elements viewed as a whole, to preserve and utilise these resources appropriately to prevent destruction and to create a comfortable environment. Governments play an important role in promoting such environmental policies. A regional environmental management plan conducted by a local government aims at identifying what an ideal regional environment is in terms of regional nature and social conditions and community intentions. It also aims at integrating pollution control, environmental preservation and improvement, and implementing relevant activities according to well-planned procedures and schedules.

Regional environmental planning aims at developing conditions necessary for the ideal environment, through mediation for organisations and communities with regard to environmental preservation and utilisation from a long-term perspective.

	A plan that describes specific conservation and protection actions that will be undertaken during project planning, construction, operation and maintenance to lessen the effects of the project on the environment and to ensure that sustainable development is achieved; it includes real-time and retroactive monitoring of project effects.
Erosion	A natural physical process where wind, waves, rain or surface water run-off loosen and remove soil particles from land surfaces that are often deposited in rivers and lakes.
Estimated time of arrival (ETA)	The calculated time given by the ship at which the ship expects to arrive at a certain point, mostly the pilot station.
Estimated time of departure (ETD)	The calculated time given by the ship at which the ship expects to be ready for sailing.
Estuarial berth	A berth with more wave exposure than tidal basins and with the maximum tidal range and current.
Fairway	Open water of navigable depth.
Flash point	The lowest temperature at which a liquid will evaporate enough fluid to form a combustible concentration of gas.
Geotextile	A synthetic textile material used in geotechnical structures.
Gravity wall	A retaining wall of heavy cross-section that resists the horizontal loads by its own deadweight.
Groin (Groyne)	A shore protection structure. Usually built perpendicular to a coastline to retard littoral transport of sedimentary materials.
Grounding area	A selected area with the best available bottom conditions for deliberately grounding the tanker in order to prevent a major disaster. Grounding areas will only be considered for use when no other alternative remains. They are nominated by the appropriate body of the country concerned.
Groundwater	All subsurface water that fills voids between highly permeable ground strata comprised of sand, gravel, broken rocks, porous rocks, etc., and moves under the influence of gravity. On the earth, besides seawater and Antarctic water, far more groundwater exists than water in rivers, lakes and marshes.
Harbour	A protected water area to provide safe and suitable accommodation for ships for the transfer of cargo, refuelling, repairs, etc. Harbours may be subdivided into: (a) Natural harbours: harbours protected from storms and waves by the natural configuration of the land. (b) Semi-natural harbours: harbours with both natural and artificial protection. (c) Artificial harbours: harbours protected from the effect of waves by means of breakwaters, or harbours created by dredging.
Hardness	The hardness of water is indicated by the content of dissolved calcium and magnesium salts. Calcium and magnesium salts that are transformed to

	insoluble salts by boiling denote temporary hardness, while calcium and magnesium salts that do not settle when boiled denote permanent hardness. The sum of temporary and permanent hardness is called the total hardness.
Hawser	A synthetic or natural fibre rope or wire rope used for mooring or towing.
Hydrocarbons	Compounds found in fossil fuels that contain carbon and hydrogen and may be carcinogenic.
Hydrography	The description and study of seas, lakes, etc.
IMO	International Maritime Organisation.
Impounded basins	Lock basins with an approximately constant water level and no waves or current.
Jetty	A berth structure that projects out into the water from the shore, or a berth structure at some distance from the shoreline.
Jetty head	A platform at the seaward side of a jetty.
Keel top distance (KTD)	The distance from the lowest point on the keel to the highest point on the entire ship.
Liquefied natural gas (LNG)	Natural gas that has been converted (chilled) to liquid form for ease of storage or transport. When natural gas is cooled to a temperature of approximately −160°C at atmospheric pressure it condenses to LNG. The volume of this liquid is about 1/600 of the volume of natural gas. LNG is only about 45% the density of water. LNG is odourless, colourless, non-corrosive and non-toxic.
Liquid petroleum gas (LPG)	Primarily propane, mixes that are primarily butane, and mixes including propane, propylene, n-butane, butylenes and iso-butane. In some countries, LPG is composed primarily of propane (up to 95%) and smaller quantities of butane. LPG can be stored as a liquid in tanks by applying pressure alone.
Loading and unloading window (LW)	A period of time set by the terminal where the berth is cleared for berthing of the tanker. The loading window should at least be 24 h longer than the time of loading.
Lo/lo	Lift-on/lift-off vessels that are loaded and discharged through the hatchways.
Marginal berth	A berth structure parallel to the shore.
Marine pollution	Among international efforts to control marine pollution, the MARPOL 73/78 Convention governs pollution by ships while the Convention on the Prevention of Marine Pollution by the Dumping of Wastes and Other Matter (Dumping Convention) governs the ocean dumping and incineration of waste materials.
Messenger line	A light line attached to the end of a main mooring line and used to assist in heaving the mooring to the shore or to another ship.
Mineral oils	Residues of natural fossil fuels.
Mooring dolphin	A dolphin structure equipped with bollards or quick-release hooks for mooring the ship.

Mooring operations (MOP)	An overall term for all activities involved in berthing and unberthing a ship.
Navigable waters	Traditionally, waters sufficiently deep and wide for navigation by all, or specified sizes of, vessels.
Oil boom	A device comprised of a large float with a suspended screen attached underneath. It is used to contain spilled oil and prevent it spreading further on the sea surface or to protect a given sea area, such as a fish farm, from pollution.
Oil separator	A device that isolates oil from water by flotation treatment.
Parts per million (ppm)	A weight/volume or weight/weight measurement used in contaminant analysis. It is interchangeable with mg/l or mg/kg in the case of liquids. Chemical dosages are often referred to in parts per million (e.g. 100 ppm of polymer). 100 ppm = 0.01%.
Passive tugboat escort (PTE)	Escort with at least one tugboat that is following close to the ship. In order to assist the ship, which will be going ahead on its own engine, towing lines have to be put in place before assistance can take place.
pH	A measure of the acidity or basicity of a solution (i.e. the negative of the logarithm of the hydrogen ion concentration); 'hydrogen ion exponent', a unit for measuring hydrogen ion concentrations. A scale (0–14) represents the acidity or alkalinity of an aqueous solution. Low pH values indicate acidity; high values indicate alkalinity. The mid-point of the scale, 7, is neutral. Substances in an aqueous solution ionise to various extents, giving different concentrations of H^+ and OH^- ions. Strong acids have an excess H^+ ions and a pH of 1–3 (HCl, pH = 1). Strong bases have an excess OH^- ions and a pH of 11–13 (NaOH, pH = 12).
Pier	A berth structure projecting out from the shore line.
Port	A sheltered place where the ship may receive or discharge cargo. It includes the harbour with its approach channels and anchorage places. Ports may be subdivided into: (a) Ocean ports: ports located on coasts, tidal estuaries or river mouths where the port can be reached directly by ocean-going ships. (b) Inland waterway ports: ports located on navigable rivers, channels and lakes.
Port clearance (PC)	Clearance that should be given by the port authorities before a ship is allowed to continue to the terminal from the AP.
Port side	The left side of the ship looking towards the bow.
ppb	Parts per billion, commonly considered equivalent to μg/l or μg/kg.
ppt	Parts per trillion, commonly considered equivalent to ng/l or ng/kg.
Precipitation	Rainfall, snowfall or any condensate.
Quay	A berth structure parallel to the shore line.

Refraction of water waves	The process by which the direction of incoming waves in shallow water is altered due to the contours of the seabed.
Relieving platform	A platform or deck structure built below the top deck level and supported on bearing piles. The main function of the platform is to reduce the lateral soil pressures over the upper portion of the sheet wall.
Requested time of arrival (RTA)	The time given by parties ashore (port authority, pilot, terminal, etc.) when the ship has to appear at a given point.
Rip-rap	A wall or foundation made of broken stones thrown together irregularly or loosely in water or on a sea bottom.
Risk assessment	In dealing with environmental problems, a certain degree of uncertainty is unavoidable, despite advances made thus far in the scientific elucidation of negative impact (or risk) on human beings and the natural environment. However, irreversible damage could occur if necessary measures were delayed until complete scientific elucidation is achieved. In such a situation, an integrated policy-making approach of two processes, namely scientifically estimating and evaluating the negative impact of human activities on humans and the environment (risk assessment) and deciding and executing rational policies for risk mitigation based on risk assessment (risk management), is becoming established. International agreement made on the protection of the ozone layer is a precedent of this approach.
Ro/ro	Roll-on/roll-off ships that are loaded and discharged by way of ramps.
Route draught (RD)	The minimum depth of the call channel at the shallowest point at low tide.
Sea island	A berth structure with no direct connection to the shore, at which ships can berth. Berthing can take place on either one or both sides of the structure.
Sediment	Materials such as sand, silt or clay suspended in or settled on the sea bottom. Solid fragmented material originating from weathering or rocks or by other processes and transported or deposited by air, water or ice, or that accumulated by other processes such as chemical precipitation from solution or secretion by organisms. The term is usually applied to material held in suspension in water or recently deposited from suspension, and to all kinds of deposits, essentially of unconsolidated materials.
Sheet wall	A retaining wall that resists loading.
Silt	Fine particulate organic and inorganic material; strictly confined to material with an average particle size intermediate between those of sands and clays, but often taken to include all material finer than sands.
Silt curtain	A curtain or screen suspended in the water to prevent silt from escaping from an aquatic construction site.
Starboard side	The right side of the ship looking towards the bow.
Tail	A short length of synthetic rope attached to the end of a wire mooring line to provide increased elasticity and also ease of handling.

Terminal safety zone (TSZ)	A specified distance from the terminal where there are restrictions on activities. Restrictions are not only for terminal activities but also for general public activities.
Terms of reference (TOR)	A statement of the specific work to be done under a consulting agreement or similar contract.
TEU	Twenty-foot equivalent unit (e.g. a standard 40 ft container = 2 TEUs).
Tidal basin	A basin with a greater than usual range of water levels.
Turning basin (TB)	A specified area for turning the ship in order to go alongside the berth before mooring or after departing.
ULCC	Ultra-large crude carrier – used for a carrier with a deadweight greater than 400 000 DWT.
VLCC	Very large crude carrier – used for a carrier with a deadweight between 140 000 and 400 000 DWT.
Yard	The yard in a port is subdivided into: (a) The primary yard, which is the section of the yard adjacent to the apron, and primarily used for the temporary storage of inbound and outbound cargo (the storage area). (b) The secondary yard, which is the section of the yard used for chassis and empty container storage, miscellaneous equipment and facilities.

REFERENCES AND FURTHER READING

BSI (British Standards Institution) (2000) BS 6349-1: 2000. Maritime structures. Code of practice for general criteria. BSI, London.

CUR (Centre for Civil Engineering Research and Codes) (2005) *Handbook of Quay Walls*. CUR, Gouda.

US Army Coastal Engineering Research Centre (1984) *Shore Protection Manual*, vols 1 and 2. Department of the Army Corps of Engineers, Washington, DC. See also later editions.

Port Designer's Handbook
ISBN 978-0-7277-6004-3

http://dx.doi.org/10.1680/pdh.60043.559

Chapter 23
Conversion factors

In this book metric units have been used. When the word 'ton' (tonne) is used, this is a metric ton.

To convert between the imperial, or the US, system and the metric system, the following conversion factors should be used.

23.1. Length

1 mm	= 0.03937 in	1 in	= 25.4 mm	
1 m	= 3.281 ft	1 ft	= 12 in	= 0.3048 m
	= 1.094 yd	1 yd	= 3 ft	= 0.9144 m
1 km	= 0.6214 mile	1 mile	= 1760 yd	= 1.609 km

1 fathom = 6 ft
1 cablelength (UK) = 100 fathoms = 1/10 nautical mile
= 185.2 m
1 nautical mile = 10 cablelengths = 6080 ft = 1852 m
1/60 of a meridian degree
1 degree = 60 nautical miles

For interest:
1 Swedish inch = 2.47 cm
1 English inch = 2.54 cm
1 Russian inch = 2.54 cm
1 Norwegian inch = 2.62 cm
1 Danish inch = 2.67 cm
1 Paris inch = 2.71 cm

23.2. Speed

1 km/h	= 0.278 m/s	= 0.62 mph	= 0.54 knots
3.60 km/h	= 1 m/s	= 2.24 mph	= 1.94 knots
1.61 km/h	= 0.45 m/s	= 1 mph	= 0.87 knots
1.85 km/h	= 0.5145 m/s	= 1.15 mph	= 1 knot

23.3. Area

1 mm	= 0.00155 in^2	1 in^2	= 645.2 mm^2
1 m^2	= 10.76 ft^2	1 ft^2	= 0.0929 m^2
	= 1.196 yd^2	1 yd^2	= 0.8361 m^2
1 ha	= 10000 m^2		
	= 2.471 acres	1 acre	= 0.4047 ha = 4047 m^2
1 km^2	= 0.3861 $mile^2$	1 $mile^2$	= 2.59 km^2

23.4. Volume

1 m^3	= 1000 litres	= 35.3147 ft^3 = 1.308 yd^3
1 ft^3	= 0.02832 m^3	
1 yd^3	= 0.7646 m^3	
1 litre	= 0.22 imperial gallons	= 0.2642 US gallons
4.546 litres	= 1 imperial gallon	= 1.201 US gallons
3.785 litres	= 0.8327 imperial gallons	= 1 US gallon
1 pint (UK)		= 0.5683 litres
1 pint (US)		= 0.4732 litres
1 US barrel	= 5.6146 ft^3	= 158.99 litres

23.5. Weight

1 kg	= 1000 g	= 2.2046 lb
1 tonne	= 1000 kg	= 2204.6 lb
	= 0.98421 long tonne	
	= 1.10231 short tonne	
1 hundredweight (1 cwt)	= 112 lb	= 50.802 kg
1 lb	= 0.4536 kg	
1 long ton	= 2240 lb	= 20 cwt = 1016 kg
1 short ton	= 2000 lb	= 907.3 kg
1 kip (US)	= 1000 lb	= 453.6 kg

23.6. Force

1 N	= 0.2248 lb	= 0.1020 kg	
4.448 N	= 1 lb	= 453.6 kg	
9.807 N	= 2.205 lb	= 1 kg	= 1 daN
1 kN	= 0.1004 long tonne	= 102.0 kg	= 0.1020 tonne
9.964 kN	= 1 long tonne	= 1016	= 1.016 tonne
9.807 kN	= 0.9842 long tonne	= 1000 kg	= 1 tonne

23.7. Force per unit length

1 N/m	= 0.06852 lb/ft	= 0.1020 kg/m
14.59 N/m	= 1 lb/ft	= 1.488 kg/m
9.807 N/m	= 0.672 lb/ft	= 1 kg/m
1 kN/m	= 0.0306 long tonne/ft	= 0.1020 tonne/m
32.69 kN/m	= 1 long tonne/ft	= 3.333 tonne/m
9.807 kN/m	= 0.3000 long tonne/ft	= 1 tonne/m

23.8. Force per unit area

1 N/mm^2	= 145.0 lb/in^2	= 10.20 kg/cm^2
0.006895 N/mm^2	= 1 lb/in^2	= 0.0703 kg/cm^2
0.09807 N/mm^2	= 14.22 lb/in^2	= 1 kg/cm^2 = 10 $tonne/m^2$
1 N/m^2	= 0.02089 lb/ft^2	= 0.102 kg/m^2
47.88 N/m^2	= 1 lb/ft^2	= 4.882 kg/m^2
9.807 N/m^2	= 0.2048 lb/ft^2	= 1 kg/m^2
10 $tonne/m^2$	= 1 kg/cm	= 0.1 MPa = 0.1 N/mm^2 = 100 kN/m^2
1 atmosphere	= 1 kg/cm^2	= 10 m water pressure

23.9. Moment

1 N m	= 8.851 lb in	= 0.7376 lb ft	= 0.1020kg m
0.1130 N m	= 1 lb in	= 0.08333 lb ft	= 0.01152 kg m
1.356 N m	= 12 lb in	= 1 lb ft	= 0.1383 kg m
9.807 N m	= 86.80 lb in	= 7.233 lb ft	= 1 kg m
1 tonne m	= 3.229 long tonne ft		
1 long tonne ft	= 0.3097 tonne m		

23.10. Temperatures

Celsius (°C) = (°F −32)/1.8 = (°F −32)5/9
Fahrenheit (°F) = 1.8°C + 32

23.11. Useful data

Standard gravitational acceleration	= 9.807 m/s^2	= 32.174 ft/s^2
Density of water	= 1000 kg/m^3	= 62.4 lb/ft^3
Weight of reinforces concrete	= 23.6 kN/m^3	= 2400 kg/m^3
		= 150 lb/ft^3
1 horsepower (HP)	= 0.746 kW	

Port Designer's Handbook
ISBN 978-0-7277-6004-3

http://dx.doi.org/10.1680/pdh.60043.563

Index

Figure locators are given in *italics* if not in main text range.